Hydrogenation of Fats and Oils

Theory and Practice

Second Edition

Hydrogenation of Fats and Oils

Theory and Practice

Second Edition

Editors
Gary R. List
Jerry W. King

AOCS PRESS
Urbana, Illinois

AOCS Mission Statement

To be a global forum to promote the exchange of ideas, information, and experience, to enhance personal excellence, and to provide high standards of quality among those with a professional interest in the science and technology of fats, oils, surfactants, and related materials.

Library of Congress Cataloging-in-Publication Data

Hydrogenation of fats and oils : theory and practice / editors, Gary R. List,
Jerry W. King. – 2nd ed.
p. cm.
Rev. ed. of: Hydrogenation of fats and oils / H.B.W. Patterson. c1994.
Includes bibliographical references and index.
ISBN 978-1-893997-93-6 (alk. paper)
1. Oils and fats, Edible. 2. Hydrogenation. I. List, Gary R. II. King, Jerry W.

TP680.P337 2011
664'.3–dc22

2009051527

Printed in the United States of America

15 14 13 12 11 5 4 3 2 1

The paper used in this book is acid-free, and falls within the guidelines established to ensure permanence and durability.

Contents

Preface to the Second Edition

Patterson's book, first published in the 1980s and revised in 1994, has long been a standard reference textbook on the hydrogenation of edible oils. Much has changed and we believe that a third revision is warranted in view of emphasis on *trans* fatty acid reduction in food products resulting from nutrition labeling regulations enacted by the Food and Drug Administration. Chapters on the use of critical fluids in hydrogenation and electrochemical hardening have been added as well as a chapter on low *trans* oils. The mechanism of *trans* fat formation is discussed. The editors and authors hope this new material will be useful to the reader.

Gary R. List
Washington, Illinois

Preface to the First Edition

The genesis of the first edition of this book is located in a paper originally given at a United Nations Industrial Development Organization (UNIDO) conference and distributed by the United Nations. As in the first edition, discussion is not confined to vegetable oils, and the hydrogenation technique is considered in detail. The "why" as well as the "how" of hydrogenation are treated. Written for the production staff who need advice on specific problems as well as for development personnel who are provided directions, if not solutions, the book gives direct practical advice as well as explanations why changes occur as they do. The glossary of technical terms contains a more detailed explanation of some features that would be inconvenient to present for readers throughout the text.

Our general understanding of the principles of the industrial hydrogenation of fats and oils has not undergone any radical change since the first edition. An immediately obvious difference in this edition is the presentation of texture measurements of hardened oils in terms of solid-fat content (SFC) instead of dilatation valuers [Conversion of dilatation to SFC where necessary was performed by the use of published tables (van den Enden et al., 1978, 1982, 1986)]. The solid-fat index (SFI) is given where appropriate. Although certain plant items or manufacturing procedures for the production of a catalyst and hydrogen as well as hardened oil are no longer popular, their descriptions are retained because they show the development of the subject, and may be useful for those readers working with an older plant. The reference list is expanded to include items which gained prominence in recent years.

Automatic measurement and computerized process control are, of course, widespread today, leading to reductions in laboratory and plant staff. Hydrogenation is exothermic, and efforts were always made to take advantage of this, but in recent years plant design emphasized energy usage more than ever.

As with several earlier publications, I wish to express my gratitude to Mrs. Marjorie Honor for her careful work in preparing the text of this edition.

H.B.W. Patterson, D.Sc.
9 The Wiend
Bebington
Merseyside L63 7RG
England

Acknowledgments

Acknowledgments and sincere thanks are hereby extended to the following individuals and organizations who have readily agreed to the use of illustrations, data, and comments already published by them in journals, books, or company brochures. This help was invaluable, and is cited in detail via the references in the text along with acknowledgments to other published sources.

R.R. Allen; J.W.E. Coenen; G. Leuteritz; B.F. Teasdale; W. Zschau; Alfa-Laval (Tumba, Sweden); American Oil Chemists' Society (Illinois, United States); Buss AG (Basel, Switzerland); Canada Packers (Toronto, Canada); Chemetron Process Equipment (Kentucky, United States); Chemistry and Industry (London, United Kingdom); EMI Corporation (Illinois, United States); Girdler Catalysts (Kentucky, United States); Harshaw Catalysts (Ohio, United States and De Meern, The Netherlands); Lightnin' Mixers Ltd. (Poynton, United Kingdom); Lurgi GmbH (Frankfurt am Main, West Germany); M and M-Schenk Filters (Hemel Hempstead, United Kingdom); P and S Filtration (Haslingden, United Kingdom); Peabody Holmes (Huddersfield, United Kingdom); Süd-Chemie AG (Munich, West Germany); Unichema International (Wormerveer, The Netherlands); UNIDO (Vienna, Austria); Western Co-operative Fertilizers (Calgary, Canada).

The author is especially grateful to Mr. A.R. Nash, who helped to update and clarify the text at certain points, and, where necessary, converted the dilatation values to SFC by the use of published tables (van den Enden et al., 1978, 1982, 1986).

The following individuals and companies are also thanked for substantial contributions they made in bringing the text up-to-date.

R. Ariaansz (Engelhard De Meern BV; Pk De Meern, The Netherlands); J. Bonnachewski (Unichema International; Emmerich, Germany); J. Erbe (Süd-Chemie AG; München, Germany); R.C. Hastert (Hastech; Cleveland, Ohio, United States); P. Haynes (Schenk Filterau GmbH; Waldstetten, Germany); Hoechst AG (Frankfurt am Main, West Germany); G.M. Leuteritz and R.F. Duveen (Buss AG; Basel, Switzerland); LurgiGmbH (Frankfurt am Main, West Germany); P.K. Page (Lightnin' Mixers Ltd.; Poynton, United Kingdom); P. Price (Calsicat Catalysts; Malinckrodt GmbH; Gennef/Sieg 1; Germany); F.G. Veldkamp (Lochem BV; Lochem BV, The Netherlands).

As well, the following individuals are additionally thanked for the substantial contributions they made in editing the second edition. Albert J. Dijkstra and Peter N. Pintauro (Vanderbilt University).

Introduction

Since publication of this book in 1994, hydrogenation has undergone much change. Once the backbone of the edible oil processing industry, health, and nutrition concerns arising from *trans* fatty acids produced during hydrogenation have impacted its use. Indeed *trans* fat labeling regulations enacted in 2006 have caused a shift from heavily hydrogenated fats and oils to liquid or lightly hydrogenated products in order to meet the less than 0.5 gm *trans* fat/serving required for a zero *trans* fat claim on nutrition labels. During the period 1999 to 2007, the number of foods reformulated to a zero *trans* claim increased from essentially zero to over 1900. Much of the reformulation was accomplished by replacing hydrogenated oils with naturally stable oils such as corn, cottonseed and sunflower as well as trait modified oils whose fatty acid composition have been altered by plant breeding. *Trans* fat reduction has also been achieved through use of both chemical and enzymatic interesterification as well as fractionation of tropical oils. On the other hand, *trans* fat issues have prompted research directed towards improving the hydrogenation process. Examples include hydrogenation in critical fluids, electrochemical hardening, and use of modified hydrogenation process variables. Other approaches include production of low *trans* fats by blending low iodine value bases with liquid oils or by use of two-stage processes first with conventional nickel catalysts followed by hardening with platinum catalysts. Several low *trans* products have been introduced commercially by use of chemically modified nickel catalysts. Complete hydrogenation will continue to be an important process because stearines offer low *trans* options in many food applications including shortenings and deep fat frying oils. Although much has changed over the past several decades, hydrogenation will continue to be an important industrial process.

Chapter 1

The Hydrogenation Reaction

H.B.W. Patterson

Economic Value of Hydrogenation

Reference is made in the Introduction to the economic value of hydrogenation resulting from its ability to modify chemical and physical behavior, so that one oil may frequently be substituted wholly or largely for another according to market conditions, and indeed sometimes to make a product of novel characteristics. To understand how this is possible, we must look briefly at the chemical nature of fats and oils.

Triglycerides

The basic units of fats or oils consist of one molecule of glycerol combined with three molecules of fatty acid. If the result is liquid at ambient temperature, it is commonly known as an oil, and if it is solid, as a fat. Coconut oil is an example of a commodity which is liquid in the tropical regions of origin but solid in temperate zones. Whereas the glycerol component is a constant feature, several related families of fatty acids exist. Where all three fatty acids in the triglyceride are of exactly the same kind, the substance is known as a simple triglyceride, but if more than one kind is present, it is known as a mixed triglyceride (Fig. 1.1).

In nature, fats are physical mixtures of various triglycerides, most of which are themselves mixed triglycerides. The proportions of the different triglycerides which go to make the complete fat will largely determine its character, just as the different kinds of fatty acid combined in any one triglyceride will affect both its chemical and physical nature. To make this point here is opportune: in the case of mixed triglycerides, the relative positions 1, 2, or 3 which different fatty-acid groups may occupy with respect to one another also have some influence on the character—especially the physical character—of the mixed triglyceride in question. For greater detail, see the notes in the Glossary. Differences between various fatty acids provide the principal reasons why one fat differs from another, and by modifying these differences where possible, hydrogenation modifies the behavior of the fat as a whole.

Fatty Acids

Fatty acids are comprised of a chain of carbon atoms combined with hydrogen and terminating in a carboxyl group:

$$H_3C\text{-}CH_2\text{-}CH_2 \text{ .... } CH_2\text{-}CH_2\text{-}COOH$$

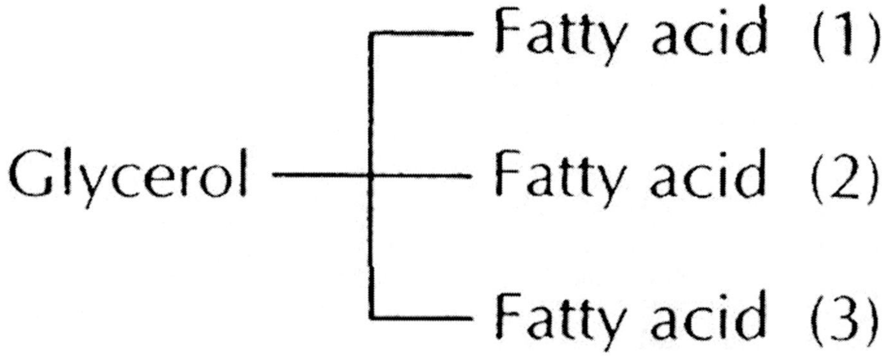

Fig. 1.1. Structure of a triglyceride. Where positions 1, 2, and 3 are all alike, the molecule is a simple triglyceride; if one differs, it is a mixed triglyceride.

A chain of carbon atoms combined with hydrogen is commonly spoken of as a hydrocarbon chain, and when all available carbon valencies for hydrogen are satisfied, the chain is said to be saturated. The first three members of the series—formic, acetic, and proprionic acids—are not at all fatty in character, but simply fit the structural pattern of the series; even the fourth member, butyric acid, only qualifies since it is found combined in butter fat, yet does not itself show the customary fatty character of refusing to mix with water. With the sixth member of the series, caproic acid, this fatty character is, however, plainly evident, and the acid is found in the oils of various species of palm. Those fatty acids which contain an even number of carbon atoms predominate greatly in natural fats (no doubt by reason of the metabolic process by which they are formed), but small amounts of some containing an odd number of carbon atoms were isolated. Owing to advances in analytical methods, several unusual fatty acids containing a short branch in the chain, a three- or five-membered ring system, keto and epoxy groups were discovered (Swern, 1979). Ricinoleic acid, containing one hydroxyl group, is the principal fatty acid of castor oil, and was known for many years. Notes on nomenclature, characteristics, and further structural details of fatty acids are given in the Glossary.

Fatty-Acid Chain Length and Unsaturation

Of special importance in determining the character of a fatty acid are its chain length and the extent to which pairs of adjoining carbon atoms may each lack one hydrogen atom, hence, forming a double bond between them, in which case the fatty acid is said to be unsaturated. We already noted that a fatty or hydrophobic character is obvious in caproic acid (C6), and this increases as chain length increases within any particular series or family of fatty acids.

The most widespread and important fatty acids are those containing 16 or 18 carbon atoms, although some containing up to 30 were identified. Among the oils of fish and marine animals, very long chains of 22 and more carbon atoms are most common. For fatty acids, all with an even number of carbon atoms in the chain, a regular increase occurs in melting point as the chain increases; similarly, for odd-numbered chains, although, as one may remark, the melting point of any particular member here is just a little less than the corresponding even-numbered member immediately before it (Luddy, 1979). Unsaturation is associated with a liquid or a lower melting condition, greater solubility, and chemical reactivity. It may be distributed in different ways along the chain, and this again has a marked effect on the physical and chemical properties both of the acid itself and of the glycerides in which it is combined. Consider first the case of a solitary bond in a chain (monounsaturation or monoethenoid). Here the two remaining hydrogen atoms, one at each side of the double bond, may each lie on the same side of the chain. This is described as the *cis* configuration, and the chain is considered as shaped here into a rigid arc with the hydrogen atoms positioned toward the outside of the arc. Alternatively, the two hydrogen atoms may be positioned on opposite sides of the chain, now seen as acquiring a kink or dog-leg effect at this point in its length; this is the *trans* configuration. For a normal single bond, complete freedom of rotation exists; at a double bond, rigidity exists, and only the two fixed positions of *cis* and *trans* are possible; at a triple bond, only one rigid configuration can exist.

$$\begin{array}{c}\quad\quad\quad\;\; H\;\; H \\ \quad\quad\quad\;\; |\;\;\; | \\ -CH_2-CH_2-C{=}C-CH_2-CH_2\end{array}$$ *cis*

$$\begin{array}{c}\quad\quad\quad H \\ -CH_2-CH_2-C{=}C-CH_2-CH_2 \\ \quad\quad\quad\quad\; H\end{array}$$ *trans*

$-CH_2-CH{=}CH-CH{=}CH-CH_2$ conjugated (1,3 unsaturation)

$-CH_2-CH{=}CH-CH_2-CH{=}CH-CH_2$ nonconjugated (1,4 unsaturation or skipped)

$CH_3-(CH_2-CH{=}CH)_n-COO-$ very common in nature; *cis* greatly predominates

$-CH_2-CH{=}C{=}CH-CH_2$ 1,2 unsaturation; unknown in fats

Where only a difference in spatial configuration between the double bonds of one unsaturated fatty acid and another (*cis* and *trans*) exists, they are geometric isomers. Where the same number of double bonds is present but not all are located at the same points in the chain, they are positional isomers. The *cis* isomer is nearly always the form found in natural fats; it is normally the lower melting and more reactive form since it distorts the chain more and does not lend itself so readily to being packed into crystals. When two double bonds are separated by a single CH_2 (methylene) group (1,4 unsaturation above), the latter is particularly active and seemingly the initial point of attack when fats containing it are exposed to atmospheric oxidation. The conjugated system (1,3 unsaturation) is also reactive, lending itself both to thermal polymerization and hydrogenation.

Conventionally, unsaturated fatty acids containing two or more double bonds are classed as polyunsaturated fatty acids (PUFAs). Possibly taken for granted is that those found in natural fats will be of *cis* configuration; however, to attain their full usefulness as starting points for building-up by the human metabolism of extremely important regulatory compounds, the double bonds must be acceptably positioned in the chain. Linoleic acid—in which *cis* double bonds occur following carbon atom numbers 9 and 12 (numbering the carboxyl group carbon atom as 1) in an 18-carbon atom chain—is an example of such a useful fatty acid. These have long been termed "essential fatty acids" (EFAs). The saturated fatty acids are sometimes referred to as SAFAs. Since they contain no double bond to distort the chain, they pack most easily into crystal form, and consequently are higher in melting point than an unsaturated fatty acid of the same chain length, more hydrophobic and less vulnerable to attack.

Nonfat Components

Several classes of compounds occur as minor components of fats and oils, especially, seemingly, in vegetable oils. If these compounds contain obvious nickel-catalyst poisons (Chapter 7) like sulfur or phosphorus, a preliminary cleaning of the oil prior to hydrogenation is normally pursued to a point where their presence is so diminished that they no longer seriously impede the attainment of the purpose of the hydrogenation. A processor may easily find himself in a position where sacrificing some catalytic activity is more economical than incurring an extraordinary expense in the further cleansing of the oil. In the past, a largely exhausted catalyst whose nickel was still capable of capturing sulfur to be used at the prehardening bleaching stage brought down the sulfur content in a marine oil. The phosphatides are the obvious source of phosphorous poisoning, but usually these are brought to a very low level in any case by the preliminary degumming, neutralization, and earth-bleaching steps, all of which have become better understood and more effective since the 1950s. Sulfur, apparently present in the form of thioglucosides, is more evident in rapeseed, crambe, and mustardseed oils. Although, as purported, deodorization should immediately follow the alkali neutralization and earth bleaching of rapeseed oil intended for hardening because the sulfur compounds are sufficiently volatile to be stripped in this way, one must calculate whether the expense of such steam stripping equals or exceeds that of the additional nickel which would otherwise be used. Pigments such as carotene, chlorophyll, and gossypol do not affect hydrogenation, although the color of an oil is greatly lightened thereby, especially if carotene is the pigment. Sterols (e.g., cholesterol), tocopherols, complex hydrocarbons (e.g., carotinoids, squalene), waxes, and phenolic compounds (sesamol) do not hinder hardening, although some may confer or reinforce useful characteristics such as resistance to oxidation. Chapter 8 concerns itself with the hydrogenation of individual oils, and there questions concerning the removal of catalyst poisons are treated if they are present in particular cases.

The Hydrogenation Reaction

In the "Fatty Acids" section and the "Fatty-Acid Chain Length and Unsaturation" section in this chapter, obviously shown is that any process which alters the shape, distribution, or number of double bonds within the fatty-acid chain is able to modify the chemical stability and the physical behavior of the fatty acid or the glyceride of which it forms a part. Hydrogenation is just such a process and very frequently does all three at the same time, hence, its importance. To avoid confusion in our attempt to evaluate what is involved in the industrial operation, distinguishing three kinds of activity is helpful: (i) the chemistry of the interaction between the molecules at the site of the reaction; (ii) the physics involved in bringing the reactants together and dispersing the products; and (iii) the engineering best suited to the safe, economical, and convenient accomplishment of the simple or complex production program required.

Chemistry and physics are, of course, simply engineering on a very small scale. Point (iii) is considered later in Chapter 4. As for point (i), the reaction itself, nothing is achieved simply by mixing the hydrogen and oil, since the direct addition of hydrogen to the double bond of an unsaturated fatty acid involves surmounting a considerable energy barrier. However, both hydrogen and unsaturated bonds are readily absorbed at the surface of a catalyst such as finely divided nickel, in which case the energy barrier is much smaller; now the reaction can be much faster and energy is liberated. Similarly, the desorption of the reaction product from the surface of the nickel requires surmounting only a second modest energy barrier before more energy is liberated. If hydrogenation—that is, the saturation of the double bond—has in fact taken place, the net liberation of energy for a drop of one unit in iodine value (IV) is sufficient to raise the temperature of the oil by close to 1.7°C; depending on the specific heat of the oil, which itself varies appreciably with temperature (*Bailey's Industrial Oil and Fat Products*), the exothermic heat of reaction was computed as 1.7 BTU/lb or 0.942 kcal/kg (specific heat 0.6 kcal/kg, say) per unit drop in IV. This liberation of heat has very important implications for the design of hydrogenation autoclaves (Chapter 4).

Isomerization

To greatly help our understanding of the several effects we can obtain from hydrogenation, we now look closely at what are understood to be the different possibilities open when either a solitary double bond is adsorbed at the catalyst surface or a polyunsaturated acid containing the common skipped group, $CH=CH-CH_2CH=CH$. At the surface of an active nickel catalyst, the nickel atoms are about 2.5 Å apart [one angstrom (Å) unit = 10^{-10} m; one micron (μm) = 10^{-6} m]. The ideal spacing of about 2.7 Å is held to fit easily with the 1.5 Å between the carbon atoms of a double bond (Waterman, 1951). Metals which catalyze the hydrogenation of these double bonds,

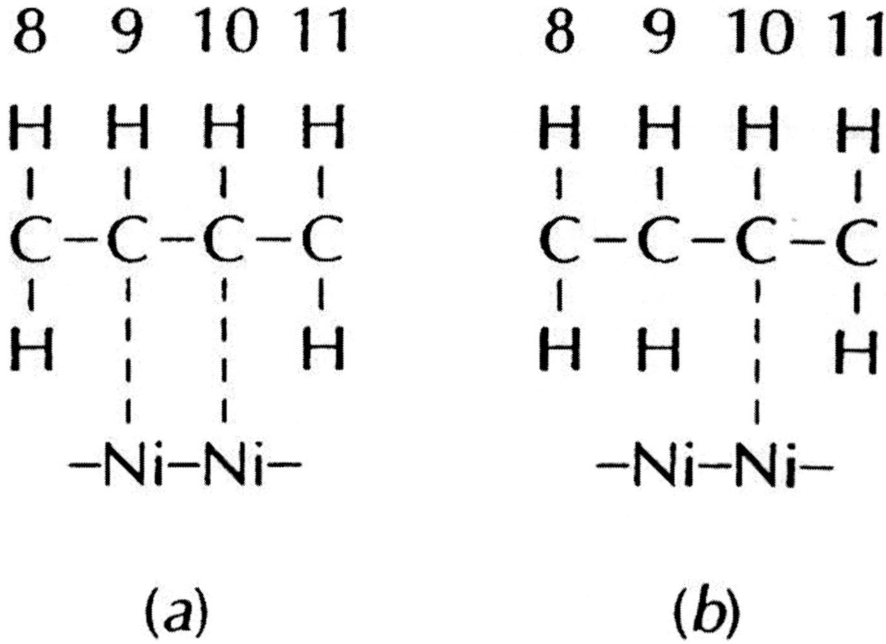

Fig. 1.2. Linkage of a double bond to catalyst atoms.

such as nickel, copper, palladium and platinum, all have atomic spacings close to 2.7 Å. When the double bond is adsorbed at the surface of active nickel, the situation is as represented in Fig. 1.2a which shows part of a carbon chain containing one *cis* double bond between the 9C and 10C atoms that has formed links to the nickel surface. From the hydrogen also adsorbed in the vicinity, one atom now links with, say, the 9C atom (Fig. 1.2b). These possibilities are now foreseen:

- The same H atom is lost by 9C before 10C can acquire one; the original 9–10C *cis* bond reforms and is desorbed: no change;
- The other H atom on 9C is lost; hence, a *trans* bond forms and is desorbed: geometric isomer;
- An H atom is lost from 11C, and a 10–11C *cis* bond forms and is desorbed: positional isomer;
- The other H atom is lost from 11C, a 10–11C *trans* bond forms and is desorbed: positional isomer; and
- The 10C captures an H atom while 9C still holds two; the bond is therefore saturated and is desorbed.

Suppose 10C first captured an H atom; possibilities comparable to those above now emerge, except that this time the migration of the double bond, if it took place, would be to the 8–9C position, *cis* or *trans*. Although other essentially similar explanations of the reaction were also offered (Beckmann, 1983; Coenen, 1978; der Boer & Wosten, 1968; Heertje et al., 1974; Larsson, 1983; Roonev et al., 1960; Sebedio et al., 1981; Van der Plank & Van Oosten, 1975) this comparatively simple one, the Horiuti-Polanyi mechanism (Horiuti & Polanyi, 1934), affords us an easy way of visualizing how the double bond may not only be hydrogenated but also can migrate in either direction and change its configuration (Albright, 1967; Allen & Kiess, 1955; Chahine et al., 1958).

Commonly, the case that, starting from a fat with all of its double bonds being *cis*, an equilibrium position is reached where two-thirds of the remaining double bonds are *trans*, after which the proportion remains the same until all are hydrogenated. However, one may select conditions which will much delay the attainment of this equilibrium, as we shall see in the "Low-Temperature Hydrogenation" section and the "Iso- or *trans*-Suppressive Hydrogenation" section in Chapter 2. The important question remains of the competition which takes place between fatty-acid chains of different degrees of unsaturation for a place on the active nickel, and the important consequences this has for the hydrogenation industry. Evidence that polyunsaturates are adsorbed preferentially and more strongly at the catalyst surface than monounsaturates has grown since the mid-1950s (Albright & Wisniak, 1962; Eldib & Albright, 1957; Wisniak & Albright, 1961). Coenen and Boerma (1968; Coenen, 1970) demonstrated this clearly when they compared the sequence of events during the hydrogenation of glyceryl trioleate (containing no polyunsaturates) and rapeseed oil which contains both linolenic and linoleic acids as well as the (*cis*) monounsaturated C22 erucic acid. From the very beginning, the trioleate showed the expected formation of saturated (stearic) groups and *trans* isomers (elaidic) of oleic acid; in the case of the rapeseed oil, only after the disappearance of about one-half of the total linoleic and linolenic acids did modest amounts of C22 saturate and C22 *trans* monounsaturates become noticeable. Further, the very character of the reaction rate during the earlier part of the hydrogenation of oils known to contain PUFAs ("pseudo zero order," Chapter 1, "Order of Reaction" section) indicates that during that time they dominate the catalyst surface (Coenen, 1960a,b; Coenen & Boerma, 1968). Now we have to consider a possible course of events when these PUFA groups in the triglyceride are adsorbed. As already mentioned, much of the unsaturation in natural fatty acids is present in the skipped or 1,4 unsaturation form, $=CH=CH-CH_2-CH=CH=$. The sequence of events following the adsorption of such a group is seen in Fig. 1.3.

In fact, the experimental evidence (Allen & Kiess, 1956; Chahine et al., 1959; Coenen, 1960a; Coenen & Boerma, 1968; de Vnes, 1963; de Vries & Jurriens, 1963) shows that conjugated systems appear, double bonds migrate, and the usual proportion of *trans* isomers can be attained; indeed, migration was recognized as long ago as 1919 (Moore, 1919). Copper appears to operate as a hydrogenation catalyst only on oils capable of forming conjugated intermediates—hence, its ability to reduce natural polyunsaturates to monoenes without producing saturates. When

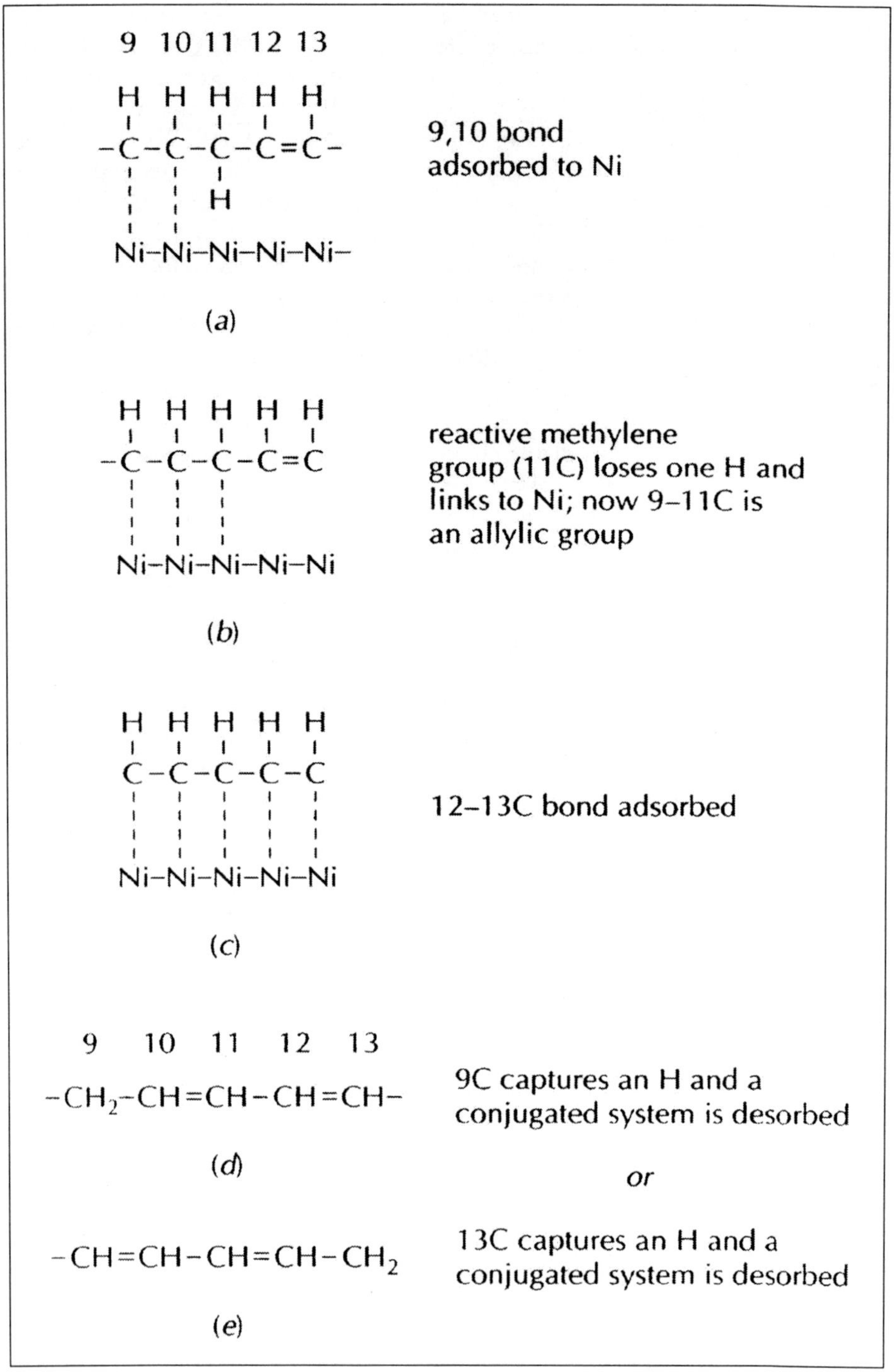

Fig. 1.3. Linkage of double bonds to catalyst atoms.

one appreciates that double bonds not only migrate in both directions along a hydrocarbon chain, forming *trans* isomers at each step, but also that the products of such isomerization then become available for further isomerization or hydrogenation, one can imagine the complexity of the result. At any moment, it represents the net effect of many changes. Evidently, a plentiful supply of hydrogen on the nickel surface promotes hydrogenation rather than isomerization; even the polyunsaturated chain, for a proportion of its contacts, may not have the chance to form the triple link of the allylic group shown at Fig. 1.3b before it is hydrogenated and desorbed. If this is the case, these are the circumstances in which the stronger linkage of the polyunsaturated system in effect becomes rather less dominant. By contrast, when the supply of hydrogen on the nickel surface drops to a much more modest level, the chances for isomerization improve markedly, and then the preferentially linked polyunsaturate best exerts its temporary dominance of the catalyst surface. Note that in the hydrogenation of very long chain polyunsaturates containing up to six double bonds, stages are reached at which the remaining two or three double bonds are comparatively isolated from one another by several intermediate $-CH_2-$ groups. In these circumstances, the whole unsaturated system is noticeably less reactive toward hydrogenation or oxidation than normal *cis, cis,* 9,12 linoleate or 9,12,15 linolenate would be, and if the latter were to have its central double bond on C12 hydrogenated, the resultant 9,15 isolinoleic group is much less reactive than normal 9,12 linoleate. Even so, partly hydrogenated chains, if fragmented by chemical (oxidative) attack near the middle of the chain, may yield C8 and C9 portions containing a *trans* group and a terminal aldehyde or ketone group which in some cases represent potent off flavors.

We have seen the more important possibilities open to the chemical reaction on site (Chapter 1, "The Hydrogenation Reaction" section, point 1 of the list) and particularly how the greater or lesser availability of hydrogen there plays such a decisive part; now, we may conveniently pass to review the different physical factors ("The Hydrogenation Reaction" section, point 2 of the list) which condition the supply of reactants to the site, thus influencing the reaction in one way or another.

Hydrogen Dispersion

The gaseous hydrogen must first dissolve in the liquid oil and then pass to the solid surface of the catalyst before hydrogenation can occur. The involvement of different states of matter classifies this as a heterogeneous catalytic reaction. Unsaturated triglycerides must also make their way to the catalyst surface, and the products of reaction must move back from the surface into the bulk oil. The successive stages are visualized as follows, and are as represented by Fig. 1.4.

1. Hydrogen of known purity, pressure, and temperature—hence, in effect, of known concentration—saturates the immediate surface of the oil to a point approaching the solubility of the gas in oil at those conditions. Supposedly, a very thin layer of less mobile gas at the gas/liquid interface is present. At a

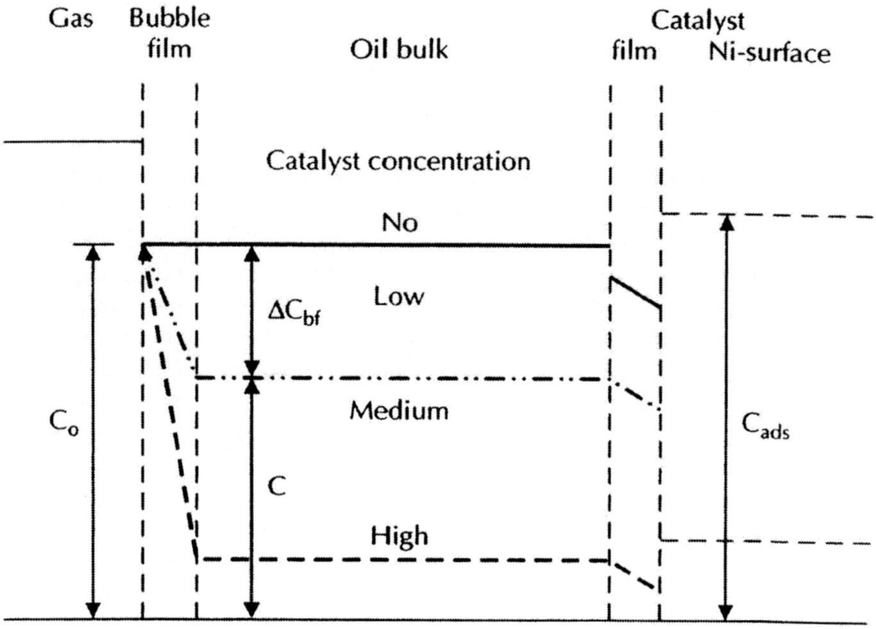

Fig. 1.4. Concentration of hydrogen at different stages in the mass transport of gas bubbles to catalyst pores (Coenen, 1976).

pressure of one atmosphere, the solubility of hydrogen in oil (v/v), S, at t °*C* is given approximately by the expression:

$$S = 0.0295 + 0.000497t \qquad \text{[Eq. 1.1]}$$

Only a slight variation is noted from one oil to another (Table 1.1). Solubility increases with temperature over the ranges at which hydrogenation is operated; this is true for other common gases such as oxygen, nitrogen, and carbon monoxide. The solubility increases in virtually linear fashion with pressure (Wisniak & Albright, 1961).

2. The dissolved hydrogen must now in effect penetrate the thin film or less-mobile layer of oil regarded as enveloping each bubble of gas. This would apply equally to a droplet of oil thrown into a space filled with gas.

TABLE 1.1
For 1000 Liters (m^3) of Oil, the Number of Liters of Hydrogen Which Will Dissolve Approximately at Certain Temperature

Temperature (°C)	Liters hydrogen per m^3 oil
25	42
100	79
150	104
180	119

3. Once in the bulk of the oil, dissolved hydrogen must travel to that layer of oil surrounding a particle of catalyst.
4. The catalyst oil layer or film must be crossed, and then the hydrogen is within the region of the catalyst pores.
5. Movement within a pore toward the nickel surface follows.
6. Hydrogen is then adsorbed onto the active portion of the nickel exposed to it. Obviously, a drop in hydrogen concentration is the driving force which brings the hydrogen from the bubble into the bulk oil; turbulent movement within this bulk evens out the concentration, but again another fall occurs as the layer surrounding the catalyst is penetrated, and still a further fall as at least some of the hydrogen moves to the interior of the pore. This latter part of the journey has special features which are made evident in Fig. 1.12 ("Operation of Selectivity" section in Chapter 1). The molecules of unsaturated triglyceride are of course only involved in steps 4 to 6 above, but following these, whether they are hydrogenated or isomerized, once or more often, by contacts with the active catalyst, they must behave as follows:
7. They must desorb finally from the surface.
8. They must retreat to the mouth of the pore.
9. They must return through the film surrounding the catalyst and rejoin the bulk oil.

The dispersion, mass transport, or migration of the hydrogen is influenced on the incoming side by the concentration gradient through the bubble film and on the outgoing side by the concentration gradient through the catalyst film. It is also markedly influenced by the transfer area available to it; thus, if agitation is increased, and this is effective in enlarging the gas/liquid interface, it assists in increasing the hydrogen concentration in the bulk oil. If the pressure in the gas phase is increased, this also increases the concentration of dissolved hydrogen, and the dissolved hydrogen is nearer to affecting the amount of hydrogen adsorbed on the nickel and hence, affecting the reaction:

$$\text{Rate of hydrogen solution} = \text{a constant (interface area)} \times (H_2 \text{ conc. in gas} - H_2 \text{ conc. in oil}) \qquad \text{[Reaction 1.1]}$$

where the constant expresses the particular character and thickness of the oil film, and concentration is synonymous with pressure.

By causing a more violent flow of oil across the surface of a gas bubble, increased agitation can diminish the effective thickness of the bubble film, and by similar reasoning, the effective thickness of the catalyst film. This reduction of resistance to dispersion assists the mass transport. While increased agitation also accelerates oil currents within the bulk oil and hence, the transit from bubble to catalyst, this movement is in any case more rapid than diffusion through a film, so unlikely will it itself ever be the rate determinant. Agitation improves heat transfer via heating/cooling coils and maintains the catalyst in suspension.

Since Normann's first design came into operation in 1906 [(German Patent 139,457 (1902); British Patent 1515 (1903)], an increasing stream of devices for stirring, blowing, and sucking hydrogen into oil, or oil into hydrogen, has continued—Chapter 4 deals with this aspect. As will be obvious, increased agitation opened the door more widely for incoming hydrogen, so the greater provision of accessible active catalyst surface opens the door for hydrogen going out into a combination with the oil ("Catalyst Action" section, Chapter 1). Conversely, increasing the agitation above the level needed to saturate the available nickel surface with hydrogen will achieve nothing.

In the end, some factor has to control the rate of reaction, and it may often be that the mass transport of the hydrogen through the oil to the catalyst is indeed the limiting factor in full-scale factory equipment, whereas in the laboratory, higher rates of agitation and dispersion are more commonplace (Albright, 1967, p. 201; Allen, 1981; Hastert, 1981). We shall return to this question during this chapter as the other factors are considered. From what has already been said, evidently, considerable amounts of heat have to be conducted away during the reaction, and if a high availability of hydrogen is maintained on the nickel surface, this can sway the reaction in a particular direction which may not always be the desired one ("Operation of Selectivity" section, Chapter 1) (i.e., the reaction tends to become "nonselective").

Hydrogen Pressure

At the site of the reaction, pressure on the catalyst surface does not play a significant role; but indirectly, by speeding up the rate of solution of hydrogen into the oil, the pressure on the system increases the hydrogen supply to the surface and therefore, in turn, the reaction rate. As mentioned already ("Hydrogen Dispersion" section, step 1), an increase in pressure also increases solubility itself on an almost linear basis. Discussion as to exactly how the hydrogenation reaction rate varies with pressure is evidently virtually as old as the process of hydrogenation, and one suspects that the validity of various conclusions is limited by the influence of the other factors discussed in this chapter on the particular experiments which were carried out by many teams of workers (Swern, 1964). For edible-oil hydrogenation, an absolute pressure of 3 atm is popular, with the facility of increasing this to 10 atm at most, if required, in a plant with a varied program. In these particular circumstances, an increase in rate with an increase in pressure will probably be somewhat less than directly proportional but higher than the square root of the increase. In plants where one cannot possibly increase the rate of agitation so as to accelerate hydrogenation, sometimes possibly one can increase pressure temporarily for this purpose. For edible-oil hardening with a nickel catalyst, successive pressure increases beyond 10 atm become less and less rewarding; for fatty-acid hardening, pressures in the range of 10–30 atm are very useful (Hastert, 1979) and ensure shorter hardening times and lower nickel expenditure. In an extensive program of tests covering the hydrogenation of soybean oil with copper catalysts, Koritala and co-workers (1980; Mounts et al., 1978) used pressures from 50 to 30,000 psig, being mainly concerned with chemical changes in the nature of the product as well as reaction mechanism ("Isomerization" section, Chapter 1); an increase of reaction rate with an increase in pressure in the 500–3000 psig range was noted although evidently as hydrogenation progressed and the triunsaturated linolenic acid disappeared, the further fall in IV became more and more dependent on the ever decreasing amount of diunsaturated linoleic acid remaining.

TABLE 1.2
In Practice, Several Temperature Levels Exist Near Which Critical Effects for the Course of the Reaction Are To Be Obtained. Although These Are Considered in More Detail Later (Chapter 2), Listing Them Briefly Here Is Useful

Temperature (°C)	*Critical Temperatures*
100–115	Partial hydrogenation of a vegetable oil to reduce majority of linolenic acid group with minimum formation of *trans* isomers.
c. 120	The first of two stages in a hydrogenation which seeks minimum solids content with flavor stability.
150	(a) Not to be exceeded until a certain IV drop has been attained on oils containing substantial amounts of linolenic acid and even more unsaturated groups so as to avoid cyclization of hydrocarbon chain. (b) The popular level at which to conduct hardening of the all-hydrogenated vegetable shortening with prolonged melting range.
160	(a) Above this temperature, nickel carbonyl is completely unstable; hence, the poisoning effect of CO on Ni ceases. (b) Above this temperature, migration of double bonds and formation of *trans* isomers are encouraged to reach their equilibrium level.
180	The usual level for edible oil hydrogenation which may follow a set amount at a lower temperature for reasons given above. If a relatively quick melting range is needed, this temperature should be used as much as possible after any other control requirements have been met. At this level, polyunsaturates diminish markedly.
200	Should not be exceeded for edible product hardening. Above this, the risk of worsening color and increase in free fatty acid grows.
210	Maximum for most technical or non-edible hydrogenations; above this temperature, hardening rate may even be increasingly retarded (Coenen, 1975).
240	Acceptable maximum for hydrogenation of dimer and trimer fatty acids.

Temperature

Seemingly, beyond dispute, the increased reaction rate on the nickel surface, which, like other chemical reactions, occurs with a rise in temperature, far outweighs the fact that the solubility of hydrogen in oil and the transport through it are also increased. Hence, when a fall in hydrogen concentration occurs at the surface, the more unsaturated groups temporarily dominate the scene, and isomerization is more probable. Again, with copper as with nickel, reaction rate and temperature increase together. In practice, several temperature levels exist near which critical effects for the course of the reaction are to be obtained, as seen in Table 1.2.

Catalyst Action

A description of catalysts—their structure and behavior—is given in Chapter 7. Here, for the moment, we shall mostly concern ourselves with their effects on the speed and course of the reaction. As the amount of surface accessible to the reactants is increased, the rate of reaction increases, but only for as long as the hydrogen supply is adequate. If a series of laboratory tests is performed by using batches of the same oil known to be relatively free from catalyst poisons and under the same conditions of temperature, pressure, and the extent of IV drop, the type of results obtained is illustrated in Fig. 1.5.

Once the reaction is underway, the hydrogenation is completed in a certain time corresponding to the average rate of the uptake of hydrogen. If the dose of catalyst is increased, the time of the hydrogenation falls (i.e., the average rate of the uptake of hydrogen has risen). After a few tests, noticeably the increased dose of catalyst is no longer as effective in speeding up the reaction, and this corresponds to the leveling out of the curve.

If a second series of tests is performed with, for example, double the agitator speed, the hydrogenation rate at the lower catalyst doses will be in line with the size of these doses and not so very different from the early results of the first series, but a useful effect from further increased doses of catalyst now continues to be obtained up to a higher level. The series of tests may be repeated at four times the agitator speed, and again, while the effect at the lower end of the scale is similar, at the upper end a still higher dose of nickel has achieved a useful increase in the hydrogenation rate. The explanation of these results is obvious. At small doses of nickel, all rates of agitation serve to saturate the surface; thus, increasing the agitation gains little. As the nickel dose increases, more hydrogen per minute can be accommodated; hence, increased agitation is worthwhile. Conversely, for any degree of agitation, comes a level of catalyst dosage beyond which the nickel is less and less actively employed. Noticeably, for the degree of agitation commonly available in commercial hydrogenation vessels (autoclaves, converters), above 0.2% of nickel/oil is such a level. Again, if the reaction is considerably accelerated, problems of heat removal may begin to appear. Many suppliers of catalyst recommend doses of 0.02–0.1% of nickel/oil for a range of hydrogenations.

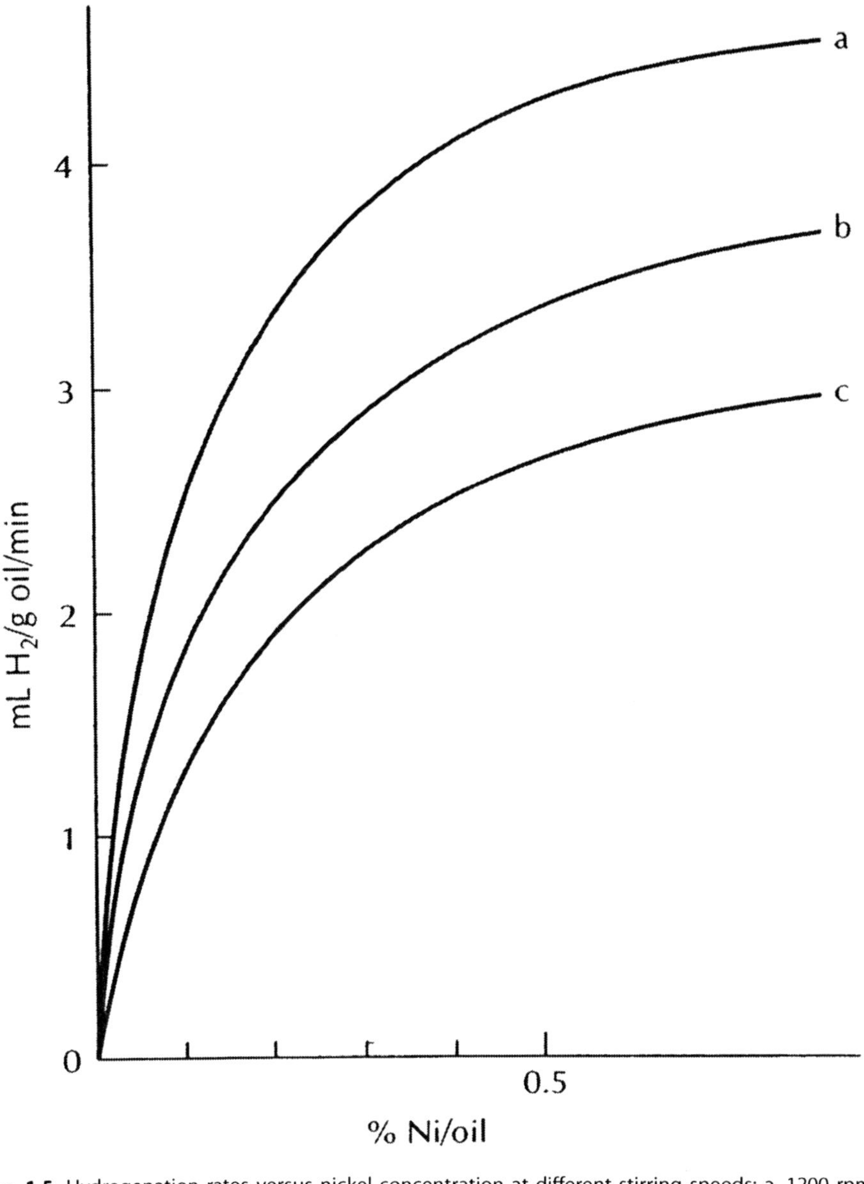

Fig. 1.5. Hydrogenation rates versus nickel concentration at different stirring speeds: a, 1200 rpm; b, 600 rpm; c, 300 rpm.

Catalyst Induction, Fatigue, and Poisoning

In the illustration of catalyst action just provided in the "Catalyst Action" section, care was taken to specify that the oil should be relatively free from catalyst poisons so that a confusing element might be largely eliminated. In practice, catalysts can suffer damage in various ways. First of all, at the time of manufacture and possibly during storage prior to use, some of the active nickel surface may be oxidized so that for some minutes very little reaction is evident at the beginning of hydrogenation. During this time, the superficial oxidation is being reduced so as to expose active catalyst. This so-called "induction period" is likely to be shorter if the temperature at the start was arranged at, say, 140°C, than if an attempt is being made to perform a low-temperature hydrogenation wholly within the 100–115°C range; in fact, a catalyst showing a considerable induction period would be best avoided in low-temperature hydrogenations.

Most nickel catalysts do not exhibit an induction period sufficiently long to pose any problem in industrial practice. When comparing the rate of hydrogenation achieved by one catalyst against another in identical circumstances, sometimes evidently one starts very well, but if the hydrogenation is longer than average—say, into the melting point range of 42°C and above—it fades more rapidly, and the last part of the hydrogenation may be obviously slower than in the case of the other. Possibly, one can attribute this falling off in performance to the thermal and mechanical attrition of the catalyst surface as much as to a degree of mechanical clogging and/or direct chemical poisoning. If this is the case, the deficiency may relate to the support, or lack of it, which was provided for the active nickel within the catalyst structure. Lastly, the question of poisoning arises, by which we mean the combination of nickel with carbon monoxide, sulfur, phosphorus, halogens, soap, free fatty acid, oxidized fat, or degradation by contact with excess moisture; no doubt, some of these act by clogging the surface as much as by chemical combination. Whereas poisoning by carbon monoxide is reversible in the sense that nickel carbonyl is unstable and by 160°C breaks down completely, the other substances may be regarded as irreversible in their action. The first and obvious effect of many poisons is that they are first neutralized by combination with valuable nickel; this amounts to a nickel cleansing process and underlines the merits of the adequate pretreatment of triglycerides or acid oils and fatty acids, since in the great majority of cases this will prove cheaper. It may easily be the case that for some impure feedstocks, 0.1–0.2% of nickel/oil is first taken up in this way, and hence, a first dose of 0.2–0.4% of nickel/oil may be necessary to ensure tolerably rapid progress to a low IV end product. This is an example of poisoning at its most potent level; often it operates merely by reducing the number of possible reuses of the catalyst. If one can not possibly stimulate the rate of a hydrogenation by purging with pure hydrogen, by applying increased pressure, or by using some other form of agitation, then almost certainly the catalyst is dead and more will be needed. Chapter 7 ("Durability and Poisoning" section) considers a range of catalyst poisons, and in Chapter 8 advice is given on the hydrogenation of individual oils, including information about the kind of pretreatment needed in cases where particular poisons such as phosphorus and sulfur are features of the oil concerned. Catalyst poisoning remains the subject of investigation and review from time to time in the literature (Coenen, 1975; Drozdowski & Gonaj-Moszora, 1980; Drozdowski & Zajac, 1977; Hastert, 1979; Ottesson, 1975).

In addition to exerting a most marked influence on the rate of hardening, the extent and manner of the poisoning have a strong effect on the access to different parts of the surface by unsaturated triglycerides, possibly on their adherence to it and certainly on whether migration and isomerization quickly attain their full possible effects as distinct from the hydrogenation of the double bonds. One can describe all of these phenomena very conveniently under the heading of selectivity, where the structure of the catalyst is shown to have much influence on them. Before doing this, something remains to be said concerning the general character of the hydrogenation reaction in the light of what was explained concerning the factors of hydrogen dispersion, pressure, temperature, and available active catalyst surface.

Order of Reaction

The "order of reaction" is an expression of the number of different molecules or atoms directly taking part in a reaction and whose concentration is therefore a factor in deciding the velocity of that reaction. One molecule is involved in the decomposition of nitrogen pentoxide, which is therefore a first-order reaction; radioactive decay of an element is another example. First-order reactions are characterized by the fact that the time taken to fall from the initial concentration, whatever that might be, to one-half of that concentration (a half-life of radioactive decay), is always the same. Put simply, if C represents concentration,

$$-\mathrm{d}C/\mathrm{d}t = kC \qquad \text{[Eq. 1.2]}$$

represents the rate of change. Where the concentrations of two different kinds of molecules govern the rate of reaction, the second order is represented. If one component of a reaction is being continuously renewed, a "pseudo order of reaction" condition is established since the usage of this component is masked by this renewal. In the case of the hydrogenation of many oils, especially the very unsaturated ones, so many double bonds are available at the beginning that the disappearance of several makes no appreciable difference to the reaction rate; also, of course, hydrogen is continuously renewed, so that initially the concentration of neither component appears important, and we have a pseudo zero-order reaction. Soon, however, as unsaturation diminishes, the supply of remaining double bonds begins to exert its expected influence, and this means that the hydrogenation reaction, for the remainder of its course, takes on the character of a first-order reaction.

Selectivity

Meaning

In its most general sense, selectivity in hydrogenation means a preference for hydrogenating one class of unsaturated substances rather than another, and in practice an ability to maintain this preference until the concentration of the preferred unsaturate is much decreased. Immediately we see selectivity is very relative to the classes of substances which we have under consideration. Further investigation also shows that the physical conditions in which the hydrogenation is performed have a powerful influence on the magnitude of the selective effect. This selectivity has both qualitative and quantitative aspects. While the above statements were probably acceptable from the earliest days of hardening, until about the 1960s some confusion existed because selectivity was not only related to the texture of the fats produced but an attempt was made to express it in terms similar to "softness" or "firmness" for a given final IV. But texture is influenced not only by saturates but by *trans* isomers (iso-acids), as explained in the "Iso- or *trans*-Promoting Hydrogenation" section in Chapter 2, and depending on the catalyst and conditions, more or fewer of these *trans* isomers are associated with selectivity. Hence, selectivity for some time continued to be discussed in qualitative and imprecise terms which did not have the same force for all concerned. This is a feature of older books and reports, but happily since the 1960s a common understanding arose that selectivity means a comparison between the rates at which two hydrogenation products are produced when measured over an accepted range of the particular hydrogenation reaction. This understanding clearly implies we are thinking in terms of the comparison of chemical reaction rates or ratios of rate constants. This approach by Bailey (1949) removes the time factor; ratios are also independent of catalyst activity and the overall hydrogenation rate. However, a product may be derived by a sequence of reactions, and, according to the purpose we have in mind, we may consider only the net end result, or we may probe more deeply to look at the intermediate steps. Similarly, we may choose to class together as the product the composite groups—such as all diunsaturated C18 fatty-acid components which were derived from more highly unsaturated groups—without concerning ourselves with whether they are predominantly *cis* or *trans* or where the double bonds are located in the chain.

These conventions are perfectly legitimate; however, in assigning numerical values to selectivities, we must bear in mind that these often represent net effects and not necessarily a one-step transition between a single species. Enough has now been said to make possible the offering of a set of definitions covering several important and recognized selectivities of different types. The terminology used by Coenen (1970, 1976) is followed; here and there a deliberate use is made of alternative expressions with the same meaning so that the reader may recognize all of these when encountered elsewhere.

Selectivity I (S_I)

This is a comparison of the rate at which oleic acid is produced from more-unsaturated fatty-acid groups with the rate at which oleic acid is itself hydrogenated further to stearic acid. The term "more-unsaturated fatty-acid groups" obviously includes linoleic acid with two double bonds, and indeed reference is sometimes made to "linoleic selectivity" in this connection; frequently, S_I is employed to refer to the net effect of polyunsaturate hydrogenation (of two and more double bonds) compared with monoene (oleic) hydrogenation—hence, the expression "selectivity ratio" or "SR" (Allen, 1966).

Selectivity II (S_{II})

The triunsaturate linolenic acid (*cis, cis, cis*, 9,12,15 octadecenoic) contains two reactive methylene groups (at 11 and 14C atoms), and is prone to atmospheric oxidation with the development of off flavors. Hence, if one can reduce it to the more stable linoleic or oleic acid, flavor stability is greatly enhanced while preserving the liquid or soft solid character of the oil of which it may be a constituent. Further, natural linoleic acid (*cis, cis*, 9,12 octadecenoic) is nutritionally very valuable; definitely avoid its reduction as much as possible while reducing the linolenic components. By comparing the rate of hydrogenation of linolenic acid to linoleic with the rate of hydrogenation of linoleic to oleic, we obtain what is classified as Selectivity II (S_{II}) or linolenate selectivity. Whenever oxidative stability in the product is being sought rather than an appreciable change in texture, S_{II} will be of value if the oil contains greater degrees of unsaturation than the two double bonds of linoleate. This increased number of double bonds is usually present as additional skipped groups ($-CH_2-CH=CH-CH_2-CH=CH$) which lend themselves to intermediate conjugation during the course of hydrogenation. Copper shows a much higher S_{II} than nickel; hence, fish oils rich in C22 and C24 chains possessing five and six double bonds may be reduced (by using a copper catalyst) to a situation where two or three isolated double bonds remain in such chains, yet the oil remains liquid.

As will be apparent, at the same time as linoleate is being hydrogenated to oleate, some is being produced by the hydrogenation of the linolenate. Assumably, the amount of linolenate converted to oleate during one visit within the pores of the catalyst is negligible, and isomers of linolenate and linoleate existing at the end of the hydrogenation period may be counted, if desired, with the normal acid. The calculation of the selectivity ratio from experimental results proceeds as described in the next section.

Specific Isomerization (S_i)

Very simply, this is the proponion of double bonds isomerized to the *trans* form to those saturated with hydrogen. Where a hydrogenated fat is required which shows a rapidly decreasing solid content with a rise in temperature (confectionery fats), a catalyst which encourages high specific isomerization is useful. Normally, a high selectivity I (S_I) will accompany a high specific isomerization (S_i), which is fortunate for the industrial operator.

Triglyceride Selectivity (S_T)

The types of selectivity which were already described—S_I, S_{II}, and S_i—are concerned with individual unsaturated fatty-acid groups, but in fats and oils these are linked together as triglycerides. If the hydrogenation that takes place is shared out in a random fashion over the various molecules in the bulk oil, one would describe this as an example of high triglyceride selectivity (S_T). If, on the other hand, during their proximity to the catalyst surface, more than one of the constituent unsaturated fatty acids of the triglyceride molecule sustain some hydrogenation, while many other molecules in the bulk oil remain slightly or not at all affected, this would be an instance of poor- or low-triglyceride selectivity. For example, if during the hydrogenation of an oil certain amounts of stearic-acid groups are formed and these are randomly distributed over the molecules present as a fraction, *s*, of all the fatty acids there, this would mean statistically that the proportion of tristearins (S_3) would be the much smaller fractions (*Margarit*, 1914). At the opposite extreme, where no random distribution of hydrogenation occurs, all the stearic-acid groups are confined to tristearin molecules; then the fraction of tristearins (S_3) is the same as the fraction of fatty acids which is stearic (*s*)—therefore, $S_3 = s$. The S_T would then have sunk to zero. A low S_T generally accompanies a low S_I. Coenen (1976, 1978) illustrates the extremes as shown in Fig. 1.6.

Miscellaneous Selectivities

All the selectivities described so far relate to some reaction with the triglycerides, but natural oils contain small amounts of other compounds, sometimes well under 100 ppm, which affect their appearance, odor, or taste, and these may be amenable to the removal or marked diminution by means of catalytic hydrogenation without the composition of the natural oil itself being seriously disturbed. Any catalyst or operating condition which displays a useful capacity for bringing about an improvement of this nature could be described as having selectivity of a certain kind. In view of the wide range of substances in various oils which could be affected to differing extents in this way, no general term for such selectivity has so far come into use.

The question remains of whether an unsaturated fatty-acid group situated in one position rather than another within a triglyceride has a greater chance of being hydrogenated, all other factors being equal. Primarily, the question has been of whether the 1 and 3 positions afforded a better chance than the 2 position. This attracted a modest amount of research since at least as early as the 1930s without the matter being resolved conclusively. This kind of situation can arise from the differences in experimental conditions as well as the limitations of the analytical techniques available. After earlier work which differed, the Hilditch school found, for example, that oleic acid hydrogenated with equal ease in all positions (Bushell & Hilditch, 1937; Hilditch & Jones, 1932; *Bailey's Industrial Oil and Fat Products*, p. 289 & 806). The equality of all positions seemingly was the commonly accepted view. However, in 1977, Drozdowski (1977) found that in hydrogenating linseed oil, interesterified soybean oil, and interesterified rapeseed oil with nickel, and cod liver oil with platinum, a preference existed for the 1 and 3 positions. In 1979, Kaimal and Lakshminarayana (1979) suggested a preferential hydrogenation of linoleic acid in the 1 and 3 positions in the nickel-catalyzed hydrogenation of cottonseed, sesame, soybean, and sunflower oils. More evidence for this kind of preference

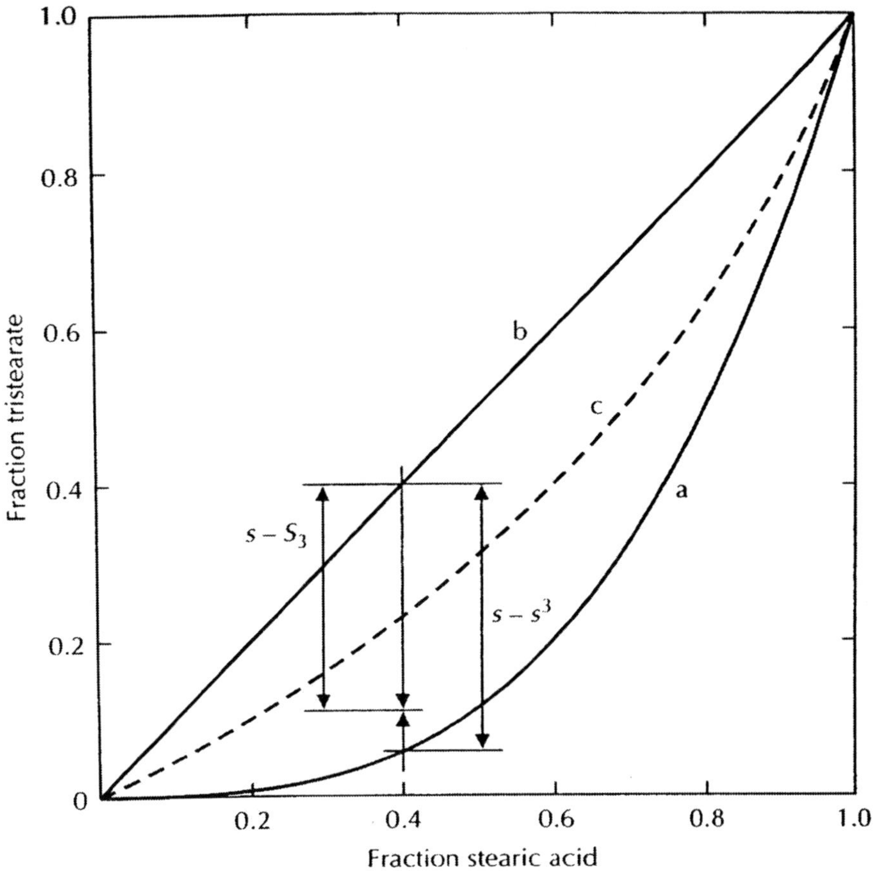

Fig. 1.6. Tristearates versus stearates at different S_T values. a, perfect triglyceride selectivity, $S_3 = s$ (*Margarit*, 1914); b, lowest possible triglyceride selectivity, $S_3 = s$. $S_T = s\text{-}S_3s\text{-}s$ (*Margarit*, 1914) equals 1 on curve a, 0 on curve b, 0.5 on curve c.

has arisen (Paulose et al., 1978), and Ackman made use of such effects in tracing the hydrogenation of herring oil (Sebedio et al., 1981).

So far as industrial practice is concerned, evidently, any selectivity must be more than marginal to be important, and unsaturated groups within the natural fats and oils offered for hydrogenation must be situated within the triglycerides so that they provide a sufficient bulk or quantity for their selective hydrogenation to account for a significant improvement in stability or texture, or both, in the chosen product. Positional selectivity of this importance appears to be absent.

Estimation of Selectivity

As previously stated, in the hydrogenation of fats, several reactions are proceeding simultaneously, and understandably we are most often concerned with the net effect at any particular stage; also, one may regard the reactions as first-order in character throughout nearly all of their duration. With these qualifications, we may consider how new analytical techniques and the use of digital and analog computers have made the estimation of selectivity vastly easier and quicker since the 1960s.

The hydrogenation of linolenate (Allen, 1967; Okkerse et al., 1967) is represented as follows:

$$\text{Linolenate} \xrightarrow{K\text{Ln}} \underset{(\text{isomers})}{\text{Linolenate}} \xrightarrow{K\text{L}} \underset{(\text{isomers})}{\text{Oleate}} \xrightarrow{K\text{O}} \text{Stearate}$$

$$C\text{Ln} \qquad\qquad C\text{L} \qquad\qquad C\text{O} \qquad\qquad C\text{S}$$

where CLn = concentration of linolenate; KLn = reaction constant of linolenate; CL = concentration of linoleate isomers; KL = net reaction constant of linoleate isomers; CO = concentration of oleate isomers; KO = net reaction constant of oleate isomers; and CS = concentration of stearate.

The disappearance of linolenate (Ln) and linoleate (L) during the reaction is given as

$$dLn/dt = -K\text{Ln}\,C\text{Ln and} \qquad \text{[Eq. 1.3]}$$

$$d\text{L}/\text{d}t = -K\text{L}C\text{L} + K\text{Ln}C\text{Ln} \qquad \text{[Eq. 1.4]}$$

which by integration gives at time t (starting from O)

$$C\text{Ln}.t = C\text{Ln.o}\exp(-K\text{Ln}.t)$$

$$C\text{L}t = C\text{Ln.o}\frac{K\text{Ln}}{K\text{L} - K\text{Ln}}(\exp(-K\text{Ln}.t) - \exp(-K\text{L}t)) + C\text{L.o}\exp(-K\text{L}t)$$

[Eq. 1.5]

If CLn.o, CLn.t, CL.o, and CL.t are found from the analyses of the original oil and after time r, then one can calculate the reaction constants KLn and KL and thus, selectivity II, since $S_{\text{II}} = K\text{Ln}/K\text{L}$. The solution of the equations may be from previously prepared graphs or by the use of a digital computer.

As early as 1965, commencing with cottonseed oil, then including groundnut, maize, soybean, and linseed oils, Albright (1965, 1967) determined initial and final concentrations of linoleate, *cis* and *trans* oleate, and stearate during hydrogenation. One could read S_{I}, in the terminology used above, $K\text{L}/K\text{O}$, from a computer-prepared series of curves on a graph comparable to Fig. 1.7 (Allen, 1966).

In Fig. 1.7, the ratio of final to initial linoleate concentrations is recorded on the x-axis. The increase in stearate is recorded on the y-axis, since this represents the hydrogenation of oleate in a simple, direct way. For example, if three-quarters of the

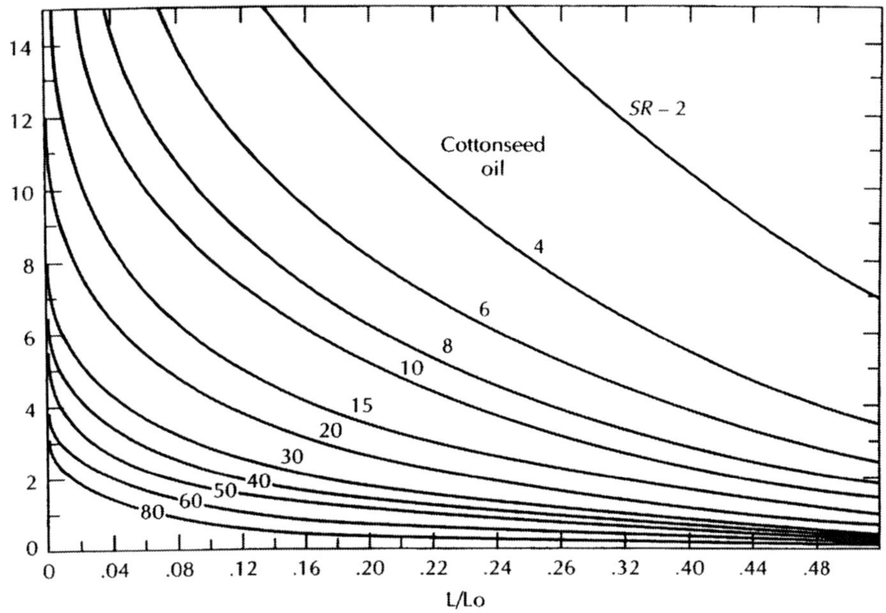

Fig. 1.7. Calculation of the selectivity ratio for the hydrogenation of cottonseed oil.

linoleate was hydrogenated by a certain time, the concentration of linoleate would have fallen such that *CL*/*CLo* = 0.25. If at the same time stearate had risen by 8%, the corresponding selectivity would be 4. This means linoleate had hydrogenated four times faster than oleate over the period in question. Had the increase in stearate been merely about 1.7%, the selectivity would be 20. The amount of linoleate hydrogenating directly to stearate (i.e., not via oleate) is taken as negligible: the increase in stearate represents the hydrogenation of both *cis* and *trans* oleate isomers. One must note: the extent of the hydrogenation reaction over which the selectivity is measured is a necessary part of the complete statement of test conditions (along with temperature, pressure, hydrogen dispersion, catalyst, etcetera—since throughout the reaction the concentrations of different groups are changing); mass action affects the operation of selectivity; isomers form and disappear; and finally, if the hydrogenation is carried far enough and the concentration of linoleates becomes low, the selectivity begins to break down, and oleate is converted to stearate in a rapidly increasing fashion (Scholfield et al., 1979). An illustration given by Coenen (1976) of S_I (Fig. 1.8) is most useful in showing the opposite extremes of this type of selectivity, assuming it is invariant during two ideally pseudo first-order reactions. The illustration is then brought to bear on the example of soybean oil hydrogenated to 95 IV under high selectivity (S_I = 50) and therefore *trans*- promoting conditions and conditions of low selectivity (S_I = 4) (Coenen, 1976). Fig. 1.9 shows the difference in the result. For salad oils, the solid content index (SCI) at 5°C should be low, coupled with adequate oxidative stability; hence, one may apply a light

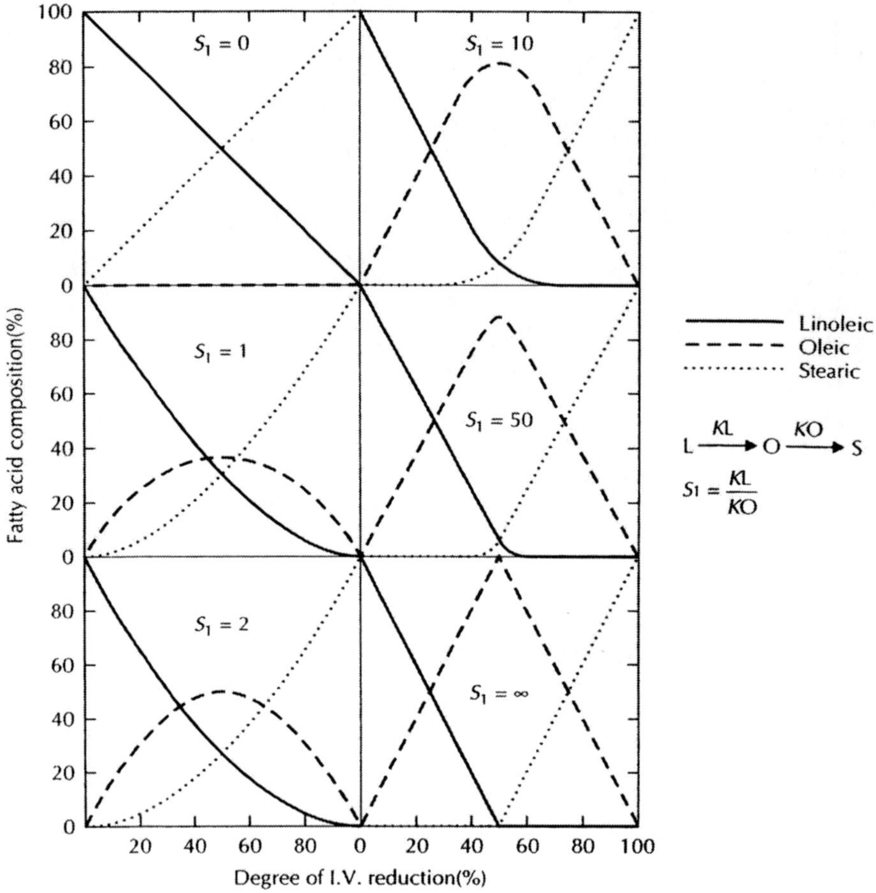

Fig. 1.8. Theoretical composition curves for the hydrogenation of linoleic esters: S_I = O, linoleates go directly to stearates; S_I = 1, linoleates have equal chances; S_I = 2, equal chances per double bond (but linoleate has twice as many); and $S_I = \infty$, no stearate is produced until the production of oleate has ceased. The first half of this section is approached when using a copper catalyst.

hydrogenation to diminish the polyunsaturates present. If this leads to more saturates than are consistent with the need for the product to remain clear in the refrigerator, winterization may reduce these. The more selective the hydrogenation, fewer solids need to be removed by winterization. For frying oils, the considerations are similar. For domestic margarines and confectionery fats, a rapid descent in SCI to a melting point below body temperature (37°C) is desirable—again, favored by high selectivity—whereas for bakery fats and shortenings, a longer plastic range is suitable, so a limited proportion of higher melting triglycerides is desirable and selectivity need not be so great.

Fig. 1.10 shows in the total shaded area what is currently achievable; for hydrogenated soybean oil, while the narrowly striated area to the right indicates what

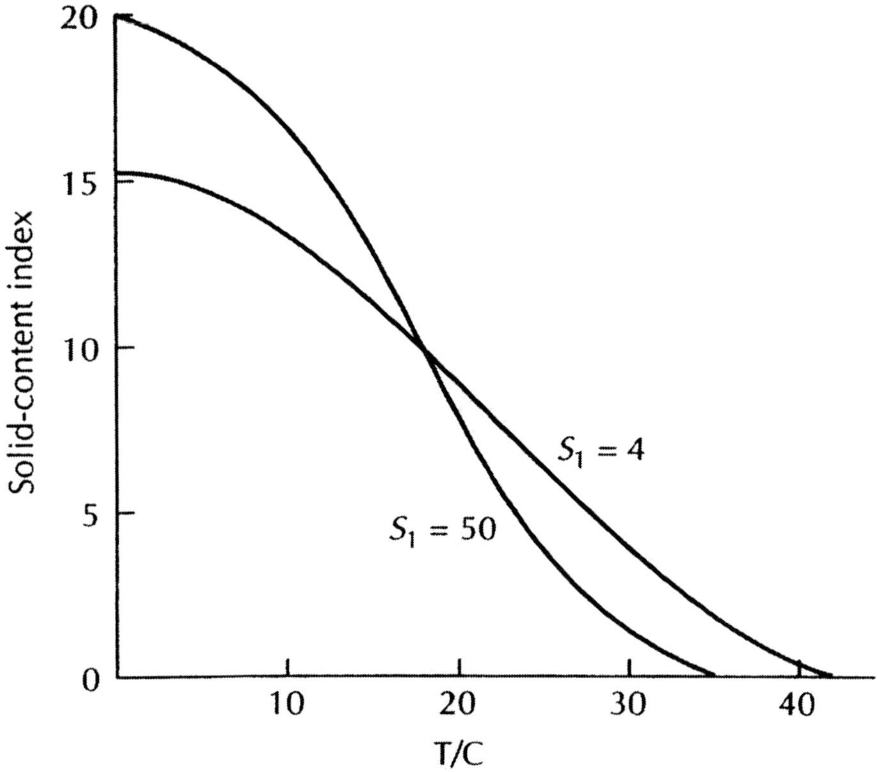

Fig. 1.9. SCI-temperature curves for soybean-oil hydrogenation to iodine value (IV) 95 under conditions of high selectivity ($S_I = 50$) and low selectivity ($S_I = 4$).

is generally the most useful SCI 20–30 region contrasted with a point typical of natural cocoa butter (Coenen, 1976).

For S_{II}, Allen (1966) uses the graph in Fig. 1.11 to illustrate how in the selective hydrogenation of soybean oil, a faster drop in the concentration of linolenates (low Ln/Lno) than is the case for the linoleates (L/Lo) means a selectivity in favor of linolenates. One should not think that, because the management of a plant lacks a ready access to a computer or because they do not wish to undertake as a routine investigation the direct comparison of hydrogenation rates of different fatty-acid groups in conjunction with the use of prepared graphical interpretations, they are unable to compare one catalyst with another regarding selectivity. Probably, for much of its production, the plant depends on a limited range of perhaps three or four crude oils which are pretreated in the plant to the standard found adequate prior to hardening. If a sample taken from such oils is then hydrogenated in the laboratory over the usual hydrogenation range and at the same conditions of temperature and pressure as are used in the plant, then the determination of solid fat content or solid fat index (SFI) at a key temperature of, say, 30°C will enable a decision to be made as

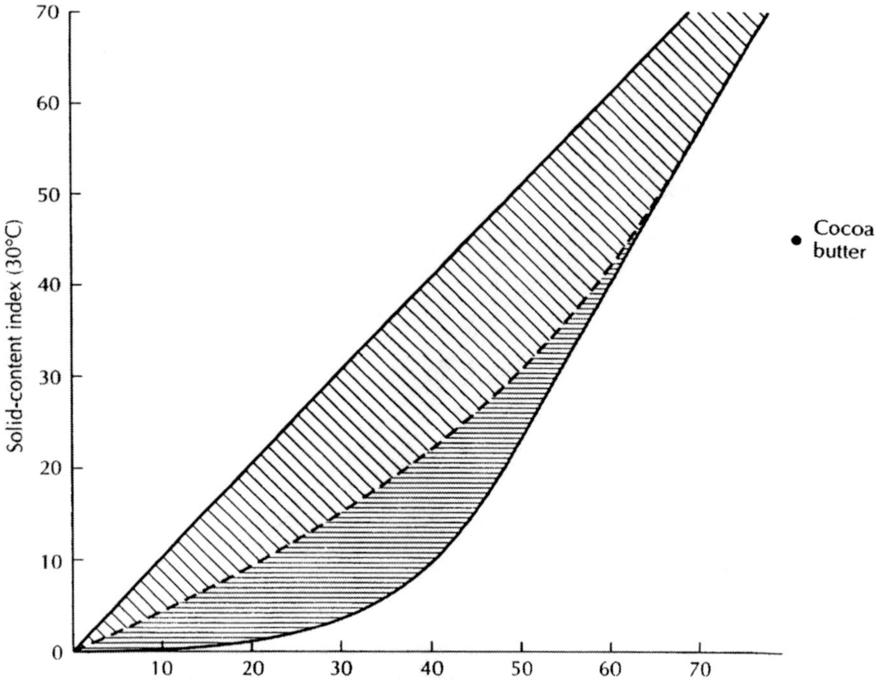

Fig. 1.10. Hydrogenated products from soybean oil accessible by present-day methods. Solid content index (SCI) at 20 and 30°C. For comparison, the SCI 20–30 point for cocoa butter is also given.

to which of the catalysts is the more selective for the intended use. Practicality may even be a reason to establish that, in given test conditions, the solid fat content or the SFI at 30°C shall not exceed some acceptable value (say, SFI 30 shall not exceed 4), which implies that the hardened fat melts completely comfortably below body temperature. Further, as pointed out by Allen (1966), since a difference between agitation speed and vessel geometry contributes to the difference in performance between laboratory autoclaves on the one hand and pilot-plant or full-scale autoclaves on the other, the speed of the laboratory agitator could be dropped to a point where, all other conditions being virtually equal, the same or very similar characteristics are obtained in laboratory test runs and operations on a larger scale. Most hardeners buy their catalyst from companies specializing in their manufacture and who usually provide selectivity data in their prospectus. This information, allied with the simple empirical test procedures outlined here, enables plant management to make a decision quite easily as to whether a catalyst is technically adequate and economical for their program.

Allen (Chen et al., 1981) published further simple mathematical models relating temperature, pressure, catalyst concentration, and agitation to reaction rate and S_I.

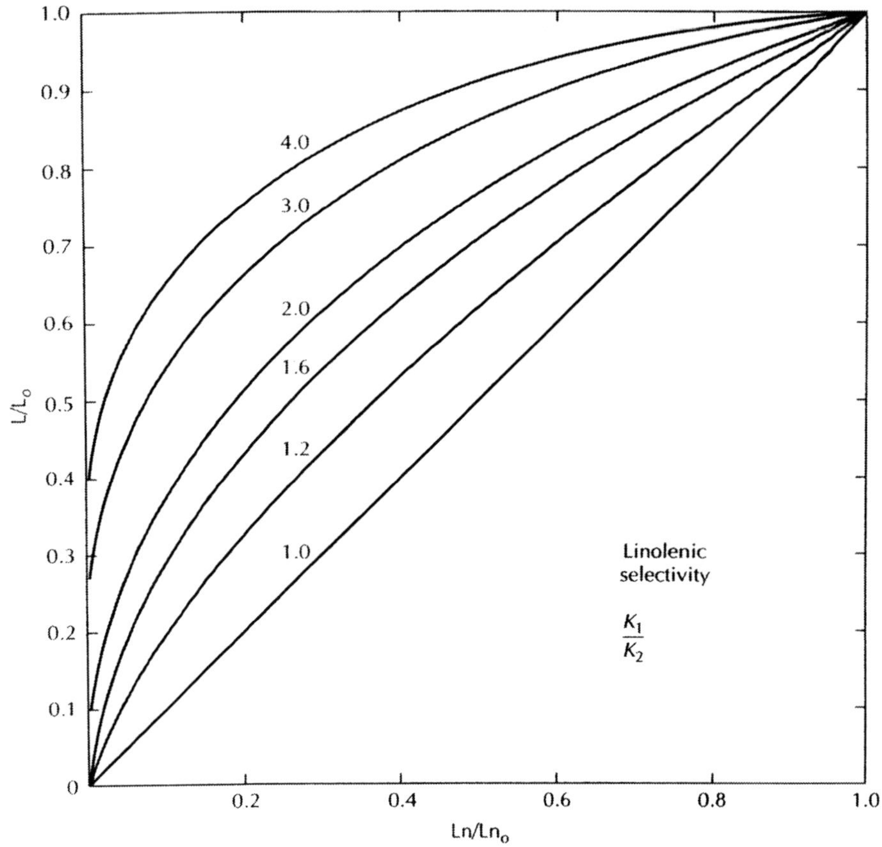

Fig. 1.11. Calculation of the linolenic selectivity ratio (S_{II}).

Operation of Selectivity

Selectivity operates very much according to conditions obtained at the catalyst surface. Although Chapter 7 is devoted to many aspects of catalyst structure and use, to describe here particular features of that surface which relate to selectivity is appropriate. The possibilities which exist when an unsaturated fatty-acid group aligns itself alongside a range of suitably spaced, catalytically active metal atoms, nickel in particular, were considered in detail in the "Isomerization" section of this chapter. Coenen (1978) suggests that the adsorbed triglycerides may cover only up to about 30% of the surface, leaving the remainder free for occupation by hydrogen if and when available. Fatty-acid groups with the most double bonds are seen as attaching themselves most firmly (Fig. 1.3a–e), therefore maintaining themselves in position slightly longer and hence, increasing their chance of obtaining a share of such hydrogen as is available. Well-known is that selectivity is higher if hydrogen is less plentiful. If the concentration of hydrogen is high, more often what may happen is that the first double bond of a polyunsaturated system to be adsorbed is hydrogenated completely before others can establish links with the surface (Fig. 1.3b,c). Thus, the molecule is desorbed quickly, and we have a mobile situation in which the chances for monoenes and polyenes to hydrogenate are more nearly equal. Again, well-known is that a rich supply of hydrogen depresses selectivity.

The distribution of intermediate positional and geometric isomers formed as hydrogenation continues supports this general picture (Coenen, 1970, 1976, 1978). The extent to which a direct linolenate-to-oleate shunt exists, as well as the routes via linoleate and isolinoleate, is discussed by Dutton (1972) along with the elucidation of the mechanism by which copper acts in the hydrogenation of soybean oil, relying on the formation of a conjugated double-bond system as an intermediate in the eventual production of monoenes; hence, the further hydrogenation of monoenes-to-saturates is inhibited. The profuse distribution of positional and geometric isomers in the nickel hydrogenation of soybean oil is also illustrated. Besides the various reactions taking place on the face of the active nickel, the conditions of the movement of molecules up to and from that surface also have a profound effect on selectivity: this time via a mass transport effect. The conventional nickel catalyst is spread out on a very porous support, usually of siliceous material, in which pore diameters may range from below 20 to several hundreds of Ångström units, all found in particles which themselves range from 1 to 10 µm or more (i.e., 10^4–10^5 Å). The dimensions of a triglyceride molecule are in the range of 15–20 Å; a hydrogen molecule, only about 2.4 Å. A schematic representation by Coenen (1970, 1976, 1978) of the state of affairs within three classes of pores—wide, medium, and narrow—during hydrogenation is given in Fig. 1.12.

In zone A, where the width is many times the dimension of the triglyceride, the molecules move into and out of from considerable depths within the pore without hindrance and mostly sustaining hydrogenation or isomerization of one double bond at most, if indeed any change occurs. Here the polyunsaturates dominate the surface, reside a fraction longer because of their firmer attachment, and take the first and much larger share of the adsorbed hydrogen. Hydrogen molecules, being much smaller, move even more freely. Selectivity in favor of the more unsaturated groups has the opportunity to operate most effectively in this situation. As the molecule penetrates more deeply into the pore, the duration of its stay is longer, and the chance of a double bond being hydrogenated increases on this account. Thus, within the pore of zone A, is a diminishing concentration of linoleate and a rising concentration of oleate. For pores of medium width (say, somewhat above 25 Å), a situation is established where nearly all the polyunsaturated molecules were hydrogenated by the time a certain depth is reached, and from there on (zone B) some oleate is being converted to stearate before it is desorbed and escapes back down the length of the pore to the bulk oil. As molecules leave zone A and penetrate zone B, the increasingly strong probability is that selectivity will have an even poorer chance to operate effectively; less and less linoleate is available, in any case, to dominate the active surface in this region. Finally, in those pores which allow a more restricted access, a zone C is recognized wherein even oleates are no longer found, and merely a semi-stagnant

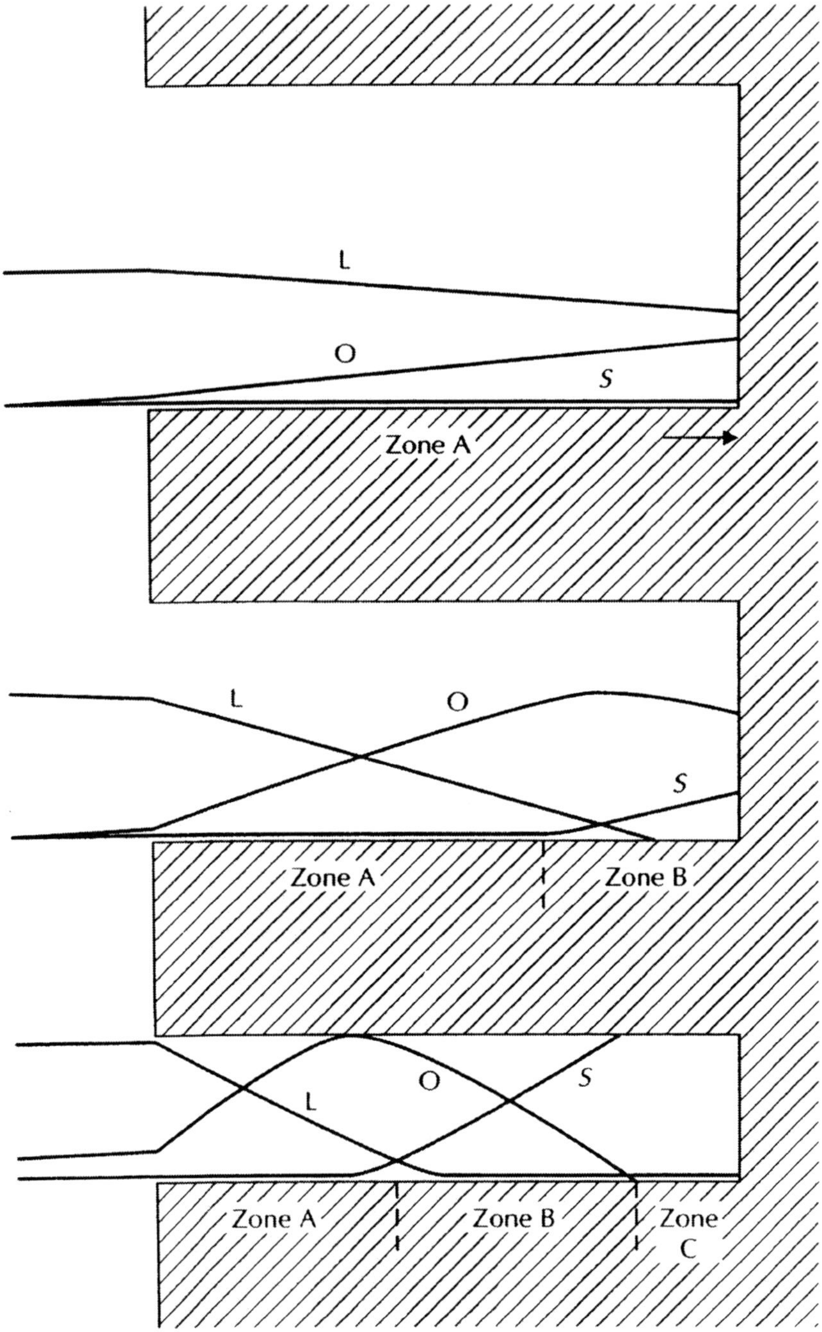

Fig. 1.12. Reactant concentration gradients in pores of different widths.

population of fully hydrogenated molecules exists. In this description, assumably, hydrogen molecules have a good opportunity to penetrate the working depth of the pore. If the hydrogen supply is diminished so that this is no longer the case, we will have a situation where, since no longer an effective concentration of hydrogen exists, even a relatively short way within the pore polyunsaturates have seized the great majority, and a selective condition exists. Again, that species of selectivity which depends on the ease or difficulty of mass transport within the catalyst pores depends therefore on the bulk of the molecules concerned. If two catalysts A and B are chosen so that A is a relatively wide-pore catalyst, whereas B possesses mainly narrow pores, and these are used to hydrogenate glyceryl trilinoleate and methyl linoleate, the results will show immediately that, in the case of the larger glyceryl molecule, catalyst A has operated selectively, at first producing much oleate and *trans* isomers before saturated stearate appears, whereas with B, stearate appears from the beginning and oleate and *trans* isomer intermediates never attain the concentration temporarily existing when using catalyst A. With the smaller molecules of the methyl ester, the results for A and B are closely similar throughout the whole hydrogenation process because the narrow pores of B no longer constitute a hindrance to movement. Evidently, catalyst poisons which mainly block the outer catalyst surface and wide pores will reduce selectivity; for substances which adhere preferably to narrow pores, a gain in selectivity will be found. For more information on this topic, consult Chapter 7.

The composition of the oil within catalyst pores is therefore, at any instant, more advanced along the route of the hydrogenation reaction than that of the bulk oil. A concentration gradient exists back toward the mouth of the pore to which the reaction products flow. Where this flow is impeded, the gradient steepens, less intermediate products escape, the reaction is less selective, a greater proportion of saturates forms, and the solid content or dilatation curve flattens (i.e., even if the melting point is elevated, it is approached more gradually). A comprehensive review of the operation of selectivity was made by Gray and Russell (1979) for nickel and other metal catalysts.

Combination of Factors Affecting Hydrogenation

Apparently, when reviewing the effect of various factors on the progress of a hydrogenation reaction, one finds that they achieve their results mainly by the influence they are able to exert on the concentration of hydrogen at the catalyst surface. When a certain objective is defined as the product of hydrogenation, its attainment is quite likely to be a matter of compromise between operating conditions which work in opposing directions—regarding the speed of reaction, selectivity, isomerization, and freedom from unwanted side reactions such as the formation of aromatic cyclic monomers intramolecularly or dimers and polymers intermolecularly when highly unsaturated groups are present. The character of the triglyceride or free fatty acid also has an influence. Hydrogenation tends to be faster when double bonds are conjugated or conjugatable, unsaturation is greater, carbon chains shorter, double bonds are *cis* rather than *trans*, and, at least in several observed cases, when the double bond is more remote from the carboxyl group.

Regarding selectivity and a nickel catalyst, which is overwhelmingly the most popular metal for industrial use, an increase occurs with unsaturation, a rise in temperature, an increase in available active nickel surface, and a decrease with greater hydrogen dispersion via faster or better agitation or rise in pressure. A series of simple test programs—covering the effect on selectivity and the simultaneous formation of *trans* isomers by changes in pressure, temperature, agitation, and the adsorption of a little purified gossypol—is reported by Allen (1966), and fits in with what is described here. The direction in which the various factors influence the results of hydrogenation is summarized in Table 1.3.

Strictly speaking, S_{II} compares the rate of hydrogenation of linolenates (and possibly even more unsaturated groups) with that of linoleates with only two double bonds. In the case of the nickel catalyst, the preference is only modest so that S_{II} is between 2 and 3, whereas linoleate selectivity (S_{II}) can reach values of 40–50. Nevertheless, S_I is included in Table 1.3 as an indication that conditions which favor S_{II}, S_{II}, and S_{II} also help S_{II}. Regarding the last item of oil unsaturation, perhaps obviously, when an extremely unsaturated oil is hydrogenated, the most unsaturated groups will be partially reduced first of all. This may lead to a situation in a long chain where two or three double bonds are isolated, that is, separated by more than two -CH_2- groups. In this case, because the opportunity of conjugation at the first adsorption is denied them, their reactivity will be low, and therefore normal linolenic and linoleic acids if present will gain a crucial advantage in the competition for hydrogen. This kind of consideration becomes particularly relevant in the important area of the hydrogenation of fish oils.

TABLE 1.3
Influence of Different Factors on Hydrogenation

	Surface hydrogen concentration	Reaction rate	S_I	S_{II}	S_i	S_T
Hydrogenation dispersion (agitation) increase	+	+	–	*	–	–
Hydrogen pressure increase	+	+	–	*	–	–
Temperature increase	–	+	+	*	+	+
Catalyst active surface increase	–	+	+	*	+	+
Oil unsaturation increase	–	+	+	+	+	+

Other Hydrogenation Routes

Various primary and secondary alcohols are well able to hydrogenate unsaturated fats and oils in the presence of conventional nickel catalysts commencing at a slow pace at room temperature, quickly near the boiling point of the alcohol, and most rapidly in the alcohol-vapor phase. Isopropyl alcohol reacts readily in this way near its own boiling point of 82°C, and it is itself converted to acetone in the process. Since acetone may readily be hydrogenated back to isopropyl alcohol, the possibilities of an industrial process begin to appear if only this route afforded technical or economic advantage. Although evidence exists of considerable S_I, and the hydrogenation can take oil to the point of complete saturation, the fact is that no industrial exploitation points to the absence of any such advantage (Patterson, 1974). This kind of reaction is known as conjugated hydrogenation or sometimes as transfer hydrogenation, and the molecules provide the hydrogen as hydrogen donors. Russian workers were among the earliest to explore this field. Their work since 1933 was briefly reviewed by Gray and Russell (1979), who provide references to the original papers and chemical abstracts. Indian researchers such as Chakrabaty et al. (1972) and Japanese workers such as Fukuzumi and Kato reported in some detail on the degree of selectivity obtained by primary and secondary alcohols and the production of conjugated dienes as part of the reduction mechanism, as well as *trans* isomers. The work of this school is by no means confined to alcohols as hydrogen donors or nickel as a catalyst; transfer hydrogenolysis (hydrogenation of carboxylic group to produce alcohols) was also investigated. Although a wide range of selectivities was demonstrated, no industrial application is known (Nishiguchi et al., 1977; Tagawa et al., 1978).

Of course, ample literature is available on the use of metals other than nickel as hydrogenation catalysts. Copper is the only one which is a serious candidate for significant industrial use, and this is remarkably slow to develop, although its activity has been the subject of research from the earliest times of fat hydrogenation. Coordination complexes of many transition metals form homogeneous hydrogenation catalysts often possessing high selectivity, but this technical characteristic has failed to secure economic justification. A description of these catalysts was therefore left to Chapter 7. Lastly, one must mention the ability of hydrazine (N_2H_4) to hydrogenate linolenates without the aid of a catalyst. Scholfield et al., in 1961 obtained no positional or geometric isomers, but they did obtain 9,12, and 15 monoenes and 9,12,9,15 and 12,15 dienes. Although of use in the laboratory, no industrial application appears likely.

Chapter 2

Hydrogenation Process Techniques

H.B.W. Patterson

Requirements

In the previous chapter, the nature of the hydrogenation reaction and the effect on it of the usual process variables, including type of catalyst, were described. This knowledge was adapted over the years as a more accurate and intimate understanding of the reaction was acquired to achieve a variety of requirements. The need for a product stable to atmospheric oxidation—and hence, able to maintain a desirable flavor (in some cases over long periods in arduous climatic conditions)—has come to the forefront. This is allied to a whole range of requirements for texture according to the intended use of the product. The requirements can often be met simply by hydrogenation; occasionally, the result of hydrogenation needs to be corrected or adjusted by a further modification technique—such as the removal of a small amount of high melting component by winterization, or the redistribution of a saturated group by interesterification of the hardened oil alone or in mixture with another oil—finally, the physical blending of the hardened oil with other soft and hardened oils to achieve a required texture over a temperature range is as old as the process itself.

In this chapter are described the conditions commonly employed to obtain various textures as closely as possible by means of hydrogenation alone and always consistent with flavor stability. These textures range from the near liquid at ambient temperatures to the firm, brittle solid. For several products, a compromise has to be devised between factors pulling in opposing directions; in these instances, the compromise may mean that for the first part of the hydrogenation one condition is used; then, before completion, an abrupt change is made, such as deliberately raising the temperature. Such a maneuver is not illogical since by the time the reaction has reached the half-way stage, we may be hydrogenating much changed material for which different conditions may not merely be tolerable but positively advantageous.

Under process conditions, one is justifiable to say something about batch and continuous operations. Traditionally, hydrogenation-process techniques were evolved on a basis of batch operation. For continuous operation, at least two kinds of consideration are involved: firstly, the possibility of readily obtaining the reaction conditions required to obtain an optimal technical specification for the product; in practice, this amounts to saying "Can the conditions of a continuous process be made sufficiently selective?"—if indeed that is an important consideration for the reaction in question. Secondly, the possibility of the production program for a particular product being sufficiently constant to afford economical uninterrupted operations of sufficient length from reasonably consistent feedstocks. A changeover from one product to another should not give rise to an amount of intermediate product which is embarrassing (i.e., the holdup must be kept to a minimum). If these conditions can be met, then, like other operators of continuous as against batch processes, the hardener is well-situated to enjoy the advantages of a savings in space, services, and labor; a more consistent quality is likely to be obtainable, especially where runs on one product extend to 24 hours or more; and because the operating unit is considerably smaller than its batch counterpart, its fabrication in more costly construction material—if this is considered desirable—becomes more economically attractive. Having said this, nevertheless remains the case that the great majority in the industry operate batch plants for triglyceride hydrogenation, and little indication is shown that this situation is about to change (Albright, 1967; Coenen, 1976; Hastert, 1981; Snyder et al., 1978). The continuous hydrogenation of fatty acids where isomerization and selectivity are less important is, on the other hand, quite popular.

A thorough cleansing of the crude oil, most often by degumming (if necessary), alkali neutralization, washing, drying, and bleaching lays the best foundation for the attainment of a hydrogenated-oil specification. This is especially the case where exists the matter of the selective removal of nearly all polyunsaturates with a minimal formation of isomers or saturated fatty acids. Particular requirements when cleansing individual crude oils prior to hydrogenation are described as each oil is considered in turn in Chapter 8. A clean, dry hydrogen is the obvious complement to clean, dry oil, and this is treated in detail in Chapter 5. Happily, the situation here has improved steadily since the 1960s in the sense that although electrolytic hydrogen of very high purity was used by some operators since the earliest days of hardening, the alternatives, which were frequently appreciably cheaper, have improved greatly in quality as the technology of hydrogen production via the steam reforming of hydrocarbons has advanced.

The fall in color during hydrogenation has always been a welcome bonus. Inevitably, if earlier process steps such as earth bleaching have lightened the color, the change due to hydrogenation will be less dramatic. Provided catalyst poisons are adequately removed, management must evaluate how heavy (in color) a bleach is really necessary if subsequent hydrogenation will itself remove much color. Obtaining the optimal effect with the greatest overall economy is the important consideration. Obviously, oils which were physically refined (i.e., free-fatty-acid-stripped by steam under reduced pressure) instead of alkali-refined are perfectly suitable for hydrogenation, since the removal of gummy material from the crude oil—normally entailing adsorptive cleansing by contact with acid-activated earth—which precedes the high-temperature steam stripping greatly reduces catalyst poisons. Again, as alkali refining has become more efficient via the use of centrifuges, this step alone, combined with washing and drying, may suffice as a pretreatment prior to hardening in a good number of cases. Do not forget that after hydrogenation a post-treatment stage follows. Hence, several ways to achieve an overall cost-savings present themselves; just how far one can take them is a matter for trial in the plant.

Batch Hydrogenation—Dead-End and Circulating

As stated previously, this is the traditional technique used in the fat-hardening industry, although several designs for continuous operation were put forward and a few operated within a decade of Normann's patent (Albright, 1967; Coenen, 1976; Hastert, 1981; Leuteritz, 1969; Snyder et al., 1978; Swern, 1964). The heart of the batch process is naturally the actual gassing time which often accounts for from one-third to one-half of the total cycle time. During this period, the temperature is certainly controlled, and may be deliberately varied according to a previously established pattern. The same is substantially true for pressure, at least in the case of so-called "dead-end" systems, that is, where hydrogen is simply fed to an internally agitated system and more can enter only as some is taken into combination by the oil. The alternative is the circulating system where the bulk of the hydrogen is passed through the oil several times, a proportion being absorbed each time; this is considered later. If pressure is to be varied, control is exercised via the gas-inlet valve situated on the gas line from the high-pressure gas store which itself is maintained at perhaps 7–10 atm for autoclaves operating up to 5 atm. Only a few plants have the facility for switching to an auxiliary high-pressure store (say, 20 atm) when their autoclaves (dead-end) are equipped to work at 10 atm. In the case of fatty-acid hardening, the normal operating pressure is likely to be in the 20–30 atm range in any case. Outside of the reaction time, the operations of filling, catalyst addition (which may be simultaneous or immediately following), preheating, final cooling, and then filtering have to be completed. In many designs, certain of these were divorced from the autoclave itself so that one may employ it for a higher proportion of its time on its essential function. Hence, the preheating may be largely accomplished as the oil enters the autoclave, through some form of heat exchange. Similarly, on the completion of hydrogenation, the oil batch may be dropped in its entirety to a receiving tank (drop tank).

Such a maneuver releases the autoclave as rapidly as possible for refilling. The cool, unhardened oil on its way to refill the autoclave may be passed through a coil in the drop tank to accomplish the preheating mentioned earlier. In some cases, the proposal was even made to discharge the hot, hardened oil from the autoclave via a heat exchanger to a filter while cool, unhardened oil passes through the other side of the heat exchanger to a second autoclave. As energy grows more costly, means of saving it will be studied more closely, and at the same time higher capital expenditure to ensure economy in its use becomes worthwhile. At the same time, remember that when numerous hydrogenation tasks are required of the plant, cycle times will vary, and therefore one should allow some flexibility in successive movements. One must therefore compare capital expenditure with energy saved and increased autoclave utilization in a realistic way; much depends on the complexity of the work program. One must provide an autoclave with cooling coils in any event (even if via external circulation of the oil) to control the exothermic reaction. To these cautionary remarks one must add another: the plant designer must always attempt to keep the possibilities of one kind of hardened oil contaminating another to a minimum;

otherwise, endless trouble ensues. The design of autoclaves and the layout of dead-end systems are discussed in Chapter 4 in more specific detail.

Hydrogen circulating systems, as just mentioned, were widely used for the first 50 years of the oil-hardening industry, probably because much of the hydrogen in use was far less pure than is now the case, and the movement of the gas from the top of the autoclave to a scrubbing train situated nearby afforded an opportunity to reduce some of the impurities before returning the gas to the base of the autoclave. The movement also provided agitation to create a gas–liquid interface and to maintain the catalyst in suspension. In several designs, this agitation was augmented by unsophisticated stirrers. With the advent of more widely available hydrogen well above 99.5% purity and agitators designed to be much more efficient in their use of power, circulating systems declined in popularity. The circulating hydrogen exports energy to the scrubbing system, and takes up energy when it is recompressed so as to rejoin the hydrogen feed; scrubbers have to be maintained. Today, at least one dead-end system provides the option of continuously sucking gas back from the headspace, condensing moisture by cooling, and then allowing it to be drawn back into the stream of fresh hydrogen entering the autoclave (Fig. 4.10). A common alternative is simply to provide the agitator with an axial-flow turbine impeller, just below the oil surface at working temperature, so that a vortex is created which sucks back hydrogen which has passed into the headspace. This amounts to a simple internal recirculation (Fig. 4.8). If a large amount of hydrogen was absorbed, as with fish-oil hardening, the buildup of moisture may be unacceptably high. In this event, the headspace may be purified by a brief purge to the atmosphere or returning some headspace gas via a condenser/scrubber (Fig. 4.1) to the hydrogen supply.

Continuous Hydrogenation—Fixed-Bed and Suspended-Catalyst

Continuous-hydrogenation processes may conveniently be divided into those employing a fixed bed of nickel catalyst and those in which the catalyst moves through the system suspended in the oil. Further subdivisions, especially of those employing the suspended catalyst, may be made, depending on whether gas is blown into oil or oil is sprayed into gas. Several designs are quite old (Ellis, 1912, 1913) and were listed by Leuteritz in his classification (1969). From the beginning of industrial hydrogenation, the usual potential advantages of a continuous process were understood by process designers. The economy in space, services, and labor with the high speed of hydrogenation, and the chance to monitor the reaction to obtain a consistent product, all provided a large incentive.

Fixed-Bed Catalyst

That the fixed-bed hydrogenation of triglycerides has not received more attention is commented on by Hastert (1981), who points out that in the petrochemical industry, for example, fixed-bed processes are the norm. He recognizes that a widespread and legitimate fear exists that impurities in the oil could quickly obscure the catalyst surface, and that in any case, even if this problem were solved, the fast-reaction rate that might well ensue would in all probability guarantee poor selectivity and perhaps only a small formation of *trans* isomers. Coenen (1976) emphasizes that the problems of the mass transport of triglyceride molecules within the pores of a supported catalyst whose particles were small enough to be suspended would become much greater if the catalyst were pelleted. The core of the pellet is foreseen to fill with stearate while only the outer shell continues to exert any effect.

If working conditions for the catalyst were made easier, one might find a useful application for fixed-bed catalysts. First of all, the refined oil could be well-cleansed of impurities. This is already often the case. The region of very light hydrogenation could first be chosen so that any gain in flavor stability would not be achieved at the expense of too great a change in texture due to either saturates or isomers being formed. Some off flavors in fats as widely different as coconut oil (probably derived from smoke-dried copra) and beef tallow may arise from mere traces of impurity. One can render a supported nickel catalyst nonpyrophoric, and then compress it into small pellets without difficulty. Even when pelleted at 25 tons/in (Normann, 1903) pressure, the catalyst is lightly abraded by the oil flow so that several ppm of nickel are found in the hydrogenated oil and must, therefore, be removed by the conventional posttreatment. To what extent this abrasion of the surface compensated for the soiling by the oil would depend on individual circumstances. One must distinguish between soiling (mechanical obstruction) and chemical poisoning by, for example, sulfur. In the latter case, well-known is that a sulfur-poisoned catalyst being employed to achieve a high *trans*-isomer content during the hydrogenation of a vegetable oil virtually free of such impurities gradually may lose its *trans*-promoting ability—sulfur content falls—and one may restore this by a controlled addition of fresh sulfur (Chapter 7).

Suspended Catalyst

As a continuous-hydrogenation system, the suspended catalyst has gained rather more acceptance than fixed-bed hydrogenation for triglycerides, and for fatty-acid hardening, it is even better placed. The same potential advantages of economy and consistency apply here with the same problems in obtaining sufficient selectivity. The immediate question is: How is best to reduce the spread in residence time, by which a minor proportion of the oil flow gets through the system with less than the average amount of hydrogenation while a corresponding amount is hydrogenated rather more? The first portion is a potential source of oxidative instability, and the latter, if it contains too much saturated fat, will cause the overall product to have a somewhat flatter saturated-fat content (SFC) curve than if hydrogenation was more evenly performed. To overcome these problems, units have taken the form of stirred-tank reactors with vigorous mixing in each. Early designs would allow three stages, but commonly, four and, better still, six are now considered necessary, as in the BUSS system (Leuteritz, 1969), where a common hydrogen headspace is sited above four or six compartments through which oil flows in a horizontal sequence, whereas Procter & Gamble patented (Albright, 1967; Hastert, 1981; Mills et al., 1950) a sequence of stirred compartments, one sited above the other in the form of a tower up which oil and hydrogen flow concurrently (see Chapter 4). Towers of this type may incorporate many more than six compartments, so the residence-time distribution is greatly narrowed, and the ideal condition of the so-called "plug flow" is approached, in which the result is what would occur if all the material flowed through a reactor at a uniform speed. Although Lurgi employees (1991; Hastert, 1979, 1981) offer their system for fatty-acid hydrogenation (see also the "Current Autoclave Agitator Designs: Loop Hydrogenation Reactor" section in Chapter 4), they have met the problem of residence-time distribution (sometimes described as back mixing) by having a tower of 16 compartments, while DRAVO at one time employed a long tubular reactor to which hydrogen could be introduced at intervals throughout its length (Hastert, 1981). In the BUSS system, the effect of residence-time distribution is opposed by the very interesting use of a high-speed venturi jet in each compartment which mixes hydrogen and oil more intimately and rapidly than any stirring action can achieve ("Current Autoclave Agitator Designs: Loop Hydrogenation Reactor" section). With all these systems, the typical hydrogenation rates of conventional factory-scale batch autoclaves of 1–3 IV units of drop/minute can be greatly exceeded, and rates of 10–30 IV units of drop/minute were demonstrated. The circulation of the reaction mixture through coolers external to the autoclave compartments or the cladding of the tower with a water jacket are devices used to contain the very considerable exothermic heat of reaction at these fast hydrogenation rates. BUSS reports (Leuteritz, 1969) that so efficient is the modern venturi design that a mass-transport barrier for hydrogen across the gas–liquid interface is virtually eliminated, with the consequence that the catalyst concentration within the limits of 0.05–0.20% of nickel/oil in their tests appeared no longer to influence selectivity and isomerization.

The capital expenditure required for a sophisticated continuous hydrogenation plant is admittedly rather higher than for a batch autoclave, so one must consider the size of the program before the payback is fully realized from operating economies. BUSS anticipates this to occur at about the 80 ton/24 hour output level maintained over at least 3 to 4 days in the week. As mentioned earlier, also the two-part question for all continuous plants is: How quickly can a change in product be achieved and what amount of intermediate product is to be expected from the transition? Research continues into the effects obtained from the use of nickel, copper, and other catalysts in various conditions of pressure and temperature. On the very small scale, tremendously high hydrogenation rates were achieved which on such a small scale presented no difficulty regarding temperature control. No dramatic improvement in Selectivity I (S_I) or Selectivity II (S_{II}), which would easily justify large-scale investment, has resulted (Illsemann & Mukherjee, 1978; Snyder et al., 1979).

Ultra-Light, Touch, Brush, or Flash Hydrogenation

Odor, flavor, and color are usually given to an oil by components present in only minute amounts. In a great many cases, these are greatly diminished by light hydrogenation, which ranges from a nominal IV drop of 1 unit to about 5; possibly the effect is gained, at least in part, by the catalyst promoting not only the hydrogenation but also the adsorption or destruction of these minor components. Texture should not be unduly affected, and if conditions of catalyst and temperature are well-chosen, the change in SFC for 20°C should be less than 5. The brevity of the contact explains the variety of names by which this technique is known among hardeners. If the crude oil concerned is known not to contain much catalyst poison, one may treat it directly; it should, of course, be dry. Lard and coconut oil are examples of this class.

Other fats already having a disagreeable odor and a deeper-than-normal color may require an adsorptive cleansing—loosely called bleaching—with from 0.5–2.0% of activated earth prior to light hydrogenation; cleansing by alkali neutralization, washing, and earth treatment is, of course, even more effective. This kind of improvement in the appearance and odor of a fat does not confer the same safeguard against future oxidation as is conferred by the selective hydrogenation of polyunsaturates in fats and oils where these are present. Sometimes simply a question of lightening the color may suffice, as in the case where palm oil is used for some technical purpose, such as soap making. If the oil or fat has suffered oxidative damage, improvement is more difficult to achieve, and this applies to other processes like refining and bleaching. Depending on the nature of the improvement being sought, whether an improvement of flavor with an absolute minimal disturbance of texture or the greatest possible lightening in color with greater tolerance for an increase in solids content, the type of catalyst may be varied from c. 0.05% of fresh nickel/oil to c. 0.50% of used nickel/oil. In this context, "used" means nickel which has lost about three-quarters of its original activity, if one knows that the fatty component of the used catalyst is not itself a source of off flavor. The temperature of the brief contact may be below 150°C or at 180°C; the lower temperature is preferred if the necessity arises to minimize the *trans*-isomer formation. In a dead-end batch autoclave, the brief contact may be arranged by raising the pressure in the headspace to 3–5 atm, closing the gas inlet, and then operating the stirrer until pressure has fallen to 1 atm. This one operation may lead to no discernible change in IV or refractive index (RI); it may, obviously, be repeated if a useful effect is gained without exceeding the amount of hardening which can be tolerated. In a hydrogen circulating system, the brief stirring is replaced by circulation for 1–2 minutes or as long as is advantageous. Although the immediate source of off flavor may sometimes be permanently removed by the rapid circulation of the oil over a column of pelleted catalyst in the presence of hydrogen, one must establish that this defect is not overcome by other process steps to which the oil is, in any case, to be subjected. Again, if only occasional deliveries of the raw material exhibit this defect and treatment in an existing batch-hydrogenation plant is readily available on the lines described above, that is the economical answer. Although a copper catalyst is found to be more selective and effective than a nickel catalyst in removing color and some off flavors, it brings with it the need to secure a stringent removal of copper in a posttreatment. The overall expense may render this answer less attractive (*Handbook of Soy Oil Processing and Utilization,* 1980).

Low-Temperature Hydrogenation

When stating low temperature, 120°C or less is intended. The aim is to bring down the proportion of polyunsaturates with three or more double bonds in the fatty-acid groups (e.g., linolenates) to under 2% while forming the minimum of solids. The low temperature certainly minimizes the production of low-melting solid isomers (iso-acids, *trans* isomers), especially if it is combined with the use of a fresh nickel catalyst at about the 0.05–0.15% of nickel/oil level and pressures of 3–5 atm. This technique applies especially to soybean oil but may be applied to other vegetable oils (Chapter 8). The extent of the IV drop depends on the amount of triunsaturates in the original oil; in the case of soybean oil, realistically aim for a drop in linolenates from c. 8% to under 2% for a fall of c. 30 IV units. The melting-point test is of little value, but gives a result in the mid-twenties in a centigrade range. As long as the catalyst remains substantially unpoisoned, its tendency to produce *trans* isomers will remain low. The low temperature is maintained throughout the hydrogenation, and at the end perhaps less than 20% of the fatty acid present will contain the *trans* form as seen by comparison with methyl elaidate via infrared spectroscopy. Naturally, the crude soft oil should be degummed (if necessary), refined, and given an adsorptive cleansing (activated earth) to get rid of catalyst poisons (Chapter 7). By the same token, less than 0.05% v/v of carbon monoxide in the hydrogen should appear. In these favorable circumstances, possibly one can even use the catalyst a second time. A vigorous dispersion of the pure hydrogen into the soft oil is also a favorable circumstance to the extent that it also minimizes isomerization even if it reduces selectivity. Evidently, from what was said of conditions for batch hydrogenation, this same class of hydrogenation can be performed with a suspended nickel catalyst in a continuous system (Mills et al., 1950), where the reaction is several times faster than in the conventional stirred-batch reactor and the residence time (during which *trans* isomers form) is correspondingly shorter. This class of hydrogenation, whether as batch or continuous, is a classic instance of the need to compromise between opposing factors to obtain an optimal result. This is applied to vegetable rather than marine oils.

Iso- or *trans*-Suppressive Hydrogenation

This kind of hydrogenation aims at a stable, rather soft-solid end result whose melting point is in the 30–40°C range. The approach to the melting point is gradual; hence, the SFC or solid-fat index (SFI) curves are rather flatter than normal, and the fat is said to have a long plastic range. This, of course, is what is required of a shortening or baking fat. For this purpose, a fresh or nearly fresh nickel catalyst is used at the usual 0.05–0.15% of nickel/oil level, up to 5 atm pressure with an upper limit of 160°C on temperature. We have now moved in the direction of greater selectivity as compared with the low-temperature hardening of the "Low-Temperature Hydrogenation" section in this chapter and with the vegetable oils concerned and the hydrogenation being pursued further, a melting point of 35–40°C is common. If the desired combination of a flat SFI curve and an adequate melting point does not arise directly from the one operation of hardening, customarily for many years one adjusted the texture in the desired direction by the addition of 3–5% of fully hardened vegetable oil. Fully hardened cottonseed or palm oil would also suffice (Patterson, 1974; Swern, 1964), and the effect is first measured by testing laboratory blends.

Normal Hydrogenation

The pattern for a great many hardenings of vegetable and animal fats is to commence after preheating to 140°C, and allow the temperature to rise to 180–200°C while maintaining a pressure of c. 3 atm: for marine oils, a pause at 150°C is indicated ("Cyclization and Polymerization" section in this chapter). True is that if a melting point of over 40°C is to be attained, 0.1–0.15% of fresh nickel/oil or three times that dosage of partly-exhausted nickel is likely to be employed, mainly to speed the reaction, since as unsaturation is very substantially reduced, whether the remaining double bonds have attained their usual equilibrium proportion of 66% of those present matters less and less; in any case, they are virtually certain to be close to it. If, on the other hand, the aim is to stop hardening at a melting point of c. 32–37°C, the texture may be influenced by whether the hydrogenation was sped on its way by a heavy dose (0.45% of nickel/oil) of repeatedly used catalyst, or a much smaller amount (0.08–0.10% of nickel/oil) of fresh catalyst which is expected to develop some isomerizing (*trans*-promoting) character from the poisoning it is sustaining from this, its first use. If no other restrictions apply, the above would be accepted by most hardeners as normal as far as a norm can be described. Sometimes the heavy dose of old catalyst has a bonus effect regarding the removal of color; again, as in the case of palm-oil hardening, where not much exothermic heat of reaction is available because of the restricted IV drop, assistance to the maximal extent from the heating coils is important. This achieves speed, lighter color, more selectivity, and maximal *trans* isomers, and if any CO is present, it will be unable to injure the catalyst at 160°C or above, since nickel carbonyl is then unstable. Obviously, if the operating temperature is designed never to exceed 150°C, no restriction as described in the "Cyclization and Polymerization" section applies.

Cyclization and Polymerization

Polyunsaturation is present in natural fats and oils most often in the skipped arrangement of the double bonds, and as a preliminary to hydrogenation at the catalyst surface, these can and do rearrange themselves into a conjugated system which is particularly reactive. For fatty-acid groups with three, four, and more double bonds in circumstances where insufficient hydrogen is present at the nickel surface, the conjugated systems in one molecule may react with those in others to form dimers or trimers, etc., but they may also form ring compounds within the same fatty-acid chain, which upon the loss of two hydrogen atoms to the nickel surface become aromatic in character (Coenen, 1970; Coenen et al., 1967). These compounds are biologically undesirable, and fortunately, a simply controlled procedure was evolved to heavily restrict the possibility of their formation.

Obviously, the greater the unsaturation, the greater the risk of cyclization at higher temperatures. Once the unsaturation level has dropped beyond a certain point, however, even at normal operating temperatures of 180°C, the risk becomes minute. If the temperature is restricted to no greater than 150°C until the drop in IV equals 0.002 (original IV) (Normann, 1903), the formation of aromatic fatty acids is reduced to under 0.1%, even with a very unsaturated oil in the conditions of poor-hydrogen availability (Coenen et al., 1967). As the drop in IV and a fall in the RI (sodium D line at 60°C) for most oils run closely together, this required drop in IV is sometimes interpreted as a corresponding fall in RI (one unit IV equivalent to one fourth-place unit in RIQ); hence, a fall in IV of 50 units would be counted as a drop from n_D^{60} 1.4650 to 1.4600 ("Oxidative and Stability" and "Unsaturation" sections in Chapter 12). Typical examples would be starting IV and units of RI drop = 0.002 (IV) (Normann, 1903): 130 and 34 (herring oil); 160 and 51 (menhaden oil); 190 and 72 (anchovy oil) and (linseed oil).

The restriction of temperature to 150°C maximum slows down hydrogenation so that less chance of a hydrogen scarcity at the catalyst surface is possible. Conversely, for autoclaves with a good dispersion of hydrogen, the risk of cyclization occurring is less; presumably this is because hydrogenation and possibly normal isomerization represent a higher probability than the sequence of rearrangements which must occur before cyclization. For oils with under 3% of linolenic groups (or groups of even higher unsaturation), the risk becomes negligible. For low-risk oils such as soybean, one can apply the rule, or one can check the circumstances of the hydrogenation by testing for aromatic fatty acid [via nonurea adducts (Coenen et al., 1967) left in a sample of the product] to establish that for that particular type of oil being hydrogenated in that autoclave in the conditions in question (temperature, pressure), the formation of aromatic fatty acids is below 0.1%, if indeed it is detectable at all.

In a continuous system, one can apply the same kind of check, and if necessary, the same restriction to within 150°C of the hydrogenation temperature for the early part of the transit. For batch and continuous systems, once the necessary hydrogenation within the 150°C limit is completed, perfectly acceptable would be to assist the temperature to rise to the 180–200°C level where it is controlled for the remainder of the reaction, since this assists selectivity and works toward a situation in which no more completely saturated fat is produced than is necessary. When hydrogenating oils with a substantial proportion of linolenates or groups of even greater unsaturation such as many marine oils, the following phenomena indicate that side reactions of the polymerization type have occurred:

1. The RI rises noticeably at the beginning of hardening, showing that conjugated systems are forming and then producing polymers. When marine oils are deliberately polymerized (300°C) for use as ingredients in wool-combing oils or for soap manufacture, a marked rise in RI is an indication that polymerization was achieved. Evidently, in hydrogenation exhibiting this behavior, because of the early use of temperature above 150°C, the polymerization is temporarily having a greater effect on refraction than the hydrogenation, which itself causes a fall.

One may arrest the effect by returning to a temperature lower than 150°C, but such polymers, etcetera that have already formed will remain in the oil.

2. If, as is often the case, the hardening is being monitored by RI, very probably the expected melting point–RI relationship was appreciably disturbed, and that some SFC or SFI figures were abnormally low because polymers, etcetera have formed. Such events, if not understood, may become a source of friction among hardening-plant staff.

3. Hydrogenated oil which has suffered the formation of polymers during hardening will probably refuse to harden to complete saturation when a sample is tested in the laboratory, even with a generous dose of fresh catalyst. The lowest attainable IV may be around 3; deliberately polymerized oils are found to cease hydrogenation at a considerably higher IV.

Two-Stage Hydrogenation

To resolve a conflict between different factors influencing the progress of the reaction, a compromise set of conditions is often imposed over virtually the entire sequence. However, once an oil is partially hydrogenated, its character is so modified that it would be logical to impose revised conditions, even to continue updating the revision, so that a constant feedback ensures the most accurate attainment of the exact end-point specification.

The control of pressure and temperature via some variety of computer was demonstrated as technically feasible as early as 1957 (Eckman & Lefkowitz, 1957), part of the expense arising from the analytical monitoring equipment. Analog computers were used in elucidating reaction mechanisms (Butterfield, 1964), and in 1972 Dutton described how a mass spectrograph was associated with computers as an investigational tool and what developments might soon be made in the way of computerized automatic control. An elementary and inexpensive approach to the compromise, amounting to no more than a switch to a new set of conditions at a predetermined point when it was considered advantageous to do so, was evidently adopted by some producers of hydrogenated shortening (Swern, 1951) in the United States many years ago. Basically, the concept was to proceed with the hydrogenation of a vegetable oil at a low temperature to suppress the *trans*-isomer type of low-melting solid, accepting meanwhile a diminished selectivity, and then to switch the reaction to a higher temperature so as to conclude it in much more selective conditions which would secure the necessary oxidative stability.

The location of the switch point as a certain IV or RI and the relatively low temperature at which it was approached no doubt varied according to the type of oil and the sought-for specification. This technique need not be confined to vegetable oils and shortenings; to indicate what limits one should accept in endeavoring to secure an advantage from this two-stage hardening is possible. First of all, 180°C for the final stage favors selectivity, and an IV drop of at least 15 units should be achieved during this stage. The time taken to rise from the switch point to 180°C will depend on the effectiveness of the heating coils and the degree of exothermic heat of reaction. In many plants, this interval will account for an IV drop of another 10 units at least, making a total of at least 25 units from the time the signal is given to raise the temperature at full speed. From the switch point onward, to maintain no more than a moderate hydrogen pressure of, say, 1 atm on the autoclave may be helpful, the better to assist selectivity.

The last decision concerns how low a temperature should be used for the first stage of hydrogenation prior to the switch point. The practical limits for this choice seem to lie between 110 and 140°C. To some extent, the slowing down of the rate of reaction and the loss of some selectivity at the lower temperature stage may be offset by employing an increased dose of catalyst of up to 0.25% of fresh nickel/oil. As some indication of likely acceptable oxidative stability, to achieve a reduction of linolenic acid to about 1% is advantageous. If greater selectivity in the two-stage process is needed, this will result by bringing in the switch point earlier in the hydrogenation, or as a second choice by lifting the first stage nearer to 140°C—or a combination of these. The price for better selectivity and greater stability will be an increase in *trans* isomers with a possible drop in the amount of saturates. Employing a catalyst very low in sulfur content will be an advantage when little room is available for maneuver in other conditions and a product is sought with a low SFC or SFI at 20°C. Obviously, the two-stage process described here is very comfortably inside the rules given ("Cyclization and Polymerization" section in this chapter) for avoiding aromatic fatty acids.

Iso- or *trans*-Promoting Hydrogenation

The need arises, especially in domestic margarines sold as a wrapped block and in confectionery fats, to employ material which melts rapidly in the mouth—thus, avoiding a fatty or gummy aftertaste—yet remains substantially solid at ambient temperatures. In temperate zones where often a gap of 20°C exists between ambient and body temperatures, a good performance in the above respect presents no serious problem, but as we move through semi-tropical to tropical zones, the gap narrows to a few degrees or vanishes, and the problem becomes ever more important. This book does not concern itself with the details of product formulations, but in this section the conditions favorable to the production of hardened fats which melt rapidly as 37°C is approached are described. This behavior means that the SFI or SFC curve in the 20–30°C zone should be steep and, ideally, the gradient will be at its steepest around 30°C. At the slip melting point, it is usual for the SFI to be about 4 or the SFC to be about 5%. A drop in SFC values between 20 and 30°C of 25–35%, depending on the oil being hardened and the final melting point, is commonly acceptable, although in the confectionery trade, much steeper drops of 50% are sought and obtained. The *trans* isomers which arise during hydrogenation are the comparatively low-melting solids which contribute to this quick-melting effect, although the special class of lauric and hardened lauric oils must be noted (coconut, palm kernel, babassu, and tucum). In their case, the preponderance of short-chain fatty acids in the triglycerides evidently secures the quick-melting effect, since even when hardened to an IV of 1, differences of 50% and more arise between SFC values at 20 and 30°C.

The hydrogenation conditions which promote isomerization also favor selectivity and delay the production of saturates. These conditions are: temperatures above 160°C and low-to-moderate concentrations of hydrogen at the nickel-catalyst surface. The point must be made immediately that if the oil being hydrogenated is one for which a risk of cyclization occurring exists as described in the "Cyclization and Polymerization" section in this chapter, the precaution of first-stage hydrogenation not above 150°C should be adopted. Where this danger is not present, the operator can commence hydrogenation at, say, 140°C, proceeding rapidly to the 180–200°C level at which most of the hardening will presumably take place with an advantage to the formation of *trans* isomers and a steep SFC curve. Some operators who are able to preheat above 140°C may do this. The "low-to-moderate" hydrogen concentrations referred to above may be interpreted as 0.5–2.0 atm pressure. Results will soon indicate the satisfactory level in balancing isomerization and speed of reaction.

Finally, the questions remain as to the type and dosage of the catalyst, to which the answers have a vital importance in securing the steepest possible SFC curve. We already know (see "Isomerization" section in Chapter 1) that *trans* isomers reach a peak concentration of some 66% of remaining double bonds after hydrogenation is underway for some time, and after that an equilibrium is established such that *cis* and *trans* isomers disappear together to form saturates; hydrogenation now progresses into the 40°C+ slip melting-point range where the SFC curve becomes less steep. Therefore, to choose a catalyst which will encourage early isomerization is vital, so that the 66% *trans* equilibrium position is reached as early as possible and the amount of *trans* isomer (iso-acid) is at its highest. For this purpose, a sulfur-poisoned catalyst is traditionally the most useful. Since the poisoning of the catalyst has reduced its activity, customarily one compensates for this by increasing the dosage if the operator finds this necessary. The common source of sulfur-poisoned catalyst in the past was the same plant in which the catalyst was used when fresh. This has the disadvantage that most of the catalyst's surface is merely clogged by a variety of substances from its previous usage, some of which happen to have combined with active nickel, whereas what we most need is a nickel surface which was modified in a way which best suits our current purpose. Catalyst poisoning is discussed more fully in Chapter 7, but these practical details will indicate the most useful approach to securing a supply of a consistently steep SFC-curve material.

Fresh, active supported nickel catalyst, if derived from nickel sulfate, may contain c. 0.3% of sulfate/nickel (but if from other nickel salts, such as chloride, no sulfur), and will be used at about the 0.1% of nickel/oil level. Eventually, its activity drops to perhaps one-sixth of what it once was, and by then it is put to work on the production of hardened oils in the 31–38°C melting-point range. An increase of 3% of sulfate/nickel is usually sufficient to reduce heavily the activity of a fat-hardening catalyst. This old or spent catalyst probably contains 2–3% of sulfate/nickel; it can be employed at even 0.5% of nickel/oil, and will be distinctly iso-promoting when first diverted for use with vegetable oil very low in sulfur and other catalyst poisons. Soon its ability to give sufficient *trans* isomers, as judged by SFI or SFC at 20°C, becomes less, although the rate of hardening has not diminished. In fact, the sulfur content has probably been lowered and the activity marginally increased. To cure this, flowers of sulfur equivalent to 3% of S/nickel should be mixed with the catalyst, and during the subsequent hydrogenation much of this will repoison the nickel. In obstinate cases of failure to secure sufficient *trans* isomers with this type of catalyst, take steps to slow the hardening to the minimal acceptable rate while always ensuring that some hydrogen continues to be available. This might mean maintaining a hydrogen pressure of only 0.5 atm in the early stages of hydrogenation or throughout, but this measure is exceptional. Positive results should appear in the batch subsequent to the one in which the sulfuring was performed, if not immediately. After further use, a resulfuring is again likely to be needed. Finally, the catalyst's filtration characteristics may become so poor, due to a combination of abrasion by mechanical handling and further soiling, that their use must be abandoned. Hydrogenation times may also have become too long (i.e., in excess of 3–5 hours of gassing as required by the plant program).

A decidedly more effective catalyst in the field of isomerization is the fresh sulfur-poisoned supported catalyst which one can purchase from suppliers (Chapter 7). If sulfuring is carried out in such a way that an even distribution of the sulfur over the nickel surface occurs, a situation is evidently created wherein the chance of an adsorbed double bond losing the first hydrogen captured before a second is taken up is very much greater. The bond desorbing is most likely to be *trans*, and the full complement of *trans* isomers is attained in time for a very steep SFC curve to be reached. One prominent supplier Harshaw Catalysts (Okonek, 1986) indicates that little difference in *trans*-content results between hardening temperatures of 160 and 200°C when hardening soybean oil with 0.1% of nickel/oil of this specially prepared catalyst, although hydrogenation is several times faster at the higher temperature and 3 atm pressure is recommended. Seemingly, this class of "tailored" catalyst is so effective that it is less dependent on high temperature and low pressure to secure its effect. Although other operators are able to work in the 1–3 atm pressure range, all prefer 180–200°C, and concentrations of nickel at 0.05–0.2% of nickel/oil could be influenced by the IV drop required and the time available, provided the texture specification can be met. Naturally, this specially poisoned catalyst is likely to sustain several cycles of use. If its iso-promoting power drops, no reason is evident why the operator should not resort to resulfuring as given earlier for the naturally poisoned catalyst. This class of hardening is more tolerant of hydrogen which contains some catalyst poison. Baltes (1970) described the extremely high *trans*-promoting quality of Ni_3S_2 and Ni/W sulfide catalysts and their durability, but their advantages have not led to commercial exploitation.

Higher-Melting and Fully-Saturated Hardened Oils

Unsaturation in these oils has become so small that the questions of selectivity and isomerization are of little importance. If a hardened oil of the melting point of a few degrees above 40°C and some appreciable unsaturation were required to have a flatter-than-normal dilatation curve, a question of hardening under nonselective conditions would exist—using a rather lower temperature than usual and, if possible, at a higher pressure with a normal dose of fresh catalyst. Otherwise, the production of hardened vegetable or marine oils to between a 40 and 50°C melting point is a means of making further use of a catalyst which has lost part of its activity and the dose of which is therefore increased two- or threefold while working mainly at 180°C. When a fully saturated oil of a melting point possibly over 60°C is sought, largely the solution is a matter of providing a dose of fresh catalyst which will complete the hydrogenation within a few hours for the convenience of the program (say, about 0.1% of fresh nickel/oil); excessively long hardening times can bring about some increase in free fatty acid (FFA) above that which is usual for this class of hardening. Normally, an increase from 0.1 to 0.3% of FFA might be experienced. If the oil being hydrogenated is liable to cyclization in the earlier stages, then they must be completed at temperatures not above 150°C as explained in the "Cyclization and Polymerization" section since the polymers, etcetera, if formed, would remain. A particularly important precaution to be observed when the specification is tight and calls for a final IV below 2 is to ensure that, in being discharged and filtered, the hardened oil is not contaminated. Fully hardened lauric oils are in a class by themselves and have rather low melting points. If quite small amounts of other vegetable oils (c. 0.5%) are introduced via the catalyst or in some other way into the lauric oil about to be hardened, this can lead to an elevation of two or three degrees in the melting point of the final hardened oil. No difficulty lies in obtaining fresh catalyst made up in lauric oils, as they provide a stable and convenient medium for this purpose. In this class of fully saturated oil production, where the headspace above the oil in a dead-end autoclave may have accumulated 20–30% of inert gases, a short purge—probably to atmosphere—will be well worthwhile in helping restore the rate of hydrogenation. Naturally, as the IV falls and fewer and fewer double bonds remain to take up the hydrogen, the speed of the reaction will slacken. In this situation, hardeners who have the facility of increasing the normal working pressure readily at hand will probably elect to do so.

Consistent Quality in Hydrogenated-Oil Deliveries

The procedure of handling and storing fats and oils before, during, and after processing so as to cause minimal damage is a worthwhile study in itself, from both a technical and an economical standpoint. All that is said in this section relates merely to stratagems whereby the hardener may succeed in keeping close to specifications in his deliveries of hydrogenated oil to another part of his factory or to customers outside.

Specifications will almost certainly include a requirement concerned with slip point, texture (flat or steep SFC curves), IV, and color. The problem of consistency arises in two ways. Firstly, a substantial program of several-hundred tons per week of one quality delivered in tank cars may exist; and secondly, one delivery of, say, 20 tons of some other unrelated product has to be made each month. In the first program, assuming the attempt is being made to adhere conscientiously to the hydrogenation regime laid down (temperature, pressure, catalyst), the ever-present problem exists of stopping the hardening on target. The specified IV may allow a total spread of 3 units. The judicious use of the refractometer (Chapter 12) where exists a similar spread of ±2 units in the fourth place will allow the hydrogenation to be halted close within the upper IV limit. The inspection of results over a period possibly reveals a fall from autoclave end point to receipt in filtered oil tanks of 1 or 2 IV. More than this suggests a leaking-gas inlet valve or that, as the finished hardened oil is dropping steadily from the autoclave, hydrogen to take its place is being allowed to bubble gently up through the receding oil instead of gas being fed in quietly to the headspace above the oil. The successive autoclave batches, or collected parcels from a continuous unit, must be checked prior to transfer to a main storage tank of 200 tons or more. Autoclave batches which were overshot seriously (by which is meant they are 5 IV too low or SFC 5% too high) may have to be diverted for use in another product or for further hardening where such an outlet exists. Equally, if the degree of under-hardening implies a lack of oxidative stability, this batch must also be returned for some brief further hydrogenation lest it destabilize a much larger quantity of oil. For the same melting point, charges which were hardened on an old catalyst will probably show rather higher SFI or SFC values at 20°C than those hardened on fresh catalyst. By the time the large storage tank is half-full, the possibility exists to see if a drift above or under specification is present and to take action accordingly, always provided that no batch is arrested at so soft a stage that the high IV jeopardizes the stability of the ultimate blend. The best way to mix a large storage tank of oil is by means of a side-mounted propeller, as in the petroleum industry; within 5 hours, 500 tons can be made uniform.

The solution to the second problem, of delivering 20 tons within specification which allows some room for maneuvering, is to complete the 20-ton parcel by hardening it as four 5-ton charges or two 10-ton charges. The tests on the first charges, added to a store of data from the production of charges of the same material on earlier occasions, place the operator in a position of adhering to specifications without attempting unacceptably wide variations to achieve a satisfactory blend.

Chapter 3

Hydrogenation Using Critical Fluids

Jerry W. King[1] and Gary R. List[2]

[1]University of Arkansas, Department of Chemical Engineering

[2]National Center for Agricultural Utilization Research, Agricultural Research Service/USDA

Introduction

Historically, hydrogenation processes have employed pressure as a variable to provide higher reaction yields and to affect better contact between the hydrogen, substrates, and catalyst components during the hydrogenation reaction. Such reaction conditions are described in the chapters in this book. The variable of pressure is also prominent in processes which employ supercritical fluids as a medium for extractions and reactions (King & List, 1996; King, 2003), particularly when supercritical carbon dioxide (SC-CO_2) is utilized with its capability of readily dissolving fat, oil, and associated lipid moieties (Friedrich, 1984; Stahl et al., 1987). Hence linked by the common variable of pressure, it could be anticipated that a merger between processes that use supercritical fluids and hydrogenation would occur in time.

Many binary gas systems are highly miscible over a large concentration range at pressures and temperatures above their critical point, thus critical fluids such as SC-CO_2 can serve to dissolve and transport hydrogen during the process of hydrogenation very effectively. Only at very low temperatures, substantially lower than the liquefaction point of the supercritical fluid, can gas-gas phase separation be observed; and this in affect is the basis of separating such gas mixtures (Cipollina et al, 2007). From a thermodynamic perspective, hydrogen considerably above its critical temperature (T_c = –240°C) and pressure (P_c = 12.8 atm), can be viewed as a supercritical fluid having very little densification. Thus the cohesive energy density of the hydrogen-supercritical fluid binary fluid is largely a function of the density of the supercritical fluid component having the higher T_c and P_c (King et al., 1995; Zhang & King, 1997).

Research and process development in supercritical fluid-based hydrogenations was largely catalyzed by the projected and verified higher mass transport rates facilitated by using supercritical fluids in conjunction with hydrogen versus conducting similar hydrogenations of lipids in their molten state or in conventional liquid solvents (Patterson, 1994). Although to date a large production plant employing supercritical hydrogenation of fats and oils has not evolved, similar pilot plant units for conducting hydrogenations exist that awaits potential scale up (Wandeler & Baiker, 2004). On the other hand, production plant facilities do exist employing supercritical fluids for the production of chemical intermediates (Licence et al., 1993; Hitzler, 1998), and a number of hydrogenation processes that have been conducted at super- and sub-critical conditions without acknowledging this fact (Wandeler & Baiker, 1994).

This chapter on hydrogenation using critical fluids follows a logical progression starting with fundamental material on why supercritical and subcritical fluids can facilitate hydrogenations advantageously, including a background discussion on characteristics of critical fluids. An initial overview of hydrogenation reactions in critical fluid media is presented which also includes reaction options on non-lipid substrates. The rationale for presenting such information is to open additional vistas to the synthetic organic chemist and/or process engineer for using hydrogenation under critical fluid conditions that have not been previously explored. We have also discussed in the space available, some equipment and processing scale up options to facilitate supercritical fluid hydrogenation.

Of course a major emphasis has been placed on supercritical fluid hydrogenation of fats and oils using either SC-CO_2 or propane and similar agents. The synthesis of oleochemicals is also presented—particularly the production of fatty alcohols under supercritical conditions. The coupling of supercritical fluid hydrogenation with other processes employing supercritical fluids, such as supercritical extraction (SFE), or coupling hydrogenation with a consecutive reaction in supercritical fluid media is discussed to the extent of the possibilities that have been demonstrated in the literature. The section on "Critical Fluids and Catalysts" is largely concerned with miscellaneous topics, such as information on catalysts and hydrogenation in compressed water, which have relevance in the context of supercritical fluid hydrogenation. A brief summary of key patents pertaining to critical fluid hydrogenation is included as a starting point for readers desiring more information on patented technologies in the field. Finally, both of the chapter's authors have been involved for three decades in supercritical fluid research involving fats and oils, as well as related technologies, largely through their affiliation with the USDA-ARS' National Center for Agricultural Utilization Research in Peoria, Illinois. This would not have been possible without the contribution of Dr. John Friedrich who developed the High Pressure Laboratory at the Peoria USDA facility, and we wish to acknowledge this in our prologue to this chapter.

Why Supercritical and Subcritical Fluids?

Research involving supercritical fluids has a long and interesting history, and the reader can trace developments and the background to the field by consulting key texts and reviews on the subject (McHugh & Krukonis, 1986; Brunner, 1994; Arai et al., 2002; Dunford et al., 2003). Supercritical fluid phenomena can be traced back into the 1800's but its utilization in applied technology fields first surfaced in understanding the dynamics of gas and liquid recovery and fractionation in the petroleum industry (Katz & Rzasa, 1946; Sandrea & Nielsen, 1974) and later in the synthesis of high polymers using high pressure, i.e., the polymerization of ethylene under supercritical conditions. In the early 1970s the emergence of classic comprehensive patent by Zosel working in Germany (Zosel, 1976) demonstrated the possibilities of using supercritical fluids in the processing of foodstuffs, particularly the decaffeination of coffee and processing of hops. This opened up the "golden era" of SFE using SC-CO_2 for many food applications—which has continued to this very day. Spurious examples for conducting reactions in critical fluid media can be found in the above texts and the patent literature, but it was probably not until the very late 1980s, and certainly in the 1990s that reaction chemistry conducted in the presence of a supercritical fluid were being widely investigated (Jessop & Leitner, 1999).

The two-dimensional phase diagram for carbon dioxide, by far the most utilized critical fluid, is shown in Fig. 3.1. Here the supercritical fluid region is generally defined as the region above CO_2's critical temperature, T_c = 31°C and critical pressure, P_c = 72 atm, i.e., the upper right quadrant of the phase diagram. Carbon dioxide can also exist as a liquefied gas or high temperature fluid over a range of temperatures and pressures, defined above the vapor-liquid (V-L) equilibrium line, between the T_c and the boiling point for CO_2. Note that a requisite minimal pressure is required (defined by the V-L curve) to prevent liquefied CO_2, i.e., subcritical CO_2, from converting to the gaseous state. Very similar phase diagrams exist for all substances which define their physical and intermediate states (e.g., water in Fig. 3.1) although the magnitude of the pressures and temperatures associated with the axes are quite different. The above is a rather bland description of a supercritical fluid based on phase equilibria relationships, however suffice to say for functional purposes, a supercritical fluid is a gas held at high pressures above its T_c which exhibits the solvent properties of a liquid and the mass transport characteristics of a gas, which are dependent on the applied pressure and temperature to which the fluid is subjected.

The tunable solvent properties of supercritical fluids such as SC-CO_2 are of course what have made them versatile extraction solvents. Coupled with their mass transport parameters such as diffusion coefficients and viscosity, this can result in very fast extraction fluxes being exhibited by targeted solutes (products) from sample

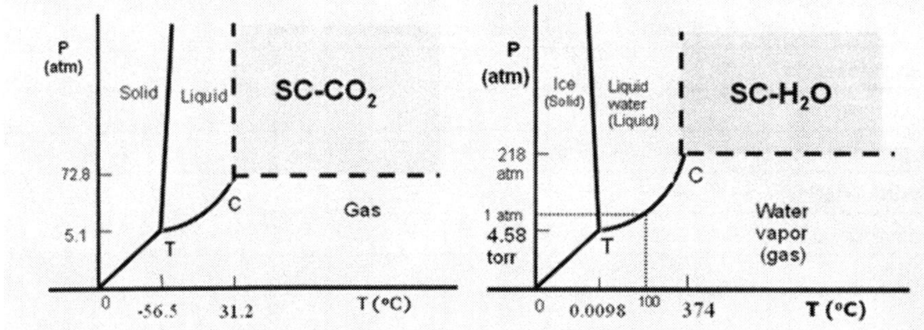

Fig. 3.1. Two dimensional phase diagrams for carbon dioxide and water.

matrices. These same properties also make critical fluids attractive media in which to conduct reactions since reactant contact with catalyst surfaces and subsequent transport of reaction products from catalyst are readily facilitated. Fluids such as SC-CO_2 if properly used are of low toxicity, thus products made in the presence of SC-CO_2 are devoid of the presence of organic solvents, an important feature for oil or fat products intended for use in the food consumer marketplace. "Solventless" hydrogenation also is attractive from a process engineering perspective since recovery and possible disposal of the solvent is avoided. The ability to design critical fluid processing in which the supercritical fluid can be recycled also avoids the objections to jettisoning for example, carbon dioxide into the environment. These attributes will be discussed in further detail in the section on "Equipment, Processing Concepts, and Scale-Up."

Table 3.1 below tabulates the types of reactions that have been conducted in the presence of critical fluid media.

Our focus is on hydrogenations in this chapter, but these reactions are also done employing both heterogeneous and homogeneous catalysis. As will be documented in the section on "Coupled Processes Using Supercritical Hydrogenation", supercritical fluid hydrogenations can be coupled with SFE and other types of reactions noted in Table 3.1 to achieve a specific product type or distribution. Products from hydrogenations conducted in the presence of critical fluids can be further fractionated using columnar or chromatographic methods utilizing the same critical fluid medium (King, 2002). Coupled with the control of reaction rates and product distribution via the supercritical fluid reaction (SFR) options, such "coupled" or tandem processing options allow a wide diversity of goals to be achieved. It is also possible to employ critical fluid media for the regeneration of catalysts used in the hydrogenation step, thus the capital equipment investment required to build a supercritical fluid hydrogenation plant, can also be utilized for other unit processing operations, thereby justifying the cost of plant construction and creating additional profits.

An excellent and fairly recent review of the entire supercritical fluid hydrogenation field has been provided by Seki, Grunwaldt, and Baiker (2008). This review is

TABLE 3.1
Types of Reactions Conducted in the Presence of Critical Fluids

Polymerization
Enzymatic
Heterogeneous or homogenous catalysis
Hydrogenation
Conversions in sub- and super-critical H_2O
Pyrolytic
Photolytic
Reactions of Analytical Significance

dominated by hydrogenation of small compounds and the characteristics of generic and specialized catalysts used in the hydrogenations. A short compilation reaction rate data for catalytic hydrogenations in supercritical fluids has been authored by Ramirez et al., (2002) from the Universitat Politècnica de Catalunya in Barcelona, Spain. The cited works deal largely with heterogeneous catalysis in the presence of supercritical fluids; however, observed reaction rates, space velocities, selectivities and apparent kinetic constants have been noted. In the case of vegetable oils, data are available on the effect of pressure and reaction conditions on the selectivity toward the preferred *cis*-isomer during linoleic hydrogenation.

Fundamentals of Critical Fluids Pertinent to Hydrogenations

Aside from the previously noted benefits of conducting hydrogenations in the presence of critical fluids, some more specific benefits are afforded by hydrogenating with the aid of critical fluids: improvement in mass transfer or reactants and products, regulation of product distributions, reduction in the amount of hydrogenation required, higher product quality, and a higher rate of hydrogenation. Aside from control of reactant-product solubility and accelerated mass transfer, higher rates of hydrogenation are inherent due to the pressure dependence of the rate constant on the activation volumes for reactants and products involved in the hydrogenation reaction. For such reactions, one can expect the change of the reaction rate constant (based either on mole-fraction or molar-concentration) with pressure at constant temperature to be:

$$-RT\left(\frac{\partial \ln k_x}{\partial P}\right)_T = \Delta V^{\ddagger} \tag{3.1}$$

where the activation volume, $\Delta V^{\ddagger}$, is:

$$\Delta V^{\ddagger} = \bar{V}_{\ddagger} - \bar{V}_A - \bar{V}_B$$

and, V denotes the partial molar volumes of the reactants and products and ‡ indicates the activation-state complex. In practice, application of Equation 3.1 can be complicated because the partial molar volumes and the activation volume can themselves be a function of pressure, which is the case for a normal or supercritical hydrogenation conducted under pressure. Reactions can increase or decrease their rates with pressure, depending on the value of $\Delta V^{\ddagger}$. As an example of the possible magnitude of the pressure effect, some organic reactions can double the reaction rate when the pressure is increased from atmospheric (0.1 MPa) to 50 MPa (which gives $\Delta V^{\ddagger} = -0.025$ L/mol). Higher product quality is often times found in the lighter color of the resultant fat or oil and after supercritical fluid hydrogenation, partly due to the prophylactic benefit of using carbon dioxide as one of the fluids in the binary fluid hydrogenation mixture.

Supercritical fluids are normally utilized as one fluid phase, or perhaps with a co-solvent consisting of organic solvent up to mole fractions of 0.15–0.20, an amount consistent with maintaining one phase with respect to the imbibed co-solvent. This co-solvent can also consist of using water, although water has a very low mole fraction solubility in SC-CO_2 (Takenouchi, 1964) which increases monotonically with temperature. For the case of the binary fluid combination of carbon dioxide and hydrogen, both fluids are usually applied at reaction conditions above their respective critical temperatures and pressures. Indeed one of the advantages of supercritical fluid hydrogenation is that both fluids are mutually miscible in all proportions at these higher reaction temperatures. The critical loci defining the one-phase and two-phase regions of hydrogen with a specific critical fluid having a higher T_c and P_c are occasionally available in the literature, but can also be estimated by computational methods using a NIST data base (NIST-Refprop, 2009). Suffice to say the terminal points of the critical loci for hydrogen—supercritical fluid pair correspond to their respective T_c and P_c, and the locus of the critical loci is not often a monotonic function of mole fraction composition of the binary pair, but will attain pressures and temperatures above each of the binary pairs respective T_c and P_c, often times reaching a maximum at a temperature and pressure above their T_c and P_c.

It is a documented fact that when utilizing a binary pair of supercritical fluids, the solvent power of the supercritical fluid having a higher T_c and P_c will be reduced by the presence of the supercritical fluid having a lower T_c and P_c (hydrogen). This effect has been demonstrated by one of the authors for the SC-CO_2 – He or N_2 system (King et al, 1995; Zhang & King, 1997) as well as others (Ruckenstein & Shulgin, 2003). This can also change the cohesive energy density of the fluid (Cipollina et al., 2007), and hence its ability to dissolve solutes (lipid materials) in comparison to using the neat supercritical fluid (SC-CO_2). The variation in solubility parameter, the square root of the cohesive density energy, for certain binary gas mixtures, such CO_2 – H_2 is shown in Fig. 3.2 (Cipollina et al., 2007).

Supercritical CO_2 can also change a substrate's physical properties or morphology when it is used for both SFE and SFR (hydrogenation). Expansion of the lipid phase by dissolved SC-CO_2 has been cited a number of times, particularly more recently in biodiesel synthesis using supercritical fluids (Wyatt & Haas, 2009). This expansion of the lipid substrate due to dissolution of SC-CO_2 can be verified experimentally by use of view cell measurements as reported in the literature (Bezanehtak et al., 2002; 2004). On a volumetric basis this expansion can be considerable, e.g., 10^3 for CO_2 in methanol. Such an expansion partially explains why hydrogenations, when H_2 is added as a third component to such a system, can be facilitated in a multiphasic system since H_2 can be readily absorbed by the expanded liquid. The dissolution of the supercritical fluid in the fat or oil substrate also affects the melting point of the lipid moiety causing a depression under its ambient melting point—hence a solid fat can be turned into a liquid when this process occurs. Absorption of the SC-CO_2 into the lipid phase also causes an increase in its fluidity potentially aiding in the transport and contact between the lipid phase-hydrogenation-catalysts

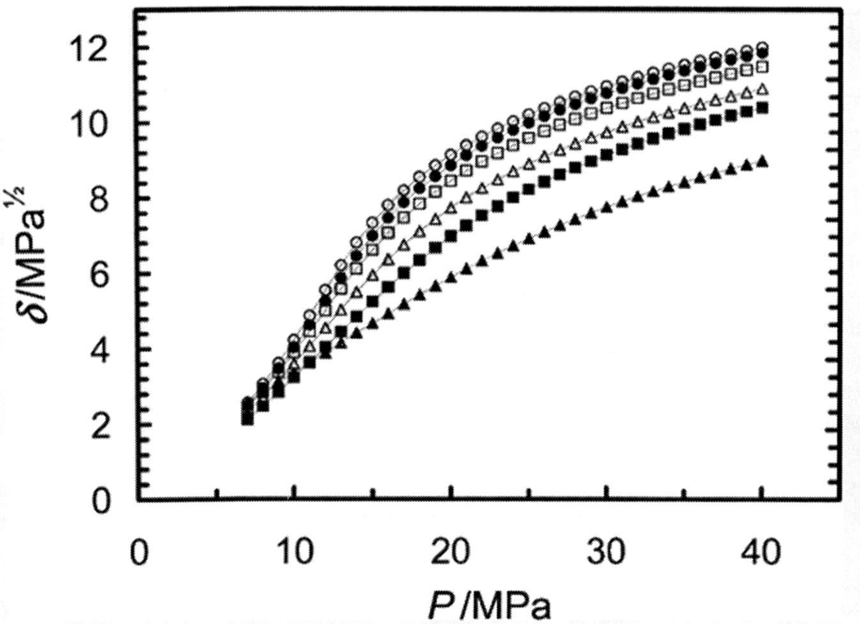

Fig. 3.2. Hildebrand parameter vs. total pressure for selected binary CO_2-CO (empty symbols) and CO_2-H_2 (filled symbols) mixtures at $T = 333$ *K*: O, $x_2 = 0.046$; 0, $x_2 = 0.089$; 4, $x_2 = 0.138$; b, $x_3 = 0.045$; 9, $x_3 = 0.135$; 2, $x_3 = 0.244$. Reprinted with permission from Cippolina, et al., (2007). Copyright 2007 American Chemical Society.

surface leading to faster reaction rates. An interesting discussion of the properties of CO_2 – expanded lipids has been recently provided by Seifried and Temelli (2010). Interestingly, SC-CO_2 can also be used as a reactant in some SFRs as reported by Beckman (2004) and Ikariya and Noyori (2003).

Hydrogenation under supercritical fluid conditions, which if applied by design or unwittingly, has seen numerous applications. This includes the synthesis of fuels (Saka, 2006), fine chemical production (Hitzler et al., 1998), fuel feedstock conversion in both the oil and coal industries, and hydrogenation of polymeric substrates (Jessop et al., 1999). Hydrogenation for transforming lipid substrates covers not only constituent triglycerides composing fats and oils, but also fatty acid or alcohols, as well as specialty lipids. In this chapter, we will cover these in more depth in the following sections: "Supercritical Fluid Hydrogenation of Fats/Oils Using CO_2 or Propane," "Oleochemical Synthesis in Supercritical Fluids," "Coupled Processes Using Supercritical Hydrogenation," and "Critical Fluids and Catalysts;" and several excellent but outdated reviews have been previously written on this subject (Ramirez et al., 2002). An excellent and fairly recent review of the entire supercritical fluid hydrogenation field has been provided in the Ph.D. thesis of Ramirez (2005) which is available on the Internet.

Hydrogenation Reactions in Critical Fluids—An Overview

Perhaps one of the basic benefits of conducting reactions, such as hydrogenations in a supercritical medium, is to facilitate contact between the catalyst, substrate, and hydrogen, a contact which is inhibited due to mass transfer through the condensed liquid state as noted below in Fig. 3.3. Here we see the barrier imposed by having to convey hydrogen gas through the liquid phase to the catalyst surface which is reduced considerably by employing a supercritical fluid that approximates a normal gas phase mass transport process. This results in facilitating considerably faster reaction rates as will be shown later in this chapter.

This section is concerned with a broad view of the possibilities of using critical fluids for hydrogenation reactions. In conducting hydrogenations using traditional methods involving gas-liquid phase systems, it is customary to increase the reaction temperature to enhance hydrogen's solubility in the starting substrates; however when using a supercritical fluid for hydrogenation, temperature can be used as an independent variable to improve selectivity without sacrificing conversion. The higher effective hydrogen solubility in the starting substrate and at the interface with the catalyst reduces the possibility of isomerization reactions from occurring during the overall hydrogenation process. However there is a limit to the extent of hydrogen's solubility in the starting substrate even in the presence of a supercritical fluid as temperature is increased, and it can actually decrease depending on the

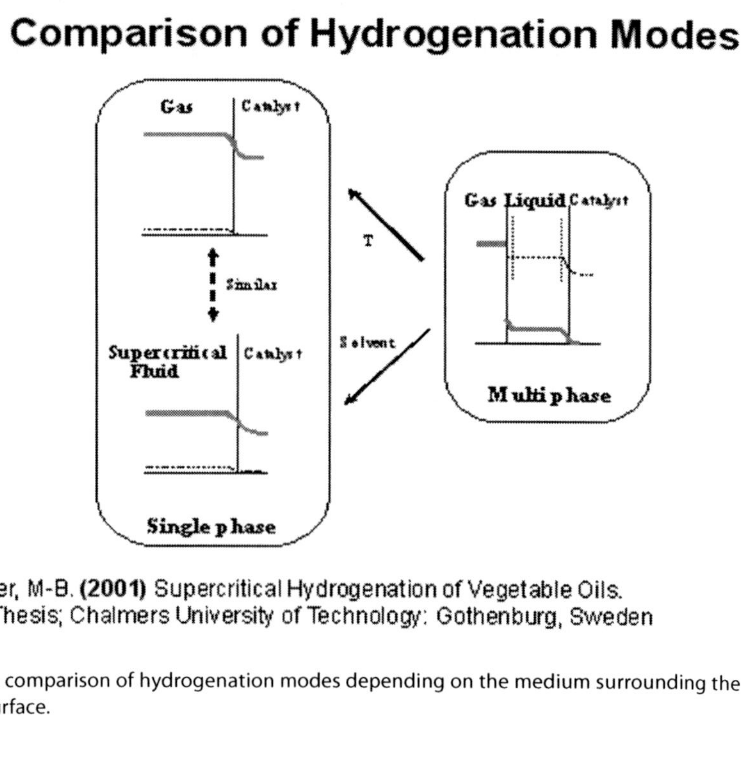

Fig. 3.3. A comparison of hydrogenation modes depending on the medium surrounding the catalyst surface.

amount of supercritical solvent employed during the reaction. An excellent example of this phenomenon occurs during the hydrogenation of α-pinene in which it was found that the hydrogenation rate was improved in a biphasic system as opposed to a one phase system (Chouchi et al., 2001; Milewska et al., 2005). This is frequently reflected in the occurrence of multiple phase formation upon the introduction of the supercritical fluid. Hence, as will be described later, phase equilibria studies conducted with the aid of a sight glass or modeled if a suitable equation of state is found to be applicable or available.

The choice of the supercritical fluid can also have an impact when conducting reactions with the aid of a supercritical fluid. The selective hydrogenation of cinnamaldehyde can be facilitated in SC-CO_2 using a Pt/Al_2O_3 catalyst system, but rather requires high pressures and a CO_2 mole fraction to achieve a one phase system in the presence of SC-CO_2 (Bhalchandra et al., 1999). This situation changes if one employs supercritical ethane in place of SC-CO_2 resulting in a much lower pressure requirement to maintain a single phase region.

The versatility of conducting hydrogenations in SC-CO_2 is largely due to the complete miscibility between H_2 and SC-CO_2 as cited by Jessop (2004) in one of his incisive reviews on the subject. For example, Jessop notes the application of asymmetrical catalysis in the presence of supercritical fluids to produce pharmaceutical intermediates and food-related compounds, and that even CO_2 alone can be hydrogenated to produce formic acid or formate derivatives using a homogeneous metal catalyst. High enantiomeric excess conversions approaching 90% and beyond can be realized by conducting the hydrogenations in SC-CO_2, often times aided by the inclusion of co-solvents such as alcohols or fluorinated alcohols. Compressed gases aside from SC-CO_2, i.e., ethylene, offer synthetic routes plagued by the formation of carbon monoxide, where CO negates hydrogenation catalyst activity. It should be appreciated that one advantage of dissolving H_2 in a supercritical fluid versus a conventional liquid solvent is that a much higher molarity for H_2 can be achieved, whereas much higher pressures of H_2 are required to achieve an equal molarity of H_2 in a liquid solvent phase.

Jessop and Leitner (1999) have commented on the important factors relevant to conducting hydrogenation reactions using organometallic catalysts. There are three basic types of metal-catalyzed reactions that can be conducted in the presence of supercritical fluids: Type 1—where the interaction between substrates, products, and catalyst occur in single SCF phase; Type II—the reaction is conducted in a single phase containing the dissolved substrate and reactants using an insoluble metal catalyst; and Type III—where the catalyst is dissolved in the liquid phase and the catalyst in the liquid phase, or vice versa. These researchers offer a simple procedure for determining the nature of the active species in such metal-catalyzed hydrogenations and the reader is referred to their classic text (Jessop & Leitner, Chapter 4.7, 1999) for further details.

Type I and II reaction scenarios noted above require some data regarding the solubility of metal complexes in supercritical fluids. A key factor in conducting reaction chemistry using the dissolved catalysts in the supercritical fluid phase is the introduction of CO_2–phillic modified catalysts produced by employing fluoroalkyl, fluoroether, and/or siloxy substituents, to make the catalysts compatible with supercritical fluids such as SC-CO_2. The above-cited reactivity between SC-CO_2 and H_2 can be used to produce formic acid or methyl formate for further synthetic purposes. Similarly other well known solvents can be produced by hydrogenation of substrates like secondary dialkyl amines in the presence of SC-CO_2 to yield for example, dimethylformaide (DMF).

Pillai and Sahle-Demessie (2003) have reported a very complete study on the hydrogenation of 4-oxoisophorone over Pd/alumina catalyst in SC-CO_2. They found that the reaction rate for this hydrogenation conducted in SC-CO_2 was similar to those recorded using liquid solvent media, however selectivities did differ depending on which medium was employed. In addition, catalysts deactivation in SC-CO_2 was found to be much less versus the same hydrogenation conducted in liquid solvents.

A significant contribution in supercritical fluid hydrogenation has been made by Poliakoff, et al., (2001, 2003) who have illustrated the general utility of supercritical hydrogenation as applied to simple compounds. Contributions from this group have ranged from the development of simple, small-scale continuous flow reactors, to unique phase equilibria measurements, and the scale up of several of these hydrogenations to commercialization (see the "Key Patents Involving Hydrogenation in the Supercritical State" section of this chapter). This group's contribution has nicely been summarized by Hitzler et al., (1998). Typical of the Poliakoff studies is the continuous hydrogenation of cyclohexene to cyclohexane in SC-CO_2.

As can be seen in Fig. 3.4, the mutual solubility of the three components involved in cyclohexene's hydrogenation as a function of temperature increases substantially, with all of the components being miscible at approximately 270°C as modeled using the Peng-Robinson equation of state. A similar trend is also found for the hydrogen-propane-cyclohexene system. Typical conditions for such a conversion were as follows: catalyst 5% Pd on APII Deloxan, pressures 120–140 bar, H_2:cyclohexene was 2:1, catalyst bed—4 mL, and experimental flow rates of 0.5–20 mL/min. Using these conditions, 95% conversions could be achieved. Similar hydrogenations have also been reported in either SC-CO_2 or propane for other diverse substrate types, such as aldehydes, ketones, nitriles, alcohols, oximes, and Schiff bases. Large scale hydrogenation commenced at the Thomas Swan plant in Consett, England in 2002 involving the hydrogenation of isophorone which could be produced in-situ without additional purification at a 99.4% purity level.

Supercritical fluid media also behave like liquid solvents in terms of their heat capacity characteristics. For example in the hydrogenation of cyclohexene, the adiabatic temperature rise is only 47°C versus 45°C in the liquid phase for an equimolar mixture in the olefin and H_2 in 90% CO_2 in a flow reactor. This is due to the fact that the heat capacity, C_p, for the above reaction is 141 J/mol/K versus 147 J/mol/K for the supercritical versus the liquid medium. One of the beneficial results of this close heat capacity to those exhibited by liquids is in eliminating hot spot development in reactors during the course of the supercritical hydrogenation.

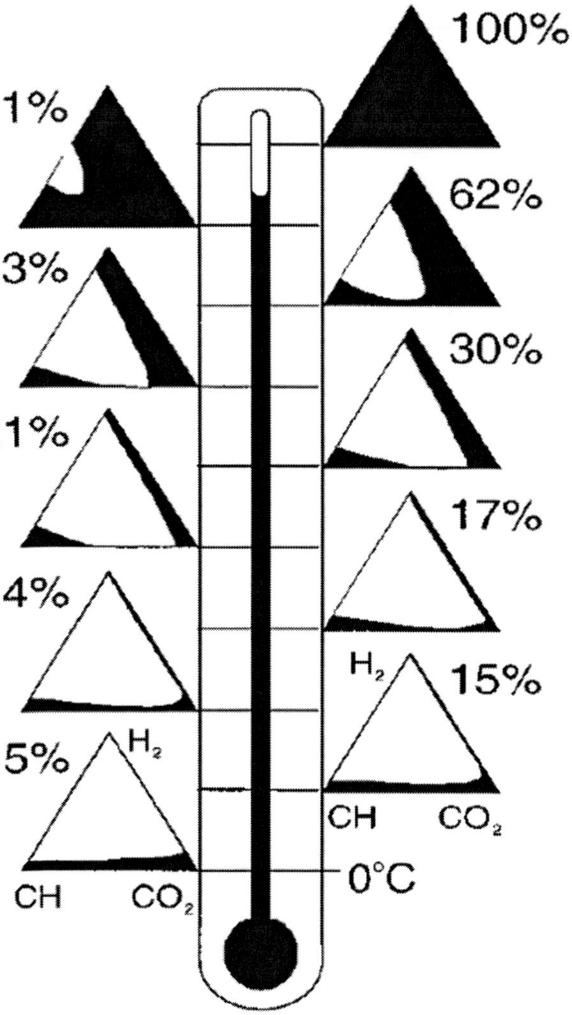

Fig. 3.4. Pictorial representation of the temperature variation of phase equilibrium for the system cyclohexane + CO_2 + H_2 using so-called "Gibbs triangles" (in steps of 30°C between 0 and 270°C). The dark areas represent homogeneous phases (totally miscible) whereas the areas of immiscibility are left white. The figures show the percentage of each triangle occupied by the homogeneous phase. The system was modeled for 120 bar using the Peng-Robinson equation of state. Similar results were obtained for the system cyclohexane + H_2 + propane. Reprinted with permission from Hitzler, et al. (1998). Copyright 1998 American Chemical Society.

Equipment, Processing Concepts, and Scale-Up

Much of the equipment utilized in supercritical fluid hydrogenations under bench scale conditions as well as scale up to pilot plant conditions, is similar if not identical, to typical high pressure autoclaves and devices designed for SFE and high pressure chemical synthesis. Therefore we will not concern ourselves in this chapter with descriptions of such equipment which is routinely utilized for such purposes since there are books and catalogues which describe the specifications of high pressure hardware. The focus in this section is on the basic approaches that have been used in conducting supercritical fluid hydrogenations.

Reactors in general fall under two basic types: conventional batch stirred reactors and continuous fixed bed reactors. Augmenting these reactor types are reactors which use a combinatorial approach to synthesis, i.e., multiple reactors and those that embody multiple fluid inputs. These basic reactor types can be configured for hydrogenations in several ways as illustrated in Fig. 3.5. As shown in Fig. 3.5, a simplistic batch reactor (A) for hydrogenations allows for the introduction of the binary supercritical fluid mixture (SF) into the top of reactor headspace where through agitation it is contacted with the reactants (R) and suspended catalyst (C). A more extended batch reactor configuration is shown in (B) where the supercritical fluids are circulated through the batch reactor and contacted with the hydrogenation catalyst held in a fixed bed using a recirculation pump. As with configuration A, both reactants and products (P) must potentially be separated at the conclusion of the reaction if unconverted reactants remain in the reactor. Reactor configuration B however has the advantage that removal of the catalyst from the reactor at the conclusion of the hydrogenation can be avoided since it is already isolated in the fixed bed. Of course reactant throughput and generation of product are dependent on the solubility of the oil-fat in the binary supercritical fluid mixture. The extractive batch reactor approach C couples the SF hydrogenation followed by SFE of the reaction products from the batch reactor. Hence by reduction of pressure and/or temperature, the products can be captured from the compressed fluid in a separator vessel, and the binary SF mixture potentially recycled back to the reactor accompanied by H_2 makeup gas.

Configurations D and E in Fig. 3.5 are continuous flow reactor options which employ some of the same features shown in options B and C. Such reactor systems are very attractive primarily due to the fact that in a packed bed reactor, the catalyst concentration is much higher facilitating better catalyst—reactants—H_2. The semi-continuous flow reactor dissolves the reactant (fat-oil) into the binary supercritical fluid mixture and transports it over the catalyst bed and eventually into a separator vessel where the supercritical fluid components are separated from the lipid product phase by adjustment of pressure and/or temperature. Configuration E is a variant of D whereby the supercritical fluids and reactants are separately pumped into the catalyst reactor bed followed by phase separation of the products from the SF. Configuration E has the additional advantage that the ratio of substrate to hydrogen and supercritical fluid can be varied as the situation

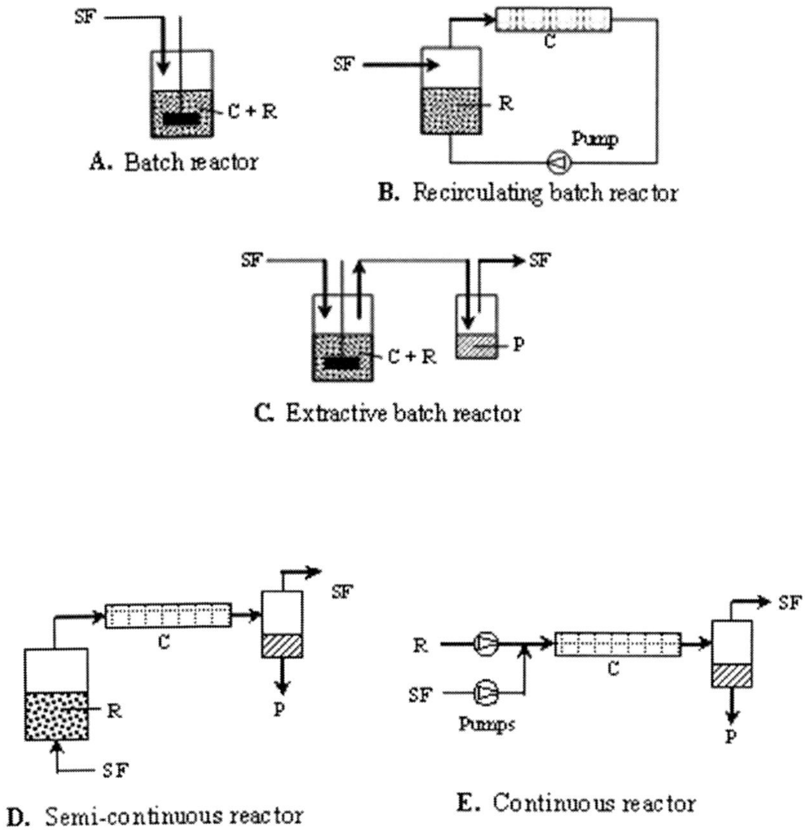

Fig. 3.5. Reactor options for conducting supercritical fluid hydrogenations.

dictates, and is perhaps the most preferred scheme for conducting supercritical fluid hydrogenations.

As remarked previously, the use of continuous supercritical hydrogenation system is preferred whenever possible. The residence time in a continuous reactor is a function of the reacting mixture density which is dependent on the mole fraction compositions and system pressure and temperature. Such data is rarely available and must be experimentally determined or modeled using appropriate equations of state and predictive phase equilibria software. A more detailed example of this approach is shown in Fig. 3.6 where the vegetable oil or lipid substrate is simultaneous fed along with the H_2 – SC-CO_2 or H_2 – n-propane into a mixer, M, whereby this mixture is then introduced into columnar reactor under optimized reaction conditions. The requisite reaction pressure is adjusted via the expansion valve while the tubular reactor is thermally-controlled. Separation of the products is facilitated by using a separator vessel after the expansion valve.

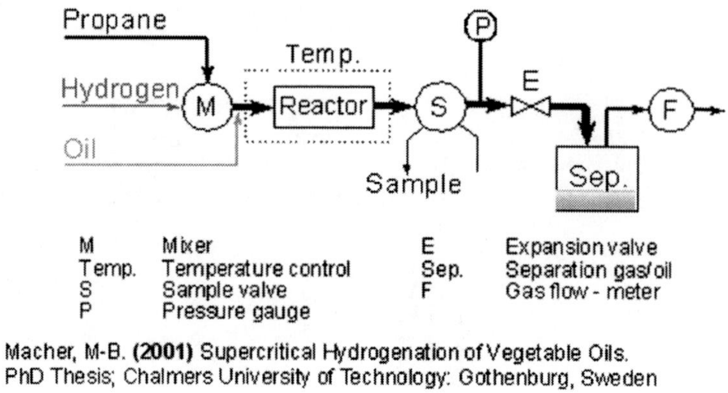

Fig. 3.6. Continuous flow system for the supercritical hydrogenation of vegetable oils.

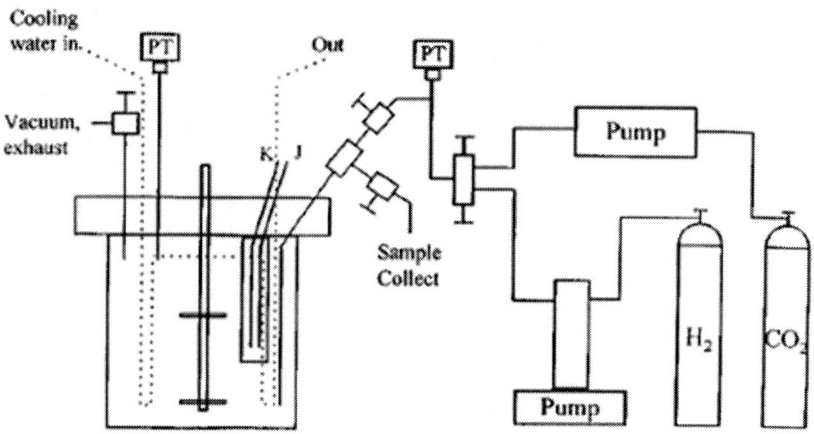

Fig. 3.7. High pressure hydrogenation system for conducting conventional and binary hydrogenations on vegetable oils. PT = pressure transducer, J, K = thermocouples.

Mixing of the hydrogen with the supercritical fluid prior to contact with the lipid substrate can be achieved by using a dual pumping system for each component followed by mixing them in a high pressure reaction vessel prior to using a gas booster pump to inject them into a mixing tee or chamber with the fat-oil substrate. A satisfactory design for mixing binary supercritical fluids has been described by Zhang and King (1997).

Designs for supercritical fluid hydrogenations conducted in batch reactors vary, but Fig. 3.7 illustrates one that the authors have used in their research, similar in concept to design A in Fig. 3.5. The binary gas hydrogenation mixture is introduced via a dog leg tube into the heated vegetable oil contained in the heated stirred autoclave under pressure by combining the two flows of volumetric pumps. Provision has been made for periodic sampling as pictured to track the progress of the hydrogenation.

Variations in batch reactors have also been reported, for example, Bertucco and coworkers (1997) hydrogenated a unsaturated ketone using an Al-supported Pd catalyst in a gradientless internal recycle reactor of the Berty-type. These researchers acknowledged the existence of multi-phases within the reactor and suggested that the observed beneficial effects of having a supercritical fluid medium was to "expand" the liquid phase thereby facilitating rapid mass transfer of hydrogen and contact between the catalyst and reactants. Further studies by Devetta of the Recasens group (1997) with these gas-expanded phases were conducted with the aid of a trickle bed reactor. This latter study is of importance because the hydrogenation reaction was scaled to a pilot plant application and made a rigorous attempt to account for radial temperature profile variations in the exothermic hydrogenation reaction.

The Recasens group in their studies involving the hydrogenation of sunflower oil used an internal recycle, gradientless-microreactor in which the catalyst is held in an annular basket made of mesh screens. The recycle flow was delivered by a variable-speed stirring-shaft, pumping "radially" through the bed. To avoid reactor-wall effects the vessel was fabricated from a nickel-free, alloy bar. This system could be mixed using stirrer shaft fan speeds used (up to 105 rad/s) to assure well-mixed conditions within the reactor. Liquefied propane was pumped using a high-pressure diaphragm pump (Milroyal® D, Dosapro Milton Roy, France) to the reactor, to provide and maintain a system downstream pressure of 18–25 MPa, which was manually set with a high-pressure regulator. The sunflower oil was pumped at a constant flow rate using a high-performance liquid chromatography (HPLC) pump and H_2 was compressed by a gas-booster system (Haskel Model AG-62) equipped with a high-pressure gas reservoir. The experimental reactor setup is shown in Fig. 3.8.

The sunflower substrate was mixed with propane in a 20-cm long, 1/4-in. Kenics Model 37-04-065 and H_2 was added downstream of mixer. The reactant mixture was preheated to the desired operating temperature before entering the reactor. This reactor is a Robinson-Mahoney type-reactor (Autoclave Engineers, Erie, PA), which creates a flow through the basket to the reactor wall for upward/downward deflection and a fixed annular catalyst basket, and that has baffles inside and outside the basket to prevent vortexing. After leaving the reactor, the effluent was continuously expanded to atmospheric pressure on an externally-heated needle valve to control the total flow of the reactor mixture. This effluent was then sent to a series of glass U-tubes, immersed in an ethylene glycol-water (40% v/v) bath held at 249 K to condense the oil from the propane and un-reacted H_2 mixture.

Although somewhat unconventional, the use of a membrane reactor is another possibility for conducting hydrogenations which has already been demonstrated for the reduction of a simple reactant. Using a rhodium-complex coupled with a fluorous phosphine ligand, a membrane reactor having a pore size of 0.6 nanometers has been successfully used to hydrogenate 1-butene in SC-CO_2 as pictured in Fig. 3.9.

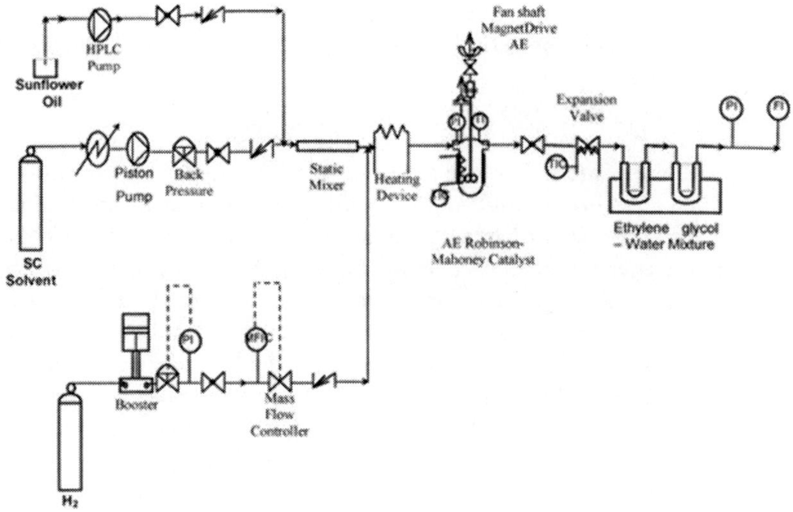

Fig. 3.8. Continuous internal recycle, gradientless-microreactor hydrogenation reactor setup. Reprinted with permission from Santana, et al., (2008). Copyright 2008 Elsevier.

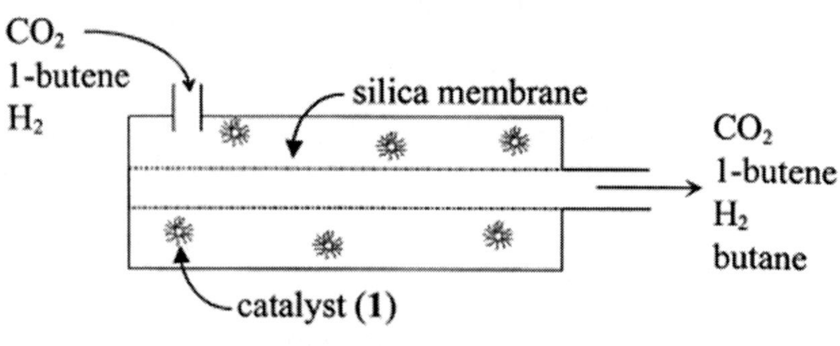

Fig. 3.9. Continuous reaction and separation concept. The membrane reactor is operated in a dead-end configuration. van den Broeke, et al.: Homogeneous Reactions in Supercritical Carbon Dioxide Using a Catalyst Immobilized by a Microporous Silica Membrane. Angew. Chem. Int. Ed. 2001. 40. 4473–4474. Copyright Wiley-VCH Verlag GmbH & Co. KGaA. Reproduced with permission.

The silica membrane is permselective to 1-butane, 1-butene, H_2, and SC-CO_2, but does not allow passage of the catalysts. A 40% yield of 1-butane was achieved and a turnover rate of $4000h^{-1}$ recorded during continuous operation using SC-CO_2 at 200 bar and 80°C. The mechanical strength of the ceramic silica membrane allows its use under relatively harsh reaction conditions. The precipitated catalyst was prepared in situ from $[RhCl(cod)]_2$ (cod=*cis*,*cis*-1,5-cyclooctadiene) and six equivalents of $P(p\text{-}(SiMe_2CH_2CH_2C_8F_{17})C_6H_4)_3$. In the batch reactor the turnover rate (TOF) at 25% conversion was found to be 9400 h^{-1}. The characteristics of the membrane reactor are given in Table 3.2.

TABLE 3.2
Characteristics of the Membrane Reactor

Property	
Thickness selective silica layer	200 nm
Length membrane	0.30
Outer diameter ceramic membrane	0.014
Membrane reactor volume	35.0 mL
Co_2 permeation[a] (200 bar, 353 K)	$3.0x10^{-3}$ mol m^{-2} s^{-1} bar^{-1}

[a]Flux divided by the pressure difference across the membrane.

Although critical fluid media are often cited as being miscible in all proportions with common reactive gases such as H_2, CO, and O_2, the addition of reactants or co-solvents have a profound effect on the associated phase equilibria, often times resulting in the formation of multiple phases during the reaction sequence. This has led to the monitoring of such reactions using pressurized view cells to observe and record the formation of multiple phases. Due to their lower T_c's, the above cited reactant gases act as anti-solvents which along with the fact that reactants can act as both co-solvents or anti-solvents leading to complex phase behavior.

Fig. 3.10 shows the type of phase equilibria data that can be obtained with a view cell. Note that here the mixture critical point (CP) is not necessarily the maximum with respect to either T or P, as it is with a single component phase diagram. As shown in Fig. 3.10, the dew point line (DPL) and bubble point line (BPL) are composition-dependent and the pressure maximum, P_{max}, and temperature maximum, T_{max}, are not consistent with the critical point for the mixture. Similarly, Poliakoff's group (Oag, 2004) have developed an apparatus for determining the critical points and phase boundaries of single component and binary mixtures using acoustic or shear-mode piezoelectric sensor methods to define bubble-point lines, dew-point lines, and critical points.

Phase behavior as well as interphase and intraparticle mass transfer may change as the reaction proceeds due to changes of the concentration of the different reaction components. This information is normally not available and consequently phase equilibria data is needed to optimize reaction conditions and to verify modeling of reaction conditions. A simple case is illustrated in Fig. 3.11 where different inlet and outlet conditions can exist during a reaction in either the two phase or single phase region, or exactly on the critical loci. In this case, the two P – T critical loci as indicated, are for the n-butane–hydrogen–methyl palmitate entering and exiting a continuous flow reactor designed to produce fatty alcohols.

In the supercritical single phase region, where gas–liquid mass transport resistance is eliminated, superposition of the dilution effect of the liquid substrate and the difference in activation volume (see the "Fundamentals of Critical Fluids Pertinent to Hydrogenations" section of this chapter) among the reaction rates that can occur in the same reaction mixture might be related to the results. Every reaction

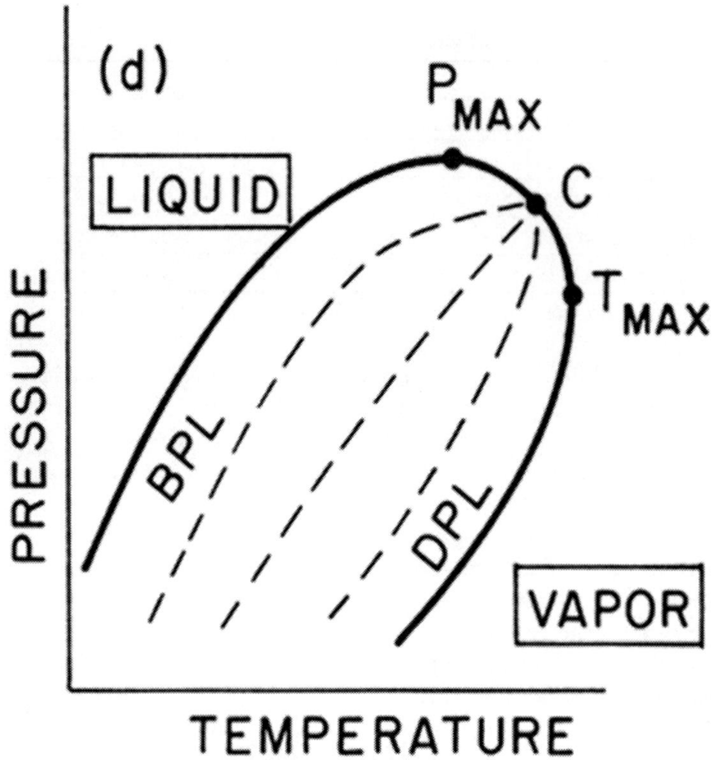

Fig. 3.10. Pressure–temperature phase diagram for a binary system.

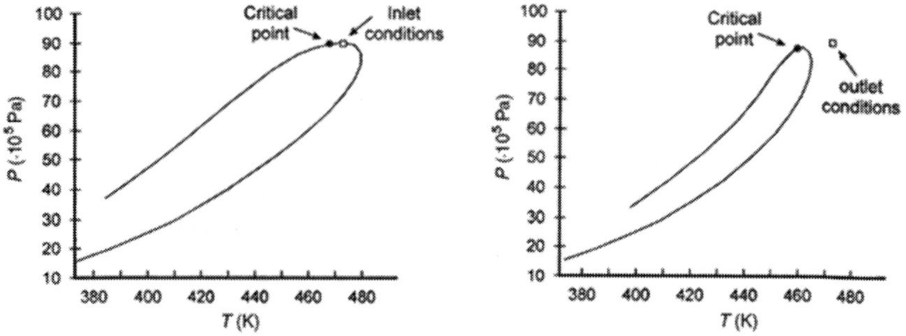

Fig. 3.11. Pressure–temperature critical loci for the reactant feed (left) and product flow (right) to/from the reactor along with critical points and process conditions for both flows. System: n-butane/hydrogen/fatty acid methyl esters.

has its own activation volume which changes with reaction conditions, and a negative activation volume results in an increase in reaction rate with increasing pressure. Savage et al. (1995) and Eckert et al. (1974) have given comprehensive descriptions on the relation between kinetic constant and activation volume.

Although homogeneous single-phase conditions seem beneficial to achieve high catalytic performance for most hydrogenations, some hydrogenations do proceed faster in two-phase systems containing an expanded liquid (Hutchenson et al., 2009). The expanded liquid-substrate contains large amounts of CO_2 and thus can dissolve sufficient amounts of hydrogen for the reaction to take place under appropriate conditions. Consequently, for the hydrogenations where the access of substrate to the catalyst is the limiting step, a biphasic system may be advantageous to achieve a higher reaction rate. This choice seems to be essential, for instance, when we use a small amount of heterogeneous catalyst in a large batch reactor. For a comprehensive review on the reaction engineering employing expanded liquids, the reader is referred to the recent excellent review of Jessop and Subramaniam (2007).

Due to the great miscibility of hydrogen and substrates in SCFs, most simple hydrogenation reactions can be performed below 100°C with high conversion and selectivity. This is attractive not only from an economical point of view but also in terms of suppression of undesired side reactions such as coke formation which is often observed in conventional gas-phase hydrogenations operated at much higher temperatures. Note that this is not always the case in the hydrogenation of lipid components, however, that even in the narrow temperature range (critical temperature to 100°C), increasing temperature sometimes greatly increases not only the desired hydrogenation rates but also the rates of undesirable side reactions. In addition, at the same total pressure, the SCFs density drastically changes with the change in temperature near the critical points. Thus the reaction temperature also must be carefully selected for beneficial use of SCFs in catalytic hydrogenation.

An excellent and comprehensive discussion of the thermodynamic and phase behavior which accompany hydrogenation of triglycerides has been provided by Weidner, Brake, and Richter (2004). Physicochemical data generation using an integrated experimental system is described in which phase behavior, viscosities, densities, and flow measurements as applied to H_2 with CO_2, propane, and dimethyl ether binary systems having soybean oil as the common lipid component. Multi-dimensional phase behavior diagrams were generated on the above systems to study the contrasting solubility behavior and component miscibilities. The relative reduction in densities associated with CO_2, propane, and dimethyl ether-soybean oil (SBO) systems was quantified, including the relative changes in each binary pair's viscosities over a similar range of pressures and temperatures. This is important in the case of coupling H_2 with dimethyl ether or propane since one can predict where the homogeneous phase equilibria exists with SBO, and the much lower pressures versus a CO_2-based system with SBO since much lower pressures are required to sustain miscibility and phase homogeneity in the two former systems.

Similarly, the higher pressures and temperatures associated with the reaction of fatty acid methyl esters (FAMES) to yield oleochemical derivatives, such as alcohols have been modeled by the above group. Since this type of hydrogenation scheme is envisioned as a continuous system in practice, and consists of five components some of which are being recycled, and/or separated, the phase behavior will be quite different than those phase relationships for the SBO-supercritical gas systems. The modeling of the phase relationships can be achieved via ASPEN and its subroutine equations of state (EOS), but each EOS must be studied to pick the optimal one that predicts the targeted phase equilibria and its agreement with limited experimental data. This primarily is a one- versus two-phase region problem for the five components making up the FAME reaction system, and includes requisite data for recompression of the system components.

Both model and real fat-oil systems have also been studied with respect to their phase equilibria characteristics by Pereda, Bottini, and Brignole (2002) group in a number of papers. For example, the fluid phase behavior of H_2 – propane – tripalmitin system, between 323–453°K up to pressures of 15 MPa as noted in Fig. 3.12. Fig. 3.12 shows the range of these conditions and shows the projection of a plane which defines the one phase region from the multi-phase region, and hence can be used to ascertain what are the correct conditions (pressure and amount of solvent) required to guarantee a homogenous reaction conditions.

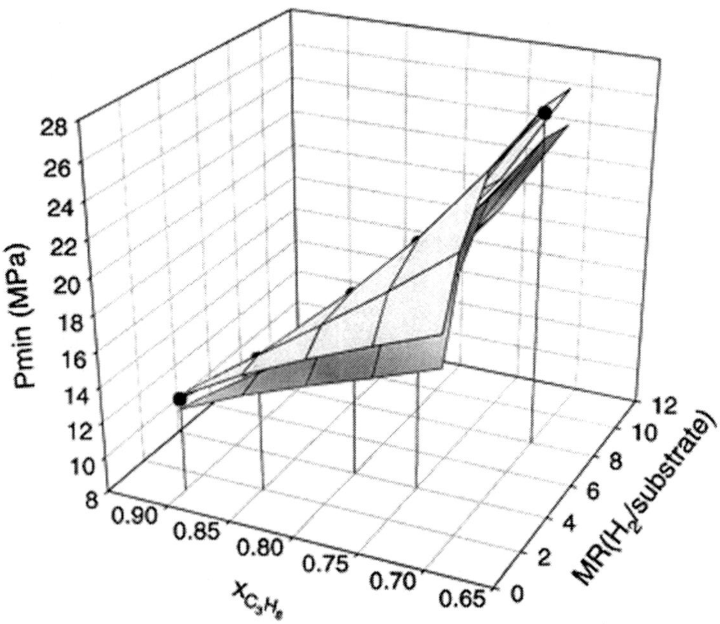

Fig 3.12. Minimum operating pressures and amount of mole reaction solvent required for the supercritical hydrogenolysis of methyl palmitate.

The phase equilibria studies principally completed by this Argentinian-based research group are impressive and seminal to anyone working in supercritical fluid hydrogenations. For this reason they are listed below as a distinct resource group in Table 3.3. A seminal contribution to the literature on phase equilibria during hydrogenation proceeds those listed in Table 3.3 is "Phase Equilibrium Engineering of Supercritical Hydrogenation Reactors" by Selva Pereda, Susana Bottini, and Esteban A. Brignole which appeared in the *AIChE Journal*, Volume 48, pages 2635–2645 in November 2002. Studies even prior to this on light hydrocarbon

TABLE 3.3
Phase Behavior Studies Related to Hydrogenation of Lipids due to Bottini-Brignole-Perada-Peters-Rovetto

Effect of phase behavior in the hydrogenation of triglycerides under supercritical and near-critical propane. C.M. Piqueras, D.E. Damiani, and S.B. Bottini, J. Supercrit Fluids, 50, 128–137, 2009.
Effect of phase behavior in the hydrogenation of triglycerides under supercritical and near-critical propane, C.M.Piqueras, D.E.Damiani, S.B.Bottini, J. of Supercrit. Fluids, 50, 128–137, 2009.
Advances in phase equilibrium engineering of supercritical reactors, S. Pereda S., E.A.Brignole, S.B.Bottini, J. of Supercrit. Fluids, 47, 336-343, 2009
High pressure phase equilibria of supercritical alcohols with triglycerides, fatty esters and cosolvents, P. Hegel, A. Andreatta, S. Pereda, S.B. Bottini, E.A. Brignole, Fluid Phase Equil., 266, 31–37, 2008.
Phase equilibria in ternary mixtures of methyl oleate, glycerol and methanol, A.E. Andreatta, L.M. Casás, P. Hegel, S.B. Bottini, E.A. Brignole, Ind. Eng. Chem. Res., 47, 5157–5164, 2008.
Sunflower oil hydrogenation on Pt catalysts: Comparison between conventional process and homogeneous phase operation using supercritical propane, C.M. Piqueras, G. Tonetto, S. Bottini D.E. Damiani, Catalysis Today, 133–135, 836–841, 2008
Sunflower oil hydrogenation on Pd/Al_2O_3 catalysts in single-phase conditions using supercritical propane, M. Piqueras, S. Bottini, D. Damiani, Applied Catalysis A: General, 313, 177–188, 2006.
Hydrogenation of sunflower oil on Pd catalysts in supercritical conditions: effect of the particle size, C.M. Piqueras, M.B. Fernández, G.M. Tonetto, S.B. Bottini, D.E. Damiani, Catalysis Commun., 7, 344–347, 2006.
Phase equilibrium modeling in the hydrogenation of vegetable oils and derivatives, S. Pereda, L. Rovetto, S.B. Bottini, E.A. Brignole, J. Am Oil Chem. Soc., 83, 461–467, 2006.
Supercritical hydrogenolysis of fatty acid methyl esters: phase equilibrium measurements on selected binary and ternary systems, L. Rovetto, S.B. Bottini, E.A. Brignole, C.J. Peters, J. Supercrit. Fluids, 35, 182–196, 2005.
Supercritical fluids and phase behavior in heterogeneous gas-liquid catalytic reactions, S. Pereda, S.B. Bottini and E.A. Brignole, Applied Catalysis A: General, 281, 129–137, 2005.
Phase equilibrium data on binary and ternary mixtures of methyl palmitate, hydrogen and propane, L.J. Rovetto, S.B. Bottini and C.J. Peters, J. of Supercrit. Fluids, 31, 111–121, 2004.

phase equilibria with triglycerides were published as early as 1997 (de la Fuente et al., 1997). For modeling purposes, P-V-T data on the lipid substrate is frequently desired. Toward that end Acostia et al. (1996) have provided such data on a variety of natural fats and oils, tripalimtin, and triolein in the temperature range from 303–353 K and from 1–150 MPa. The P-V-T data was correlated by means of the well known Tait equation.

Not all phase equilibria studies of relevance to lipid materials are focused on triglycerides or their methyl ester analogues. Bunnner et al., (2009) for example have conducted a fundamental phase equilibrium study on the system, hydrogen-carbon dioxide-squalene-squalane. The hydrogenation of squalene can be of some importance since it occurs in many vegetable oils, but is particular ubiquitous in marine animals such as shark liver.

The SCFs required for the hydrogenations are usually provided from high-pressure gas cylinders using compressors. The necessity of such special apparati for high-pressure conditions can be a barrier to the implementation of hydrogenations in SCFs, particularly for non-specialists and for those who work in laboratories where the working space is narrow and thus only limited number of gas cylinders can be set for safety reasons. An interesting alternative is the one reported by Poliakoff and co-workers (Aiekn & Poliakoff, 2009) who developed practical continuous supercritical hydrogenation processes "without gases", using the in situ decomposition of formic acid and ethyl formate. Formic acid is a source for both H_2 and CO_2, while the decomposition of ethyl formate is performed to generate C_2H_6 that dilutes hydrogen concentration. Thus formic acid and ethyl formate are separately fed into the first reactor containing 5% Pt catalyst to give the CO_2-C_2H_6-H_2 fluid, which is subsequently mixed with a substrate fluid and fed into the second hydrogenation reactor, giving the corresponding product downstream. During the catalytic decomposition, the reverse water–gas shift reaction also takes place to give carbon monoxide which poisons hydrogenation catalysts. However, by careful setting of the conditions, the CO concentration can be suppressed to levels that do not affect the performance of hydrogenation catalysts markedly. The other product, water, also deteriorates the performance of some hydrogenation catalysts by dissolving the active metals, but this can be solved by drying the gas mixture (e.g., insertion of a drying agent-packed column).

Supercritical Fluid Hydrogenation of Fats/Oils Using CO_2 or Propane

The supercritical fluid hydrogenation of fats and oils has been studied extensively by several groups, namely Härröd and Möoller and associates; King and List and colleagues; and the Recasens group in Barcelona, Spain, among others. Although their experimental approaches have varied from using batch, stirred reactors to continuous flow reactors, they exploit the advantages discussed in previous sections associated with combining hydrogenation with supercritical fluids. Historically gas phase hydrogenations have been applied with success using reactants that are low molecular weight; hydrogenating larger molecules requires that the hydrogenation be conducted at temperatures above the melting points of the liquid reactants, which compromises the dissolution of hydrogen in such liquid reactant media, and requires that hydrogen be forced into the liquid phase at high pressures. The use of supercritical fluids alleviates this problem by allowing either one phase reaction conditions to be attained with respect to H_2 and the reactants (fats and oils) or the use of a multi-phase medium under pressure that still promotes high contact between dissolved hydrogenation and the reactants. Both types of phase equilibria usually result in conferring superior mass transport properties with respect to hydrogen and the reactants, although a one phase system is usually optimal in promoting a fast reaction rate.

King et al. (2001) studied the hydrogenation of soybean oil using pure hydrogen mixed with SC-CO_2 at 14 MPa and 393–413 K, in a conventional nickel catalyst in a slurry reactor. They found that at constant temperature, the mixture of SC-CO_2 + H_2 exhibits a slower reaction rate than that using pure hydrogen, and that reaction conditions had a strong influence on the characteristics of the final product. In Table 3.4, the hydrogenated oil products, using the listed gas compositions, were characterized with respect to their IV, percentage *trans* fatty acid content, and overall fatty acid composition for 2- and 4-h reaction times. Also included in Table 3.4 are the results from conducting hydrogenation with just H_2 at 50 psi for the same time period. Here the extent of hydrogenation decreases as the overall pressure of the binary fluid system decreases as indicated in the corresponding IV values. The percentage of *trans* fatty acid content at either 2- or 4-h sampling periods also decreased as the overall system pressure was decreased, as did the saturated fatty acid content of the resultant oil (i.e., stearic acid). These two results suggest that nonselective hydrogenation is taking place under these conditions, yielding oils that have quite different properties from the nonselective hydrogenated product produced at 50 psi.

Additional hydrogenations were run without the second supercritical fluid component, i.e., SC-CO_2, using a nearly equivalent total pressure to that used in the

TABLE 3.4
Properties of Soybean Oil Hydrogenated Using Binary Fluid Mixtures of Carbon Dioxide and Hydrogen

	Non-selective 50psi H_2		1000 psi CO_2 1000 psi H_2		500 psi CO_2 500 psi H_2		250 psi CO_2 250 psi H_2	
Time (hrs)	2	4	2	4	2	4	2	4
IV	105	69	116	82	118	96	122	109
% trans	7.1	23.3	1.9	6.4	1.5	5.0	1.4	3.8
% 18:0	5.8	16.8	8.3	23.5	7.3	17.0	5.7	10.7
% 18:1	42.6	61.4	28.2	35.6	27.6	33.2	26.6	31.4
% 18:2	33.8	6.4	45.2	25.2	46.6	33.3	48.9	40.3
% 18:3	2.8	0	4.8	2.0	5.0	3.4	5.3	4.2

above-described CO_2/H_2 pressure ladder. These results are tabulated in Table 3.5 for 2- and 4-h sampling intervals so as to compare them against the nonselective, lower-pressure hydrogenation results. In this case, the IV of the resultant hydrogenated oils is similar to the IV exhibited by the product from nonselective low-pressure hydrogenation. The percentage of *trans* fatty acid content of the hydrogenated oils produced using higher hydrogen pressures is significantly lower compared to the oil hydrogenated at 50 psi, while the stearic acid content increased using a higher hydrogenation pressure. These results point to a set of reaction conditions that can produce low *trans* fatty acid levels, but that is inherently nonselective with respect to the mode of hydrogenation.

Experiments were also conducted to see what effect the reaction temperature would have on the results of the hydrogenation (Table 3.6). In this case hydrogenations of soybean oil were performed with a binary fluid system of 500 psi for both CO_2 and H_2, as well as in pure H_2 at a 500 psi level, but the reaction temperature

TABLE 3.5
Properties of Hydrogenated Soybean Oils Produced Using Pure Hydrogen

	Non-selective 50 psi H_2		1900 psi H_2		1000 psi H_2		500 psi H_2	
Time (hrs)	2	4	2	4	2	4	2	4
IV	105	69	108	75	108	72	110	76
% trans	7.1	23.3	2.7	7.0	3.1	7.4	3.4	8.6
% 18:0	5.8	16.8	11.7	27.8	12.6	30.8	11.0	26.9
% 18:1	43.6	61.4	29.2	35.3	29.1	33.0	29.7	35.7
% 18:2	33.8	6.4	41.2	21.7	40.9	21.2	41.5	22.2
% 18:3	2.8	0	3.5	1.3	3.6	1.8	4.3	1.9

TABLE 3.6
Properties of Soybean Oils Hydrogenated at a Higher Temperature (140°C)

	Non-selective 50 psi H_2 120C		2000 psi CO_2 100 psi H_2		500 psi CO_2 500 psi H_2		500 psi H_2	
Time (hrs)	2	4	1	3	1	2	1	2
IV	105	69	104	65	88	39	91	48
% trans	7.1	23.3	9.0	25.4	7.2	12.0	6.9	12.3
% 18:0	5.8	16.8	5.3	16.1	19.3	49.0	19.8	43.8
% 18:1	43.6	61.4	46.3	69.7	36.7	33.1	33.2	33.9
% 18:2	33.8	6.4	31.1	2.2	28.0	4.4	30.5	8.9
% 18:3	2.8	0	2.2	0.2	2.7	0.3	2.9	0.6

was 140°C, rather than 120°C. Here we observed a difference from the oil properties achieved previously when comparing the results from hydrogenating with a binary fluid mixture vs. pure hydrogen at 500 psi. In this case, both the IV and percentage of *trans* fatty acids, as well as the stearic and oleic acids contents, are very close for the two above reactions.

Fig. 3.13 is a plot of reaction time vs. IV value for the hydrogenated soybean oil products. In Fig. 3.13, the symbol codes have been grouped together: the first three representing hydrogenations conducted at elevated pressures in a pure hydrogen atmosphere, the next three representing the binary fluid mixtures at 120°C followed by a conventional low pressure hydrogenation at 50 psi, and then hydrogenations done at 140°C. An increase in pressure increases the reaction rate for hydrogenation (steeper IV vs. time plots) for the CO_2/H_2 mixtures. Fig. 3.13 leads one to the conclusion that the binary gas mixtures are retarding the hydrogenation reaction relative to pure H_2 at 120°C. The use of higher pressures with pure H_2 yields no apparent advantage above 500 psi in terms of reaction rate and yields results similar to those obtained at 50 psi. Two of the hydrogenations conducted at 140°C (500 psi CO_2/500 psi H_2 and 500 psi H_2) show a rapid drop in IV with reaction time compared to hydrogenations conducted under the above-described conditions. This trend can be partially ascribed to the increase in reaction temperature (II) but requires that enough H_2 be available to contact with the catalyst/oil (note the result for the 2.000 psi CO_2/100 psi H_2 mixture).

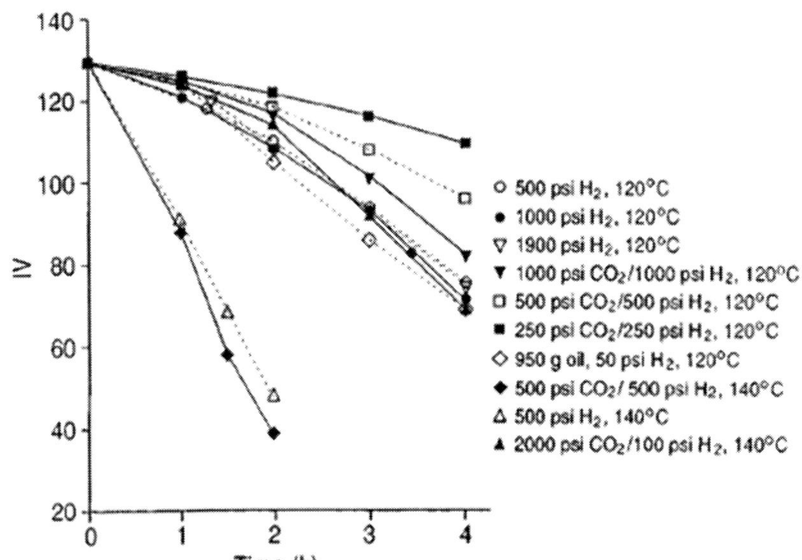

Fig. 3.13. Iodine value (IV) versus reaction time for the hydrogenation on soybean oil under various experimental conditions

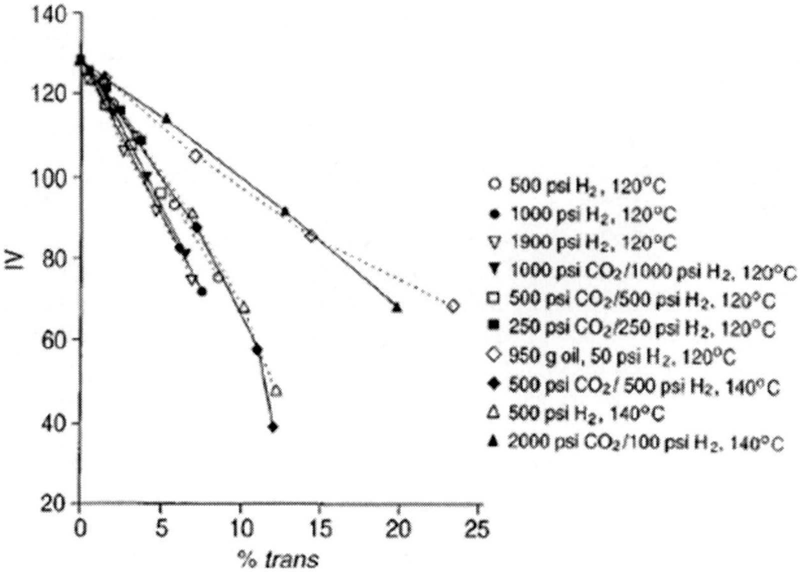

Fig. 3.14. Iodine value versus percent *trans* fatty acids for soybean oil hydrogenation under supercritical conditions.

Fig. 3.14 shows the relationship between the percentage of *trans* fatty acid vs. IV value of the resultant soybean oil. In this case, the *trans* fatty acid content-IV relationship is linear and independent of reaction pressure, the lone exception being the 2.000 psi CO_2/100 psi H_2 hydrogenation conducted at 140°C which mimics the conventional low-pressure hydrogenation at 50 psi. Thus, the binary fluid compositions and pure H_2 atmospheres at higher pressures yield lower *trans* fatty acid content having similar IV values when compared with the traditional low-pressure hydrogenation conditions. For most of the hydrogenations in Fig. 3.14, the percentage *trans* fatty acid content is 30% lower than that usually found in hydrogenated soybean oil with an IV value of 70. Also note that reaction conditions which yield the lower *trans* fatty acid content at similar IV continue to head downward in Fig. 3.14, which suggests that the trans fatty acid content will continue to remain low as the reaction proceeds under the described conditions.

In Fig. 3.15, the percentage of stearic acid content vs. IV shows similar loci for all the reported reaction conditions except for the conventional low-pressure hydrogenation and the 2,000 psi CO_2/100 psi H_2 result at 140°C. These two hydrogenations yield a lower percentage of stearic acid in the final product that has a similar IV to that obtained under the other eight hydrogenation conditions. This corroborates the trends for *trans* fatty acid production shown in previous figures and indicates the nonselective nature of these hydrogenations.

The results obtained from the described experimental hydrogenation runs have potential application in the food industry since the properties of the resultant oils closely approximate the IV, *trans* fatty acid content, and solid fat content (% 18:0)

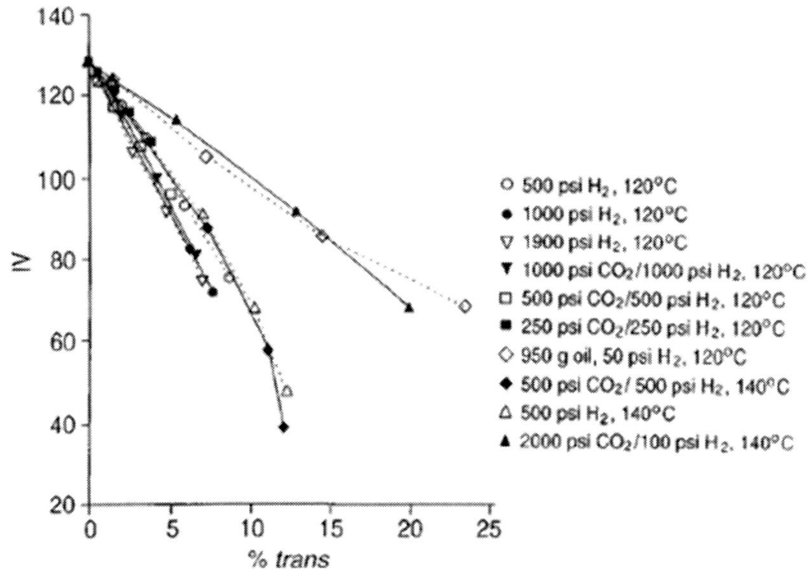

Fig. 3.15. Iodine value (IV) versus percentage of stearic acid for soybean oil hydrogenated at various conditions.

of margarine and shortening basestocks having similar DP. A conventional margarine basestock will usually have a DP of 32–39°C and will exhibit the properties listed in Table 3.7. The hydrogenated oils obtained in this study have a slightly higher IV than that observed for a commercial margarine basestock, the percentage of *trans* fatty acid being a decade lower than that found in a commercial sample. Table 3.7 shows that the stearic acid content of both conventional margarine basestock and the hydrogenated oils is almost identical. It is also possible to obtain a lower *trans* fatty acid content in the hydrogenated products compared to that found in a conventional shortening basestock (DP = 45–52°C) having a similar IV range (see Table 3.7). Table 3.7 also shows that the experimental hydrogenated products tend to have a slightly elevated level of stearic acid relative to levels found in commercial shortening

TABLE 3.7
Comparison of Experimental Hydrogenated Soybean Oils Versus Conventional Margarine or Shortening Basestocks

	Margarine Basestock (D.P. 32–39°C)		Shortening Basestock (D.P. 45–52°C)	
	Conventional	Experimental	Conventional	Experimental
% 18:0	6–9	7–11	11–13	13–24
% Trans	11–30	1–3	15–20	3–8
IV	90–110	108–114	85–90	88–102

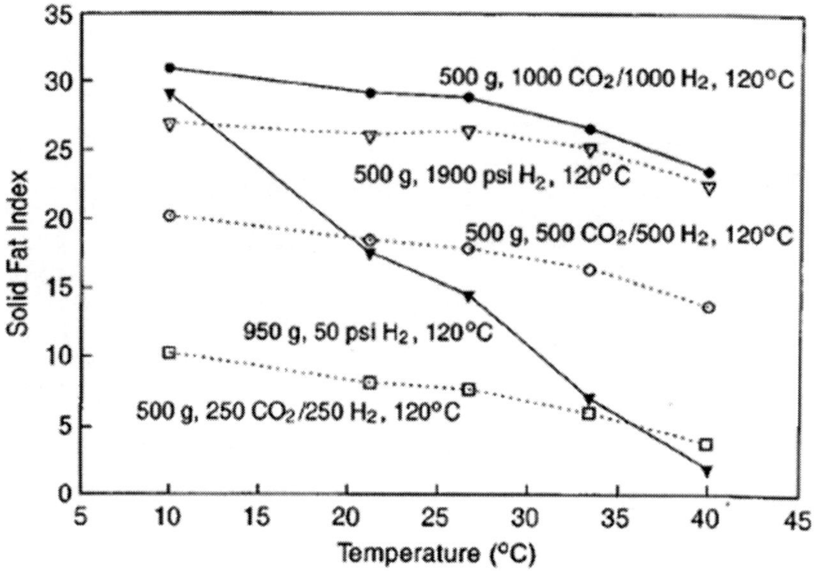

Fig. 3.16. Solid fat index (SFI) vs. temperature for five hydrogenated soybean oils using binary mixtures of CO_2/H_2 as well as under conventional hydrogenation conditions.

basestocks. Both of the previous results suggest that the hydrogenation conditions described in the study offer considerable versatility in designing an appropriate basestock for margarine or shortening use, since lower *trans* fatty acid levels are desired.

Solid fat indices (SFI) were measured for several of the hydrogenated oils synthesized in this study. Five of these products after 4 h of hydrogenation time were characterized by their SFI vs. temperature plots shown in Fig. 3.16. For the binary fluid mixtures and oil hydrogenated with pure H_2 at 1,900 psi, the temperature dependence of the SFI is a weak function of temperature. This is similar to *trans-suppressive* hydrogenation, which yields a long plastic range of melting behavior desired for shortenings, (but contrasts markedly with the SFI vs. temperature curve) for the conventional hydrogenation conducted at 50 psi H_2 pressure. The SFI results in Fig. 3.16 were all determined on hydrogenated oils having IV in the range of 60–70; however, the observed differences in their SFI vs. temperature curves reflect different saturated fatty acid content. For the four oils displaying an invariant SFI versus temperature curve, it should be possible to stop the hydrogenation reaction earlier before the saturated fatty acid level increases to yield different SFI values.

The commercial utility of the products obtained by using the above hydrogenation methods is worth noting. Fig. 3.17 illustrates the relationship between the percentage of *trans* fatty acid content of soybean oil hydrogenated at various conditions along with their respective DP vs. temperature curves. For both the pure hydrogen and CO_2/H_2 mixtures, the curves for percentage of *trans* fatty acid vs. DP are very similar, which is encouraging since these DP can be achieved with oils having a low-*trans* fatty acid content.

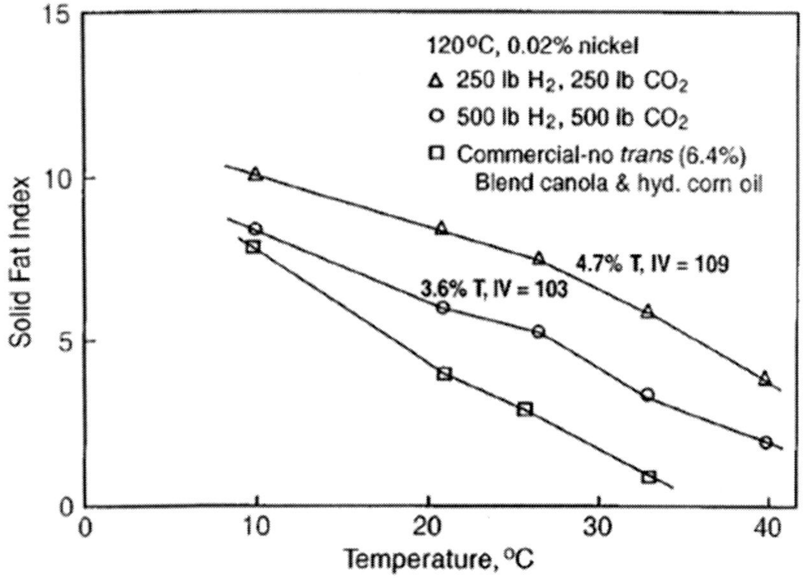

Fig. 3.17. Comparison of the solid fat index (SFI) vs. temperature curves for potential low trans-shortening base stocks derived from supercritical fluid hydrogenations with a low-trans fatty acid blended oil mixture (T = % *trans* fatty acid content; hyd. = hydrogenated).

By adjustment of the hydrogenation conditions, it is also possible to produce an oil having a low-*trans* fatty acid content (% T) at higher IV that behaves similarly to blending oil mixtures [e.g., a low-*trans* (6.4%) mixture consisting of canola and hydrogenated corn oils]. Such a comparison is made in Fig. 3.18 between two hydrogenated soybean oil products produced using binary fluid mixtures of CO_2 and H_2 at elevated pressures and the above-mentioned blend of oils. The two hydrogenated soybean oil exhibits somewhat similar SFI versus temperature curves as does the oil blend, making them good substitutes for such margarine basestocks that have low-*trans* fatty acid content.

The results obtained on the above study have been modeled by Holser et al., (2002) using a series of 1st order irreversible rate equations which correlate pressure, temperature, hydrogen pressure, and catalysts concentration. The modeling parameters were estimated from the experimental data and used to simulate anticipated results in the pressure range of 0.069–6.9 MPa. The rate expressions for a particular triglyceride were written as the sum and differences of the overall rate constant, K, triglyceride concentration, hydrogen pressure, and catalyst concentration for the reactant and product species. Further simplification was achieved by grouping the hydrogen pressure and the catalyst concentration with the rate constant. This new constant takes the form, $K' = K (P/P_0)p(C/C_0)c$, with P, the pressure; P_0, a reference pressure; C, catalyst concentration; and C_0, a reference catalyst concentration. The material balance equations were written on the basis of the reactants and products

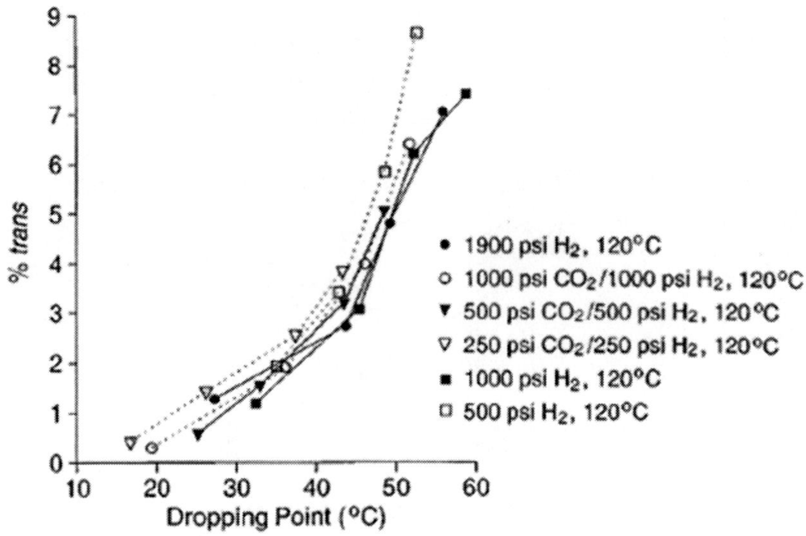

Fig. 3.18. Percent trans fatty acid content of soybean oil hydrogenated using binary CO_2/H_2 mixtures and conventional hydrogenated oil vs. dropping point (DP).

of the triglyceride species. This model assumed that positional isomers possess comparable chemical reactivity, although they could be treated as unique species and introduced through material balance equations. An example of a simulation for the changing triglyceride composition as a function of reaction time is shown below under the stated hydrogenation conditions. The estimated rate constants using this approach are given in Table 3.8.

TABLE 3.8
Estimated Rate Constant and Their Standard Deviations from Experimental Data for the Hydrogenation Reactions at 120°C

Estimated Rate Constants for Triglyceride Reactant to Product Conversions		
Reactant → Product	K(h-1)0.34 MPa	K(h-1)3.4MPa
L LL → LLO	0.120	0.191
LLO → OOS	0.204	0.290
OOS → OSS	0.153	0.797
OSS → SSS	0.110	0.299
LLP → LOP	0.637	0.171
LOP → OOP	0.544	0.292
OOP → OSP	0.335	0.411
OSP → SSP	0.213	0.644

Holser, et al., J. Agric. Food Chem., 50, 7111–7113 (2002)

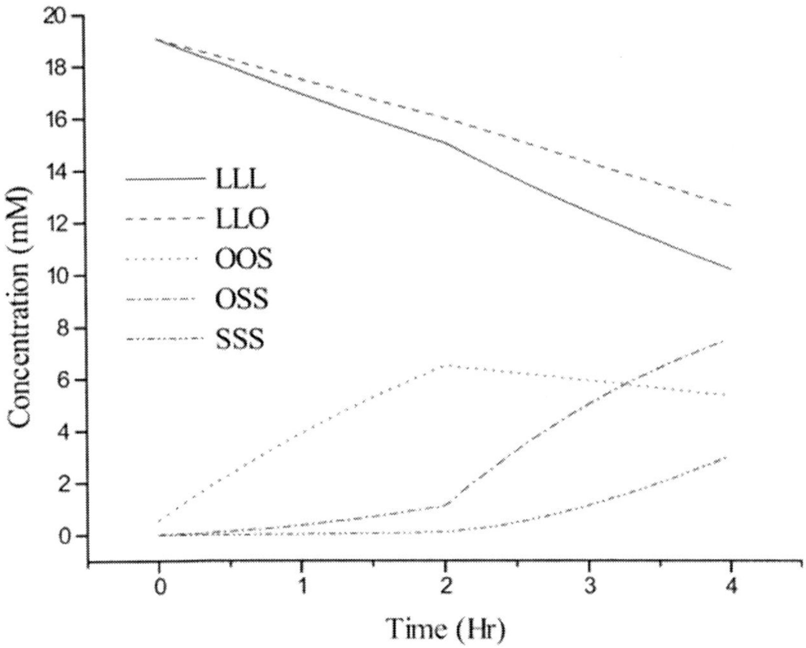

Fig. 3.19. Simulation of changing triglyceride distribution with increasing pressure. Reaction was initiated at 0.34 MPa, and pressure was increased to 3.4 MPa after 2 h.

The model could be extended to include geometrical isomers formed during hydrogenation. Triglyceride data were analyzed using the simulation software, SCOP (Simulation Resources, Inc., Berrien Springs, MI). A set of differential equations was formulated to represent possible hydrogenation reactions of the triglyceride species. Solutions were obtained by numerically integrating the equations using a fourth-order Runge-Kutta method. The principal axis method was used to estimate parameter values from experimental data. Variables such as pressure and temperature and their influence on the hydrogenation reaction were then correlated and used to predict the resulting triglyceride distributions. Fig. 3.19 shows the results of a simulation where hydrogenation is initiated at a pressure of 0.34 MPa, which is increased to 3.4 MPa after 2 h. This simulation indicates how the triglyceride distributions could be directed by controlling the pressure. This would be particularly advantageous in the partial hydrogenation of soybean oil, where a product of a specific composition is sought to provide particular material properties. The model equations were developed from a mechanistic description of the physical and chemical processes. This provides a fundamental basis to relate the change in triglyceride composition to temperature, pressure, catalyst concentration, and mass transfer. The model could be extended to include the formation of geometrical isomers in a straightforward manner if provided with additional experimental data.

Similar hydrogenation studies have been conducted by the Recasens group (Ramirez, 2004) using sunflower oil and hydrogenating it using a Pd/C catalyst in supercritical propane. An internal re-cyclic, radial flow and packed bed microreactor ($50\ cm^3$) over the temperature range of 428–488 K at a sunflower oil hourly space velocity of LHSV= 30–70. The hydrogen mole fraction was varied from 0.02–0.10 using stirrer speeds starting at 52 rad/s up to 250 rad/s. A central composite four-variable, two level, design was used to predict the effect of process variables on the iodine value (IV) and on *trans* $C_{18:1}$ fatty acid content. Both the feed and product were at a composition which permitted a single phase to exist during the hydrogenation. The total system pressure, the molar oil concentration and the catalyst mass were held constant at 20 MPa, 1 mol %, and 0.1085 g, respectively. Use of a quadratic-form response-surface model allowed the fitting of the experimental results and allowed one to ascertain the process could be operated to obtain a certain product I_2 value and a minimum *trans* - C18:1 content.

To ensure a single-phase vapor phase, the operating pressure and temperature for the reactor were held above the mixture critical values estimated using the Chueh-Prausnitz approximation. For a typical reacting mixture composition, H_2 = 9%, oil = 1%, C_3H_8 = 90%), the critical values were predicted to be T_c = 389 K and P_c = 6.2 MPa. In studying the effect of the operating variables on IV reduction and *trans* fat content, it was found that different sets of reaction conditions can lead to both an increased conversion while at the same time lowering the *trans* C18:1 isomer content. For example, one set of reaction conditions would be to operate the reactor on the high-temperature range and % H_2 range together with a high space velocity (LHSV), yet in contrast, a second combination could also be to use low LHSV, lower the temperature, and still keep the % H_2 < 4%.

Investigations by Tacke and coworkers (1996) describe the use of SC-CO_2 and propane in a fixed bed reactor in hydrogenating oils and fats, free fatty acids, and fatty acid esters. Using Degussa's Deloxan polysiloxane-supported Pd and Pt catalysts, they have reported a 18-fold increase in productivity versus the conventional Ni/kieselguhr supported catalyst used in conventional fat hardening processes. These researchers felt that the increase in hydrogenated products was due to the reduced viscosity of SC-CO_2–reactant mixtures which once again facilitated faster H_2 mass transfer rates. Tacke and coworkers tested a wide variety of experimental conditions which were as follows: reactor temperatures 40–190°C, CO_2 flow rates of 50–200 L/h, propane flow rates of 50–150 L/h, hydrogen flow rates of 5–100 L/h, catalyst volumes from 2–30 mL, total operating pressures of 2.5–20 MPa, and liquid hourly space velocities (LHSV) through their reactor of 5–240 L/h. They also explored both trickle bed and batch stirred reactors in performing the above spectrum of hydrogenation conditions. Catalyst screening studies which consisted of evaluating immobilized metals supported on activated carbon, alumina, silica, and titania, and the above-mentioned Deloxan. These have been summarized in an excellent review by these researchers (Tacke et al., 2003). Tacke et al., found that besides the Deloxan-supported catalysts being twice as active as traditional supported precious metal catalysts, they provided superior linoleate selectivity and lower *cis/trans* isomerization when using the ethyl ester of linoleic acid.

Extension of these model compound optimization studies were then transferred to vegetable oil hardening studies in both liquid, near-critical and super-critical carbon dioxide; propane, and SC-CO_2–propane mixtures using both discontinuous, batch stirred tank reactors as well as continuous tickle bed operation. When using conventional Ni catalysts in the temperature range of 373–425 K with supercritical propane, the reaction rate was found to increase 10^3 fold with respect to rates achieved using classical conditions. Supercritical propane far outperforms SC-CO_2 in part due to the much higher mutual miscibility of the triglyceride-based oils with propane relative to SC-CO_2. The propane requirement for operation for hydrogenations conducted in the homogenous supercritical state can be ascertained through the development of 3-component phase diagrams either from experimental measurements or modeling. Such studies indicate that there will be a specific mole fraction of propane for every specific pressure and temperature combination. Batch reactor experimental results were characterized by low space time yields and the production of larger amounts of *trans* fatty acids. In contrast, the continuous hydrogenation experiments using Deloxan AP II/1 wt % Pd on a fixed bed support gave higher space-time yields, however the linoleate selectivity remained low. Selectivity definitions employed by Tacke and coworkers to assess different hydrogenation results were as follows:

Linolenate selectivity: $S_{Ln} = k_3/k_2$
Linoleate selectivity: S_{Lo}
Specific isomerization: $S_i = \dfrac{\text{[No. of trans double bonds formed]}}{\text{[No. of hydrogenated double bonds]}}$

where k_3 was the rate constant associated with the hydrogenation of tri- to di-unsaturated fatty acids, k_2 is a measure of the hydrogenation of di- to mono-unsaturated fatty acids, and k_1 is the rate associated with mono-unsaturated fatty acid conversion to the corresponding saturated fatty acids. The above results indicate that reaction is strongly hydrogen mass-transferred controlled and that an increase in the hydrogen partial pressures increases the space-time yields.

An example of hardening using the conditions above was performed on free tallow fatty acids in SC-CO_2 using a space velocity of 15 h^{-1}. When compared to a trickle bed hardening using activated carbon and TiO_2—supported 2 weight percent Pd mixed bed catalysts, 6–15 times higher space time yields were obtained using SC-CO_2, and the hydrogenation could be done even faster by employing propane. Another advantage of the supercritical hydrogenation is the extension of the catalyst lifetime; the Deloxan—supported 1 weight percent Pd—fixed bed catalyst lasted longer relative to its use in a trickle bed hydrogenation. Such reactions will be noted again in the section, Key Patents Involving Hydrogenation in the Supercritical State, concerning the patent citations on supercritical fluid hydrogenation.

Beginning in approximately 1996–1997, Härröd and Möeller initiated the development of a technology platform in continuous hydrogenation using supercritical fluids exclusively devoted to lipid substrates. Due partly to the desire to protect intellectual property, their developments were largely discussed to a limited extent at select technology conferences, and then followed in the next few years by a series of theses, patents, and eventually publications. The intellectual property in their patents will be summarized in the "Key Patents Involving Hydrogenation in the Supercritical State" section, however the research and development of Härröd and his associates has resulted in a pilot plant facility devoted to the hydrogenation of oils and fats largely in supercritical propane and carbon dioxide. The pilot-plant is designed for hydrogenation of fatty acid methyl esters (FAMEs) to fatty alcohols. The production rate of product is 10 kg/h, which utilizes propane at a rate of 40 kg/h. The pilot plant has provisions for hydrogen recovery. The pilot-plant was commissioned in October 2002.

Van den Hark and Härröd (2001) studied the supercritical hydrogenation of fatty acid methyl esters at 15 MPa and 553.15 K using propane as reaction solvent. These authors found that at high substrate concentrations (2 mol% of oil and 20 mol% of hydrogen), a rapid fall of the reaction rate was observed due to the split of the supercritical homogeneous reaction mixture into two phases. In this case, a large excess of hydrogen is available to the catalyst, if the reaction mixture having hydrogen ratios around 10 or lower are of interest and the minimum required propane feed five to six times the product weight in order to create the necessary single–phase conditions. They note that if the process is operated under conditions comparable to those of the conventional processes, the advantages of the propane addition are lost. These investigators suggest than one of most simple methods of determining single–phase conditions is to observe the reaction rate, since they found that the difference in reaction rate between two–phase and single–phase conditions is very significant.

Macher et al., (1999) studied the partial hydrogenation of rapeseed fatty acid methyl esters under near-critical and supercritical conditions. Experimental variables such as temperature, residence time, hydrogen pressure, and catalyst life were varied systematically, using a statistical experimental design, in order to elucidate reaction rate and *trans* fatty acid formation as functions of the above variables. The experiments were carried out in a microscale fixed-bed reactor, using a 3% Pd-on-aminopolysiloxane catalyst. At 92°C, a hydrogen pressure of 4 bar, and a residence time of 40 ms a *trans* content of 3.8 ± 1.7% was obtained on the hydrogenated rapeseed which had an iodine value of 70. The obtained results support the findings from traditional processes that at a constant iodine value (IV), the *trans* content decreases with decreasing temperature, increasing pressure of hydrogen, and also increasing residence time. The reaction rate at our best conditions was ~500 times higher than could be obtained using traditional batch hydrogenation.

Macher and Holmqvist (2001) carried out the hydrogenation of palm oil in near-critical and supercritical propane using a small (0.5 cm^3) continuous fixed reactor and 1% Pd/C as catalyst, temperature (338–408 K), H_2/triglyceride mol ratio (4–50), and residence time (0.2–2 s) to assess the iodine value (IV) as a function of the operating variables. The authors observed high reaction rates [a residence time of 2 seconds (!) is sufficient at 393 K], which indicates that the reaction could also be run successfully at lower temperatures. To correlate the trends they observed from their hydrogenation experiments, response surface analysis plots were developed.

Fig. 3.20 shows the IV of the resultant products as a function of temperature and the H_2/TG ratio at a constant residence time. The IV was found to decrease with increasing temperature, as could be expected, while the effect of hydrogen is not significant. A possible explanation can be found by looking at Fig. 3.21, where the effect of H_2/TG and residence time has been plotted. Although the IV decreases strongly with increasing residence time, it shows no significant response to hydrogen.

However, what can already be seen is that the best results, i.e., the lowest IV, are found under rather extreme reaction conditions within the experimental range (see upper right corner of the plot). At 120°C and a residence time of 2.0 s, IV< 10 can be obtained.

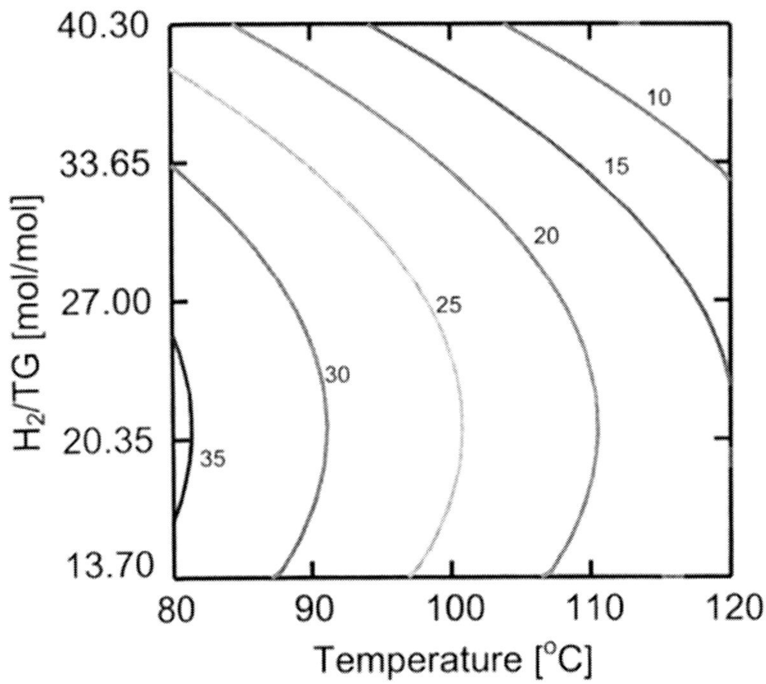

Fig. 3.20. Iodine value (IV) as a function of temperature and H_2/triglyceride (TG) ratio at a residence time of 2000 milliseconds and SEE = 8.68. Reprinted with permission from Ramirez, et al., (2004). Copyright 2004 John Wiley and Sons.

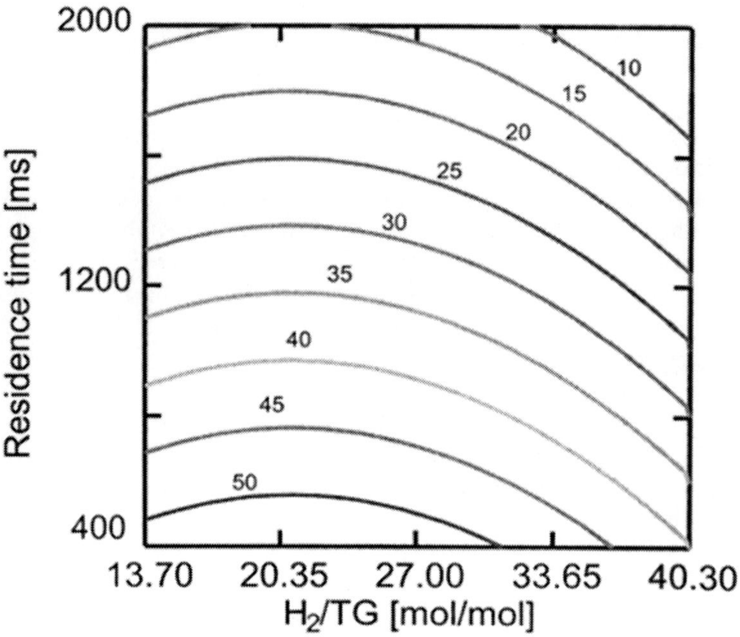

Fig. 3.21. Iodine value as a function of H_2/triglyceride ratio and residence time. Temperature = 120°C; SEE = 8.68. Reprinted with permission from Ramirez, et al. (2004). Copyright 2004 John Wiley and Sons.

To complete the picture, Fig. 3.22 shows the effect of temperature and residence time at a constant hydrogen concentration. For short residence times, the temperature does not have any effect at all, whereas it is marginally significant for long residence times. This can be explained by the kinetics of the reaction: at low temperature the reaction rate is low; although an increase in residence time does have an effect, the overall change in IV is still low. At high temperatures, the reaction rate is high, which means that an increase in residence time automatically causes higher conversion of the substrate, i.e. a lower IV. However, a certain minimum time is required: if the amount of catalyst is too small, the activating effect of temperature alone is not sufficient to obtain an acceptable degree of hydrogenation.

The above results can also be re-interpretated in terms of the fatty composition of the resultant products and how they correlate with IV. The trend is shown in Fig. 3.23 below where the saturated fatty acid content continues to grow reaching a maximum about one-half way through the reaction which is in line with the *trans* content. Macher (2001) remarks that very little information can be obtained on S_{Lo} since the disaturate content of the starting material. Hence at lower I, the *trans* content decreases more slowly than the corresponding *cis* fatty acids. These trends are consistent over a number of experimental conditions that were employed in conducting these palm oil hydrogenations as shown in Fig. 3.24 when using a new catalyst charge.

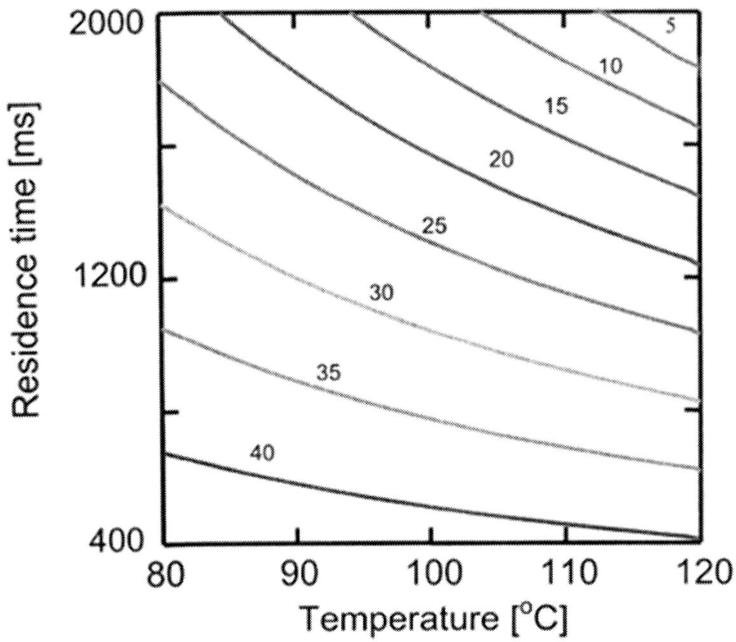

Fig. 3.22. Iodine value as a function of reaction temperature and residence time at a H_2/triglyceride mole ratio = 40.3 and SEE = 8.68. Macher & Holmqvist: Hydrogenation of Palm Oil in Near-Critical and Supercritical Propane. Eur. J. Lipid Sci. Technol. 2001. 103. 81–84. Copyright Wiley-VCH Verlag GmbH & Co. KGaA. Reproduced with permission.

Oleochemical Synthesis in Supercritical Fluids

This section is concerned with oleochemical synthesis employing both hydrogenation and supercritical fluids as opposed to the previous section which was concerned with the hydrogenation of fats and oils in critical fluid media. Both applications of hydrogenation coupled with supercritical fluids share a commonality with respect to the equipment and methods used to study them, and these techniques have been described in a proceeding section. Suffice to say that the synthesis of oleohemicals in supercritical fluid medium is perhaps more amenable to the use of continuous flow systems based on reports in the literature to date.

The one oleochemical synthesis that has been studied the most is the exhaustive hydrogenation of fatty acid methyl esters, FAMES, to fatty alcohol mixtures. Fatty alcohols and their derivatives are important in many industrial processes where they are used as raw materials for surfactants and lubricants. Commercially, fatty alcohols are produced by one of three processes: the Ziegler process, the Oxo process or by a high pressure hydrogenation of fatty acids or esters. The latter process is the only process that uses natural fats or oils whereas the two first processes utilize petrochemical feedstocks. Depending on their application, fatty alcohols are divided into subgroups. Fatty alcohols having eleven or more carbon atoms are usually called detergent-range alcohols because they are used in the detergent industry as sulfate,

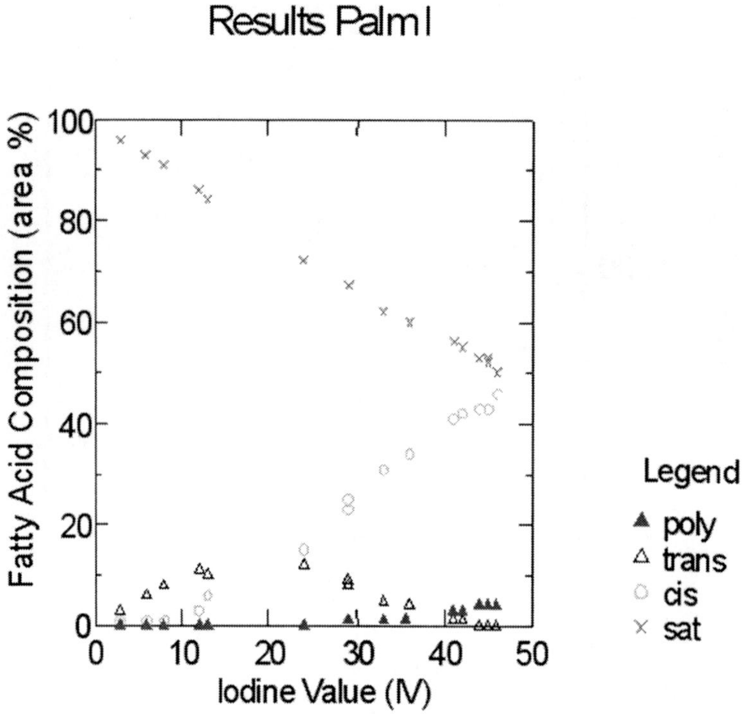

Fig. 3.23. Fatty acid composition versus IV for palm oil hydrogenated according to 1st experimental design (Macher, 2001). Macher & Holmqvist: Hydrogenation of Palm Oil in Near-Critical and Supercritical Propane. Eur. J. Lipid Sci. Technol. 2001. 103. 81–84. Copyright Wiley-VCH Verlag GmbH & Co. KGaA. Reproduced with permission.

ethoxylate or ethoxy sulfate derivatives. Fatty alcohols with less than eleven carbon atoms are called plasticizer range alcohols, and they are used in the polymer and lubricant industries mainly in the form of their ester derivatives.

Tacke and coworkers (1996) have applied similar methodology as described in the section on Supercritical Fluid Hydrogenation of Fats/Oils Using CO_2 or Propane to harden fatty acids, such as tallow derived fatty acid in SC-CO_2 and achieved iodine values (IV) below 1.0 g I_2/100 g-product at 15 h^{-1}. A summary of various fixed bed processes is available in the literature (Tacke et al., 2003). Compared to the trickle bed approach which use C- and Ti—supported Pd catalysts (2% Pd), the fixed bed approach allowed the realization of 6–15 higher space—time yields using Deloxan AP II / 1% Pd-supported fixed bed catalyst in SC-CO_2. For both the trickle- and fixed-bed processes, the partial pressures (2.5MPa) of H_2 were equivalent. They note that since the process is carried out at lower temperatures, the acid value of the fatty acids remains at a high level. The lifetime of Deloxan AP/1% Pd supported was found to be three times when using fixed bed hydrogenation versus trickle bed technology using AC-CO_2.

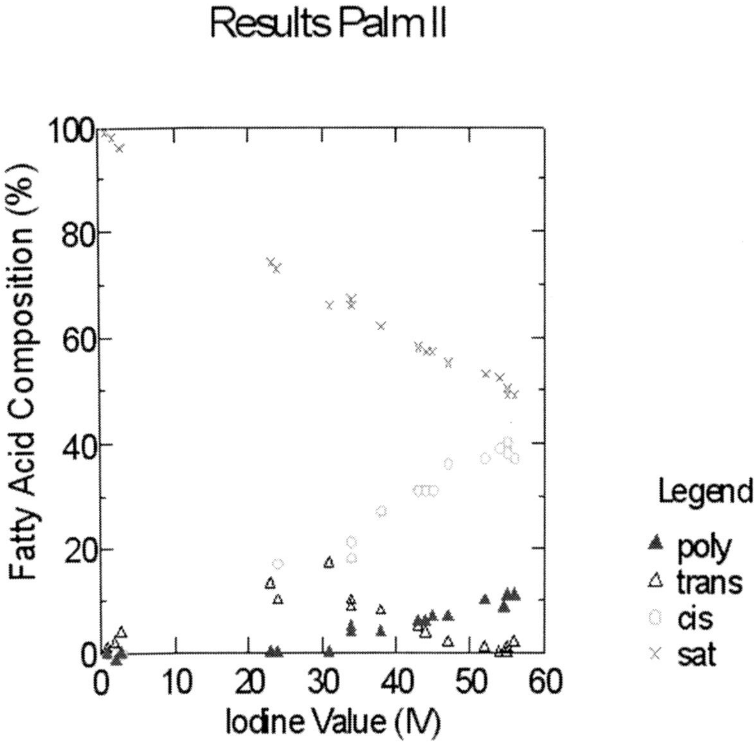

Fig. 3.24. Fatty acid composition versus IV for palm oil hydrogenated according to 2nd experimental design (Macher, 2001). Macher & Holmqvist: Hydrogenation of Palm Oil in Near-Critical and Supercritical Propane. Eur. J. Lipid Sci. Technol. 2001. 103. 81-84. Copyright Wiley-VCH Verlag GmbH & Co. KGaA. Reproduced with permission.

Early studies by Härröd and Möller (1996) illustrated the relatively phenomenal results that could be obtained by hydrogenating FAMEs in a packed bed reactor using Cu-Cr catalysts over the temperature range of 473–573 K, where productivity rates of 7.0×10^{-5} kg/m^{3}h were attained. Härröd and Möller remarked that such a high productivity, a reactor of 3 m^3, could well serve the entire worldwide production of hydrogenated oils. Over 98% conversions have been reported with reaction rates being over 250 times faster than achieved using traditional gas-liquid processes—particularly when propane is employed.

Van den Hark (2000) initially studied the model hydrogenation of methyl palmitate to establish the operating conditions which would be generally applicable for the hydrogenation of fatty acid methyl esters (FAMES). He surmised that to operate in the homogeneous phase region, temperatures in excess of 530 K would suffice and always at pressures exceeding 16 MPa for a 0.826 mole fraction of propane and 0.154 mole fraction of hydrogen for a 0.02 FAME mole fraction. The critical temperatures and pressures will of course vary as indicated previously as the mole fractions of the solvent and reactants change. An example of the three-dimensional phase equilibria relevant to the above hydrogenation conditions is shown in Fig. 3.25.

Van den Hark and Härröd investigated the hydrogenation of methylated sunflower oil (1999) to the corresponding fatty alcohols in SC-C_3H_8 using a fixed-bed reactor system consisting of two reactors. The substrate oil was first introduced into the first reactor containing 2% Pd/zeolite for the saturation, followed by the second reactor containing a Cu-based catalyst, in which the saturated substrate was converted into the corresponding fatty alcohols. A flow schematic of this reactor setup is shown in Fig 3.26.

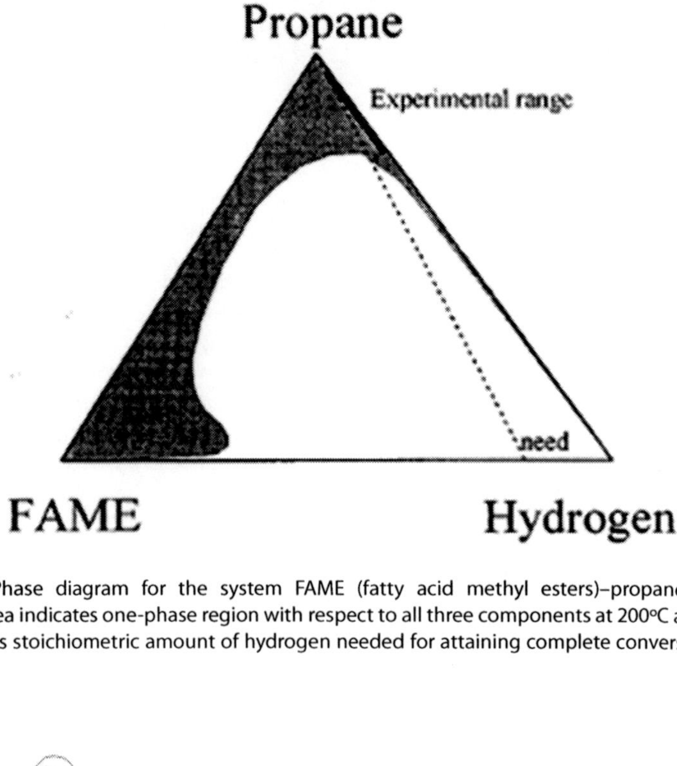

Fig. 3.25. Phase diagram for the system FAME (fatty acid methyl esters)–propane-hydrogen. Darkened area indicates one-phase region with respect to all three components at 200°C and 100 bar; dashed line is stoichiometric amount of hydrogen needed for attaining complete conversion to fatty alcohols.

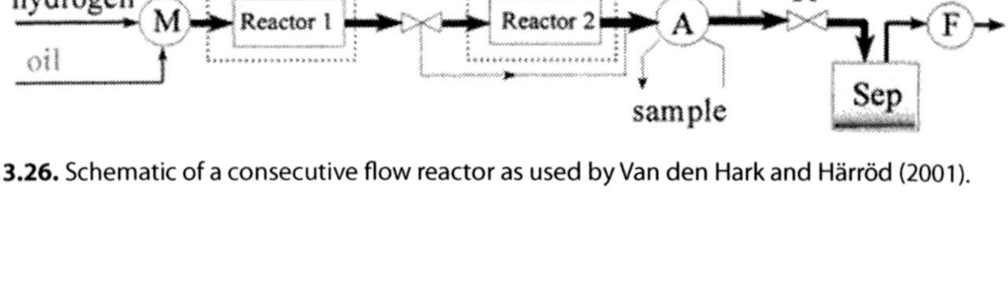

Fig. 3.26. Schematic of a consecutive flow reactor as used by Van den Hark and Härröd (2001).

Van den Hark and Härröd (2001) showed at high substrate concentrations, a rapid drop of the reaction rate was observed, and the benefits of the propane addition were completely lost. This drop was attributed to a split of the supercritical reaction mixture into two phases, i.e., a substrate-rich phase and a hydrogen-rich phase. When this phase split occurred using small catalyst particles (32 μm), the pressure drop over the catalyst bed increased sharply, because the formed liquid droplets blocked the void space in the porous catalyst bed. These two phenomena, i.e., decrease the reaction rate and pressure, and could be used to deduce the product and substrate solubility in the reaction mixture. The product, i.e., fatty alcohols, showed the most unfavorable solubility among the components in the reaction mixture. The solubilities increased with increasing pressure and decreased with increasing temperature and in the presence of hydrogen (Fig. 3.27). Under the process conditions (15 MPa, 280°C, and 20 mol% hydrogen), a single phase was observed up to 2 mol% (i.e., 15% by mass) substrate/product. Besides the minimum pressure in the catalyst bed, substrate oil, and transport limitation was shown to be an important factor in process optimization. This situation is illustrated in Fig. 3.28.

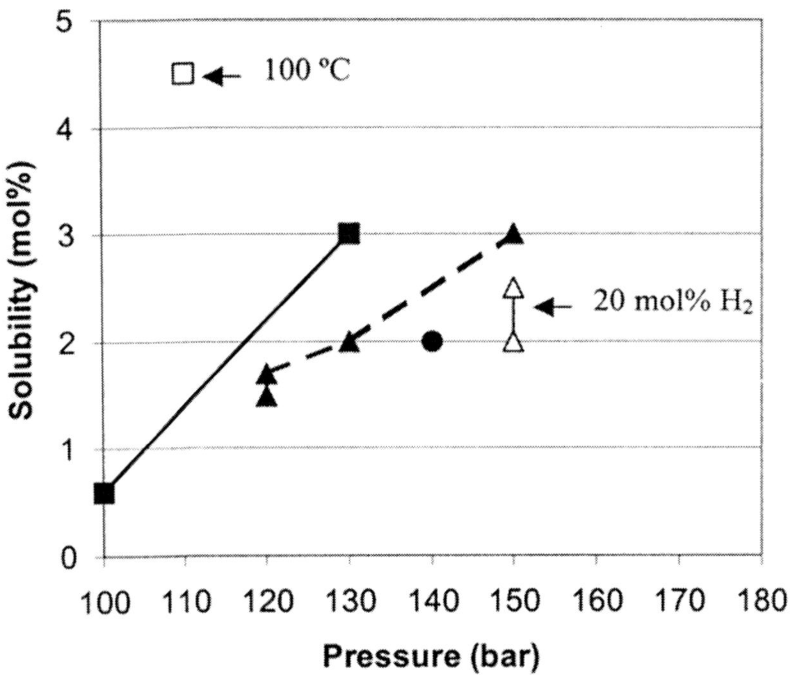

Fig. 3.27. Solubility of FAME (■), FAME-FOH (▲), and FOH (●) in the reaction mixture with 10 mol% hydrogen at 280°C, except open symbols (□) FAME at 100°C and 10 mol% hydrogen; (Δ) FAME-FOH at 20 mol% hydrogen and 280°C). Together with FOH, an equal molar amount of methanol is formed. (FAME—fatty acid methyl esters, FOH = fatty alcohols).

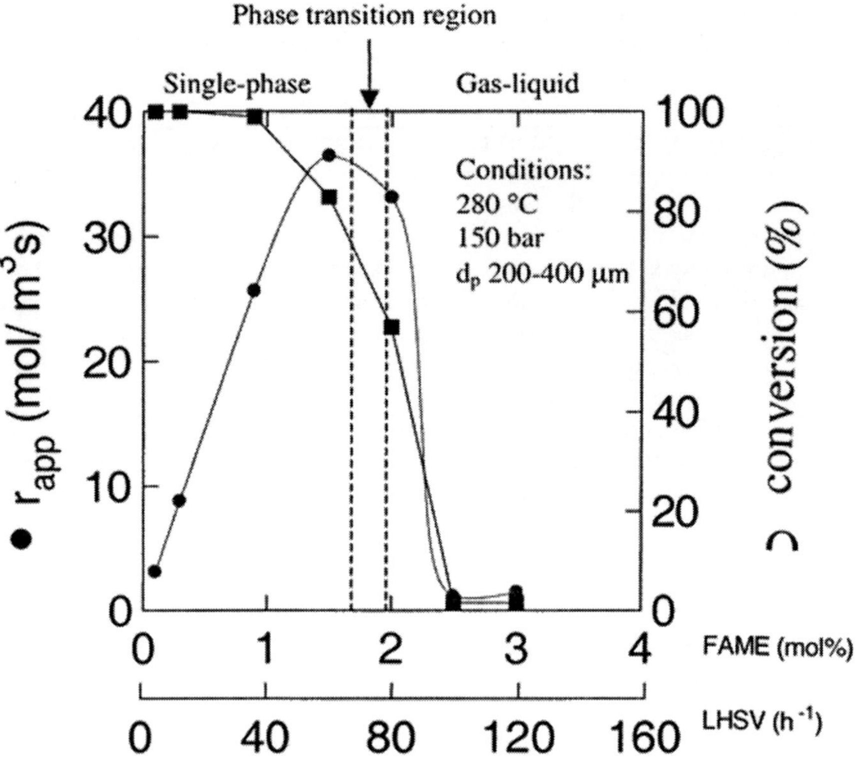

Fig. 3.28. Apparent volumetric reaction rate (r_{app}) and conversion versus substrate (FAME) concentration in the reaction mixture. Second *X* axes indicated the substrate-based space velocity (conditions: 20 mol% hydrogen, residence time) 800 ms, flow rate), 120 mmol/min, i.e., 32 mL/min at reaction conditions). Reprinted with permission from van den Hark & Härröd (2001). Copyright 2001 American Chemical Society.

•(r_{app}) = flow (mol/s) X_ FAME (mol%) × conversion (%)/reactor volume (m^3) (mol/m^3)

The large excess of hydrogen available to the catalyst, if the reaction mixture forms a single phase, turns the substrate, not hydrogen, into the limiting factor. Because process settings such as pressure and catalyst particle size have opposite effects, e.g., on substrate access to the catalyst–solvent requirement, fluid compression, and phase equilibria, they have to be carefully balanced. Besides kinetic data, precise solubility data and physical properties, such as viscosity and diffusivity of the supercritical reaction mixture, are needed to further optimize this promising supercritical hydrogenation.

Macher et al. (1999) also studied continuous partial hydrogenation of methylated rapeseed oil (fatty acid methyl esters) in SC-C_3H_8 using 3% Pd supported on aminopolysiloxane. The goal of this investigation was to obtain partially hydrogenated oil product with a low trans content (<5% at the iodine value range 50–80) as well as to achieve high reaction rate. An HPLC tube (inner diameter, 2.1 mm; length, 50 mm) was employed as microscale fixed-bed reactor, into which the mixture of substrate oil, H_2, and CO_2 was introduced. The single-phase operation performed at 100°C, total pressure of 10 MPa, and a residence time of 400 ms in the presence of 2.4 wt % (0.35 mol%) of the substrate oil and 0.40 MPa of H_2 afforded iodine value of 60 and lipid hourly space velocity of 50,000 $kg_{FAME}\ m_{Reactor}^{-3}\ h^{-1}$, which is approximately a factor 125 higher than that obtained for traditional triglyceride hydrogenation in batch processes. In addition, by increasing the substrate oil concentration and H_2 pressure up to 15 wt % and 2.3 MPa, respectively, the space velocity could be further increased to 213,000 $kg_{FAME}\ m_{Reactor}^{-3}\ h^{-1}$, which is roughly 530 times higher than that of the traditional processes; the iodine value also increased to 70. These results clearly indicate that elimination of gas–liquid transport resistance under supercritical conditions accelerates the hydrogenation. Concerning the product quality, the lowest trans concentration of 3.8% could be achieved at an iodine value of 70 when the hydrogenation was carried out at 92°C, total pressure of 10 MPa, and residence time of 400 ms with 0.4 MPa of H_2. The concentration was far lower than that obtained with a batch process under optimized conditions. The tendency observed in classical processes, namely, at constant iodine value the *trans* content decreases with decreasing temperature, increasing H_2 pressure, and increasing residence time, was observed also under the supercritical conditions.

Van den Hark and Härröd (2001) reported that selectivity in the hydrogenation of FAMES could be achieved by selective choice of the hydrogenation catalyst so that the double bonds in the fatty acid chain are retained while the ester group is reduced, thus yielding unsaturated fatty alcohols. Conditions of 15 MPa and temperatures from 558–573 K using a Zn-Cr catalyst proved optimal for achieving these results. It was found that by using propane under these conditions, that 10-fold less propane was required to maintain homogeneous conditions. Using a starting mixture of FAMES obtained from sunflower oil having an IV = 113 and the above conditions permitted the production of fatty alcohols having IVs ranging from 61 to 78.

Using a somewhat different supercritical fluid, Brands et al., (2001, 2002) investigated the hydrogenation of methyl palmitate in supercritical n-butane as a model study for the production of fatty alcohols. The study was divided into two parts: (1) concerned with an estimation of thermodynamic and phase equilbria for the system, and (2) a flowsheet analysis accompanied by process design with economic cost analysis. Reliable estimates of the supercritical mixture properties were obtained by fitting experimental vapor-liquid equilibrium data with the Schwartzentruber-Renon cubic equation of state. The reaction mixture is supercritical to a maximum pressure of 9 MPa and 470K for the H_2–methyl palmitate system. These values were determined for a hydrogen mole fraction of H_2 and methyl palmitate mole fraction of 0.025. Under these conditions the equilibrium conversion was found to 99.2 mol%. These values indicate the practical upper values of reactant concentrations in supercritical butane.

TABLE 3.9
Utilities for a 50 Kilo Ton Per Year Fatty Alcohol Plant Using Supercritical n-Butane[a]

Stream	Cold/Hot	Utility Name	Supply T(K)	Target T(K)	Duty (kW)
1	C	Cooling water	293.2	298.2	2793.0
2	H	LP-steam	417.2	417.1	401.4
3	H	MP-steam	453.2	453.1	823.0
4	H	HP-steam	513.2	513.1	694.7
5	H	Superheated steam	563.2	553.2	1037.7
		Total cold utilities			2793.0
		Total hot utilities			2956.8

[a]LP, low pressure; MP, medium pressure; HP, high pressure.

In the second publication, Brands et al., (2002) performed a computer simulation study to optimize the operation of a fatty alcohol production facility based on the above vapor-liquid equilibrium data. Both separation and recycle options were evaluated and it was indicated that at a hydrogen to ester molar ratio of 4:1 in the feed is achievable, which compares favorably to existing liquid- and gas-phase processes, and allows recycle streams to be reduced. Data excerpted from the publication with respect to process utility costs is shown in Table 3.9 and the relevant flow sheet is reproduced in Fig. 3.29. The results of this simulation indicated that an economically efficient hydrogenation for fatty alcohol production can be conducted using a supercritical n-butane.

Coupled Processes Using Supercritical Hydrogenation

As cited previously, Anderson and coworkers (2000) studied the reduction of fatty acid methyl esters (FAME) to fatty alcohol mixtures in two different types of supercritical media (H_2/CO_2 and H_2/C_3H_8) using two different hydrogenation catalysts. They reported high and rapid conversions are achieved at 250°C at a hydrogenation mole fraction of 0.25. The hydrogenation step can then be coupled with an enzymatic-catalyzed transesterification to yield a novel sequential two step synthesis of fatty alcohols from soybean oil derived-FAME. This synthesis also emphasized the use of "green" chemistry in devising the method since it was not only based on the use of supercritical carbon dioxide as an environmentally benign solvent, but employed a lipase catalyst in the transesterification step and a chromium-free catalyst for the hydrogenation step. The overall concept and flow schematic are shown below in Fig. 3.30. One of the products from the exhaustive hydrogenation of the resultant FAMES was methanol that can be recycled back into the transesterification stage. High yields of saturated alcohols are realized (*c.* 90% steryl alcohol) at pressures and temperatures above 150 bar and 210°C, respectively.

The basic experimental apparatus used in this hydrogenation of FAMES study under supercritical conditions is shown in Fig. 3.31. The individual gases (fluids) were

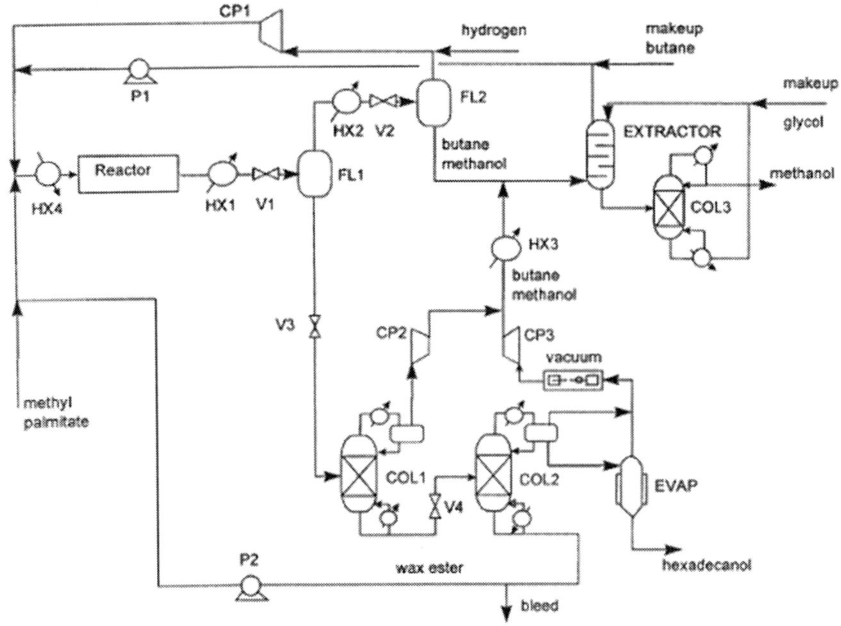

Fig. 3.29. Flowsheet for the fatty ester hydrogenation using supercritical n-butane. EVAP = evaporator; COL = distillation column; V = valve; HX = heat exchanger; P = pump; CP = compressor; FL = flash tank. Reprinted with Permission from van den Hark & Härröd (2001). Copyright 2001 American Chemical Society.

metered into a 1-liter capacity Parr Instruments Co. high pressure, stirred autoclave to assure homogeneity. The resultant binary fluid mixture was then compressed to the desired reaction pressure using a Haskel gas booster compressor and introduced into the reaction vessel containing either a copper chromite catalyst (E-406TU-Engelhard Inc., Erie, PA) or a chromium-free catalyst (T-4489-United Catalysts Inc., Louisville, KY). For the hydrogenations, the cell was contained in the thermostatted oven of a Spe-ed supercritical fluid extraction unit (Applied Separations Inc., Allentown, PA). The synthesized alcohol mixtures were collected in a vial after decompression of the binary fluid mixture through the micrometering valves on the Spe-ed unit. The FAME feedstock was fed into the SC-H_2/CO_2 or SC-H_2/C_3H_8 stream using an Isco syringe pump (Isco, Inc., Lincoln, NE). The FAME-SC-H_2/CO_2 or SC-H_2/C_3H_8 mixture was then transported over the hydrogenation catalyst contained in a small volume reactor vessel (1–4.5 mL cells).

The system used to study the sequential, two-step reaction conversion of soybean oil to fatty alcohol mixtures coupled an Isco SFX-2-10 extractor module with flanking Isco syringe pumps that were used to deliver the oil, 4 μL min^{-1}, and methanol, 5 μL min^{-1}, respectively, into the SC-CO_2 prior to the first reactor cell.

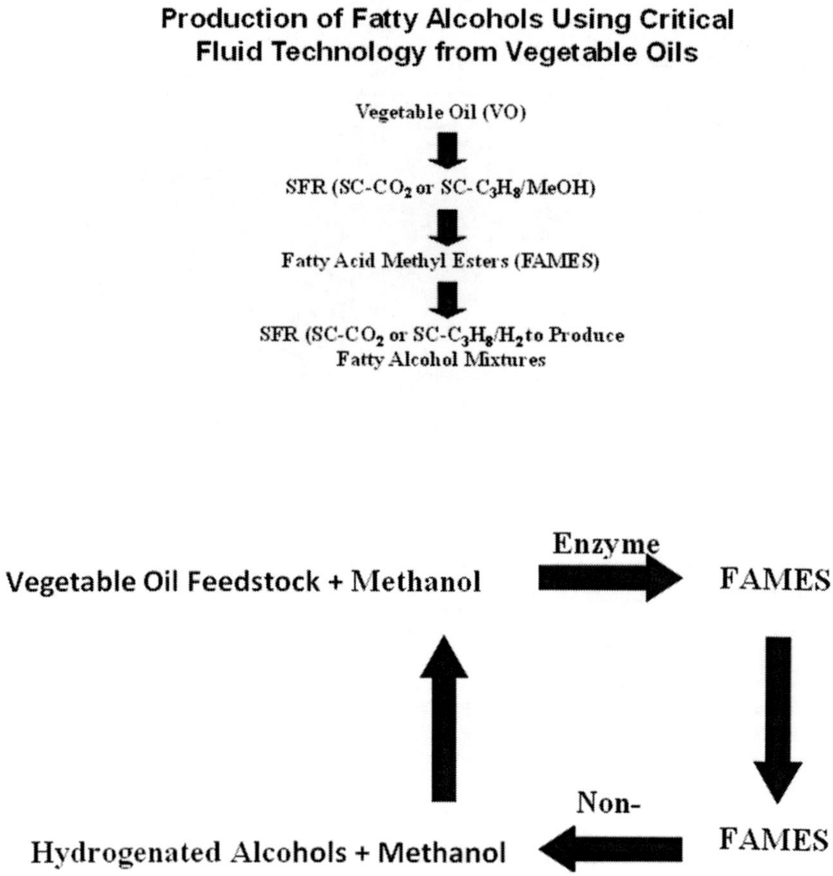

Fig. 3.30. Production of fatty alcohol mixtures using a (a) SFR-SFR sequential reaction scheme, (b) feedback of methanol into the SFR process and the green nature of the process.

Transesterification of soybean oil was accomplished using the reported enzymatic synthesis of FAMES as described by Jackson and King (1997). Utilizing a pressure of 170 bar and a temperature of 50°C, a supported-enzyme catalyst, Novozym SP 435 (Novozyme, Danbury, CT) contained in a 2.5 mL reaction vessel was used to perform the FAME conversion. The resultant FAMES were then transferred into the apparatus shown in Fig. 3.31 where either SC-H_2/CO_2 or SC-H_2/C_3H_8 were used to hydrogenate the synthesized fatty acid methyl esters.

Here in Fig. 3.31, the hydrogenation step uses the identical experimental apparatus previously noted for the critical fluid hydrogenations. A multiple syringe pump system (A–D) was used to deliver the appropriate amounts of CO_2 (A–B), soybean oil (C), and methanol (D) into the extraction (reaction) cell of an Isco SFX-2-10 module (E), containing the Novozym SP 435 catalyst. The resultant

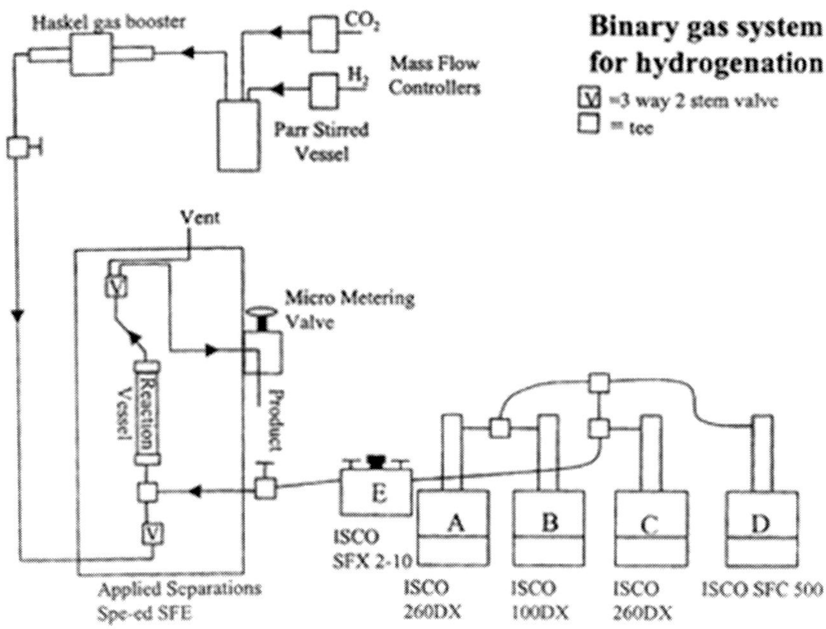

Fig. 3.31. Reaction system for hydrogenating FAMES to alcohols using binary fluid mixtures of SC-H_2/CO_2 or SC-H_2/C_3H_8.

product from the transesterification was then fed into the flow stream as designated in the hydrogenation sequence of the overall reaction.

For the SC-H_2/CO_2 as well as for the SC-H_2/C_3H_8 system only two of the experimental variables were found to have a significant impact on the hydrogenated alcohol purity. These were temperature and hydrogen content in the supercritical fluid. As shown in Figs. 3.32. a-b, these variables should be kept at higher levels as is clearly evident from the response surfaces. Substrate flow rate, pressure and residence time were not found to have a significant effect on the hydrogenation process within the investigated range of experimental parameters. As an example, a response plot obtained with the chromium-free catalyst and SC-H_2/C_3H_8 mixtures is shown in Fig. 3.32b. When the hydrogenation was performed under optimal conditions the two catalysts gave similar results.

The response surface for the SC-H_2/C_3H_8 system indicates that high alcohol conversions can be accomplished also when using lower mole fractions of hydrogen in propane than for the corresponding SC-H_2/CO_2 system (*c.* 50%). Response surfaces for the copper chromite catalyst, Figs. 3.32.c and d, follow a similar pattern but the trends are not as clear. At high mole fractions of H_2 the SC-H_2/CO_2 system is superior in terms of product yield to the SC-H_2/C_3H_8 system. However, at low mole fractions of H_2, both the SC-H_2/CO_2 and the SC-H_2/C_3H_8 systems give approximately the same yield, i.e., 80% conversion. This

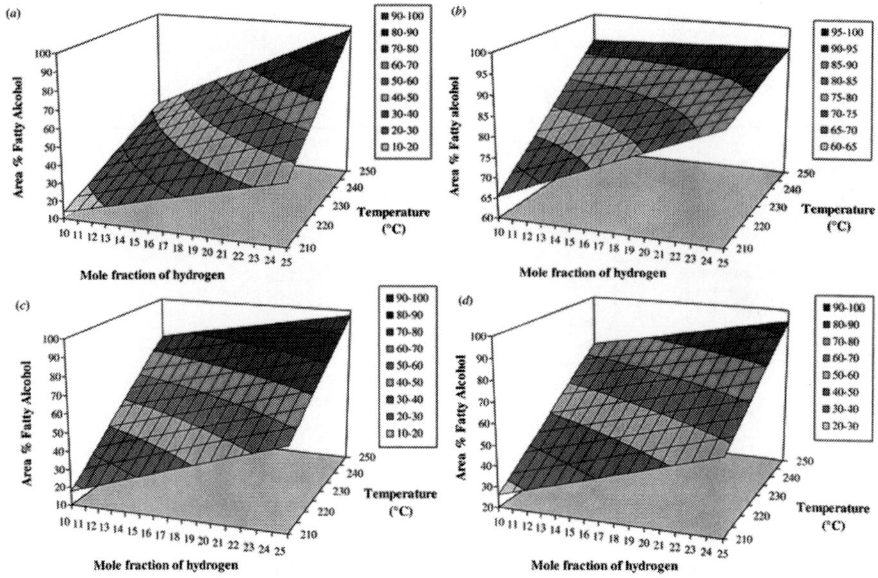

Fig. 3.32. a-d. (*a*) Response surface plot for production of fatty alcohols using chromium-free catalyst in SC-H_2/CO_2. Approximate residence time, 5.5 s; substrate flow, 37.5 μL min^{-1}; pressure, 200 bar. (*b*) Response surface plot for production of fatty alcohols using chromium-free catalyst in sc-H_2/C_3H_8. Approximate residence time, 5.5 s; substrate flow, 150 μL min^{-1}; pressure, 200 bar. (*c*) Response surface plot for production of fatty alcohols using copper chromite catalyst in SC-H_2/CO_2. Approximate residence time, substrate flow and pressure as in Fig. 3.32(*a*) and (*d*) Response surface plot for production of fatty alcohols using copper chromite catalyst inSC-H_2/C_3H_8. Residence time, substrate flow and pressure as in Fig. 3.32 (*b*).

indicates the importance of catalyst selection for the overall yield and the reaction conditions that are required.

The gas chromatographic profiles of the hydrogenated product mixtures derived from the reaction in the SC-H_2/CO_2 and SC-H_2/C_3H_8 systems are shown in Figs. 3.33. a and b. Fig. 3.33a indicates a high degree of conversion (97.2%) for both methods used in preparing the hydrogenated fatty alcohols (steryl and palmityl alcohols) from soybean oil. Only trace levels of unconverted FAMES remain. However, Fig. 3.33b which shows the composition of the product obtained with the SC-H_2/C_3H_8 system indicates not only the appearance of the fatty alcohols, but also significant amounts of C_{16} and C_{18} n-alkanes as by-products. In this case, the two alcohols constitute 95% of the product, while the *n*-alkane components were 4.6% of the total product yield. Both product mixtures were generated at 250 bar, 250°C, using a hydrogen mole fraction of 0.25 and residence time of 9 seconds, using chromium-free catalyst. Using a copper chromite catalyst for the two binary gas mixtures under basically the same conditions yielded a similar result.

There are tradeoffs in employing the SC-H_2/CO_2 or SC-H_2/C_3H_8 system; the first system allows the highest potential conversion to alcohols to be achieved, while

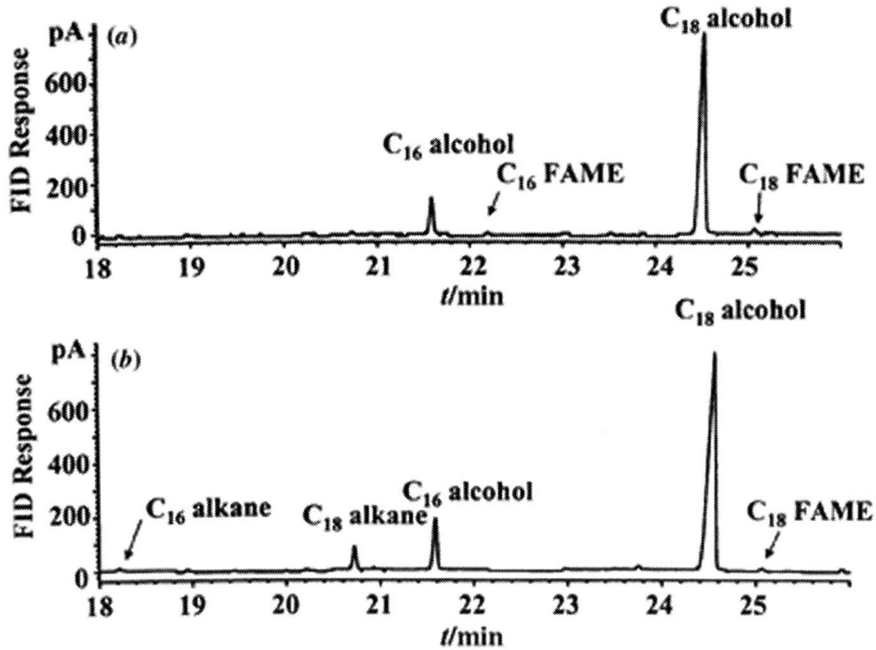

Fig. 3.33. (*a*) GC profile of products from critical fluid hydrogenation (SC-H_2/CO_2) of FAMES using chromium-free catalyst. Reaction conditions: pressure 250 bar, temperature 250°C, mole fraction of H_2 0.25, residence time 9 s; substrate flow rate 50 μL min^{-1}. (*b*) GC profile of products from critical fluid hydrogenation (SC-H_2/C_3H_8) of FAMES using chromium-free catalyst. Reaction conditions: pressure 250 bar, temperature 250°C, mole fraction H_2 0.25, residence time, 9 s; substrate flow rate 250 μL min^{-1}.

the second binary fluid mixture (H_2/C_3H_8) permits a higher throughput. The rate of conversion of FAMES to alcohols is higher for the SC-H_2/C_3H_8 mixture than for the SC-H_2/CO_2 binary fluid. For example, three times more alcohol can be synthesized in the SC-H_2/C_3H_8 mixture in two-thirds of the time as compared to what can be accomplished under similar conditions in the SC-H_2/CO_2 system.

Fig. 3.34 illustrates the GC analysis of the product mixture obtained when using enzymatic catalysis to form the FAMES from soybean oil followed by a hydrogenation over a chromium-free catalyst. High conversion to alcohols, 96.5% yield, with only traces of the original FAMES and *n*-alkanes were detected in the chromatogram. This demonstrates that the two-stage reaction sequence performs well when using the same conditions as applied for the reactions leading to the product composition illustrated in Fig. 3.34. The conversion of the soybean oil over Novozym SP 435 was examined using capillary supercritical fluid chromatography, SFC. As indicated by the SFC, the reaction was relatively complete; C_{16}, C_{18} FAME and only minor amounts of mono-, di- and tri-glyceride appeared in the chromatogram as shown in Fig. 3.35.

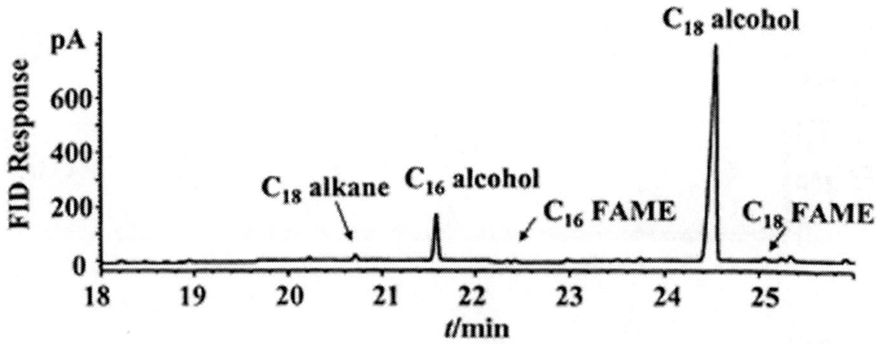

Fig. 3.34. GC profile of products from SFR (transesterification)/ SFR (hydrogenation-SC-H_2/CO_2) of soybean oil. Reaction conditions for the transesterification step: pressure 170 bar, temperature 50°C; oil flow rate 4 μL min^{-1}, methanol flow rate 5 μL min^{-1} CO_2 flow rate 1 mL min^{-1}. Reaction conditions for the hydrogenation step: pressure 250 bar, temperature 230°C, mole fraction of H_2 0.25, residence time 9 s.

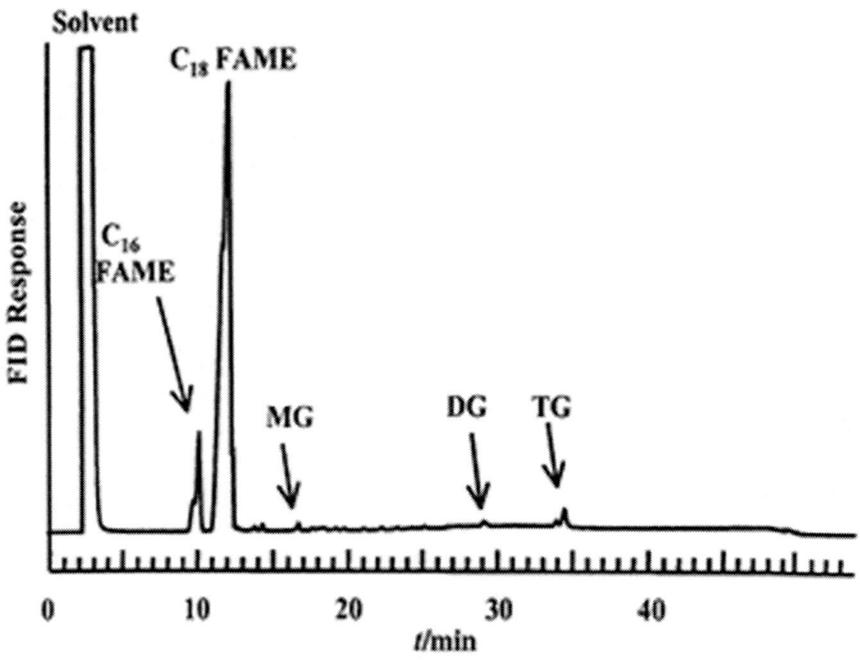

Fig. 3.35. SFC profile of the transesterification step using Novozym SP 435. MG, mono-glyceride; DG, di-glyceride; TG, tri-glyceride.

It should be emphasized that by using the above-described sequential reaction method, that conversions of FAMEs via the described enzymatic synthesis route gave yields in excess of 98% conversion. Conversions for the second reaction step, i.e., hydrogenation, on the average yielded 96.5% conversion yielding mixtures based on a soybean FAME feed of 90% steryl and 8% palmityl alcohol. The hydrogenation rate is fast—between 4 to 9 seconds and the catalyst lifetime on this laboratory scale reactor system lasted two months. Here, as throughout the cited FAME hydrogenation studies in this chapter, the use of FAMEs as opposed to neat fatty acids decreased the corrosion on the stainless components. King remarked that further fractionation of the resultant hydrogenated fatty acid mixtures could be achieved by inserting a supercritical fluid fractionation (SFF) column after the hydrogenation step to achieve high purity free fatty acids.

Hobbs et al., (2006) carried out the Pd-catalyzed hydrogenation of acetophenone and subsequent enzyme-catalyzed kinetic resolution of the corresponding (R/S)-alcohol product with vinyl acetate as shown in Fig. 3.36. They connected a hydrogenation reactor and an enzyme reactor so that the hydrogenation-kinetic resolution could be performed in series. Thus a mixture of acetophenone, H_2, and CO_2 was first fed into the hydrogenation reactor containing 2% Pd/SiO_2–Al_2O_3. Then the product downstream was mixed with vinyl acetate and fed to the enzyme reactor which contained cross-linked *Candida antarctica* lipase B aggregate as catalyst. The conversion in the hydrogenation reached a maximum of 82% even without optimization and the subsequent kinetic resolution took place well to give the corresponding products in excellent enantioselectivity (maximum >99%) enantiomeric excess (ee) at reasonable conversions. Unfortunately, the selectivity to 1-phenylethanol in the hydrogenation was not reported, although it might be not important for the outcome of the kinetic resolution. Note that the group had previously performed the same hydrogenation using a Deloxan-aminopolysiloxane-

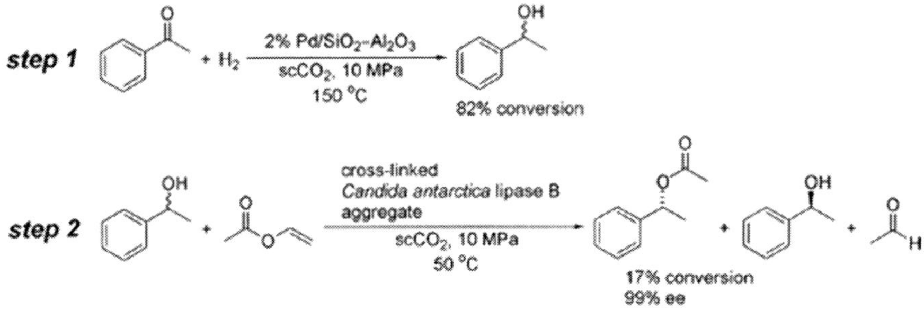

Fig. 3.36. Hydrogenation of acetophenone and subsequent kinetic resolution of the corresponding alcohol product with vinyl acetate (These reactions were performed in series in continuous-flowing SC-CO_2, using a hydrogenation reactor and an enzyme reactor. The conversions and ee were obtained under the following conditions: CO_2 flow rate, 1 mL min^{-1}; acetophenone flow rate, 0.1 mL min^{-1}; acetophenone:H_2 = 1:4).

supported 5% Pd catalyst. The catalyst then afforded ethylbenzene, 1-cyclohexylethanol, and ethylcyclohexane in addition to 1-phenylethanol, and the product selectivities could be controlled by changing reactor temperature.

Supercritical fluid hydrogenation can also be combined with supercritical fluid extraction (SFE). For example Tacke et al., (2003) reported the coupling of the countercurrent mode of SFE in tandem with a high pressure hydrogenation in SC-CO_2 or propane. Typically pressure of 20 MPa and temperatures between 60–100°C were used to fractionate the fatty acids having acid numbers of approximately 190 mg KOH/g. The latter value is comparable to that obtained using conventional distillation in the production of free fatty acids. Mass balance studies indicated that over a 95% yield could be obtained, a yield that would permit and integration of supercritical hydrogenation after the fractionation step. Typical hydrogenations temperatures were at 140°C and were increased if evidence of catalyst deactivation was recorded. The lowest IV values were recorded using SC-CO_2/propane mixtures. It should be appreciated that coupling the hydrogenation step with SFE or a similar fractionation technique can provide a superior feedstock for hydrogenation and thereby increase catalyst usage lifetimes.

Another relative unknown coupling of SFE with supercritical hydrogenation was the extraction of squalene from the waste stream associated with olive oil production followed by supercritical hydrogenation to produce squalane. These studies were part of a FAIR consortium effort in the European Union in which the partners showed that relative high yields (70–78%) of squalene could be extracted from saponified olive oil residue using SC-CO_2. Further studies were conducted that coupled supercritical fluid chromatography (SFC) with this initial SFE enrichment stage, resulting in squalene of 90% purity. An alternative secondary enrichment process is to utilize a fractionating columnar approach in the countercurrent mode which provides 80% pure squalene. This can be even further improved by methylating the olive oils residue to enhance the separation factor using the packed columnar fractionating column in the countercurrent mode. The final tour de force was to combine SFE-a fractionating enrichment method with the hydrogenation step in SC-CO_2. It was found that hydrogenation conducted in SC-CO_2 was somewhat faster than in condensed n-hexane, although the system was not totally optimized.

Critical Fluids and Catalysts

It is not possible within the scope of this chapter to cover on the aspects of catalysts use and their design relative to supercritical fluids. The reader is referred to a very excellent and illustrative review by Wandeler and Baiker (2000) and in the context of heterogenous catalysis in fine chemicals synthesis (Ciriminna et al., 2008; Hutchings, 2009), as an introduction to the subject. As remarked previously, the existence of multiple phases when conducting hydrogenations on liquid substrates coupled with the inclusion of a heterogeneous catalyst (solid) suspended in a liquid substrate makes hydrogen transport to the catalyst surface difficult. Aside from the low solubility of H_2 in such liquid phases, there exist large concentration gradients at both the gas-liquid and liquid-catalyst interfaces as noted by Härröd and coworkers (2001). Part of this is due to the fact that there is not sufficient H_2 at the catalyst surface and transport into the catalyst pores is also inhibited, particularly so in gas-liquid phase systems. Such a system is the gas-liquid hydrogenation fat hardening process which requires as noted in other chapters of this book, large batch reactors where reactants, H_2, and catalyst are mixed for long times at high temperatures (140–200°C) and low pressures (1-3 bar). To make a hydrogenated fat containing 30–40% trans fat content requires that the batch reactor be scaled to yield approximately 400 $kg_{oil}/m^3_{reactor}h$.

Various catalytic metals have been tested for the hydrogenations in SCFs. The most active metal for the hydrogenation of olefinic double bonds in SC-CO_2 is palladium, and the reaction can be performed without affecting the double bonds. This feature has led to successful selective double bonds bond hydrogenation of α, β-unsaturated carbonyl compounds. Platinum exhibits interesting catalytic behavior for hydrogenations in SC-CO_2 and its activity strongly depends on the structure of substrates. Typically platinum is effective in SC-CO_2 not only for the olefinic double bonds bond hydrogenations (Milewska, 2005) but also for the selective hydrogenation of α, β-unsaturated aldehydes to unsaturated alcohols and aryl nitro compounds to arylamines. However, in spite of its high activity for the hydrogenation of the double bond in α, β-unsaturated aldehydes in SC-CO_2, platinum is ineffective for double bonds bond hydrogenation of α-ketoesters like ethyl pyruvate in the same medium. The unique property of platinum in SC-CO_2 has not been elucidated entirely, but several researchers observed the formation of carbon monoxide and its adsorption on platinum by IR spectroscopy. It has been suggested that the carbon monoxide is formed by the reverse water–gas shift reaction catalyzed by platinum. The adsorbed CO can block the platinum sites active for both the desired and undesired reaction. An interesting example where adsorbed CO originating from the above reaction poisons sites active for an undesired reaction is the selective hydrogenation of halogenated nitrobenzenes to the corresponding halogenated anilines, in which the platinum sites active for the dechlorination were shown to be blocked by the in situ formed carbon monoxide.

As the studies of Härröd and Macher have shown, even after short running times, the catalysts showed significant signs of deactivation. The IV of the reduced fat or oil increased continuously, and sometimes the activity disappeared completely before the end of their experiments (i.e., after 1 kg oil/g Pd had passed the catalyst). Reasons for this could be impurities in the substrate, e.g., poisons, oxidation products, etc., or the substrate itself, which are/is blocking the catalyst pores or polymerization of double bonds and subsequent coking. An indication of impurities was found in an experiment where the oil was pretreated with catalyst at 100°C, prior to the hydrogenation. This resulted in a slightly slower deactivation of the catalyst. A similar effect has already been reported for Ni-catalysts. The purification procedure probably needs to be optimized to remove all impurities (since the deactivation was not completely inhibited), however the result is a promising first step.

The second possible reason for the catalyst deactivation seems initially less likely, since palm oil is a highly saturated oil and the risk of coke formation is related mainly to the presence of polyunsaturates, but this has not yet been investigated. An analysis of deactivated catalyst showed that most of the small pores had disappeared, most likely due to blocking by oil. It seems that polymerization is not necessarily occurring, but that even unaltered oil can simply fill the pores and not be able remove itself again. This is related to the shape of the pores and also to the type of carrier material, especially carbon, which is considered to adsorb different reactants very strongly. The research at Chalmers University in Sweden showed that high concentrations of hydrogen can prevent the deactivation of the catalyst, which would support the coking theory. However, no studies on the substrate/catalyst combination used in the present study have been done to date. Initial experiments indicated a slight delay in the deactivation at hydrogen concentrations as high as 10 mol%.

As previously discussed, in fatty alcohol synthesis it should be noted that at high substrate concentrations, a rapid fall of the reaction rate is observed, and the benefits of using propane are completely lost. This fall in the hydrogenation reaction rate depends on a split of the supercritical reaction mixture into two phases (a substrate-rich and a hydrogen-rich phase). If this phase split occurred by using small catalyst particles (≤32 μm), the pressure drop over the catalyst bed increased sharply due to the formation of liquid droplets which blocked the void space in the porous catalyst bed.

These two phenomena can be used to deduce the product and substrate solubility in the reaction mixture. The product showed the most unfavorable solubility which increased with higher pressure. Under such process conditions (150 bar, 280°C, and 11 mol% hydrogen), a single phase was observed up to 2 mol% (i.e., 15% by mass) product. Besides the minimum pressure in the catalyst bed, substrate transport limitation could be shown to be an important factor in process optimization. Therefore, egg-shell catalysts or fine catalyst particles (100–300 μm) should preferably be used in the continuous supercritical reactors. Optimized, continuous flow reactors can produce very large production rates ~ $2.4 \times 10^3\ kg_{FAME}/m^3_{reactor}h$, which is 500 times higher than production rates in stirred batch reactors.

As reported in the "Supercritical Fluid Hydrogenation of Fats/Oils Using CO_2 or Propane" and "Oleochemical Synthesis in Supercritical Fluids" sections, polysiloxane-supported noble metal catalysts are compatible with both SC-CO_2 and propane in continuous flow reactors. Hitzler and Poliakoff (1997; Hitzler et al., 1998) using similar catalysts have reported the hydrogenation of a variety of organic functionalities, e.g., the hydrogenation of cyclohexene to cyclohexane. They attained yields of cyclohexane between 95–98% using SC-CO_2 at 120 bar and propane at 60–80 bar. Among the other conversions they reported is the hydrogenation acetophenone which yielded a mixture of four products depending on the temperature, pressure and hydrogen concentration used. Extension of this work utilizing the aforementioned Deloxan series of catalysts have shown these to be versatile as applied to alkynes, epoxides, phenols, cyclic ethers, oximes, nitrobenzenes, and nitriles. It was found that propane was the best supercritical fluid for processing amine-containing substrates, since the use of SC-CO_2 and its reactivity to form carbamates was problematic.

Successful hydrogenations have also employed other catalysts besides polysiloxane supports. Studies on the hydrogenation of α, β–unsaturated aldehydes in SC-CO_2 using Pt/Al_2O_3 supported catalyst show increasing yields and selectivity with increasing SC-CO_2 pressures. This trend is justified due to the adjustable solvent polarity of SC-CO_2 and electronic state of Pt which promotes preferential hydrogenation of carbonyl bonds. Hydrogenation to produce isophytol can be carried out in SC-CO_2 employing a flow reactor packed with Pd-Si catalyst. Similarly, maleic anhydride can be hydrogenated to γ–butyrolactone using Pd/Al_2O_3. Although these reactions might not seem highly relevant to the hydrogenation of lipids, they point out the possibilities of conducting hydrogenations of different functionalities which could also be transferred into lipid reaction chemistry.

Nanoparticles with catalytic activity have been used with SC-CO_2 to perform hydrogenations. Polymeric–supported colloidal Pd nanoparticles efficiently hydrogenate 1-hexyne at a high turnover rate ($4.0 \times 10^6 h^{-1}$) in SC-CO_2 at only 15 bar pressure and 50°C. It has also been reported that Pd nanoparticles stabilized by water in CO_2 microemulsion aided by rapid dispersion in SC-CO_2 can efficiently hydrogenate olefins. A sample synthesis using this approach would be the conversion of 4-methoxycinnamic acid to 4-methoxyhydrocinnamic acid in 20 seconds at 50°C using SC-CO_2. Rhenium nanoparticles have also been used in conjunction with SC-CO_2 for hydrogenating phenol and naphthalene.

Although it is beyond the scope of this review on supercritical hydrogenation, it should be noted that reference is made in the literature as to the synthesis of catalysts in supercritical fluid media on even the nano-particulate scale (Zhang & Erkay, 2006) and using the principles of "green" synthesis (Hutchings, 2009). Likewise, supercritical fluids have been cited for their potential to reactivate catalysts (Trabelsi et al., 2000) using supercritical fluid extraction on Pd/activated carbon catalysts (Zhang et al., 2009).

Hydrogenations in Compressed Water

Although water would not be thought to be considered as a conventional solvent for reaction chemistry on lipid substrates except at very high temperatures (King, 2000; Adams et al., 2004), the ability to use water in multi-phasic systems in conjunction with common supercritical fluids such as SC-CO_2, offers some interesting considerations for the synthetic lipid chemist. Water soluble catalysts, such as water-soluble phosphine complexes bearing sulfonic acid groups have been used to achieve hydrogenations in mixed SC-CO_2/H_2O systems. An example of using this approach is the carbonyl-selective hydrogenation of cinnamaldehyde which allows the easy separation of the catalyst from the reaction products.

The addition of surfactants accelerates reactions taking place in biphasic media, for example the hydrogenation of styrene in a SC-CO_2/H_2O emulsion mixture. These emulsions are sometimes troublesome with respect to their separation from the reaction products at the conclusion of the reaction, however when employing SC-CO_2/H_2O emulsions, the biphasic system readily separates by depressurization allowing easier product recovery. A rather novel reduction of ketone moieties can be affected using a flow reactor with alcohol dehydrogenase immobilized on a hydrophilic polymer (Rao et al., 2009). It is theorized that the reported conversion takes place in the aqueous phase similar to what occurs in a SC-CO_2/H_2O biphasic system.

High temperature pressure water provides an advantage for such reactions. Water can act as a solvent for both gases and organic substrates providing a single phase reaction which overcomes mass transfer limitations providing rapid reaction rates. However, both processes require the use of gases such as hydrogen and oxygen posing significant problems. It has been found that compressed hydrogen and oxygen, although possible, is expensive, requires safety precautions and it is difficult to control on the small scale required for bench-work. One practical solution to this problem was noted in the section on "Equipment, Processing Concepts, and Scale-Up," is to generate oxygen or hydrogen by thermal decomposition of the right precursors. Thus, hydrogen peroxide can be used as source of oxygen and formic acid or related formates to generate hydrogen. The "gasless" hydrogenation of aromatic compounds in near-critical water was carried out using the formic acid (HCO_2H), sodium formate ($NaCO_2H$) or ammonium formate (NH_4CO_2H) aqueous solutions as a hydrogen source by thermal decomposition. Hence no catalyst was required and the reduction of different cyclic and aromatic ketones, olefins, and aldehydes could be accomplished using a green process achievable with very simple equipment using pressures between 15 and 20 MPa, temperatures between 413.15 and 563.15 K, and residence times between 6 and 30 s in a continuous flow reactor. Conversions up to 80% combined with mass recoveries ~ 99% have been obtained so far using this type of reaction.

An interesting application that could have some application in controlling the degree of unsaturation in vegetable oils has been reported Aydogan et al., (2006) and is included here because it is conducted in aqueous medium with the aid of SC-CO_2. The soybean oil is reacted with $KMnO_4$ in the presence of water and dense CO_2 to facilitate contact between the soybean oil and aqueous $KMnO_4$ solution. Experiments were done over the temperature range of 10–50°C and 2.5–16 MPa, with and without $NaHCO_3$ addition, while the amount of $KMnO_4$ amount was varied. While only 40% on the double bond were reacted at the near critical conditions of CO_2, the addition of $NaHCO_3$ enhanced the conversion to more stauration.

Key Patents Involving Hydrogenation in the Supercritical State

This section will provide only a brief overview of the seminal patents primarily related to the supercritical fluid hydrogenations of fats and oils, as well as some related patents. It is somewhat interesting to note that in one of the original Zosel patents, namely, "Process for the Simultaneous Hydrogenation and Deodorisation of Fats and Oils" (Zosel, 1976) advocated the use of a near-critical solvent for the hydrogenation of triglycerides using SC-CO_2. This old issuance was concerned with the simultaneous hydrogenation and deodorization of at least one product from the group consisting of fats and oils, wherein said product is treated with carbon dioxide at a temperature of from 100 to 250°C and a pressure of from 150 to 300 atmospheres in the presence of a hydrogenation catalyst, and hydrogen. The fats and oils hydrogenated and deodorized in this process were to be used in the manufacture of margarine. Over the intervening years, increasing technical interest in this reaction technology is reflected by a number of patents that have been filed and the development of industrial plants for continuous hydrogenation of fats and oils as well as other moieties. Specifically, supercritical fluid reaction chemistry has been scaled up to plant production by Hoffmann-La Roche, the Härröd group in Sweden, as well as the University of Nottingham/Thomas Swan and the University of Göttingen/Schering AG groups.

The initial Möeller and Härröd patent entitled, "Hydrogenation of Substrate and Products Manufactured According to the Process" was filed in 1996 as WO 96/01304 (Möeller and Härröd, 1996). The eventual issuance of this patent involving the great improvement in reaction rate due to conducting hydrogenations in the homogeneous supercritical phase was eventually covered under U.S. Patent 5,962,711 which was issued in 1999 (Härröd and Möller, 1999). Both patents primarily are concerned with double bond hydrogenation in lipids, reduction of ester linkages to the corresponding fatty alcohols, as well direct hydrogenation of oxygen to H_2O_2. Another key patent issuance to Härröd and Möller (2001) is described in U.S. Patent 6,265,596, July 24, 2001, (Härröd and Möller, 2001). Here partial hydrogenation of fats and oils is described in the presence of supercritical fluids to produce a reduced *trans* fat–containing product. Specific relationships are stated indicating what IV and *trans* fatty acid content can be expected by employing this method.

The "Tacke" patent portfolio has quite a long history beginning in 1996 as WO95/22591 concerning the "Hardening of Unsaturated Fats, Fatty Acids, or Fatty Acid Esters" which was issued as U.S. Patent 5,734,070 (Tacke et al., 1998). The focus here as with the Möeller and Härröd fillings is on fats and oils reduction using hydrogen in the presence of a near-critical or supercritical fluid medium, including comparison with trickle bed catalysis using the same catalytic medium. Aside from the specifications on the conditions for conducting supercritical hydrogenations, data is also provided as to what specific catalysts are to be preferred such Pt or Pd supported on a variety of carriers. Another world patent, WO 97/38955 and U.S. Patent 6,156,933 (Poliakoff et al., 1997; 2000) simply entitled "Supercritical Hydrogenation" is focused more on the hydrogenation of simple aliphatic and aromatic organic compounds. The catalysis is heterogeneous involving continuous flow reactors as applied to numerous classical reactions. The patent issued to Thomas

Swan & Company in England and Degussa is based on the studies of Poliakoff and coworkers and has wide applicability for hydrogenation in supercritical fluid media. An example of its industrial application would be the continuous hydrogenation of isophorone to trimethylcyclohexanone in SC-CO_2 using a supported Pd catalyst which is now being produced on an industrial scale. In this conversion the yield is high with little by-product being formed thus requiring no purification of the products after hydrogenation has been performed. The above patent is a joint filing between the companies; Thomas Swan in England and Degussa in Germany, and involves the latter's Deloxan-type of catalysts. Interestingly, these catalysts are not currently offered commercially.

A closely related patent which is connected with the above citations is, "Three Stage Processes for the Separation of Supercritical or Near-critical Mixtures" as U.S. Patent 7,084,313 (Gutsche et al., 2006). Here processes were described for separating supercritical or near-critical mixtures containing hydrogen, a solvent gas, methanol and fatty alcohols under an initial pressure of from 100 to 300 bar are described, wherein the processes comprise: (a) reducing the pressure of such a mixture in a first stage to a pressure of from 50 to 150 bar to form a first recycle gas and a first partially-separated intermediate mixture, wherein the reduced pressure in the first stage is at least below the initial pressure; (b) reducing the pressure of the first partially-separated intermediate mixture in a second stage to a pressure of from 10 to 50 bar to form a second recycle gas and a second partially-separated intermediate mixture; and (c) reducing the pressure of the second partially-separated intermediate mixture in a third stage to a pressure of from 1 to 10 bar to form a third recycle gas and a fatty alcohol product. Although not directly related to the hydrogenation of fats and oils with the aid of supercritical fluids, the patent was issued to Subramaniam et al., (1997). "In situ Mitigation of Coke Buildup in Porous Catalysts by Pretreatment of Hydrocarbon Feed to Reduce Peroxides and Oxygen Impurities," U.S. Patent 5,690,809 is worth noting since it is concerned with coke buildup in porous catalysts during hydrogenation can be minimized by using critical fluids.

Concluding Remarks and a Salute to Supercritical Fluid Research at NCAUR

We have attempted to show in this chapter the merits and advances in using supercritical fluids for primarily the hydrogenation of fats or oils and additional lipid moieties. Unlike some applications of supercritical fluids, very high pressures are not necessarily required to affect a successful hydrogenation when using SC-CO_2, the lower compressed alkanes, or dimethyl ether as hydrogenation media. Aside from offering some essential preliminary introductory material to the field of supercritical fluid technology, our focus has been on hydrogenation reactions, including some of those not involving lipid substrates in the hope that researchers will see additional applications for this technological platform. Space limitations have prevented a detailed discussion of all of the relevant equipment and conditions, and we have chosen to elaborate on those studies we are most familiar with. Included within the reference section of this chapter are some general texts that will provide the reader with additional information however this is by no means the extent of books on supercritical fluids which definitely number over 50 as of this writing.

The role of supercritical fluids in the reduction of industrial oleochemicals has featured primarily a platform built around fatty alcohol synthesis. It is interesting to note that there are other applications of supercritical fluids to oleochemical derivatives—of particular note is the rapid synthesis of biodiesel using near or supercritical methanol (Kusdiana & Saka, 2001). The section on coupled processes does not do justice to an integrated extraction-reaction platform embracing critical fluid media as a green and sustainable platform; many other tandem processes have been conceptualized, and this is the subject of an extensive review by one of the authors in a recent issue of the *Journal of Supercritical Fluids* (King & Srinivas, 2009).

A current and on-going research focus in the area of critical fluids research is in the synthesis of micro- and nano-particulates, including catalytic media (Zhang & Erkey, 2006; Guo et al., 2009). Compressed water technology, i.e., subcritical water has been an active area of research since the 1980s; however, it is receiving increased emphasis particularly in the areas of biofuels and material science. Subcritical water along with SC-CO_2 can be advantageously combined for successive extraction, fraction, or reaction unit operations (King, 2003) and are viewed as mutually complimentary "green" fluids. In this context, the authors anticipate that additional research will be conducted in coupling hydrogenation in compressed water.

Finally the litmus test of the success of a technology can be measured by its use in the real world of industrial manufacturing. Recently several larger scale production plants utilizing supercritical fluids for synthesis have come "on-line", including the large DuPont facility in North Carolina for fluorinated polymer production. As noted in this chapter, two production facilities devoted to hydrogenation have been in operation since 2002: one large pilot plant in Göteborg, Sweden (Härröd Research AB) and one industrial plant in Consett Co. Durham, United Kingdom (Thomas Swan & Co, Ltd). We have noted several of the products originating out of the Swan plant in England; however, the former plant hydrogenates fatty acid methyl esters to fatty alcohols (10 kg-alcohols/h, 40 kg-propane/h, maximum pressure 30 MPa). Several pictures (Figs. 3.37 and 3.38) of the Swan chemical intermediates plant and Härröd facility in Sweden are shown on the following page.

In concluding this review of hydrogenation conducted with the aid of critical fluids, the authors should like to pay tribute to Dr. John P. Friedrich of the Northern Regional Research Center (NRRC)—now the National Center for Agricultural Utilization Research—whom we were both privileged to collaborate with while we were with the Agricultural Research Service of the USDA in Peoria, Illinois. Dr. Friedrich developed the original "high pressure" laboratory for the express purpose of conducting hydrogenation studies under pressure. Among Dr. Friedrich's contribution were studies on continuous slurry hydrogenation of soybean oil using copper-chromite catalysts (Koritala et al., 1980), synthesis of cyclic monomers from vegetable oil feedstocks (Friedrich, 1967) are but a few examples of his contributions in applied hydrogenation studies to oil and fat chemistry.

Fig. 3.37. The Thomas Swan company production plant in Consett, Durham, England.

Fig. 3.38. The Härröd AB pilot plant in Goteborg, Sweden.

Dr. Friedrich was also responsible for initiating the program in supercritical fluid extraction (SFE) of vegetable oils at Peoria, Illinois laboratory in the early 1980s and worked extensively with one of the authors (G. List) which resulted in a series of publications (List et al., 1989) dealing not only with the characterization of the properties and stability of the SC-CO_2 – extracted oils, but also the residual protein- and carbohydrate-containing meals for use in foods. Several patents also resulted from these SFE studies, namely U.S. Patents 4,466,923; 4,495,207; and 4,493,854 (Friedrich, 1984; Christianson & Friedrich, 1985; Friedrich & Eldridge, 1985).

These initial studies by Friedrich and coworkers provided a scaffold upon which one of the authors (J. King) continued to build upon which involved diversification from the original SFE studies of oils and fats. These studies included the use of SC-CO_2 and similar fluids for fractionating lipids and conducting various reactions in such media, including the hydrogenation studies cited in this chapter. King and colleagues (King, 1998) also applied many of the concepts and principles gained from Friedrich's initial research in reaction chemistry and SFE into the field of analytical chemistry involving sample preparation and chromatography to "green" the government regulatory chemistry laboratories. A nice summary of these studies including the state of the art of the field as applied to lipid chemistry and processing through the turn of the century is provided in the tome by King and List (1996).

Chapter 4

Hydrogenation Facility

H. B. W. Patterson

General Considerations

In considering the layout and the components of a hydrogenation facility ponder this worthwhile reminder again—the conditions at the catalyst surface determine the course of the reaction. Various means exist to create these physical conditions for bringing together oil, hydrogen, and catalyst; naturally, the manufacturer seeks to find those means which give the best effect for the least cost. His choice is influenced by what is most readily available to him—expensive, very pure hydrogen or cheaper, less pure gas; easily available skilled labor or an almost complete lack of it; and so on. Also, his choice is strongly affected by whether he needs to produce, at most, only five or six hardened oils from two or three vegetable oils, or thirty to forty different hardened fats in one week from a dozen raw materials covering vegetable and marine oils and animal fats. At the outset, a pertinent remark is that a facility designed to perform a special task with the greatest cost efficiency is—by inference—less cost-effective when put to work on substantially different tasks. For a varied production, this clearly implies some additional capital expenditure so that all relevant classes of hydrogenation can be performed effectively and the customers enjoy a feeling of confidence that their specifications will be regularly met. A large part of such expenditure will relate to the efforts made to segregate different groups of oils, and this becomes particularly relevant when work is progressing simultaneously on lauric oils, other vegetable and marine oils, and animal fats.

Quite properly, the suppliers of complete hydrogenation units will offer a basic system which contains one hydrogenation autoclave, or possibly two, and ancillary equipment such as heat exchangers, gas recirculators, and oil filters, which are proposed as the means to achieve economy in energy, maximal utilization of expensive components, and convenience in operation. The management of the hardening facility and the equipment supplier will discuss how this basic system is to be developed, if necessary, to cope with larger and more complex programs. Even in a small facility, some degree of flexibility is a good investment; for example, in a facility which is planned to possess only one 10-ton autoclave, provision could be made for the autoclave occasionally to be able to hydrogenate 5-ton charges effectively. In building a new facility or extending an existing one, one must harmonize the productive capacity of sections preparing the oil for hydrogenation, producing hydrogen, and post-treating the hydrogenated oil on its way to the final products. In this connection, pay attention to the possible respective tonnages of oils with a low- and high-hydrogen demand per ton which the facility is expected to hydrogenate within 24 hours. If evidently, during one shift of 8 hours, the hourly demand for hydrogen will be 50% greater than the hourly demand for the remainder of the day, the difference may be accommodated by an adequate hydrogen storage which enables the hydrogen-generating equipment to run steadily at over 90% of its rated capacity. Prudently arrange that the hydrogen store is large enough to permit at least a 4-hour normal operation of the hardening facility in the event of the hydrogen generators being unavailable because they are being tested or subjected to a minor repair. Larger interruptions are best covered by the ability to hold increased stocks of hardened oils. A most useful commentary on the operation of a hardening facility was given by Allen (1978).

Hydrogen Distribution: Circulation Systems

To create a gas–liquid interface by simply bubbling a stream of hydrogen up through a cylinder of oil containing suspended nickel catalyst was the aim of many early hardening-facility designs. The disturbance of the oil was adequate to keep the catalyst in suspension, to mix the oil content as hydrogenation progressed, to promote heat transfer from the coils (during the final cooling and filtering of the hardened oil, this form of agitation could not be used, of course), and easily to give rise to more than an adequate movement of oil layers across the surface of the catalyst particles. If the hydrogen itself was relatively dry, this circulation also helped strip some moisture from the oil in the first part of the hardening where necessary, and, as argued, some off flavors from the oil by the time hardening were complete. The rate of reaction could be controlled within an upper limit by the rate of circulation, any temperature within the usual acceptable limits could be used, but different operating pressures were more difficult to arrange; hence, 0.3–0.5 atm above atmospheric was common. The holes in any sparger pipe should be small, numerous, and face downward to the floor of the autoclave, partly to keep the latter clear by the rush of hydrogen, but also to prevent solids from settling into them. Not only does a question arise of hydrogen absorption, however, but also important is to maintain a uniform degree of physical hydrogen solution in the oil and indeed, in many classes of hydrogenation, a uniform distribution of saturation by the chemical combination of the hydrogen throughout the whole oil mass. Hence, earlier errors of tall, narrow autoclaves were long ago abandoned.

A substantial gas space of about 30% of the total vessel volume was popular to reduce entrainment. Stirrers were added (Waterman, 1951) to improve gas–liquid contact, and proved useful in this respect while still in the range of 20–60 rpm, which meant that gland seals in the shaft were not too difficult to arrange and maintain. The central advantage of gas circulating systems was the opportunity to clean and dry the emerging gas, usually by a simple fat catcher followed by a cooler, before recompressing it for return to the sparger at the foot of the autoclave. Again, in the earlier years, much more elaborate scrubber trains containing alkali washes and even activated carbon were known (Swern, 1964). Several disadvantages are attached to a hydrogen recirculation system, but what finally rendered it obsolete was the increased availability of hydrogen at 99.8% of purity (dry basis) and better, not only from electrolytic facilities where this had long been easily attained, but also from hydrocarbon reforming units and in several cases from other chemical facilities as a by-product. Any scrubbing train is an expense to install, operate, and maintain; it is a source of leaks—hydrogen going out or, in faulty operation, air being drawn in.

Although some impurities are removed by the scrubbers, inert impurities such as nitrogen, methane, and carbon monoxide accumulate in the gas holding system so that purges must be made. These are commonly vented to the atmosphere at the end of a hardening cycle in a particular autoclave, accompanied inevitably by the hydrogen also present. Once inerts reach a level around 25% v/v, the depressing effect on hydrogenation rate becomes noticeable. If this level of inert impurity were reached by the hydrogen holding system as a whole, then a continuing purge to the atmosphere would be needed for more than that at the end of one hardening cycle, and this would therefore be continued until the contents of the gas holding system had been appreciably improved by the introduction of an increased proportion of fresh hydrogen. Found from facility measurements in circulation systems was that four times as much compressor capacity was used as the amount of gas taken up; not only do circulation systems require this energy for the compressors, but they also require cooling water. Again, hydrogen has easily the highest specific heat of all gases, so that the flow of hot gas from the autoclave has a cooling effect which in some circumstances has to be made good by the heating coils in the autoclave, while the issuing gas itself has to be cooled before recompression. A small percentage loss of hydrogen occurs even from a well-maintained compressor. Circulation systems are the most tolerant of variations in the level of autoclave filling.

Hydrogen Distribution: Dead-End Systems

Already by 1967, Albright (1967) remarked, "Probably all batch processes for hydrogenation that have been built within the last few years in the U.S. have dead end units." The dead-end concept is simple. Compress hydrogen into the autoclave ensuring that it is extremely well-mixed with the oil charge; more hydrogen will enter as some already introduced combines with the oil; inerts and water vapor will increase in the gas space of the autoclave, where they dilute the hydrogen atmosphere and, in effect, lower the hydrogen partial pressure. If the oil is well-dried before hydrogenation commences (say, 0.05% of H_2O/oil by weight maximum) and the hydrogen by compression (7 atm) and cooling (10°C) has been reduced to 0.1% of H_2O/H_2 v/v, the buildup of water vapor in the gas space will be slow. If the purity of the hydrogen on a dry basis is better than 99.5% of H_2, the remainder being inert (99.8% is now common), we are in a strong position to complete hydrogenations with quite substantial IV drops of c. 100 when the gas space occupies 40% of the free internal space of the autoclave. If the rate of hydrogenation falls owing to the accumulation of some inerts in the headspace by the later stages, a brief purge to the atmosphere will soon improve the rate, as seen by the temperature again starting to rise. The following approximate calculation illustrates the position. Taking one m^3 of hydrogen per unit drop in IV per ton of oil, a 10-ton charge requires 1000 m^3 (15°C, 760 mm of Hg) to drop its IV by 100 units. At 0.5% of inerts, the residue amounts to 5 m^3. Corrected to the conditions in the headspace of 3 atm and 180°C, this then amounts to about 2.6 m^3, but if this is not to amount to more than 30% of the gas space, 8.7 m^3 are needed. If the 10 tons of oil occupy 11.36 m^3, then the gas space would amount to about 43% of the total internal free volume. For many facilities, a 100-unit drop in IV is exceptionally large; some moisture will be present in the gas and oil; hydrogen may be 99.9% pure (dry basis). These various factors may easily cancel out one another, but even if they do not, the tactic of purging the gas space in cases of difficulty always remains. After due consideration of the future program and the good quality of the hydrogen available, the designers of a hardening facility may decide that a gas space equal to 30% of total internal volume is adequate (see also "Autoclave Design Features of General Importance" and "Hydrogen Distribution: Dead-End Systems" sections in this chapter).

Other vital matters on which decisions are needed are the questions: What is the means of creating a good gas–liquid interface within the closed autoclave so that hydrogen has a repeated opportunity to dissolve in the oil? What is a method not only of raising the oil to the hardening temperature specified, but also of controlling it there? This implies a removal of the heat of reaction when the IV of a very unsaturated oil, such as fish oil, is dropping rapidly in the earlier part of hardening. Both of these questions are part of autoclave design ("Hydrogen Distribution: Mixed Dead End-Circulating Systems" section in this chapter).

Hydrogen Distribution: Mixed Dead-End Circulating Systems

To achieve any benefits arising from the possibility of circulating headspace gas back directly to a sparger, or alternatively to the low-pressure hydrogen holder, some designs include this facility—a mixed dead-end circulating system, even if the recirculation capacity installed is limited to merely 20% of what is taken as the normal maximal hydrogen flow. Firstly in the absence of a vacuum service, the initial circulation of dry hydrogen through the damp oil before the addition of a catalyst will eventually bring down the moisture content to the 0.05% of H_2O/oil level recommended here as satisfactory for a subsequent dead-end operation, but such a maneuver is scarcely acceptable as a routine measure. Secondly, if so much hydrogen pressure builds up in the headspace that the subsequent entry of more fresh hydrogen via the main sparger is seriously hampered, this calls for a close critical examination of the design and/or functioning efficiency of the internal agitation system of the autoclave, various aspects of which are discussed in the "Autoclave (Converter, Hardening Vessel) Design: Early Systems," "Current Autoclave Agitator Design: Radial and Axial Flows," "Current Autoclave Designs: Loop Hydrogenation Reactor," and "Autoclave Design Features of General Importance" sections in Chapter 1. In Fig. 4.11 and in the caption notes, the basic organization of a dead-end system is shown with the facilities of the so-called "mixed system" indicated as an optional extra for those who feel their particular situation justifies them.

The autoclave agitator illustrated in Fig. 4.1 carries an upper axial-flow turbine and a middle and lower flat-blade turbine. No wall baffles are shown. Safety

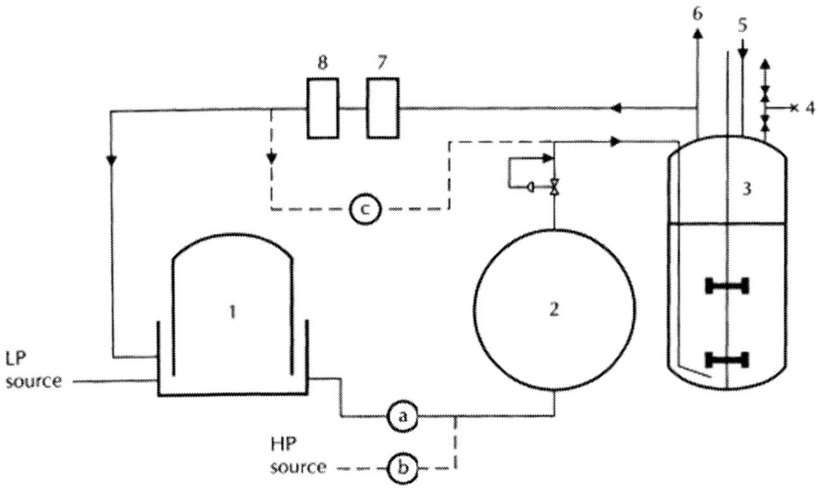

Fig. 4.1. Dead-end hydrogenation with optional recirculation. 1, low-pressure hydrogen holder; 2, high-pressure hydrogen holder; 3, autoclave; 4, vacuum connection including small vent cock to the atmosphere between two stop valves; 5, balance gas connection to replace hardened oil drawn from autoclave [this may be hydrogen drawn directly from an LP holder or an inert gas such as nitrogen]; 6, vent to atmosphere—if a recirculation line to an LP holder is included, this must include control valves; 7, fat catcher; and 8, cooler scrubber, a, compression and cooling; b, compression and cooling; either or both LP and HP sources may be available to the facility [the HP store will be designed for at least 10 atm and maybe considerably more]; if hydrogen supplies are also received by trailer, the mobile cylinder pressure will have to be reduced as it is discharged into the HP store ("Security" section in Chapter 7); c, the gas-circulating function shown by the dotted line is an optional extra, especially if no return of headspace gas to an LP holder is possible and if the operator, on occasion, prefers not to vent to the atmosphere. If headspace gas is returned immediately to the autoclave, it may join the main fresh hydrogen feed as shown, or it may feed an independent sparger.

pressure-release valve, bursting disc, sight and illumination glasses, oil and catalyst feed lines, sampling and instrumentation points, exit valve, manhole, and heating–cooling coils are also not shown.

The LP holder may have a capacity of only one-half an hour's maximal hydrogen generation, and the main hydrogen reserve (say, a 4-hour hardening facility operation) may be located in the HP store. This is a good arrangement. Some facilities are committed to holding the bulk of their hydrogen in LP holders and relying on compressors, keeping pace with demand via a relatively small HP store.

Hydrogen Distribution: Limitation of Uses

Although in many hardening facilities use is made of the pressure of hydrogen in the autoclave to assist in pumping or merely following the hardened oil to the catalyst filter, far preferable is to avoid the flow of hydrogen through the filter as the last of the charge passes out. One can do this quite simply by inserting a nitrogen feed into the filter line just under the autoclave after the oil exit valve. With the exit valve closed, the filter line, pump, and press may be blown clear.

Equally undesirable is to use hydrogen to clear soft oil lines and catalyst suspension lines which feed the autoclaves. Where vacuum is available on the autoclaves, such lines may safely be sucked clear. No connections should ever be made between a compressed air system and pipework in the hydrogen-handling system.

Autoclave (Converter, Hardening Vessel) Design: Early Systems

Normann's early design (1902, 1903) consisted of a cylindrical vessel equipped with simple blade stirrers. Hydrogen was dispersed from the foot of the vessel, collected in the gas space, and could be recirculated. The system operated at a little above atmospheric pressure, and temperature control was exercised by steam–cooling–water coils within the vessel (Waterman, 1951). The basic principles of two designs by Wilbuschewitsch are shown in Figs. 4.2 and 4.3.

In the design shown in Fig. 4.2, heating was provided by a steam jacket. Three similar autoclaves could be linked in a series to form a continuous unit, or the oil–catalyst mixture could be simply circulated from the bottom to the top of one vessel, which then acted as a dead-end hydrogenation unit until the desired end point was reached. Working pressure was about 3 atm. Oil level was not critical; a preferred system is not oil droplets in a gas space.

The versatile dead-end design shown in Fig. 4.3 operated successfully over many years. Provided that the top of the mixing tube is not flooded, a very wide tolerance concerning oil level exists. Temperature control was exercised by the normal steam–cooling–water coils. In this design, one can see the forerunner of modern high-speed hydrogenation developments; gas is dispersed into oil, and this is preferred; normal working pressure was up to 5 atm.

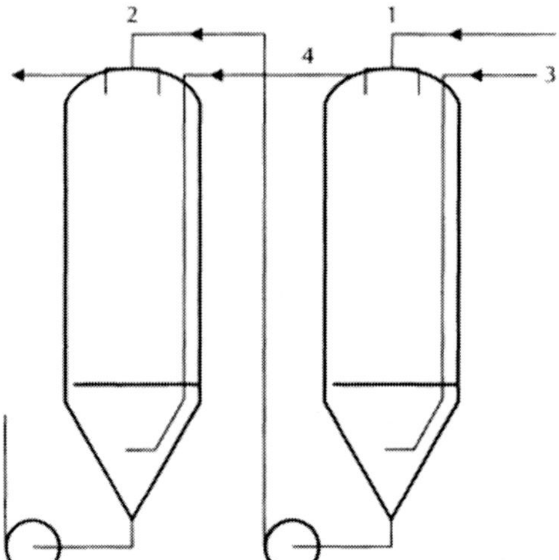

Fig. 4.2. Wilbuschewitsch oil spray (Swern, 1951). 1, oil and catalyst sprays; 2, oil and catalyst sprays; 3, hydrogen feed to sparger; 4, hydrogen exit to next stage.

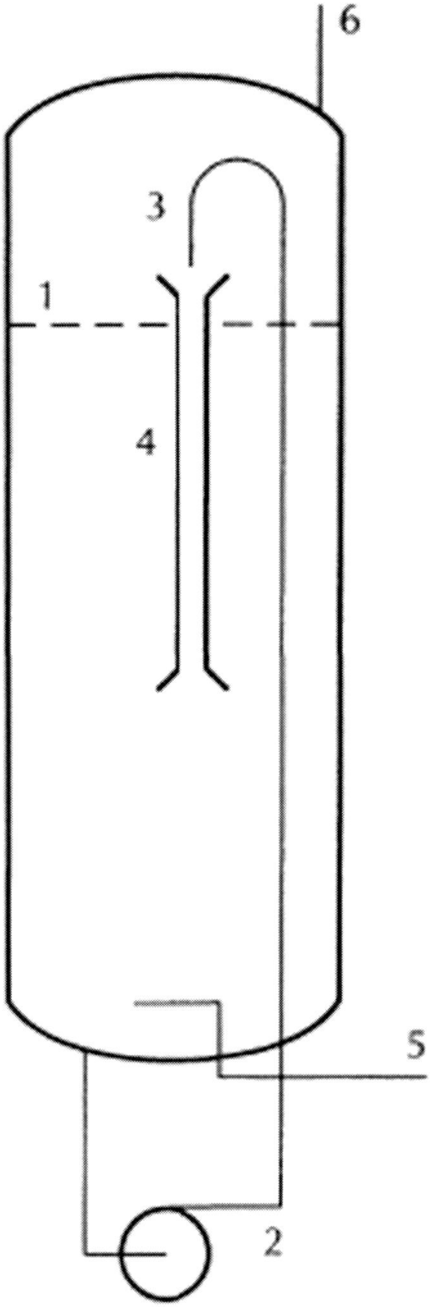

Fig. 4.3. Wilbuschewitsch mixing jet. 1, normal oil level; 2, oil–catalyst circulating pump, gland sealed with oil compatible with charge; 3, ejector; 4, oil/hydrogen mixing tube; 5, bottom gas connection to sparger; 6, top gas (balance gas).

Fig. 4.4 shows a Norwegian design, successful for many years in the United Kingdom and South Africa as a batch dead-end autoclave, although the dispersion of oil droplets through a gas space by the impact of the propeller-driven oil stream against a baffle is not highly efficient as a gas–liquid contactor. Temperature control was by means of steam–cooling–water coils; normal working pressure was 3 atm.

Fig. 4.5 shows a classic design which relies heavily on the ability of the large, top-stirring blades to disrupt the oil surface ("Autoclave Design Features of General Importance" and "Hydrogen Distribution: Limitation of Uses" sections, Chapter 4), so that hydrogen which eluded the action of the intermediate agitators is given a further chance to dissolve in the surface oil and to be promptly mixed downward. The bottom stirrer discourages the accumulation of settled catalyst. A normal dead-end operation is at 3 atm. Virtually, one can adapt the same design to gas circulation. For the power-per-ton oil charge employed, this is not as efficient as the turbo mixer. One can achieve satisfactory hydrogenation rates at stirring rates of under 60 rpm, provided attention is paid to obtaining the surface disturbance mentioned.

Many other early designs are found in the textbooks of the period (Ellis, 1912, 1913), but their differences are not fundamentally important. Those briefly described here are only a representative selection. Allen (1982) also describes a variety of systems, both batch and continuous.

Current Autoclave Agitator Design: Radial and Axial Flows

Just as batch hydrogenation via a dead-end system reached a dominant position in the fat-hardening industry by the middle of this century, the development of the turbine mixer during the second half made marked improvements in the efficiency of bringing together the three components of oil, gas, and catalyst. This meant the improved mass transfer of gas to oil for power consumed per ton of oil, and rendered the turbine the most efficient kind of stirrer (Lightnin Mixers Ltd.). The improvement arises from the correct understanding that thin layers of liquid moving rapidly across one another cause shearing, which disrupts gas bubbles and facilitates their physical solution; at the same time, a rapid flow and dispersion of gas-laden oil streams are created throughout the bulk of the liquid to obtain an enhanced and uniform effect (Mahony, 1964). What the turbine achieved has currently reached a heightened efficiency by a particular design of the axial-flow impeller (Hastert, 1990; Lally, 1987).

As distinct from a stirrer, the modern mixing jet (Leuteritz, 1971; Urosevic, 1988; van Dierendonck & Leuteritz, 1988)—which works on the same basic principle as the familiar laboratory suction apparatus attached to a water supply—secures a spectacularly effective mass transfer of surrounding hydrogen into an oil stream, and this in turn allows very high rates of hydrogenation ("Current Autoclave Designs: Loop Hydrogenation Reactor" section in this chapter). The flow effect is, of course, achieved by the circulating pump.

Returning to the turbine mixer, a brief account of its progress will help in the understanding of the advantages it offers. The radial-flow flat-blade turbine was

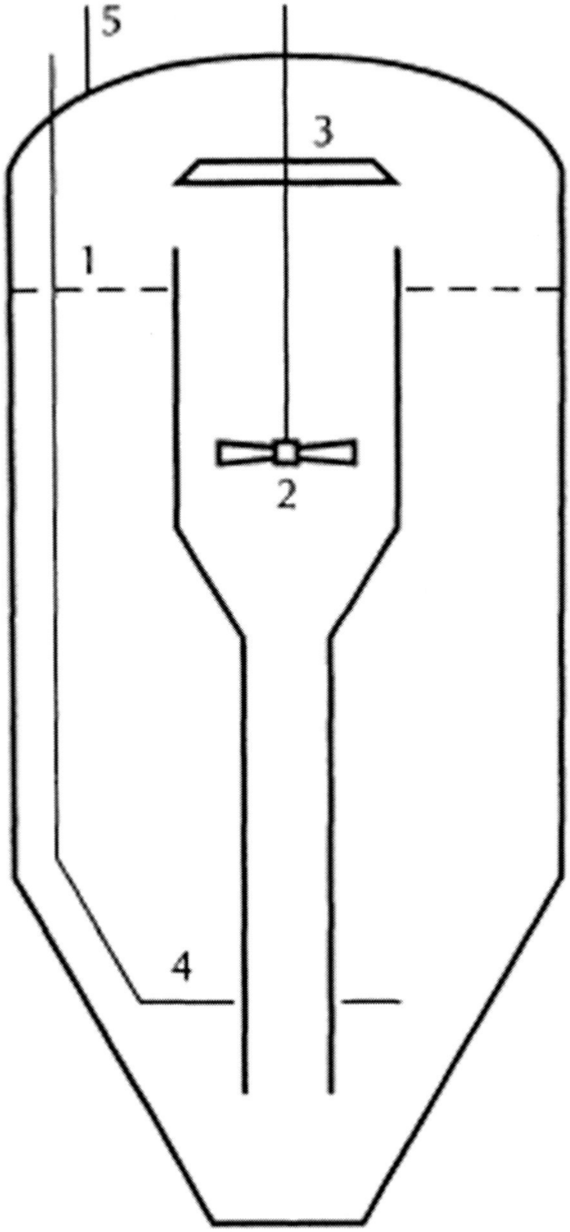

Fig. 4.4. Holmboe oil spray. 1, oil level (not severely critical); 2, oil propeller; 3, baffle; 4, hydrogen sparger; 5, top gas (balance gas).

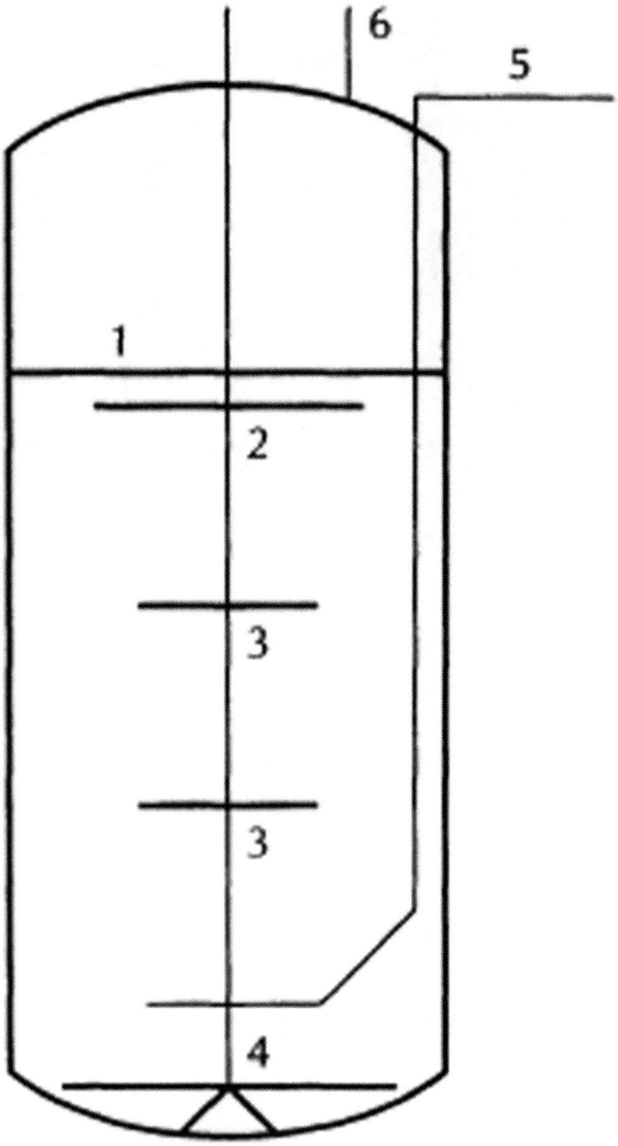

Fig. 4.5. Dead-end paddle-blade stirrer. 1, oil level (critical); 2, large surface agitator (down stirring); 3, intermediate agitators (up stirring); 4, large bottom agitator; 5, hydrogen feed to sparger; 6, top gas (balance gas).

developed by Rushton, and the commercial version was offered by the Mixing Equipment Co. Inc. in 1949 (Oldshue, 1983) and listed as their R100 model. As shown in Fig. 4.6, numerous flat blades are mounted vertically on a flat disc. The disc is bolted to the central rotating shaft, the end of which rests on a specially

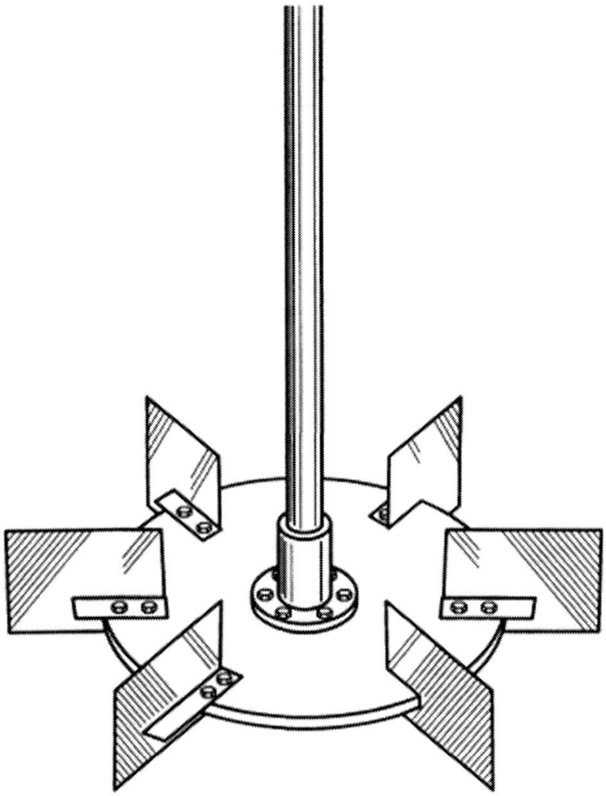

Fig. 4.6. Radial-flow flat-blade turbine.

hardened metal footstool. Provided the mechanical integrity of the impeller shaft system is within certain limits for critical speed, one may fit a stabilizing ring on the shaft below the lowest turbine. The number and size of blades may be varied when desired to meet changed circumstances, and the special gear train mounted on the top of the autoclave can very quickly be changed to cater for different speeds. Even the disc itself may be taken through a manhole in two halves for the assembly of the turbine inside the autoclave. Fig. 4.7 indicates the flow pattern in an open tank and when the usual baffles are fitted at the walls to oppose swirling and to direct oil flow both up and down. Power is consumed by (a) forcing gas through the sparger and expanding it into the autoclave and (b) driving the turbine. As the power applied to the turbine becomes several times that required to pump in the gas, the rate of gas absorption becomes high, and finally, the bubble volume in the oil space may increase to 20–30%. The oil is, of course, a non-foaming liquid in this context. Remember that the efficient mixing of liquids and gas into liquids is of major importance in many industries, and improvements in technique can naturally be passed from one industry to another. Not only are

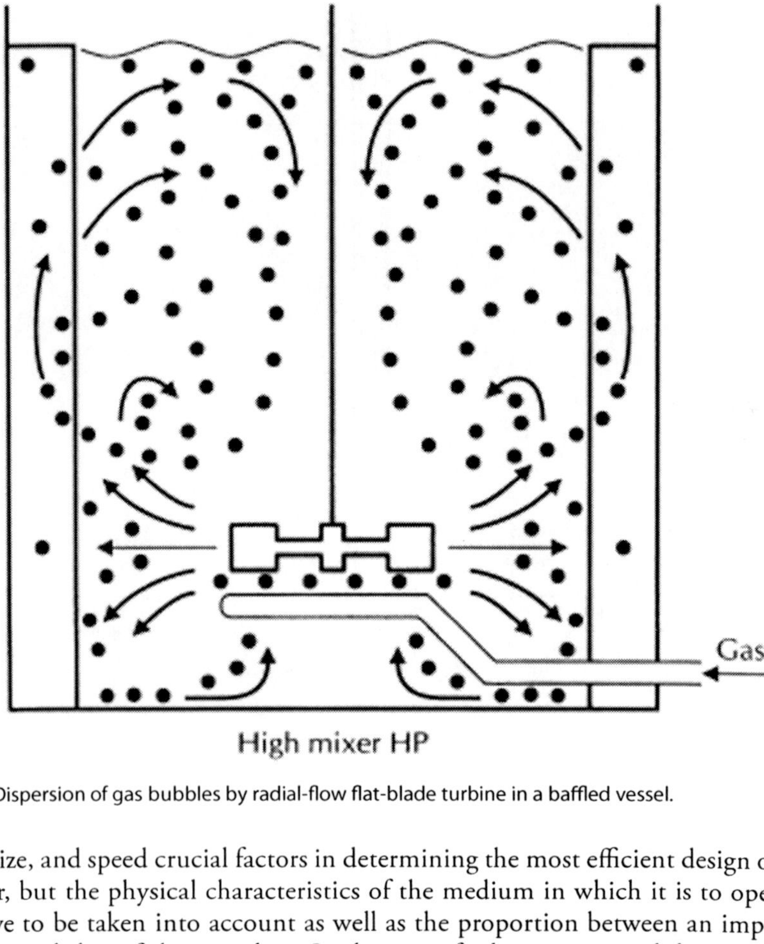

Fig. 4.7. Dispersion of gas bubbles by radial-flow flat-blade turbine in a baffled vessel.

shape, size, and speed crucial factors in determining the most efficient design of an impeller, but the physical characteristics of the medium in which it is to operate also have to be taken into account as well as the proportion between an impeller diameter and that of the autoclave. In the case of oil, viscosity, and density, both decrease with a rise in temperature, and very noticeable is that once gassing commences, the oil level in the autoclave rises appreciably. This means that the turbine is then beginning to operate in a medium of still lower overall density as compared with the ungassed state. While installing about 2 kW of turbine drive power per ton of oil charge was sensible—nearly all of which could be drawn upon in stirring the cool, ungassed oil—by the time the temperature was raised and gassing commenced, only about one-half of this power was needed. For a laboratory or pilot-facility autoclave up to about a 5-ton oil capacity, one horizontal-flow turbo mixer would both disperse gas and oil and create surface agitation to re-entrain hydrogen from the headspace. For taller autoclaves, one could install an upper and a lower turbo mixer (Garibay, 1981). Further studies (Oldshue, 1983) showed that employing a pitched-blade axial-flow turbine was advantageous on a full-facility scale (Fig. 4.8) a short way below the working surface of the oil to increase the

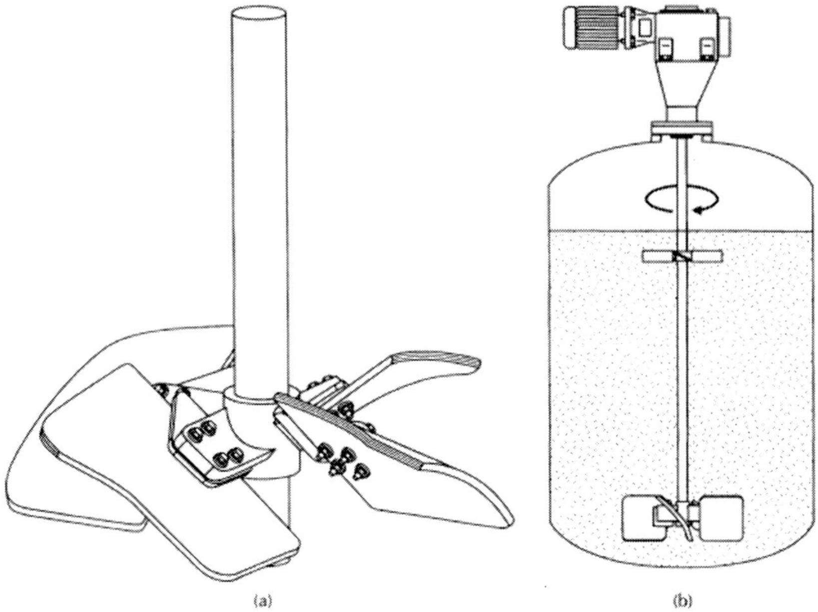

Fig. 4.8. Axial-flow flat-blade.

re-entrainment of hydrogen from the headspace. Rectangular blades are inclined at 45° to the vertical. The development of improved versions of this pitched-blade axial-flow turbine continued from the 1960s, and the current model is listed as the A200. In the 1980s, arising from more sophisticated means of measuring air flow when designing airplane wings, the possibility arose to design (Hastert, 1990; Oldshue, 1987) an axial-flow turbine which combined an adequate shearing effect with a very high capacity for dispersing the oil flow around the autoclave. This latest design is the A315 (Lally, 1987). Fig. 4.8a shows the broad blades and their high "solidity"—that is, the high ratio between the total blade area and the area of the inscribed circle of the impeller diameter. This design handles a maximal gas-flow rate by using a gas-sparger ring of greater diameter than the impeller.

The advantages of the A315 are:

1. High mass-transfer rate; therefore, reduced gassing time. The improvement in mass-transfer rate is between 25 and 30%.
2. Savings in power used.
3. Marked ability to maintain solid catalyst particles in suspension and to permit the use of smaller catalyst doses when preferred.
4. Ability to disperse very high gas rates.
5. Improved heat-transfer rate, in some cases by 10%.

For full-scale autoclaves, the A200 is still used to draw back hydrogen from the headspace (Fig. 4.8b). A315 impellers are now widely used throughout the world—not only for hydrogenation, but also in hundreds of applications involving gas dispersion, such as fermentation and various chemical processes.

Commonly, one A315 with an upper A200 suffices for facility-scale autoclaves, but when an oil depth-to-vessel diameter exceeds 2:1, two A315s may be fitted as well as the uppermost A200.

In a different design (*Bailey's Industrial Oil and Fat Products*, 1964), a suction sleeve around the shaft leading from the headspace to immediately above the top-turbine impeller was provided to facilitate the continuous withdrawal of hydrogen down into the oil.

Current Autoclave Design: Loop Hydrogenation Reactor

The most moderm design of a venturi mixing jet (Fig. 4.9) was employed, in particular, by BUSS of Switzerland (Hastert, 1981; Leuteritz, 1969, 1971; Reimann, 1978; Urosevic, 1988; van Dierendonck & Leuteritz, 1988) in the chemical industry to promote intimate contact between hydrogen and various organic liquids; these naturally include fats, oils, and fatty acids which carry a suspension of nickel catalyst: the amount of gas–liquid interface per unit volume of liquid created in this system is said to be about double that of classical mixing methods (*Bailey's Industrial Oil and Fat Products*, 1964; van Dierendonck & Leuteritz, 1988), and consequently, high rates of hydrogenation become feasible.

The old mixing nozzle system, whereby hydrogen was fed into the heart of the assembly by a separate duct and the emerging stream of oil and bubbles impacted against a plate, was abandoned long ago.

The mixing jet is incorporated in a loop reactor system as shown in Fig. 4.10 (Urosevic, 1988).

The refined oil or fatty acid passes through a heat exchanger (Harshaw Catalysts) where only a moderate rise in temperature (to 105–110°C) is allowed to avoid oxidation or polymerization. The oil next passes to a vessel (Hoechst Aktiengesellschaft/Mallinckrodt Specialty Chemicals Co.) where air and moisture are withdrawn, after which the temperature is raised in a second heat exchanger (Ottesen & Jensen, 1980). The oil, now ready for hydrogenation, passes into a reservoir (Swern, 1964).

As soon as the previous charge is discharged from the reactor (autoclave) (Horiuti & Pollanyi, 1934), enough oil is withdrawn from the reservoir for another charge, and hydrogenation commences by the circulation of this oil (Chahine et al., 1958) through the heat exchanger (Albright, 1967) back to the head of the venturi (Allen & Kiess, 1955). Not only does the suction effect at the top of the venturi draw in fresh dry hydrogen from the supply, but also headspace gas can be drawn from the reactor around the cooling/condensation system (26.1, 26.2) at the same time if desired. Especially in a lengthy IV drop, such as in fish-oil hardening, the gas-space moisture content can be kept under rigorous control,

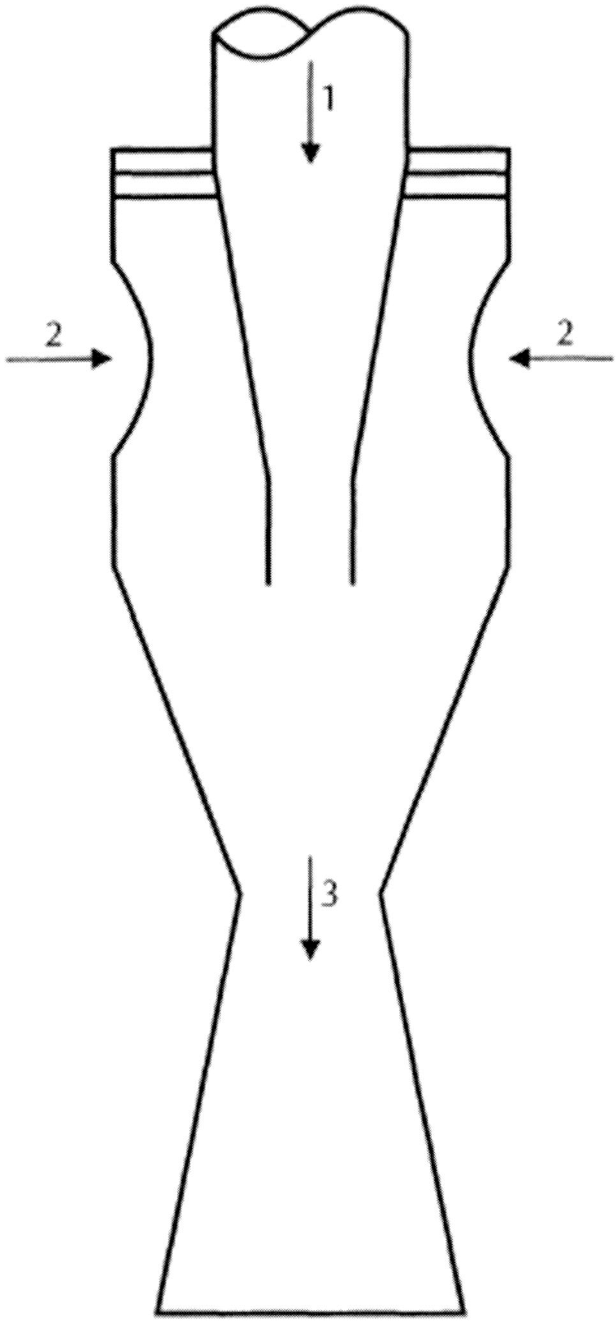

Fig. 4.9. Modern venturi mixing jet (Leuteritz, 1971). 1, oil–catalyst stream; 2, hydrogen entry ports; 3, zone of intense shear.

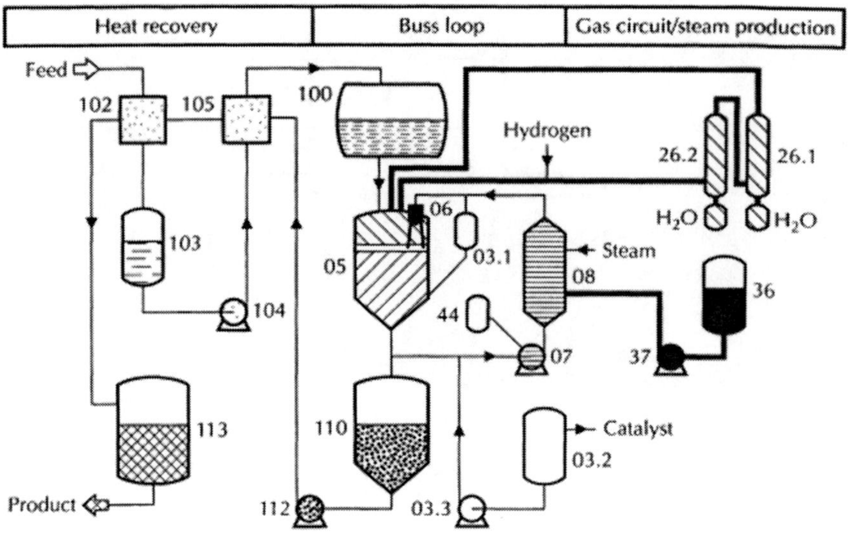

Fig. 4.10. Loop hydrogenator reactor.

and the amount of purging to the atmosphere, common in some older systems, can be avoided.

Circulation through the heat exchanger (Albright, 1967) controls the temperature to within 1.5°C of the sequence programmed, and surplus heat is collected from the reaction at this point. When the reaction is complete, the product is dropped to the container (Cowan et al., 1970) and pumped via the two-stage heat-exchange (Ottesen & Jensen, 1980) (Harshaw Catalysts) system mentioned above to the receiver (Koritala & Dutton, 1969) ready for filtration. When the IV drops of successive batches are adequate, not only is the system self-supporting regarding heat, but also a surplus can be exported for use elsewhere. A wide choice is available of pressures and temperatures at which the system can be operated. For slow reactions at low pressure, the advantage probably lies with the stirred-tank type of autoclave, but where higher reaction rates are acceptable and higher pressures and temperatures are used, the advantage goes to the loop reactor (van Dierendonck & Leuteritz, 1988).

Although the hydrogenation rates in this type of reactor can be several times greater than in conventional stirred-tank reactors and this diminishes the correlation between selectivity on the one hand and catalyst concentration and temperature on the other (Leuteritz, 1969), evidently possible is to adopt conditions, including pressure and types of nickel catalyst, which yield all types of texture in the hardened product. As with stirred autoclaves, the sulfur-promoted nickel catalyst is effective in giving products with steep melting curves. For edible-oil hydrogenation, 0.5–2.5 atm (gauge) of hydrogen pressure is common; for fatty-acid hardening, 20 atm is usual. Very economical usages of nickel are experienced. Reactor sizes range from 5.3 to 34 tons, and at least for the easier types of hydrogenation, owing to the rapid turnaround, as many as 24 batches were completed in a reactor within 24 hours (Urosevic, 1988).

A continuous system of four or six loop reactors connected in sequence was produced especially for the benefit of long hydrogenation runs by using very similar feedstock and producing virtually the same product. Even in this application, the batch reactor is successful.

Characteristics of hydrogenated marine and vegetable oils produced in a loop reactor (Fig. 4.10) and the consumption of utilities were published (Freier, 1967; Leuteritz, 1969; Urosevic, 1988).

Current Autoclave Design—Lurgi System

The layout of the Lurgi batch-hydrogenation system (*Fatty Acid Technology*, 1991) is shown in Fig. 4.11. The fatty-acid feedstock (preferably distilled grade) is charged to the autoclave, and then heated under vacuum to remove moisture. The vacuum is then closed, catalyst is added, and the introduction of hydrogen at the jet as shown causes oil to flow up the central tube of the reactor, down on the outer side, and thence, to the external heater/cooler; unabsorbed hydrogen separates in the gas space, and then recirculates as indicated. On completion of hardening, the charge is cooled externally and driven into a surge vessel prior to filtration. The next batch of fatty acid is introduced, and the cycle is repeated, normally six times in 24 hours. The pressure used is c. 25 atm, and temperatures range up to 220°C. The Lurgi continuous-hydrogenation system is shown in Fig. 4.12.

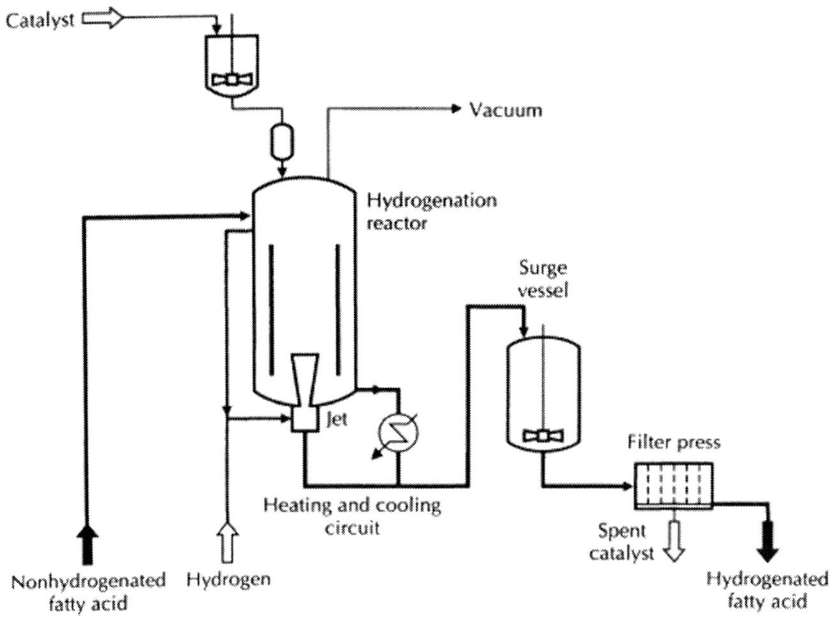

Fig. 4.11. Discontinuous hydrogenation.

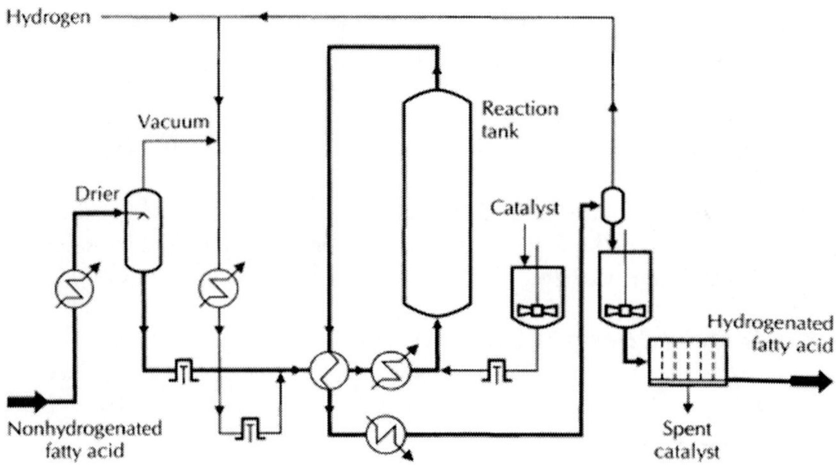

Fig. 4.12. Continuous hydrogenation.

Here the distilled fatty-acid feedstock is heated, and then dried under vacuum; the pressure is raised to c. 25 atm, and hydrogen is introduced before the feedstock passes through a heat exchanger as shown. Where desirable, the temperature may be further raised in a second heater before continuous dosing with a catalyst in suspension takes place. Temperatures up to 220°C are used. The hot reaction mixture now passes up through the reaction tower, avoiding back mixing. The hot product flows through the heat exchanger already mentioned, preheating oncoming feedstock, and may be further cooled before entering a small vessel in which hydrogen separates for reuse while the cooled catalyst/hydrogenated fatty acid passes on to a filter. As one filter fills, the stream is directed to a second while the first is cleaned. The smallest unit is capable of 80 tons/24-hour output, and as might be expected, is more economical in hydrogen and nickel than batch operation. Both types of facilities can be adapted for use with triglyceride oils on the advice from Lurgi.

Current Autoclave Design—Advanced Gas Reactor

The essential features of the advanced gas reactor of Linde Division, Praxair Inc., are shown in Fig. 4.13 (Weise & Delaney, 1992). The agitator is a downward-pumping helical screw impeller situated in a draft tube. As the oil is expelled, hydrogen is drawn in from the headspace above. Very obvious vortices form within the oil. Vertical baffles at the inlet and outlet of the draft tube suppress liquid rotation or swirling in the reactor as a whole; this increases pumping capacity or flow. The somewhat conical shape of the inlet also helps, in this regard, by reducing flow resistance. The high turbulence in the gas/oil/catalyst mixture flowing down the draft tube is further supported by a small flat-blade turbine fixed on the bottom of the impeller shaft (Fig. 4.13). This completes the fine dispersion of the hydrogen in the oil as it commences its journey through the final length of the draft tube, which is also

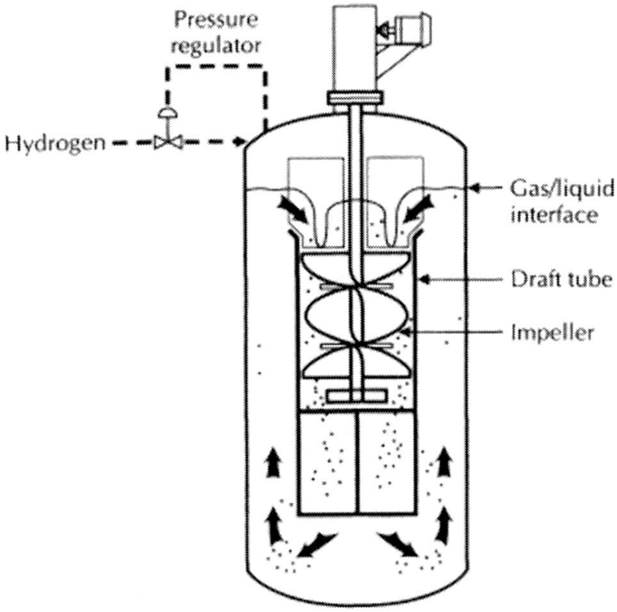

Fig. 4.13. Advanced gas reactor.

baffled. Gas which separates beneath the draft tube promptly rises to the headspace and is recirculated. The design of reactors intended for commercial use allows for batch sizes up to 45 tons of oil, and the entire volume of oil can circulate through the draft tube 0.5–3.0 times per minute. The outstanding feature of this design is a high mass-transfer rate of hydrogen to oil (Allen, 1982; Koseoglu & Weise, 1991; Weise & Delaney, 1992). This heightened mass transfer is claimed to reduce catalyst usage, batch-hydrogenation times, and maintain high selectivity. The results of the hydrogenation of soybean oil were published (Weise & Delaney, 1992). Further use of this type of reactor will establish the kind and extent of its advantages.

Autoclave Design Features of General Importance

Several design features occupy a position of importance in a high proportion of different autoclave designs; naturally, not all features are present in every design: for those who are considering the modification or extension of an existing facility, as well as those planning a new one, the following information will be useful.

Vacuum

This facility (vacuum) is present in probably a substantial majority of dead-end systems (Allen, 1978), and was regularly used in large hydrogenation facilities which date back to the early years of this century. Fears as to the safety of its use must be regarded as idiosyncratic in origin, since if it posed a danger this would have long ago become apparent, nor would it be featured in the majority of current designs. Generally recognized is that vacuum is best provided by a steam ejector, since this is the safest way of handling hydrogen withdrawn from an autoclave. Easily one can place two block valves on the hydrogen feed to the autoclave with a vent to the atmosphere cock between them, which should be open except when intentionally the autoclave should be under hydrogen pressure. Similarly, two block valves with an intermediate vent to the atmosphere cocks can be provided on the vacuum line as a precaution against hydrogen being sucked from an autoclave under pressure through a faulty vacuum-control valve. The action of these valves and cocks can be interrelated. Where nitrogen is readily available—and possibly cheaper than hydrogen—it can be employed in intermediate stages between having a facility under vacuum and under hydrogen pressure, although this further precaution was not demonstrated to be necessary in normal operation. A capacity to achieve a pressure of about 50 mm of Hg is adequate. The most economical means of raising and maintaining vacuum by steam jet were reviewed in detail by Gering (1980) at the 1980 ISF/AOCS World Congress.

Ratio Oil Depth/Autoclave Diameter

Tall, narrow vessels for batch dead-end hardening are no longer made. The amount of gas–liquid interface depends on the agitation system, but includes the disturbance at the top layer of oil when the latter is at its working temperature. This may be as low as 115°C for some procedures and as high as 200°C for others; in a great many cases, 180°C is normal. The design must allow sufficient room for the inclusion of heating–cooling coils in particular, unless these functions are to be performed by the external circulation of the oil; some space is also taken up by the agitator and baffles when present. Whereas at 180°C one ton of oil may itself occupy only about 0.9 m^3 in the autoclave, it may well occupy about 1.3 m^3 of the internal total volume when allowance is made for the displacement due to fixtures such as coils, etcetera. In these circumstances, the oil head at 180°C need not be greater than 1.5 times the diameter of the vessel, and should not be less than 1.2 times the diameter. A good mixing of oil throughout the autoclave as well as a fine dispersion of hydrogen into it can then be obtained without difficulty.

Gas Space

From what is said in the "Hydrogen Distribution: Dead-End Systems" section in this chapter, evidently, in a dead-end system, considerable latitude is available in fixing the size of the gas space, depending on the purity of the hydrogen and the volume of it that each ton of oil is expected to require [i.e., large or small iodine value (IV) drop]. Hence, with fish oils entailing IV drops of 100–140 units, a gas space of 40–45% of total internal volume, or roughly 75% of the oil volume, would be feasible as a means of keeping venting during operation to a minimum, whereas at the other extreme of vegetable-oil hardening with an IV drop of 50–60 units or even less, a gas space of only 23% of total internal volume, or roughly 30% of oil volume, would suffice. Faced with a varied program and the need to keep construction costs in close relationship to operational needs, a compromise of a gas space equal to some 30% of total volume (c. 43% of oil volume) would be sensible.

Temperature Control

The principal consideration in estimating the size of the temperature-control equipment for an autoclave is what will be needed to remove the heat of reaction from a strongly exothermic reaction, so that the temperature is easily prevented from reaching a level which damages the quality of the oil, or which in special cases defeats the aim of some process technique, such as low-temperature hydrogenation. The rate of heating the oil prior to hydrogenation, or cooling it afterward, proves to be of secondary importance. The rate at which heat is evolved is proportional to the rate at which the IV is falling; various estimations give 0.942–1.100 kcal/kg (1.6–1.98 BTU/lb), for specific heats 0.55–0.64, per unit IV drop. A heat transfer of 350 kcal/m^2 hour of °C can be relied upon where a cooling coil is not scaled and where forced flow exists across its surface arising from the circulation of the oil. Whereas in temperate zones cooling water may be at 10°C, in the tropics it is often 30°C, and may rise to 50°C. The common working temperature for edible-oil hardening is 180°C, which means a possible temperature gradient exists of 170 or 130°C. In theory, this means a provision of 5 m^2/ton of oil of coil surface could cope with a hardening rate of 1 IV/minute, but this area of coil is near the limit of what can conveniently be fitted into an autoclave if it has to be provided with stirrers, baffles, supports, etcetera; hence, hardening is organized to attain about a 2 IV/minute drop maximum, which probably corresponds to an overall rate of 1 IV/minute, and for this situation 4.5 m^2 of coil surface gives a good measure of control. Of course, external cooling (Fig. 4.10) avoids such problems.

The dose of catalyst and the hydrogen pressure applied to the autoclave will go a long way to predetermine the rate at which the ensuing hydrogenation will occur. If the flow of cooling water is failing to prevent a continued rise in temperature, then the rate of supply of hydrogen to the autoclave has to be diminished until evidently control is regained. This is an undesirable situation since, when the temperature is around 180°C or more and a distinct scarcity of hydrogen is available, the chance of polymerization by highly unsaturated groups increases. If too drastic a restriction in the hydrogen flow was avoided, polymerization will probably not have taken place. Of course, the highly unsaturated oils are most likely to hydrogenate rapidly; experience with one or two batches will quickly establish the catalyst dose which is both convenient and safe in this context (Coenen, 1976). In small- and medium-sized autoclaves, up to 10 tons in capacity, the coil can normally take the form of a large helix set near the wall, but in larger autoclaves, to site a row of small helical coils having a common header near the wall may be easier. Obviously, space must be left in the center for the agitators. When baffles are provided, they sit against the wall between the agitators and the coils. Hydrogenation rates several times faster than those envisaged here may be attained, but these bring with them the necessity of enhanced heat-removal equipment such as the heat exchanger of the loop reactor ("Current Autoclave Designs: Loop Hydrogenation Reactor" section in this chapter).

Flat-Blade Stirrers

Although this older classical shape of stirrer is unlikely to be fitted to any new autoclave, some detail here may help the many hardeners who continue to operate them. The basic functions are performed as indicated in Fig. 4.5. The stirring which disperses the hydrogen bubbles is adequate to keep the catalyst in suspension, to circulate the oil charge, and to promote heat transfer. The large top blade which stretches across most of the vessel is to disturb the surface; hence, it should not be covered by more than about 5 cm of oil at the normal working temperature of 180°C. Suppose, in the case of a larger autoclave, the blade of the stirrer has a vertical depth of 10 cm; this means that the lower edge of the stirrer is some 15 cm below the surface. If a change in oil temperature of 10°C alters the oil level by 6 cm, then at 150°C, the top stirrer rotates clear of the oil surface. To overcome any disadvantage this may bring, the last few outer centimeters of the blade can be fabricated as a paddle which extends the downward contact by another 12 cm so that some considerable disturbance occurs around the periphery of the surface even at 130°C. The rotational speed of these stirrers has produced a good effect at 40–60 rpm; tip speeds will vary upward with an increase in the size of the autoclave, but 5 m/second is about the maximum.

While devices of this kind may fall short of sophisticated agitator design, the hardening rate of older autoclaves in this class was doubled by paying critical attention to their surface agitation. In converting some tall hydrogen-circulating autoclaves to dead-end use with much purer hydrogen, paddle-type ends with a long vertical displacement were added to the top stirrer. To restrict the load on the agitator motor, the paddles were fabricated from heavy-gauge mesh. The brisk tearing of the oil surface thus obtained ensured a satisfactory hardening rate of three charges per 24 hours. Where half-charges are occasionally to be hydrogenated, one can easily arrange that one of the intermediate stirrers is not far below the working surface so that some disturbance is obtained, however crude. In general, a moderate tolerance exists regarding the level of filling of about ±5%.

Turbine Agitator Mixers

As already emphasized in the "Current Autoclave Agitator Design: Radial and Axial Flows" section in this chapter, turbine mixers have a wide application in many industries, and a large volume exists of specialist literature about their different uses. Hydrogen–oil contacting is only a small part of such use. An obvious basic consideration is how much of the task relates to mixing or circulating liquids and how much relates to shear forces to disperse and dissolve a gas in a liquid. A mixing or pumping effect is achieved notably by the large diameter, relatively slowly rotating impeller, especially if the liquid is viscous, since a high-speed, small-diameter turbine in this application would waste energy. On the other hand, to achieve high-shear forces, a slow-moving, large-diameter turbine would be less effective, and would promote much more pumping–circulating effect in a low-viscosity medium than was needed. The physical characteristics of the medium, such as density and viscosity, therefore enter into the choice of an agitator. Finally, the relationship of the agitator, including its position, with its environment is noted. Hence, in a cylindrical autoclave, the diameter of the impeller as a proportion of the diameter of the vessel is considered, and wall baffles are used to increase turbulent and vertical flow patterns. Such baffles stop short of the surface when an inclined-blade axial-flow turbine (A200) is installed. Thus, a vortex is able to form. Characteristics of different-sized turbines and autoclaves (30 liters, 500 liters, and 25 m^3) and scaling-up considerations were well-described by Bern et al. (1976).

Material of Construction

For autoclaves working on the hydrogenation of fats and oils, carbon steel conforming to local regulations has proved adequate. A temperature range up to 200°C and pressure from full vacuum up to the locally required excess above normal working maximum must be allowed (Chapter 11). Because of their more arduous duty, one may decide to employ stainless-steel coils in the heating–cooling system. The coils and any associated headers should be stress-relieved and annealed prior to being fitted. As expected, if fatty acids are being hydrogenated, one must employ the correct grade of stainless steel to resist their attack. Rice (1979) reviewed the qualities of stainless steel needed in the processing of fatty acids in equipment such as hydrogenation autoclaves, fat splitters, and stills, as well as storage tanks, taking into account whether the duty is to be at, under, or above 150°C. At working temperatures above 150°C, Type 316 is recommended, with the proviso that when welding is to be done, one should employ either a steel with less than 0.03% of carbon or an alloy containing an ingredient to inhibit carbide precipitation in the weld area to avoid corrosion. For pipe lines and store tanks when the duty is below 150°C, Type 304 is regarded as satisfactory even if some welding is done during fabrication. The additional point is that when the special resistant alloy is bonded to plates of carbon steel, not only does this save expense, but also enhances tensile strength because of the tough nature of the carbon-steel base.

Oil Segregation

The development cost for a new hydrogenated oil may amount to very little in terms of engineering facilities. Often, it can depend on the discerning combination of starting material and modified process conditions, with perhaps the employment of a catalyst which most favors the type of hydrogenation to be used. In the unlikely event that the product is the only one being made in the facility, the accurate control of the process to ensure that specification is met is the principal concern of operators, and for this purpose, sophisticated automation offers more and more help. When numerous hydrogenated oils have to be produced during the same daily and weekly program, possibly from fats and oils themselves markedly different in character, a new dimension is added to the control problem. One must evaluate the importance of segregating one oil from another. Firstly, we need to recognize the qualitative grounds on which segregation is needed; secondly, to what degree it must be ensured; and finally, how cheaply one can find a realistic answer. The information given in this section is general in nature, but it may point the way to just such an answer in many individual cases.

The crude oils fall naturally into a limited number of groups. Contamination in one direction may be more serious than in the reverse; some groups have greater compatibility with one another, especially as hydrogenation is to follow; the end-point specification for one product may be so tight that contamination by what in ordinary circumstances would be of the same size as a trace is then sufficient to spoil the parcel. Money to provide facilities for the production of sophisticated modified fats, as well as for the raw material itself, is wasted if products are allowed a degree of contact which involves a high risk of them failing to meet specifications. One must adopt a policy or strategy which avoids this; customer relations are then much better; reworked material falls to a minimum. The strategy may be a static one in the sense that separate pipe lines and process units (autoclaves, filters, tanks) may be provided for certain tasks, or it may be dynamic to the extent that a sequence of operations is followed in the production program such that the acquisition by one oil of another (usually one which it is following) sinks to insignificant proportions. This kind of so-called overflow or cross-flow has both technical and financial aspects, so the efficient facility manager regards both. Keep in mind a further point—the more modern methods of analysis, such as gas–liquid chromatography (GLC), may be able to disclose traces of contaminants (long- or short-chain fatty acids) which the specifications ignore by implication only—that is, it deals with IV, color, free fatty acid (FFA), and moisture only—and presume the oil is derived from a certain raw material from which various items are absent. The first approach, that of providing a separate track through the facility, is obviously costly, but may be justified when the volume of specialized items of a particular group, such as the lauric oils, is large enough to provide work throughout most of the week. A variation of the second or dynamic approach is to allow for a small leading proportion of the oncoming material to take up contamination at stages where it is most likely to occur, and then divert it from the main bulk. This also has its cost, and one must consider not only in which fat or fat blend the diverted portion is technically acceptable, but also where it is most valuable.

The three principal groups which recommend themselves for completely separate handling in a large facility with a continuing complex program are the lauric oils, other vegetable oils, and marine oils. Some further division may be feasible within the vegetable-oils group, but this is most likely to be achieved by what was described above as the dynamic approach, wherein certain hardened-oil parcels are scheduled for certain times in the week and possibly in relation to what immediately precedes them. Animal fats may require a rigorous segregation if at least some of them are completely unacceptable to customers by reason of their origin. Where this requirement does not arise, including them in a way which fits in with the specification of some products in the marine-oils zone may prove quite practical. So, if lightly hardened lard and medium hardened marine oil are being used as alternatives or jointly in a blend, no great problem arises. If, on the other hand, once a month a fully-hardened tallow is to be produced with a final maximum of 2 IV, this parcel has to be safeguarded by the rigorous application of the dynamic approach, since permanent static segregation is not practical.

For facility management, segregation begins when raw material bought under contract is off-loaded into their crude-oil tank farm. This means that when a store tank or pipeline changes from one use to another, it must be capable of being emptied; that adequate means must be provided for the person responsible to do this; and that the control system must bring to light deficiencies via samples and tests. Obviously, tanks must drain to a low point or sump adjacent to the transfer pump; pipelines should slope by at least 1:100; if a vertical leg of a large-diameter oil pipeline presents a drainage problem, a parallel line of very much smaller diameter may be sited next to it and connected to a pump suction so that at the conclusion of pumping, when the large line is full, it is able to drain to the feed end of the narrow pipe, which is then employed to empty it. To blow the narrow line clear or to drain it is easy. Center points should appear at the crude-oil reception, and should figure largely in oil transfer from then on; they were employed in the very earliest hardening facilities; they have grown more efficient in design and convenient in operation; they guard against mechanical failure of valves and go part way to reducing human error. The rest of the way is secured by employing male coupling components which are compatible with only certain female sockets. A faulty transfer from one pipeline system to another can then only be obtained with engineering cooperation in an unauthorized modification of the facility. Automatic self-closing of the line as the coupling is undone can be provided when this is an advantage (e.g., when the line does not need to be emptied is acceptable).

The center-point system was used very successfully in charging autoclaves, discharging them to the authorized filter group, and connecting the filter to the correct hardened-oil store tank. Oil used in mixing fresh catalyst or for holding in suspension catalyst awaiting reuse must be compatible with the main charge. This requires a similar zoning in the catalyst handling system as exists in the main soft-oil feed lines; this handling system includes the filters themselves. As providing a special filter for a parcel which is produced once a week is quite impractical, planning will ensure the charge is always filtered on a freshly cleaned filter. This remains true for many countries, especially in the developing areas of the world—hydrogenation facilities are very often concerned with one group of oils, such as the vegetable oils, and if a second group is involved, it is only on a minor scale. In these circumstances, the burden of segregation is less arduous.

Oil Protection

Hydrogenation is impeded by a variety of catalyst poisons, and the aim of pretreatment is to diminish these to a negligible level. One notable poison is oxidized oil. Ottesson (1975) and Hastert (1979) even place oxy-polymers in fish oil on the same level as sulfur in exerting hydrogenation-inhibiting effects. The poisoning effect of 1 ppm of oxy-polymer destroyed the activity of 0.0047% of nickel/oil and for 1 ppm of sulfur 0.004% of nickel/oil. Drozdowski and Zajac (1977) studied the effects of an increase in peroxide value (PV) on oils supplied for hydrogenation. One unsupported (formate) and two supported catalysts were tested; these varied in their resistance regarding the length of induction period before the reaction commenced and the subsequent rate, but a PV of c. 30 meq. of O_2/kg had a measurable bad effect, and this became more marked for 60 meq. of O_2/kg. Fortunately, pretreatment, especially if it includes activated earth, appears to curtail the inhibiting effects of oil oxidation on the catalyst. As always, the question remains of how flavor stability is affected by exposure (i) of the pretreated oil to air prior to hydrogenation and (ii) of the hydrogenated oil itself during filtering and storage prior to post-refining. In modern methods of neutralizing, washing, drying, and earth bleaching, the process is carried on with total or virtually total exclusion of air, and the oil is sometimes stored prior to hydrogenation under an inert atmosphere which may even take the form of a store tank with a floating roof under which nitrogen is occasionally introduced. The state of oxidation will then be low, and if oil for hydrogenation is normally drawn from such a stock, problems will not arise. In many facilities, oil waiting to be hydrogenated is in contact with air. Each of the following precautions is useful, and, taken together, should ensure sufficient protection from oxidation:

- Movement of oil should not entrain air or involve much splashing.
- Oil should be dry—normally no problem at this stage.
- Rate of oxidation is very dependent on the degree of unsaturation and temperature. Oils with an IV above 100 oxidize many times more quickly at their filtration temperature of around 90°C than at an ambient temperature of 25°C. With a decrease in IV, this sensitivity falls rapidly. Since heat is wasted in cooling filtered oil to ambient and then reheating it, the waiting interval for hot, very unsaturated oil should be kept short: say, within about 8 hours. Many new hardening facilities include an element of heat exchange between the outgoing hardened oil and the following incoming batch.
- Oil is normally stored in the dark, and this helps since light, especially ultraviolet, promotes oxidation.
- After adsorptive bleaching, the natural antioxidants of an oil are at their lowest concentration, and the oilmay then be most vulnerable, hence, the advisability of not leaving it standing for many hours before hydrogenation.
- Silt must be removed from intermediate holding tanks at intervals, since dirt is likely to contain prooxidative catalysts.

After hydrogenation, the oil gained in its resistance to atmospheric oxidation, depending on how far its unsaturation was reduced. Much the same protective considerations apply after hydrogenation as before. However, remember that, if appropriate off-flavor precursors are now present, possibly as a result of isomerization during hardening, some of these can yield off flavors on oxidation which are detectable in minute concentrations. Important is that hardened oil stocks are not wetted by leaking steam coils or by the incautious steam blowing of pipelines. Once hydrolytic splitting has commenced, the FFA content rises with an accelerating rate over two or three days. Advisably hold stocks if possible at not more than about 10°C above the melting point. In very general terms, possibly, if the PV has not risen above 3 by the time the hardened oil comes to be postrefined, the flavor stability will prove satisfactory.

If through storage at unnecessarily high temperatures the PV has become more than double this figure, off-flavor problems are more probable. A sequence of tests on successive parcels continued over one or two months may establish for a manufacturer how his different hardened oils are placed, in this regard. Similarly, monitoring an oil handling system for an increase in PV at two or three different stages may show in less than a month when the risk of air contact is greatest and when supplying some degree of protection would be most rewarding. Once formed, peroxides decay to aldehyde and ketone groups; hence, the total oxidation value needs to be considered in assessing the quality and history of a fat (Chapter 12). One method of protecting oils and other liquids from the attack of dissolved oxygen is to sparge a stream of very small good-quality nitrogen bubbles into the stream of liquid as it flows along a pipeline. The volume of nitrogen required is about one-half the volume of oil being treated, and this strips dissolved oxygen from the oil; the nitrogen is then allowed to accumulate in the headspace of the holding tank into which the oil is being pumped. Although such a precaution would first come to mind in connection with a refined and rather unsaturated vegetable oil, to be used eventually as such, one could apply it to lightly hydrogenated oils where the risk of attack by air was high. Citric acid in solution is an effective sequestrant of prooxidant metal catalysts. To add it at any time following the filtration of hardened oil is feasible, especially if the latter is being shipped to some other factory for post-refining. A general use of citric acid occurs among refiners at the deodorization stage; a dose of between 0.005 and 0.01% of citric acid/oil is added as an aqueous solution early in deodorization and at the end. Butyl hydroxytoluene (BHT) and butyl hydroxyanisole (BHA), generally limited to 200 ppm in total, were used successfully as antioxidants for years in deodorized oils, but remember that a hardening of flavor, such as the linolenic hardening flavor, is detectable at concentrations of 10^{-10}, which is too low to be effectively controlled by antioxidants (Patterson, 1974).

Energy Conservation

Since hydrogenation is exothermic, the ideal at which to aim is that not only should the process provide its own heating needs (for incoming soft oil), but also that surplus heat should be exported, possibly to preheat crude oil on its way to the adjacent refinery, or to provide hot-water washes or indeed any other heat requirement, providing the energy conserved always pays for the means to conserve it.

Continuous-hydrogenation systems for oils or fatty acids will now include heat exchange as a matter of course (*Fatty-Acid Technology,* 1991). For batch processes, numerous variants are offered; the hardened oil may be dropped immediately into a holding tank, and then the same autoclave filled with soft oil passes through a coil in the same tank before it is pumped out to the filter (*Flow Sheet SFS-164*). If two autoclaves are working in a roughly reciprocal fashion, incoming soft oil for one is preheated on its way in by the hot hardened oil leaving the second (*Semi-continuous Hydrogenation System*). Another more flexible arrangement provides a hot hardened-oil drop tank underneath the autoclave and three hot soft-oil holding tanks (nitrogen-protected) (*Hydrogenation Plant with Post-Treatment*). Hot, hardened oil is dropped immediately from the autoclave to the tank below. The autoclave is then filled with hot soft oil from one of the three holding tanks, while the other soft oil fills another holding tank having obtained its heat via an exchanger from the hardened oil, which has now commenced to filter. This system does cater to variations in hardening and filtration times within certain limits. In all arrangements, answer these questions: Does the flexibility match the foreseeable process-cycle time variations, and is the outlay recoverable at local energy costs? Also important is that all heat exchangers are capable of being kept clean, that catalyst deposits do not lodge in them, and that their use does not bring about a significant contamination of one oil by another.

Filtration

In the majority of filtration systems, the filter cake becomes the filter medium, and very often completes perhaps more than 90% of the separation task, as the following will make evident. Active nickel catalysts contain nickel crystallites of a 50–100 Å size (i.e., 0.005–0.01 μm) which, used as such, would present a hopeless filtration task; but if these are produced already intimately bound in an inert porous support such as the siliceous deposit kieselguhr, the particles are then in the 10 μm range (Coenen, 1970; Patterson, 1974). This brings the task of filtration much nearer that of filtering activated bleaching earths. In these, some 30–40% is under 20 μm in size, and another 20–30% is in the 20–40 μm range. The so-called "wet-reduced" nickel catalyst made by the decomposition of nickel formate in oil at 250°C contains a considerable proportion of its nickel in particles of 1 μm and less, which gives rise to serious filtration problems, but if an inert filter aid of the kieselguhr type is included at the time of decomposition, or shortly afterward, a more manageable product is obtained (Patterson, 1974, 1976, 1979).

The apertures in a filter membrane should be between two and four times the size of particles to be retained, because a good chance then arises of many particles colliding in the aperture and forming a bridge through which liquor flows and against which other particles build up, so that even smaller particles soon become trapped in the labyrinth. The type of weave adopted for wire cloth or gauze was frequently the classic Plain Dutch or Hollander weave, known in German as the Tressengewebe. Some explanation of the descriptive terms used in weaving is needed here. First, as the fabric or gauze moves steadily out from the loom, the wires or filaments running along its length form the warp, and their spacing depends on the notches cut in the reed at the head of the loom. Single-warp wires or small groups of them may be lifted momentarily above their neighbors, thus allowing a shuttle to drag a weft wire across the gauze, passing alternatively over and under the warp or groups of warp wires. Both warp and weft wires may be loosely or tightly spaced, and, if desired, may even be of different thicknesses and of different compositions. Where both warp and weft are spaced loosely and of similar thicknesses, passing alternatively over one another from top to bottom and side to side, "one up, one down," a virtual square opening with straight-through passage results. This, the simplest weave, is of course mechanically weak, and if the opening is relatively large, the retentivity is poor; the flow rate will be high, and the tendency to blind low. In a twill weave (köper bindung), the weft passes over pairs of warp, but a progressive shift occurs diagonally regarding over which warp pairs the weft passes. Openings are still of the straight-through type, but a useful gain in fabric strength and stability occurs; a medium retentivity combined with resistance to blinding results. Twill weaves were and are very popular in textile filter cloths.

By convention in specifying a gauze weave, the distance from the center of one wire to the center of the neighboring wire running in the same direction is known as the mesh. The mesh size or number is the number of such distances found in a lineal inch of 25.4 mm. Warp is quoted first then weft. After this, the respective warp and wire diameters are quoted in millimeters (mm) and in the same order. What will be appreciated is that if warp wires are substantially thicker and loosely spaced but weft are thinner and closely spaced, then instead of a square or possibly a rectangular straight-through opening, the gap becomes approximately triangular and oblique to the surface of the gauze. The metal-gauze weave popular over many years for filtering bleaching earths was a plain Dutch weave of 24 x 110; 0.314/0.254, but its retentivity was limited (i.e., single particles up to 140 µm in size could pass until bridges reduced the gap). In the 1970s, the armored Hollander weave, or so-called Panzertressengewebe, was developed, which had the much improved retentivity of 80 µ. This was specified as 132 x 36; 0.193/0.396. Here is the novel feature that 132 x 0.193-mm thin warp wires wrap themselves around 36 x 0.396-mm thicker weft wires, and a strong stable structure results. The aperture is then a tortuous channel which permits rapid liquid flow, but quickly traps particles which then begin to form bridges across the gaps. Truly, such monofilament weaves gained slower acceptance in catalyst separation than in removing bleaching earths, possibly because of a higher proportion of smaller particles in the former. Even so, where normal proportions of filter acid/precoat were acceptable, the filtration of hardened oils and hardened fatty acids was feasible. For example, when hardened fatty acids have to be separated from a guhr-supported catalyst over an 80 µ of Panzertressengewebe gauze in a Schenk automatic self-discharging filter, the addition of one part of Dicalite filter aid to two parts of catalyst solids easily met a guaranteed filter rate of 150 kg of oil/m^2/hour, and in favorable circumstances, only one-half of this amount of filter aid was necessary. The same filter manufacturers also used 100 µ Hollander weave gauze with a 1 kg/m^2 of precoat of Dicalite 740 or Speed Plus, and found this combination successful in separating nickel catalyst. Other manufacturers of similar automatic filters will no doubt provide advice as to what gauze they recommend for nickel-catalyst filtration. The penalty attached to the use of any filter aid is that it increases the amount of solids to be filtered, and if the dead catalyst comes to be sold for the recovery of its nickel, the depressed nickel content of the solids lowers the value. Suppliers of supported nickel catalysts might reasonably be asked by a fat hardener whether the use of a filter aid was necessary in all cases of metal-gauze membranes of different aperture size. The brewing industry has a high interest in efficient filtration, and advances made in the interest of one trade can naturally be used to the advantage of another. The gauze manufacturers Gebr. Kufferath of Duren contributed notably. Since the 1970s, further advances were made, and now available is an armored twill weave, the Köper Panzertressengewebe 55 μ (KPZ 55), which not only shows better retentivity, as the name suggests, but also higher flow rate and greater strength.

On the nomenclature given above, the KPZ 55 would be described as 170 x 46; 0.15 x 0.30. Here, again, the warp bends around thicker weft, and the twill weave adds to the structural stability. Not only does the KPZ 55 have an improved retentivity because of the smaller apertures, but because the number of these has so increased, providing a greater total open space per square meter, the effective flow rate per square meter (porosity) increases substantially. These advances have heightened the performance of automatic horizontal self-cleaning filters, and examples up to 200 m^2 are available (Schenk Filterbau GmbH). When wires are given a shallow U-shaped deformation so as to pass more easily over another, the technique is called crimping. At the crimped points, the wire then forms knuckles which might create focal points of pressure, metal fatigue, and wear at intersurface contacts. This tendency is reduced by a controlled rolling between steel rollers, with the knuckles then becoming flatter. This operation is known as calendering. Crimping and calendering are not allowed to distort the essential weave pattern. Normally, a filter gauze is supported underneath by a coarser, stronger gauze, the whole disc then being capable of being handled separately when necessary without the structure being distorted.

Once a permeable layer is established on the surface of the filter membrane, the continuing deposit takes over the filtration task. This initial layer may be as thin as 0.25 mm or as thick as 3 mm, depending on the circumstances of the task. The ensuing rate and effectiveness of the filtration are greatly influenced by the care with which this layer is made. Do not unduly compact it by employing too high a pressure; this is not likely to be needed, in any event, since the resistance at this stage is so low that the flow rate is well above average. Avoid sudden variations in pressure which collapse the particle bridges: an open texture must continue to apply throughout the cake as it increases, so that flow rate is not severely decreased, and so that fine particles still have the opportunity to be caught in the growing labyrinth rather than compacting on the surface.

The traditional filter membrane for the separation of nickel catalyst, which is still used in a great majority of facilities, is the filter cloth. A cloth which is woven from a single smooth fine filament would be unsuitable for catalyst filtration, and is most likely to be employed where solids are coarse and the fluid filters very freely. On the other hand, when a yarn is made from several continuous filaments or numerous short-length staples and then woven to a cloth, the fine particles are trapped not only between one length of yarn and another, but are enmeshed in the fibers which make up the yarn itself. This provides a porous matrix on which the cake builds. One such cloth of cotton or polyester staple double-shrunk twill, made to a weight of 680 g/m^2 and with an air permeability at 25 mm of WG pressure of 3 m^3/m^2 minutes, will give good filtration results with guhr catalysts (P&S Filtration, Ltd.). Further, if two thicknesses of such cloth are used, and the filtration properly is controlled as described below, the nickel content of the hardened oil will normally be restrained to under 5 ppm, which means under 25 ppm of total solids. A matter of trial could discover if the top heavy cloth could be replaced by a much lighter (240 g/m^2) close-plain-weave polyester of similar permeability while retaining the good filtration performance. Nylon–cotton mixtures in place of the all-cotton cloths may prove an economic alternative by reason of greater durability. With filter presses, a filtration rate of c. 200 kg of oil/m^2hours is acceptable; for automatic enclosed filter units, an average of 350 kg of oil/m^2hours seems realistic; this implies that the supported catalyst has filtration behavior resembling a siliceous earth (Patterson, 1979).

Good filtration control takes into account the following:

- While the flow rate during a precoating operation may be quite high, this must not arise from the use of a high pressure comparable with what will be used later in the filtration.
- Filtration proper, especially on a clean filter, commences slowly. Recycle the dark oil from the first runnings, usually at the end of the batch if it is held separately. As the filtrate clears, usually in about 10 minutes, the retention of dark oil is canceled, and the clear oil is turned to the filtered-oil tank. From now on, the rate may be steadily increased, provided no dark oil reappears.
- Avoid sudden pressure changes.
- The rate at which flow rate is increased can usually safely be made greater when the filter already has in it some cake from a previous batch. This assumes the operation is under manual control.
- A marked rise in pressure drop (resistance to flow) across the filter is a strong indication of blinding and that the routine cleaning of the membrane is due.

For a long time, one could obtain filter presses with a closed off-take so that contact with air is prevented. This usually implied only one sight glass for the inspection of the whole filtrate stream, and consequent uncertainty as to where any leak or dark oil might be taking place. To connect each chamber of a filter press to the common offtake by means of a flexible length of pipe—such as a braided hose—which carries in it a small sight glass is possible. If a leak of dark oil commences, the faulty chamber can then be identified and closed within seconds. The length of hose allows for movement during press cleaning.

Conventional European post-treatment for any hydrogenated oil normally is comprised of dilute alkali wash, mild adsorptive bleach, and deodorization. In the United States, where the increase in FFA might well leave the FFA of the filtered crude hardened vegetable oil still comfortably under 0.1%, a light bleach of the filtered oil followed immediately by deodorization has for a long time been the most common procedure (Albright, 1967; Allen, 1978). A good standard is to ensure that the deodorized final hardened oil contains less than 0.2 ppm of nickel (Katalysatoren der Süd-Chemie AG, *Removal of Metal Traces from Edible Oils*). Now combined FFA stripping and deodorization are being ever more widely used, and this must surely apply to hydrogenated oils which are obvious candidates for such a procedure. If small particles of nickel metal are present in an otherwise white fat at 5 ppm and above, they begin to impart a pearly-grey cast, best recognized by a comparison side-by-side with a nickel-free sample; similarly, if the nickel is present as soap, the cast at about 10 ppm is a dull pale green.

If the normal postbleach of filtered crude hardened oil does not succeed in bringing the Ni content below 0.2 ppm, then an addition of citric acid of about 0.05% weight of oil along with the activated earth should achieve this; even one-fifth of this dose of citric acid may be effective—it is added as a 50% w/w solution in water. Another method of getting rid of traces of nickel is to give the hardened oil a wash with very dilute phosphoric acid followed by a water wash prior to postbleaching, but this is an exceptional measure.

Whatever post-treatment of the hydrogenated oil may be adopted, the questions of a polishing filtration remain— Is the hardened oil to be delivered to a customer for further processing? and After being deodorized, is it about to be incorporated in a margarine blend or other edible fat? The eye can see particles down to 40 µm, but particles of 10–20 µm can create a haze. A careful first filtration of a guhr-supported catalyst (or bleaching earth) will easily yield a filtrate containing much less than 100 ppm of total solids, perhaps only 25 ppm. Polishing filters are graded according to their ability to retain particles down to a certain size, say, for example, 10 µm. Although after the first filtration the size distribution of solid particles in the fluid has altered markedly, so that regarding numbers of particles much more than 50% are in the 2–5 µm range, this still allows that less than 50% by weight are of this size. Thus, the polisher graded as retaining 10 µm particles may well succeed in retaining 50% by weight and above of 25 ppm of solids. This polishing step is more concerned with bleaching-earth particles, since it comes after measures specifically designed to remove nickel. Paper, natural, and man-made textiles are all employed in polishing tasks in the form of filter presses, candles, bags, etcetera. A good-quality polishing paper will show characteristics similar to these: weight 220 g/m^2, air resistance 5 kPa, dry-bursting strength 390 kPa, and wet-bursting strength 160 kPa (Patterson, 1979). An expendable fine polishing candle/cartridge may have its working life extended several times by inserting it in a close-fitting sock of coarser material, such as the 680 g/m^2 double-shrunk twill described above, which absorbs some of the material which would blind the surface of the polishing candle/cartridge itself. The frequency with which the socks and the candles/cartridges need to be replaced has to be found by trial.

Catalyst Handling and Economy

In some facilities, the catalyst is mixed into the complete charge before transfer to the autoclave, but in what is probably a considerable majority, the catalyst is mixed independently (Albright, 1967; *Drawing No*; *Fatty Acid Technology*, 1991; *Flow Sheet SFS-164*; Hastert, 1981; *Hydrogenation Plant with Posttreatment*; Leuteritz, 1971; *Semi-continuous Hydrogenation System*), and may, therefore, be added after the filling operation. This allows the bulk of the charge to be heated—probably under vacuum—to just below the temperature at which hydrogenation is to commence. It is, therefore, dry as required, and the interval the catalyst spends in heated oil prior to commencing its work is a minimum. Where more than one group of oils (laurics, other vegetable, marine, and animal oils) is being regularly processed, one must, of course, observe the same segregation of qualities in the catalyst handling system as in the main oil handling system, even if the oil filling and catalyst addition lines share the same immediate point of entry to the autoclave. When a program of catalyst reuse is in operation—again, normally the case—this strengthens the case for an independent catalyst mixing system. When neutralized bleached oil is moved to the point at which catalyst mixing is to take place, to have available a second, quite small tank adjacent to the mixing tank into which some of the oil is placed may be advantageous. This small tank can then be drawn upon for some oil with which to flush through the catalyst addition line after the catalyst dose is added to the autoclave and just before the line is sucked clear.

Two or three routines for the use of a catalyst are fashionable, the adherents of each being concerned in obtaining the desired results as consistently as possible with minimal costs to themselves. These may be summarized thus:

1. Work with the minimal amount of new catalyst every time, and then discard it.
2. Reuse catalyst repeatedly, always discarding a small amount (10–20%) and adding the same amount of new.
3. Program new and reused catalyst by the activity and size of the dose from one class of a hydrogenation task to another; discard as a failure to attain results appears probable or because hardening time is too long.

As is not difficult to imagine, the choice of routine is governed by the variety of hydrogenation tasks which are commonly featured in the weekly production program; therefore, not surprisingly, different hardening programs give rise to different catalyst-handling routines. These comments on the above three routines show how one could best apply them:

1. In the partial hydrogenation of a vegetable oil (soybean oil, for example), a selective removal of the most unsaturated fatty-acid group (linolenic) is desired with the minimal formation of *trans* isomers (lower melting solids) and solid saturates (stearic).

   Equally, one may desire to produce a hardened fish oil with a maximal conversion of highly unsaturated fatty acids or those whose double bonds are so dispersed that they very obviously cease to act as conjugate or skipped unsaturates, while *trans* isomers are at a minimum, and saturates have increased only slightly.

Such hydrogenated oils, if winterized, afford the possibility of producing a table oil from the liquid fraction, or, when hydrogenation is continued a little further, a margarine which retains plasticity at low temperatures. The activity of a good catalyst in combination with a well-refined oil and sufficiently pure hydrogen (Chapter 5) will allow one reuse on this type of hydrogenation task. Its activity is then far from destroyed: it could be used on further tasks if they existed in the program. Similarly, if a vegetable shortening is needed in which exists a long melting range, and some saturates are quite acceptable if the *trans* isomers are at a minimum, this use of a new catalyst would be indicated.

2. When a hardened vegetable or marine oil with a steep SFI or SFC curve is the dominant, even exclusive, requirement of the program, this approach is popular since the make-up of the catalyst mix remains very much the same from batch to batch, and this is an important element in maintaining closely similar textures, provided the end point of the hydrogenation is controlled fairly tightly, if possible within a total spread of 3 IV. The reason why a tightly controlled end point enables the operator to keep close to the texture desired is that the hardening conditions of temperature and pressure, as well as the dose and condition of catalyst, are consistently similar.

3. This situation of a large mixed program is one in which the big European facility commonly finds itself. Certain special tasks, such as the hydrogenation of lauric oils, require the segregation of their own catalyst. The production of stable, soft-textured, hydrogenated vegetable oils requires a new catalyst with limited reuse: stable, soft-textured, hydrogenated marine oils require a similar approach; the catalyst from these sources is very far from exhausted. By stepping up the dose from the 0.05–0.15% of the nickel/oil level of its earlier use to two or three times this range, a longer hardening task, such as the production of a hardened vegetable or marine oil of above 45°C mp, can be confidently undertaken.

   Progressively lower down the melting-point range and into an area where selectivity (S_I) and *trans* promotion are welcomed, these catalysts, which may still retain one-third of their original activity, are most useful. Here the operator may decide to accumulate a filter-full, then split into quarters or thirds for reuse, then on the next round to halves, and, finally, to whole contents of the filter if hydrogenation time and filter cycle time will tolerate it. At this stage, the Ni/oil content will have risen to 1%.

From the literature which commercial catalyst suppliers provide, evidently they recognize the hydrogenations for which a fresh catalyst is best, but for certain programs, the replacement of a portion of the recovered catalyst (10–20%) with a new catalyst, on each recycle, is feasible and popular. Unichema (1987, 1990) explicitly recognizes these different situations, and adds that if difficulty is experienced in obtaining products with sufficiently steep SFI curves ("hard consistency"), a strongly *trans*-promoting nickel catalyst capable of considerable reuse is available which can easily take the place of one which was naturally poisoned through repeated use. Regarding fish-oil hydrogenation and what amounts to the situation of the routine (Normann, 1903) above, Osinga and Balemans, after various trials, conclude that having completed some six hydrogenations with 0.05% of fresh nickel/oil to achieve a dose of 0.30% of nickel/oil by accumulation; thereafter the equivalent of 0.05% of nickel should be taken out after each cycle and 0.05% of new nickel added. Variations on this theme are described and the results compared. Girdler (now Süd-Chemie) recognizes a situation in which some of its catalysts, after four to five selective and low-temperature hardenings, continue to many more reuses at a rather high temperature. Süd-Chemie envisages a replacement level of 10–20% on repeated use (Nickel Catalyst KE-NF20 and KE-FS40). Reusability is a feature to which Engelhard also pays attention; they also supply a sulfured catalyst for high *trans*-isomer promotion capable of several reuses, provided it is not mixed with other catalysts.

Commonly, in the hydrogenation of fatty acids, the reuse of catalyst is not feasible. In their batch-hardening unit for edible oils, Lurgi allows for reuse as a normal procedure (*Fatty Acid Technology*, 1991).

Evidently, in choosing a catalyst routine, the operator must look first at consistent adherence to specification in the product. Pertinently, consider how much the whole filtration task may be multiplied by the repeated circulation of the considerable hulk of an old catalyst and whether the cost of this may be reduced to some combination of routines 2 and 3 (as listed above) without losing adherence to hardened-oil specifications.

Filling, Controlling, and Emptying an Autoclave

The suppliers of hydrogenation equipment supply written operating instructions, and will almost certainly assist commissioning. The following notes should, therefore, act as a checklist in the preliminary discussion between client and supplier. Some detail will be relevant or not, according to the precise layout of the unit and the extent of the facilities provided—for example, whether heat-exchange arrangements are provided between incoming soft oil and outgoing hardened oil, and whether these include any intermediary tanks. A dead-end system with vacuum facilities is presumed (Fig. 4.11 or a similar design).

Filling

1. The filling of the main bulk of oil (less catalyst mixture) can be performed to an advantage with the vacuum applied to the autoclave; the drying effect of the vacuum opposes the subsequent hydrolysis of the fat and the attack on active nickel. The heating of the oil commences at the discretion of facility management if coils in the autoclave are used. If a heat exchanger is operating, this will be in the action from the commencement of filling. As a general rule, the

charge is preheated at least to 120°C unless a low-temperature hydrogenation (105–115°C) is planned. If no need arises to perform the early part of the hydrogenation at a reduced temperature (140–150°C) to avoid the cyclization of polyunsaturates or to minimize *trans*-isomer formation, the preheating may reach 140°C.

2. The catalyst dose should be added, and a very few minutes are allowed with stirring and under vacuum. Stirring then ceases.

Controlling

1. The vacuum connection is closed; vacuum in the autoclave is broken with hydrogen and, if desired, hydrogen may be vented to the atmosphere very briefly (c. 30 seconds) as a final purification of the headspace. Hydrogen pressure in the vessel is allowed to rise to the required level, and stirring recommences.

2. Hydrogenation then proceeds after the steam is turned off the heating coils well in advance (say, 20°C before) of the set hydrogenation temperature. Automatic cooling arrangements are now set for action so that the hydrogenation temperature is maintained within ±5°C. If, later in the hardening, the reaction rate decreases and a decline in temperature becomes apparent, the first action is to purge the headspace by venting for a few minutes. A plate bearing a 3-mm-diameter hole placed in the vent line allows this purging to be controlled in an acceptably steady fashion. If the moisture content of the headspace is unusually high, this may be indicated by much condensation in the first vented gas as it strikes the air outside of the building. This purging usually restores the reaction rate, and may have to be repeated before the end point is reached. If a slow leak from a steam coil is present, how easily the end point can be attained by repeated venting is questionable, and the charge may have to be cooled and pumped out.

3. If the leak from the coil is larger, pressure may begin to increase in the vessel to the point where the safety valve will begin to operate, even though intermittently. Now, indubitably, the charge must be withdrawn, and the supply of hydrogen, steam, or water must be stopped to curtail rising pressure.

4. If these troubles are absent, and no reason exists to suspect mechanical failure of the agitator, possibly the catalyst was damaged. The reaction rate can often be moderately improved in these circumstances by increasing the operating pressure to whatever is permissible. For example, an autoclave usually operated at 3–5 atm but designed to allow 10 atm could be so operated.

5. In the event that a decision is made to make a moderate addition of fresh catalyst, which must be done at a rather late stage in the hardening, the oil in which the catalyst is mixed must not be of a very unsaturated type (e.g., fish oil), since this may not be hydrogenated subsequently to a point of acceptable stability;

marked flavor reversion could then occur in an oil whose test characteristics appeared normal.

Filtering

1. Before contact with air is allowed, the hydrogenated oil should be cooled below 100°C. As the oil recedes from the autoclave, its place is usually taken by hydrogen. Never allow this to bubble in from the foot of the autoclave, since this will destroy the accuracy of end-point control by encouraging further hydrogenation. The hydrogen should enter quietly from the top via the so-called "breather" or balance gas connection. If nitrogen is used for this purpose, this difficulty does not arise. The recommendation is that the filter line be finally blown clear with nitrogen and loose hardened oil expelled from the filter by the same means.

Chapter 5

Hydrogenation

H. B. W. Patterson

Quality

Two classes of impurity in hydrogen are recognized: (i) One is chemically inert, and merely serves to dilute the hydrogen, thereby lowering its partial pressure as a constituent of the gas phase. Since the rate at which hydrogen dissolves in the oil depends, among other things, on the difference in concentration (pressure) in the gas phase and the concentration in solution, anything which diminishes this difference lowers the rate of solution and, eventually, the rate of hardening. This effect can become noticeable when the concentration of inerts in the hydrogen is 25%, yet hardening may continue at an appreciable rate until they reach 50%. (ii) The second class of impurity poisons the catalyst, and some of these, such as moisture and fatty acid, encourage the splitting of the neutral oil.

The most common inert impurities are nitrogen and methane, which, while they may only be present in the fresh hydrogen at a small fraction of 1% v/v, nevertheless accumulate in the gas system until purged. If the oxygen of small amounts of air which was allowed to enter the system at intervals is converted to water, the residual nitrogen will accumulate. Fortunately, modern methods of hydrogen production normally give a gas of more than 99.5% of purity (dry basis) and frequently above 99.9%. A routine purging of the gas system at opportune intervals easily keeps inert impurities to a harmless level, even when the iodine value (IV) drop (hydrogen adsorbed per ton oil) of the program is as high as 100 units.

If oil is dried to below 0–0.5% w/w of H_2O/oil, this amount of moisture, even if not all removed during the filling and preliminary heating of the autoclave under vacuum, will not pose a serious threat to either the rate of hydrogenation or the increase in free fatty acid (FFA). It, too, can be lost by venting/purging the gas space. The common practice of compressing and cooling hydrogen to c. 7 atm before allowing it to expand to c. 3 atm for use in the autoclave will bring down the moisture level of the gas to about 0.1% v/v (see Fig.5.1), and this is an acceptable level.

Carbon monoxide (CO) poisons the catalyst by forming nickel carbonyl [$Ni(CO)_4$] which, however, becomes unstable at 110°C, and by 160°C breaks down completely. Hence, for hydrogenation purposes in general, to limit CO to under 0.5% of CO is desirable; for low-temperature (120°C) hydrogenations, less than 0.03% of CO is probably an advantage. Purportedly, at 150°C, 0.1% of CO is a considerable hindrance, but by 210°C, even 0.5% of CO has little effect (Patterson, 1974; Swern, 1964). CO should be absent from hydrogen made by the electrolysis of caustic-soda or caustic-potash solutions; it might be found up to 0.5%, depending on conditions used, when reacting water with the sodium amalgam derived from the electrolysis of brine, but this seemingly was not a problem. Modern hydrocarbon reformer plants purify their hydrogen very successfully to not exceed

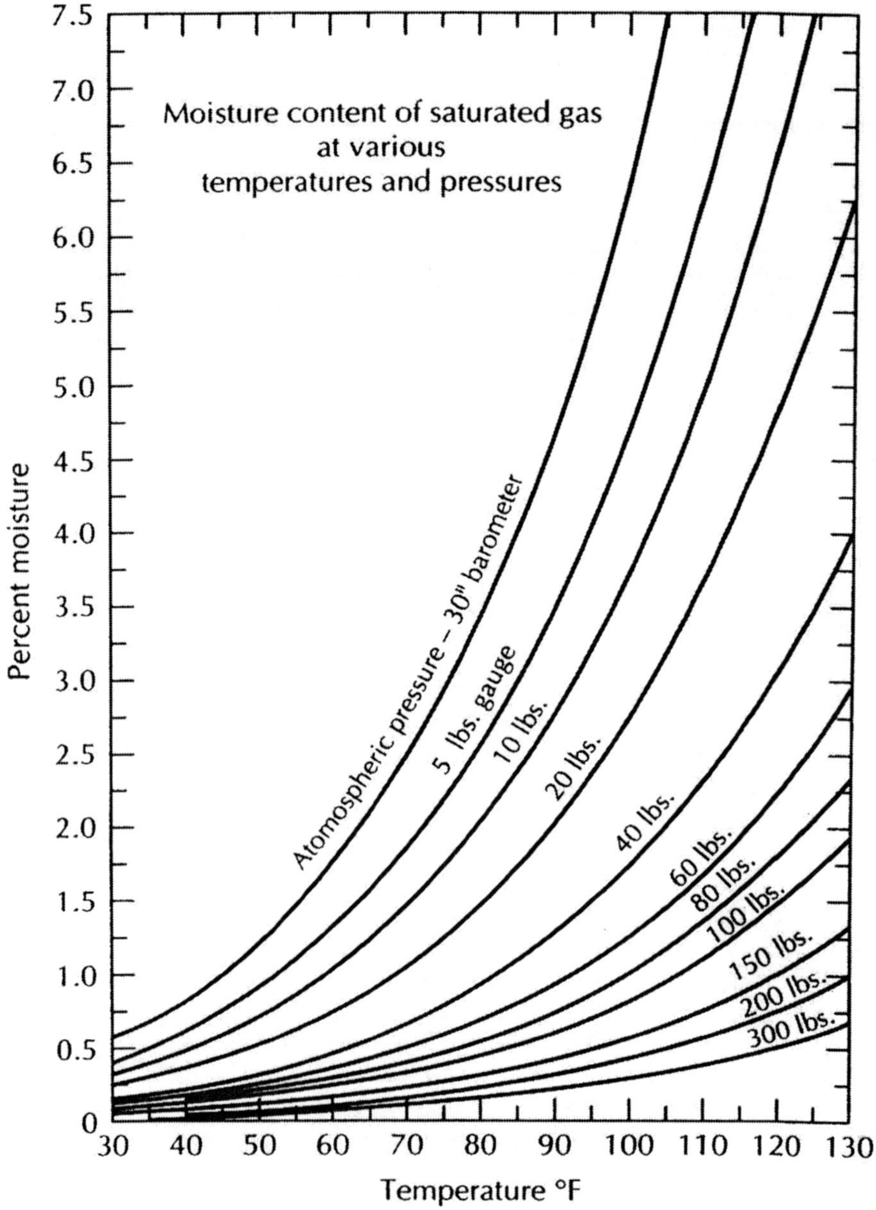

Fig. 5.1. Moisture content of saturated gas at various temperatures and pressures (*Gas Volume Correction Factors Including Properties of Gases*).

10 ppm (0.001%) of CO. Hydrogen derived from the by-product gases of petroleum refining and ammonia synthesis can also have its CO content reduced economically to well within the limit mentioned here.

The decomposition of the amalgam mentioned above yields, in the first place, a hydrogen containing mercury vapor, but the act of compression throws out much of the greater part of this. As a final stage of purification, the hydrogen is passed through a dry scrubber of activated carbon-iodized carbon-activated carbon, resulting in less than 10 μg of Hg/m^3H_2 remaining; commonly, the content runs at c. 4 μg of Hg/m^3H_2. In this connection, one may remark that the safe upper limit for the mercury content of the atmosphere of a workplace was set at 100 μg of Hg/m^3. Further, if an oil were to be hydrogenated with hydrogen containing 200 μg of Hg/m^3 for an IV drop of 90 units and all the mercury transferred to the hardened oil, the latter would contain, at most, 0.015 ppm of Hg, whereas various common natural foodstuffs contain up to 0.06 ppm of Hg. Sulfur, which is one of the best-known and effective catalyst poisons, is for practical purposes absent from electrolytic and reformed hydrocarbon commercial hydrogen. One can arrange the desulfurization of hydrogen from other sources, such as petroleum refining, without great cost. The following example shows how secure the position is. As a catalyst passes through several uses with sulfur-containing oil, an increase from 0.5 to 3.0% of sulfur/nickel greatly lowers activity and increases isomer promotion. If a typical dose of active nickel is taken as 0.1% of nickel/oil, then if the sulfur were derived from the oil, this would amount to 2.5% × 0.1% = 0.0025% or 25 ppm of sulfur/oil.

If the oil were hardened employing a drop of 127 IV, this would entail a gain in weight of 1%. If the sulfur came, not from the oil, but the hydrogen, this would imply in the gas a sulfur content of 0.025%/1% = 25% or 2500 ppm of sulfur/gas. If the 0.25% of sulfur in gas were present as H_2S, this would be equivalent to about 0.019% of H_2S by volume.

Normally, the IV drop is not nearly as much as 127 units, especially in low-temperature hardenings, so the opportunity for poisoning would be much less. On the other hand, alteration in the behavior of the catalyst would occur for gains in sulfur of much less than the rise from 0.5 to 3.0%, perhaps at the 1.0% level. Even if the maximal tolerable level of sulfur were to be fixed at one-tenth of the above value (i.e., 250 ppm of sulfur/gas), no difficulty would arise in staying well below such a figure.

Oxygen is not a significant impurity in electrolytic hydrogen; since the earliest days of commercial electrolytic-hydrogen manufacturing, passing the gas through a bed of palladium catalyst has effectively converted any oxygen present to water. Data accumulated on the quality of hydrogen derived ultimately from the initial electrolysis of brine in mercury cathode cells suggest a level of only 50–150 ppm of O_2/H_2. Similarly, oxygen is absent from reformed hydrocarbon bases.

Table 5.1 summarizes the above comments on hydrogen quality, and applies to a dead-end hardening system. Possibly, very cheap hydrogen rich in inerts is available for fat hardening. This implies that unusually high proportions of hydrogen would be lost by venting the inerts to maintain an economical rate of hardening. At a low cost for hydrogen, this could be acceptable. For example, if a cheap 85:15 ratio of a hydrogen/nitrogen mixture is in use, this could be employed until it became a

TABLE 5.1
Practical Limits for Hydrogen Impurities

Hydrogen	A 99.5% (dry basis) minimum purity is easily attainable; 99.9% is common
Inerts (N_2, CH_4, etc.)	Under 0.5%
Water vapor	Reduced by compression-cooling to less than 0.1% v/v
Carbon monoxide	0.05% v/v Maximum 0.03% v/v Preferable-easily attained
Mercury	Even 200 μg Hg/m^3H_2 fail to give a mercury content of hardened oil as great as some common natural foods.
Sulfur	250 ppm S/H_2 is about one-tenth the amount needed to inactivate nickel catalyst in one long hydrogenation
Oxygen	Negligible in commercial hydrogen (See Chapter 11 on explosive limits with air)
Halogens	Negligible in commercial hydrogen

15:15 mixture of much higher density and then vented automatically; this equals an 82% utilization of the cheap hydrogen.

For normal-quality hydrogen with efficient operation, the loss (compression, venting, etc.) lies between 5 and 8%.

Steam Iron Hydrogen

At one time, a very popular method for the preparation of hydrogen on a large scale was the so-called steam iron process, in which iron removed oxygen from steam to produce hydrogen, being itself converted to oxide; the oxide would then be immediately reduced to iron by a mixture of CO and hydrogen obtained by blowing steam over hot coke. Principal impurities in the hydrogen produced were carbon dioxide, CO, hydrogen sulfide, small amounts of more complex sulfur compounds, nitrogen, and possibly some oxygen. Formerly, the purity of the hydrogen was around 98.5% (dry basis), and this was improved to 99.5%. Compared with other processes, much of the work associated with operating and maintaining the furnaces was heavy and dirty.

Electrolytic Hydrogen

As early as 1913, a large dead-end batch-hardening plant came into action at Bromborough, England, drawing its hydrogen from the adjacent first large-scale producer of oxygen and hydrogen by using the electrolysis of a caustic-soda solution. Hydrogen purity was a minimum of 99.8%. Not only does this purity facilitate any class of hydrogenation, but also nickel consumption is halved as compared with that obtained with steam iron hydrogen. This early, very large electrolyzer was of the simple unipolar or tank type: this type, with varying degrees of sophistication, continues to be used all over the world. In the electrolytic decomposition of water into its elements, a basic minimum of 2400 A flowing for 1 hour is needed to produce 1 m^3 of dry hydrogen (0°C, 760 mm of Hg); or, put another way, 2180 Ah are needed to produce the hydrogen contained in 1 m^3 of moisture-saturated gas (20°C, 760 mm of Hg). To the minimal practical decomposition voltage of about 1.5 V, must be added an allowance to overcome anode, electrolyte, and cathode resistances, so that a final voltage across each cell will be 1.8–2.6 V, depending on cell design, type, and the concentration of electrolyte, temperature, and pressure. Caustic-soda solutions are usually 20% of concentration working at 60°C, and caustic-potash solutions are 28% working at 80°C. The energy requirement to produce the 1 m^3 of saturated H_2 (20°C, 760 mm of Hg) will, therefore, range between

$$1.8 \times 2.18 = 3.92 \text{ kWh/m}^3 \text{ of } H_2 \qquad \text{[Eq. 5-1]}$$

and

$$2.6 \times 2.18 = 5.68 \text{ kWh/m}^3 \text{ of } H_2 \qquad \text{[Eq. 5-2]}$$

Up to a certain point, the heat developed may be considered as taken up by the system, but beyond that, provision has to be made for the cooling of the electrolyzer. Depending on the other energy needs of the plant, this heat can be exchanged to some useful purpose and costs saved. Obviously, oxygen can be collected and sold; the larger the electrolyzer, the more important is the operator's evaluation of this market.

Unipolar Electrolyzers

In the simplest unipolar design, electrodes are immersed in a caustic-soda solution, and held in a tank between 1 and 2 m in depth and about 1.5 m × 1.5 m in area. Each electrode is hung inside a housing of a membrane impervious to gas; hence, hydrogen and oxygen are collected separately: electrodes in each tank are connected in parallel and tanks in a row are connected in a series. Several rows may exist. Some designs have one layer of tanks supported above another. At 2.1 V drop per cell, a consumption of 4.58 kWh/m^3 of produced H_2 occurs, which can be moderately improved by a more sophisticated design. The current density is around 500 A/m^2, and mechanical maintenance is well within the capabilities of a factory-engineering staff. The outstanding advantage of this type of installation is that, by reason of its simple modular arrangement, 5 out of 100 cells, say, may be isolated, both electrically and mechanically (gas-collecting pipes) for repair at any one time, and the remaining 95 continue functioning without any difficulty. One overhaul of the electrolyzer in eight years would be usual.

Bipolar Electrolyzers

In the bipolar design of an electrolyzer, the face of one electrolyte is all anode, and the reverse face is the cathode of the next cell. In appearance, some are likened to the plate-and-frame filter press on a larger scale. An asbestos diaphragm gripped at every edge in the frame of the electrolyzer separates the anodic and cathodic compartments from which oxygen and hydrogen are led by separate gas-collecting manifolds. Under load, the caustic-potash electrolyte surges up into the manifolds, and after giving off its gas is pumped through a filter to return to the foot of each cell via an electrolyte-distributing manifold. The current flows in series through the assembly of cells; pressure is applied evenly to the framework by tightening screws located at each end to prevent leaks. The liquor flow detaches bubbles from the surface of the electrodes, thus increasing electrical efficiency. In the case of the bipolar electrolyzer, the total cell voltage is about 2.02 V, and the specific (power) consumption is 4.1 kWh/m^3 of H_2. Current densities in different designs vary from 2000 to 3000 A/m^2. All bipolar electrolyzers are closed, so unlike the common unipolar arrangement, no contact is made between air and electrolyte. The electrical efficiency is higher than the unipolar type; an operating pressure of c. 40 cm of WG is common; the electrolyzer may occupy less floor space than a unipolar one of the same capacity, although a Canadian high current density design of the latter has minimized this difference. An overhaul is a specialist task normally undertaken by the supplier about every four years, and will not be completed in under two weeks even in the most favorable circumstances; it may last longer— until the replacement of parts is necessary. This being the case, advisably discuss with the suppliers of such electrolyzers the wisdom of installing the required hydrogen-generating capacity—at least two units, each contributing one-half—hence, avoiding complete closure.

Bipolar electrolyzers are produced in a high-pressure form working at 30 atm; efficiency is greater since the higher pressure reduces bubble size, and hence, their obstruction of the electrode surface. Cell voltage of 1.84 V and a specific consumption of 4.02 kWh/m^3 of H_2 then become attainable. The electrical efficiency and maintenance requirements of an electrolyzer must obviously be checked with a prospective supplier before purchase.

Water Supply

To minimize corrosive attack, the water fed to an electrolyzer should be free from chlorides, sulfates, and carbon dioxide in particular, and in general terms its conductivity should not exceed 6.67×10^{-6} reciprocal ohms (mho) per cm (i.e., the specific resistance will then be not less than 150,000 ohms/cm). Dry residue from one liter should not be greater than 10 mg/liter. Although one cannot rely on the ordinary condensate in a factory to meet this specification consistently, one can buy simple stills which will give a condensate whose residue is 4 mg/liter (4×10^{-6} mho/cm or 250,000 ohms/cm). One may also purchase double-effect stills and demineralizing units. The supplier of any electrolyzer will advise on the water purity required (Patterson, 1974).

Security

Hydrogen tends to migrate more readily into the oxygen produced by an electrolyzer than the reverse; 99.8% of hydrogen is probably accompanied by slightly more than 99.0% of oxygen. A falling oxygen purity is one of the most likely signs of electrolyzer trouble; as an oxygen purity of 95.6% is approached, this is an increasingly strong warning that operation should cease and checks should be made of the fabric of the unit. Today, gas purity in both hydrogen and oxygen output is probably monitored continuously by recording instruments, but if not, tests should be done every shift. Since hydrogen has a thermal conductivity seven times greater than any other gas likely to be present, its cooling effect after being dried is compared with that of a pure standard by using a heated platinum element in an instrument called a katharometer. This instrument may be used in concert with others specifically designed to detect impurities such as CO or mercury. The same instruments can be made to activate alarms and other actions, such as the venting of a feed line. Oxygen is highly paramagnetic in comparison to the great majority of gases; hence, one may monitor its presence on this basis continuously.

In the case of the unipolar cell, if (b) and (c) tend to change places in their magnitude, this indicates trouble; for the bipolar cell, the examples quoted (0.35 V + 1.65 V) relate to the total 2 V; when, however, repeated tests are obtained which show less than 0.35 V or greater than 1.65 V, this warns of likely failure. These checks on the purity of gases and cell voltages are of a routine nature common to all electrolyzers, and their makers will provide data on safe operating limits. Besides these checks, are those on the quality of the feed water, the operating temperature, electrolyte strength and purity, internal pressures where applicable, the operation of a cooling system and a water feed, and the inspection of the electrolyzer for visible signs of fabric decay and leaks.

Over and above the security precautions which relate directly to the electrolyzer itself, other obvious ones of a more general nature are taken, at least some of which will relate to any hydrogen-generating plant, whether electrolytic or of some other type. If the electrolyzer is delivering hydrogen into a low-pressure holder, an alarm should be given as this approaches the full mark so that the staff may reduce the electrical load; an alarm at a further stage may be installed so that the load on the electrolyzer is automatically reduced below the voltage at which gas is emitted, or it may be shut off entirely. In the latter case, on starting up again, the procedure should include a short purge to the atmosphere to establish that the normal gas purity is again

TABLE 5.2
An Early Warning of Deteriorating Cell Conditions Is Given When Individual Cell Voltages Depart from the Following

	Unipolar		Bipolar	
(a)	Anode-cathode	1.99–2.3 V	Anode-cathode	2.0–2.25 V
(b)	Anode-gas bell	0.6–0.9 V	Anode-diaphragm	0.35–1.65 V
(c)	Gas bell-cathode	1.39–1.4 V	Diaphragm-cathode	1.65–0.35 V

present before it is transferred to the holder—this is the same sort of procedure which would be followed after a routine longer closure. Equally, an alarm should be given as a set low level on the gas holder is approached so that staff may either reduce the demand for hydrogen or step up production. Later, if the holder continues to fall, the compressors taking hydrogen from the holder may be switched off automatically; some holders include a device whereby, when the ultimate low level is reached, a stopper mounted on the underside of the crown fits into the mouth of the gas off-take pipe which faces upward. An alarm should not merely be audible but easily visible. Normally, the suction line to a compressor is fitted with a pressure-sensitive switch such that, if pressure in the line falls below a preset minimum, the compressor is switched off and therefore does not attempt to suck in air from the surroundings.

Hydrocarbon Reforming

Since the 1950s, the technique of obtaining hydrogen from hydrocarbons has made large advances in the field, thermal efficiency, purity of product, and the ability to provide generating units for small demands of around 50 m^3/hour at costs which compete with small electrolyzers (Patterson, 1974). For the industrial use of hydrogen, in general, five main factors were identified (Daum, 1993; Kuberka et al., 1989; Twist & Sagar, 1965) as affecting the choice of method by which it should be produced. These are: capacity required, purity, type and amount of acceptable impurity, employment pressure, and feedstock cost and availability (including electricity). The first four of these were already discussed in this and earlier chapters; the last is variable according to location and time.

Three sections of a reforming unit are common:

1. The hydrocarbon is treated—probably over a fixed-bed catalyst—to reduce its sulfur content to 1 ppm, thus protecting a later nickel catalyst.

2. $C_nH_{2n+2} + {}_nH_2O \rightarrow nCO + (2n + 1)H_2$
   This is the reforming reaction: steam is present in excess; temperatures (630–1100°C), catalyst, and pressure have varied in different designs; some CO_2, and CH_4 are also produced; heat is required.

3. $nCO + nH_2O \rightarrow nCO_2 + nH_2$
   This is the so-called "shift reaction," in which CO is converted by steam at a much lower temperature (c. 380°C) and with the aid of nickel catalyst to CO_2; the CO_2 is next scrubbed out by a reagent such as methanolamine, which itself can be continuously regenerated: a repetition of this conversion step was used; and also the shift reaction may be accomplished by the use of an iron oxide–chromium oxide catalyst. The theoretical maximal hydrogen yield will now be seen to be $(3n + 1)$ H_2. The hydrogen content will now be about 99.5%, but some CO remains (possibly c. 0.2% v/v) along with inert CH_4; hence, a further purification is performed:

$$CO + 3H_2 = CH_4\ H_2O \qquad \text{[Eq. 5-3]}$$

The scrubbed gas, on leaving the final conversion (shift reaction) stage, is reheated and passed over a nickel catalyst in what is known as methanation, as above. The hydrogen content is now in the 99.5–99.9% v/v range, and, significantly, the CO is down to a very safe level of 10 ppm (Patterson, 1974; Twist & Sagar, 1965).

Reforming plants have become versatile in the sense that they can be reduced to 25% of load when required: they may also be self-sufficient regarding steam, and the hydrogen probably need not be compromised to serve the hydrogenation of fats (Minet, 1979).

Purification to a very much higher degree was well-established since the 1960s by the use of palladium–silver membranes impervious to all gases but hydrogen and giving a 99.9999% purity. This high level of purity is not required for oil hydrogenation, but is useful in other applications (Charlesworth & Schmidt, 1965; Hack & Hall, 1965; Serfass & Silman, 1965). The use of molecular sieves between the shift reaction and the methanation stage was also described in detail (Priddy, 1971). This is of particular interest to the small operator who is looking for minimal equipment costs rather than the optimum of operating costs. Apart from the widely used reforming of the hydrocarbons method, the dissociation of methanol and the cracking of ammonia were employed in the smaller-scale manufacture of hydrogen, followed by purification steps which matched purity requirements (Charlesworth & Schmidt, 1965). In all, some 25 methods of making hydrogen were listed (Brownlie, 1938), most of which are of minor importance and are viable only because of some local circumstance (Daum, 1993; Kuberka et al., 1989). If a hydrogenation plant is located within a few miles of a petroleum refinery, the possibility of obtaining a regular supply of hydrogen recovered from refinery gas should not be overlooked. Adequate purity is unlikely to present a problem, but the scale of operation and the regularity of supply may do so (Charlesworth, 1965). The standard of purity to be anticipated would be: H_2 (c. 99.7%), CO, and H_2S (each nil to a few ppm), a majority of the remainder inert (Patterson, 1974). Hydrogen recovered from ammonia synthesis purged gas may also meet a hardener's requirements since the main impurity would be nitrogen (Charlesworth, 1965). The employment of liquid hydrogen shipped into the hardening plant as an alternative to receiving the gas highly compressed in cylinders on a mobile trailer is now a practical possibility where the liquid hydrogen is produced for other larger users (Newton, 1967).

Purchase of Hydrogen

When the purchased hydrogen is being brought into the plant by pipeline, the same monitoring of quality as described in the "Quality" section of this chapter can be applied, but now the flow rate of gas has to be recorded and integrated, and a correction factor for temperature and pressure applied. The purchase contract will also stipulate that the dew point of the gas shall not exceed a certain temperature, thus placing an upper limit on the water vapor carried. Generally acceptable is that this should be checked only intermittently. Although pressure-differential flow-rate indicator–integrator meters are available, their accuracy is generally substantially poorer than the positive displacement meter where the gas flow rotates two interlocking elements so that each rotation is equivalent to a set volume. Temperature and pressure corrections have to be applied. Several sizes of meter are available; some are designed to work at higher pressures. It is generally acceptable to keep the corrected gas receipts on a 24-hour total basis. In this way, an accuracy of about 0.5% is feasible; the very small slip factor (amount of gas passing around the moving closure edges of the elements) may be checked by the meter supplier each year. If the same make of meter is in use at each end of the pipeline, this is a help in resolving any differences which may arise.

Where hydrogen is received via a high-pressure cylinder trailer, the volume delivered is measured by a pressure drop indicated on the trailer's instruments.

Hydrogen Requirements

Since the IV of an oil merely expresses the amount of iodine with which an oil combines as a percentage of its own weight, evidently, the fall in IV when an oil is hydrogenated, if divided by 127, represents the percentage gain in weight as hydrogen by the oil (i.e., IV drop/127). Conveniently have readily at hand the volume of dry hydrogen measured at certain set conditions to drop the IV by one unit for a standard weight of oil. This is shown in Table 5.3 as the theoretical figure, and then is shown the amount including a hydrogen loss of 5%, which represents a so-called hydrogen factor of 1.05. In normal circumstances, the hydrogen factor for a hardening plant will lie between 1.05 and 1.10.

In making quick estimates, a value of one cubic meter of hydrogen for a fall of one IV unit for 1000 kg is useful, and corresponds with a factor of 1.07 at 15°C and 760 mm of Hg.

TABLE 5.3
Hydrogen Required To Drop IV by One Unit

Weight of oil	H_2 (0°C, 760 mm)	H_2 (15°C, 760 mm)
1000 kg (2204.6 lb)	0.8835 m^3 (31.19 ft^3)	0.9319 m^3 (32.90 ft^3)
1 long ton (2240 lb)	31.69 ft^3	33.5 ft^3
Allowing 5% loss		
1000 kg (2204.6 lb)	0.9277 m^3 (32.75 ft^3)	0.9785 m^3 (34.55 ft^3)
1 long ton (2240 lb)	33.27 ft^3	35.18 ft^3

Chapter 6

Isomer Formation During Hydrogenation

Albert J. Dijkstra

Introduction

The hydrogenation of triglyceride oils involves mixing a small amount of catalyst with the oil, adjusting the oil temperature if necessary and then dissolving a continuous stream of hydrogen in the oil. The hydrogen molecules diffuse to the catalyst surface where they may be adsorbed. Triglyceride oil molecules also diffuse to the catalyst surface and may also be adsorbed and react with the hydrogen, provided a double bond is present in one of the fatty acid carbon chains.

This reaction is quite complex for a number of reasons. The triglyceride molecule may contain only one unsaturated fatty acid or two or even three unsaturated fatty acids. One would expect the number of unsaturated fatty acids to affect the likelihood of the triglyceride molecule being adsorbed onto the catalyst surface. In addition, the fatty acid moieties may contain a single double bond or more than one double bond. This is also expected to affect the triglyceride adsorption and in particular the position of the adsorption equilibrium.

Nickel-catalyzed hydrogenation reactions follow the Horiuti-Polanyi mechanism (Horiuti & Polanyi, 1934) according to which the hydrogen addition to the double bond takes place in two steps. A first hydrogen atom is attached to a carbon atom at one end of a double bond, thereby forming a half-hydrogenated intermediate, and then a second hydrogen atom is added to the carbon atom at the other end of the former double bond. If the addition of the first atom was irreversible, it would not matter to which of the two carbon atoms containing the double bond the first hydrogen atom attached itself. The former double bond would become saturated. However, the addition of the first hydrogen atom is reversible and the hydrogen atom that leaves the half-hydrogenated intermediate need not necessarily be the same atom as the atom that was added. As illustrated below, the molecule that results after the dissociation of the half-hydrogenated intermediate may have become an isomer of the original molecule. The double bond may have shifted along the carbon chain, a phenomenon that is generally referred to as positional isomerisation.

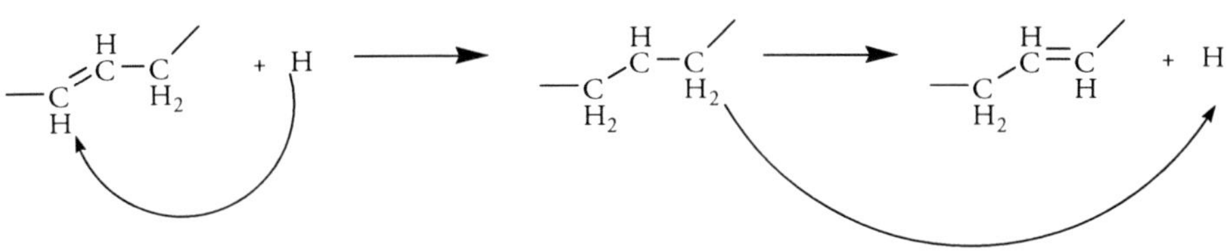

When the shift along the carbon chain happens in a methylene interrupted polyunsaturated fatty acid, a conjugated double bond system may result after dissociation. Such a system will have a different affinity for the catalyst surface and will thus affect the ratio of the various fatty acids that are adsorbed. In its turn this will affect which intermediates are generated, how they react; and finally, which products are formed.

Another form of isomerisation is observed when part of the half-hydrogenated intermediate rotates around the original double bond before the hydrogen atom that was added to one of the carbon atoms at the end of the double bond leaves again. Then the original *cis* configuration of the double bond may change into a *trans* configuration. This is referred to as geometrical isomerisation and is illustrated below.

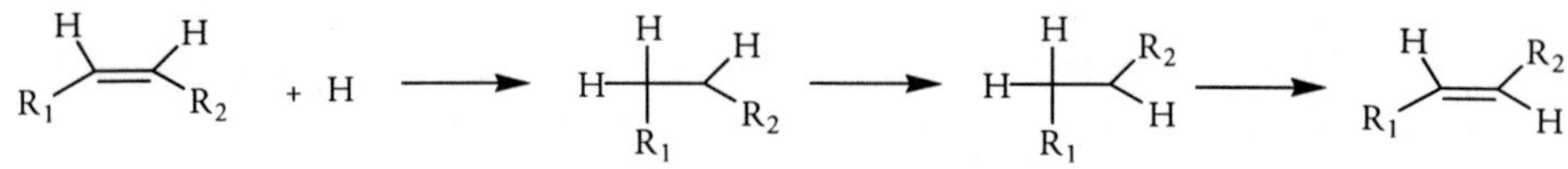

Positional isomerisation may also lead to geometrical isomerisation since after the addition of a hydrogen atom, the original double bond has become a single bond and this permits rotation. This leads to a double bond with a different configuration in a different position. Double bonds need not shift just one position along the fatty acid carbon chain: in conjugated double bond systems, the hydrogen atom leaving can be further away from the carbon atom to which the first hydrogen was attached. This is illustrated below.

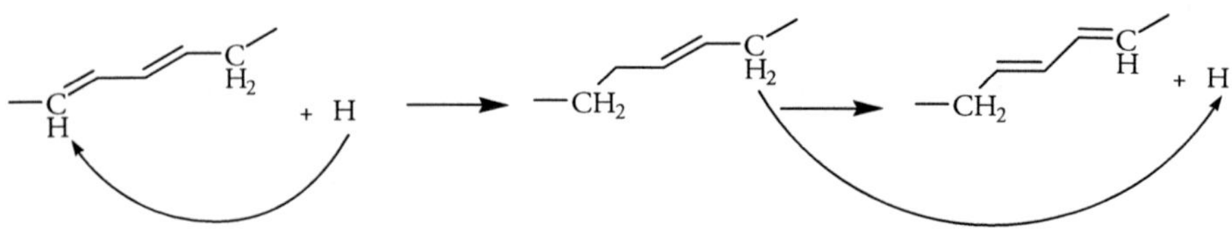

In this isomerisation sequence, the allylic hydrogen plays an essential role. That is even more so in the case of copper-catalyzed hydrogenation reactions. Whereas, the Horiuti-Polanyi mechanism of the nickel-catalyzed reaction starts with the addition of a hydrogen to a carbon atom adjoining the double bond, the copper-catalyzed reaction commences in all likelihood with the abstraction of an allylic hydrogen (Dijkstra, 2002). This facilitates conjugation and since copper catalysts only facilitate the addition of hydrogen to conjugated double bond systems, this conjugation starting with the abstraction of a hydrogen atom is a prerequisite for subsequent hydrogen addition.

History

The nickel-catalyzed hydrogenation process for edible oils has been invented by Normann (1903). At that time, molecular hydrogen was hardly used on an industrial scale so when a hydrogenation plant was built by Crossfields & Sons Ltd. in Warrington, UK, the hydrogen was manufactured by reacting steam with glowing iron shavings (Kaufmann, 1939). Because of the carbon present in the iron, this hydrogen would have an unpleasant smell, not unlike petroleum (Ostwald, 1912). Fifty years later, the steam-iron ore process was still in use (Hughes, 1953). In this process, a low sulfur coke is used to produce water gas that reduces iron ore to metallic iron and the residual water gas is mixed with air and used to superheat steam that is then allowed to react with the reduced iron to produce hydrogen.

According to Hilditch (1937), it was already evident in the early 1920s that the hydrogenation of triglycerides or mixtures of fatty acid esters showed selectivity in that the reduction to monoenes had to be almost complete before the monoenes themselves were hardened. Hilditch and co-workers also established that the 'iso-oleic acids' formed during hydrogenation consisted of *trans* isomers and positional isomers. Moreover, factors affecting the extent of isomerisation and the selectivity of the reaction had also been identified: amount of catalyst, temperature, catalyst poisons, etc. Hydrogenation reactions were followed by measuring the IV or refractive index of samples taken near the supposed end of the reaction and melting point determinations were also carried out on these samples.

In fact, proper quality control demands that more than one independent property is measured to characterize the sample because they can vary independently of each other. A 'non-selective' hydrogenation that causes monounsaturated acids to become saturated when the concentration of polyunsaturated fatty acids is still high, will cause the melting point to be much higher for a given IV than a selective hydrogenation reaction. Accordingly, measuring both the IV or refractive index and the melting point gives an idea of the selectivity of the hydrogenation. It does not give any information about the extent of *trans* isomerization. To get an idea of the latter, it is best to measure the Solid Fat Content (SFC) of the sample and a fast method to determine this characteristic has been published (Rutledge et al., 1988).

In an attempt to facilitate the understanding of the kinetics of the hydrogenation process, Bailey and Fischer (1946) introduced the concept of the 'common fatty acid pool' by suggesting that the rate of reaction of a fatty acid in a triglyceride does not depend on the chemical nature of the other fatty acids present in this triglyceride. This concept allowed various selectivities to be defined (Boelhouwer et al., 1956; Allen, 1967) such as for instance the linoleic acid selectivity. This was defined as the ratio of the rate constant of the hydrogenation of linoleic acid and that of the hydrogenation of oleic acid whereby it was assumed that both reactions depended in an identical manner on the concentration of the hydrogen that is dissolved in the oil. Similarly, the linolenic selectivity was defined as the ratio between the rate constants for the linolenic acid and linoleic acid reactions; it made the same assumption regarding the hydrogen concentration.

In order to characterize the rate of geometrical isomerization, the isomerization index or *trans* selectivity was introduced. It was defined as the increase in *trans* content divided by the decrease in iodine value because in a *trans*-IV plot, a straight line is generally observed and the slope of this line could therefore readily be determined. Actually, the fact that a straight line is observed is quite remarkable. When the IV decreases during hydrogenation, the number of double bonds still present in the oil decreases but the straight line implies that the decrease of this number of double bonds does not affect the rate at which *trans* isomers are formed. Moreover, not only does the number of double bonds decrease with decreasing IV, the number of *cis* bonds decreases even more because of the *trans* bonds formed and some of these *trans* bonds will also isomerize back to form *cis* bonds and even so, a straight line is observed.

Not all authors have been sufficiently aware of the *cis-trans* equilibrium. Baltes for instance claimed impossibly high *trans* contents in hydrogenation products when using a nickel subsulfide catalyst (Baltes, 1970; 1972). In fact, the enthalpy difference of the *cis-trans* equilibrium has been determined at $\Delta H_{iso} = -4$ kJ/mole (Veldsink et al., 1997) which means that at normal hydrogenation temperatures of 180–220°C, the *trans* isomer content is less than 75% of the double bonds (Dijkstra, 2007).

At one stage, it was thought that the rate of reaction of a fatty acid depended on its position in the triglyceride molecule: positional selectivity. By hydrogenating a randomized substrate and analyzing the overall fatty acid composition and comparing this with the fatty acid composition of the 2-position, this positional selectivity was shown not to exist (Beyens & Dijkstra, 1983). Finally, a triglyceride selectivity was defined (Coenen, 1976; 1978) to quantify the phenomenon that in some hydrogenations an unexpectedly high amount of trisaturated triglycerides is formed.

Although these selectivities had their use in characterizing different hydrogenation reactions, their assumptions turned out to be incorrect. Even the concept of the 'common fatty acid pool' was found to be untenable (Dijkstra, 1997) in the light of experiments carried out by Bushell and Hilditch (1937) and especially Schilling (1978) who showed that a linolenic acid moiety reacts more slowly when it is part of trilinolenate than when it is part of monolinolenate. The observation that a linoleic acid moiety in a triglyceride containing more linoleic acid moieties reacted more slowly than a linoleic acid moiety accompanied by medium chain fatty acids that was first of all explained by stereochemical arguments (Beyens & Dijkstra, 1983) could also be better explained by rejecting the 'common fatty acid pool' suggestion.

The definitions of linoleic acid selectivity and linolenic acid selectivity must also be rejected since calculating consecutive values of the linoleic acid selectivity during a hydrogenation experiment (Dijkstra, 1997) showed that the 'ratio of two rate constants' was not a constant. The assumption that all hydrogenation reactions depend on the hydrogen concentration in an identical manner is apparently incorrect.

Up till the 1970s, the literature regularly mentioned 'shunt reactions' taking place during the hydrogenation of polyunsaturated triglycerides. The 'oleate shunt', being a 'direct-through' reaction of linolenic acid to oleic/elaidic acid, was suggested by Bailey (1949). Subsequent authors (Mounts & Dutton, 1967) even suggested stearate shunts in which linolenic acid and linoleic acid would react straight through to stearic acid. Their conclusion that these reaction paths existed was based on otherwise inexplicable deviations from kinetic models. In retrospect, the validity of these models is doubtful, since they are based on the 'common fatty acid pool' suggestion and do not take into account that the reactions of the various fatty acid moieties have a different order with respect to hydrogen. Besides, the only difference in reaction rate between fatty acid isomers taken into account is between linoleic acid and isolinoleic acid (Δ^9, Δ^{15}-octadecadienoic acid). Accordingly, the models used are oversimplifications that do not provide valid evidence of the existence of shunt reactions.

There has also been some confusion about from which fatty acids the *trans* isomers originate during hydrogenation. Elaidic acid can be formed by the partial hydrogenation of linoleic acid but it can also be formed by the isomerisation of oleic acid. In an otherwise elegant experiment, Coenen and Boerma (1968) hydrogenated high erucic acid rapeseed oil (HEAR) and used the erucic acid as an internal marker. If brassidic acid (the *trans* isomer of erucic acid) were to be observed, this would indicate that elaidic acid would also originate from oleic acid. They observed that brassidic acid was only formed when behenic acid (the saturation product of erucic acid) was formed as well and concluded that monoenes cannot isomerize without some of them being saturated as well.

However, they performed their experiment at 100°C and thereby encouraged the saturation of monoenes. When the experiment was repeated at a much higher temperature and under selective conditions (W.L.J. Meeussen, personal communication), brassidic acid was formed without any behenic acid being formed. Similarly, the gadoleic acid (eicosenoic acid) present in the HEAR formed its *trans* isomer in the same ratio as the erucic acid did without being saturated to arachidonic acid. Accordingly, the conclusion that monoenes cannot isomerize without being hardened is an undue generalization and incorrect.

Hydrogenation Mechanism

Although cows and other ruminants are capable of hydrogenating triglyceride oils at body temperature, the industrial hydrogenation process employs elevated temperatures and also requires a catalyst to proceed. In industrial practice, the catalyst is invariably nickel. The nickel metal surface interacts with both reagents (the hydrogen and the oil or rather the double bonds in the fatty acid moieties) and causes the reagents to reach a state in which they can react with each other.

During the hydrogenation process, hydrogen is consumed. It therefore has to be supplied externally and dispersed in the oil by the use of a powerful agitator that preferably forces the gas that has collected in the autoclave headspace to re-enter the oil (Farr, 2001), or by using a venturi tube (Duveen & Leuteritz, 1982). This mechanical means constitutes one of driving forces for the dissolution of the gas. The other driving force is the difference between the hydrogen concentration in the oil and its solubility. Values for the latter have been reported by Wisniak (1974) and Andersson (1974).

Once the hydrogen has been dissolved, it can move towards the catalyst surface. As demonstrated by Koetsier (1997), the rate constant pertaining to the hydrogen transfer towards the catalyst particles is an order of magnitude larger than the volumetric liquid-side mass transfer coefficient $k_L a$ governing the rate of hydrogen dissolution. Accordingly, in industrial hydrogenations, "the effect of mass transfer rate from the bulk of the oil to the catalyst particles can therefore be neglected!"

As illustrated in Fig. 6.1, the molecular hydrogen can be adsorbed on the catalyst surface in reaction 1, which is reversible. The reverse reaction −1 therefore indicates the desorption and an asterisk (*) indicates that the species involved is adsorbed on the catalyst metal surface. Once adsorbed, the molecular hydrogen can dissociate (Reaction 2) into two adsorbed hydrogen atoms, which can then take part in further reactions with fatty acid moieties.

The concentration of the hydrogen atoms on the catalyst surface [H*] will depend on a number of factors such as the concentration [H_2] of the molecular hydrogen that is dissolved in the oil. If this is increased, Equilibrium 1 will cause the concentration of the adsorbed molecular hydrogen [H_2*] to increase and Equilibrium 2 will cause the concentration of the adsorbed atomic hydrogen [H*] to increase. Another factor affecting [H*] is the temperature since the equilibrium constants

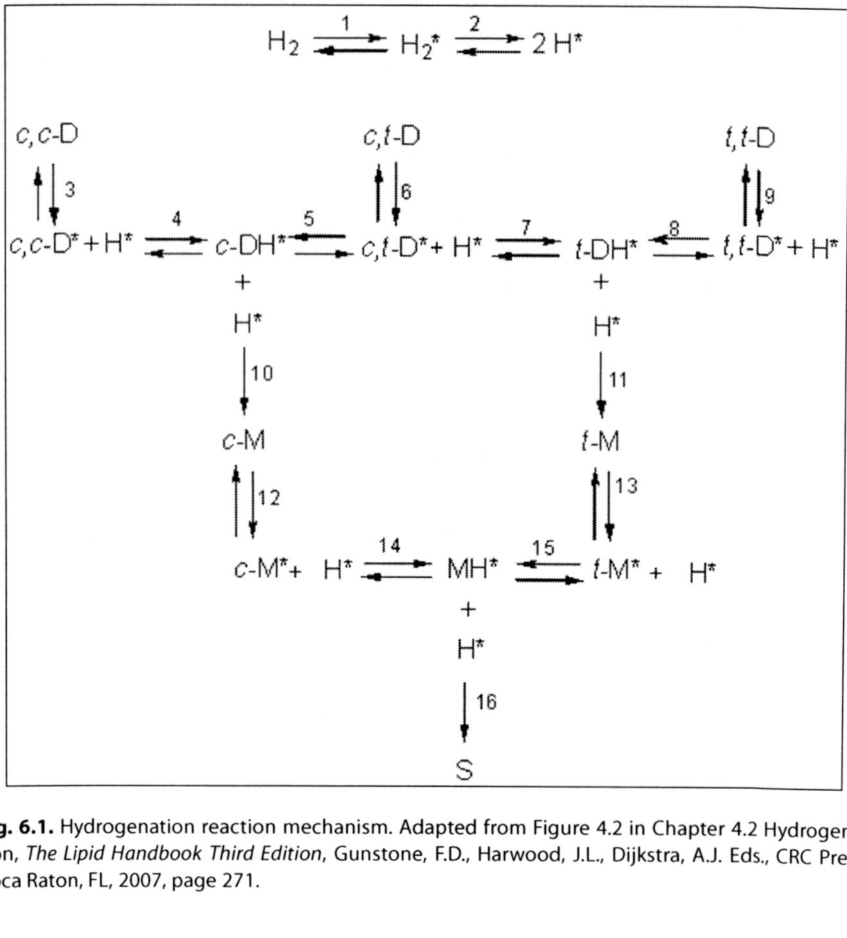

Fig. 6.1. Hydrogenation reaction mechanism. Adapted from Figure 4.2 in Chapter 4.2 Hydrogenation, *The Lipid Handbook Third Edition*, Gunstone, F.D., Harwood, J.L., Dijkstra, A.J. Eds., CRC Press, Boca Raton, FL, 2007, page 271.

of Equilibria 1 and 2 will be temperature-dependent. However, the major factor affecting [H*] will be the catalyst in that an 'active catalyst' will show a relatively high value of [H*] and that catalyst poisoning will cause the value of [H*] to decrease.

When triglyceride oils are hydrogenated, their molecules are first of all adsorbed onto the catalyst surface and this raises the question of how their fatty acid composition affects their affinity for the catalyst surface. If we denote a diene such as linoleic acid with D, a monoene such as oleic acid with M and a saturated fatty acid like stearic acid with S, it is to be expected that triglyceride S_2D will have a greater affinity than S_2M since a diene with two double bonds will have more affinity than a monoene. Similarly, SD_2 will have a greater affinity than S_2D but will it have an affinity that is twice as large because there are two dienes in SD_2? According to the 'common fatty acid pool' concept (Bailey & Fisher, 1946), this factor of two would be likely; but in practice, it may well be less than two (Schilling, 1968). Describing the hydrogenation reaction mathematically requires the affinities of each triglyceride to be known in quantitative terms and this is not (yet?) possible (see also Dijkstra, 2010). Accordingly, it will have to be described in qualitative terms and the description in Fig. 6.1 is therefore limited to what happens to a fatty acid.

In the top left hand corner of Fig. 6.1, the fatty acid *c,c*-D (*cis,cis*-linoleic acid) is denoted. In Reaction 3, it becomes adsorbed but since Reaction 3 is reversible, the fatty acid can desorb again and it is not unlikely that in an actual hydrogenation run, the value of [*c,c*-D*] reflects the equilibrium concentration. When the adsorbed linoleic acid (*c,c*-D*) reacts with an adsorbed hydrogen atom in Reaction 4, a half-hydrogenated intermediate (*c*-DH) is formed and since it is still adsorbed on the catalyst surface it has been denoted as *c*-DH*. In this notation, the asterisk indicates that this intermediate is adsorbed and the *c* that the intermediate has one double bond left that has the *cis* configuration.

This half-hydrogenated intermediate can do one of three things. It can dissociate and form the original linoleic acid *cc*-D* via Reaction -4. It can dissociate after rotation of the original double bond and form a *trans* isomer of linoleic acid *c,t*-D* via Reaction 5. This isomer is still adsorbed and can in its turn desorb via Reaction 6 to yield the isomer *c,t*-D which can be detected by analysing a sample of the oil. The third possible reaction of the half-hydrogenated intermediate *c*-DH* is with an adsorbed hydrogen atom to yield a partially saturated *cis* monoene *c*-M via Reaction 10. This monoene will desorb immediately because the addition of the second hydrogen to the half-hydrogenated intermediate will release the reaction heat of saturation so that the reaction product has a high kinetic energy. Reaction 10 is to be regarded as irreversible.

The adsorbed *trans* isomer of linoleic acid *c,t*-D* that was formed in Reaction 5 and that could desorb to yield the isomer *c,t*-D, could also react with an adsorbed hydrogen atom via Reaction 7 to give another half-hydrogenated product *t*-DH* which just like its *cis* isomer *c*-DH* can do one of three things: dissociate while going back via Reaction -7, dissociate after internal rotation to yield *t,t*-D via Reaction 8, or react with an adsorbed hydrogen atom via Reaction 11 to form a *trans* monoene *t*-M.

The monoenes *c*-M and *t*-M can both adsorb again via the reversible Reactions 12 and 13, respectively. The adsorbed monoenes can then react with an adsorbed hydrogen to yield a half-hydrogenated monoene MH* via the reversible Reaction 14 and 15, respectively; and finally, this half-hydrogenated monoene MH* can react with another adsorbed hydrogen atom to form a saturated fatty acid S via the irreversible Reaction 16.

So linoleic acid can form *cis,trans* and *trans,trans* isomers, these can form *cis* and *trans* oleic acid and positional isomers and these monoenes can form stearic acid. The pathways are clear but what has now to be discussed is the relative importance of the various paths and what factors affect their importance, in short: the kinetics of the system.

As is clear from the pathways, hydrogen plays an important role in the kinetics. This role is also illustrated by the observation that the linoleic acid selectivity decreases in the course of a hydrogenation run (Dijkstra, 1997) which means that (Dijkstra, 2007) the rates of reaction of linoleic acid and oleic acid do not depend in an identical way on the hydrogen concentration. So a study of this concentration is therefore needed. It is clear that the hydrogen concentration must be lower than its solubility to enable hydrogen to dissolve but the question is how much lower.

In this context, reference is made to a hydrogenation experiment with sunflower seed oil (Dijkstra, 1997) in which the rate of agitation was controlled in such a way as to ensure that all hydrogen that was supplied was also dissolved and reacted. It turned out that in the early stages of the hydrogenation run, the rate of agitation required hardly any adjustment to achieve a balance between supply and demand. Only when the iodine value of the reaction mixture had fallen by 60 units or at a residual linoleic acid content of close to 10% was it necessary to increase the rate of agitation.

The necessity of this increase in agitation implies that the other driving force for hydrogen dissolution had decreased and since the solubility had not changed, it follows that the concentration of the hydrogen had increased. It was also noted that after the increase in agitation, the linoleic acid selectivity decreased and that stearic acid started to be formed in appreciable amounts. Apparently, an increase in hydrogen concentration favors saturation of monoenes or in other words, the rate of saturation of monoenes depends more strongly on the hydrogen concentration than the saturation rate of polyenes.

What about the period of time before the rate of agitation had to be increased? During this period, the IV of the reaction mixture decreased from about 145 to 85, *i.e.* by a factor of almost two; and nevertheless, the rate of agitation hardly required any adjustment to balance hydrogen supply and demand. Because of the decrease in IV, the reactivity of the reaction mixture decreased considerably and the fact that it continued to react at a constant rate can only be explained by assuming that the concentration of the other reagent, hydrogen, had increased. After all, it was known that a decrease in hydrogen concentration decreases the rate of hydrogenation since a decrease in rate of agitation caused the pressure in the autoclave to go up.

All of the preceding information can be explained by assuming that the hydrogen concentration at the start of the experiment equals only a few percent of its solubility and that its exact value is the result of matching supply (rate of agitation) and demand (reactivity of the reaction mixture). When the reactivity of the mixture decreases, the hydrogen concentration compensates this decrease in reactivity by arriving at a new equilibrium value that is only slightly higher in absolute terms. In relative terms it may well be quite a bit higher to have any affect on the rate but since hardly any adjustment of the rate of agitation was needed, the difference between solubility and concentration must hardly have changed; this can only be explained by assuming the absolute level to be low.

So at the start of the experiment, when the linoleic acid concentration is high, the hydrogen concentration is low. It increases gradually when the linoleic acid concentration decreases but when this has fallen below some 15%, the rate of agitation has to be increased considerably to ensure that all hydrogen that is supplied also reacts. When this happens, stearic acid also starts to be formed. However, the hydrogen consumption only provides information about the saturation of double bonds; it does not say anything about their isomerisation.

One aspect of this isomerisation is the formation of *trans* isomers, which has been observed to be proportional to the decrease in iodine value. During hydrogenation, *cis* monoenes are formed because linoleic acid is being hardened but they also disappear since the *cis* monoenes isomerize to *trans* monoenes. Consequently, the concentration of the *cis* monoenes could be fairly constant for quite some time and thereby form *trans* monoenes at a fairly constant rate. At a certain stage in the hydrogenation, when the hydrogen concentration has to be increased and stearic acid starts being formed, the monoene concentration starts to fall and the rate of *trans* is no longer proportional to the decrease in IV. In fact, the *trans* content starts to fall and reaches zero when the oil is fully hydrogenated.

By combining the pathways with the kinetics, the following picture emerges. When a liquid oil such as sunflower seed oil is hardened, the only unsaturated fatty acids to start with are *cis,cis*-linoleic acid (*c*,c-D) and *cis*-oleic acid (*c*-M). They are both adsorbed onto the nickel catalyst's surface as demonstrated by the observation that the linoleic acid is hardened and the oleic acid is isomerized. It is not clear how much more likely a triglyceride with two unsaturated fatty acids is to adsorb than a triglyceride containing only one unsaturated fatty acid. According to the 'common fatty acid pool' concept, the likelihood should be twice as large but in fact it is less. From the available experimental data it is not clear either whether a diene is more likely to adsorb than a monoene but on theoretical grounds, there may be a slight preference for the diene.

Both the adsorbed diene and the adsorbed monoene can then react with an adsorbed hydrogen atom according to Reactions 4 and 14, respectively, yielding the half-hydrogenated intermediates *c*-DH* and MH*. These two intermediates differ in a way that is highly significant in the present context: *c*-DH* still has a double bond, whereas MH* does not. Accordingly, the linoleic acid adduct *c*-DH* has a much higher likelihood of remaining adsorbed than the oleic acid adduct. The latter lacks this residual double bond and as a result, it will be less firmly adsorbed to the nickel metal surface than the linoleic acid adduct. Consequently, the oleic acid adduct will leave the catalyst surface quite readily, either as the *cis* isomer via Reaction -14 or as the *trans* isomer via Reaction 15.

The previously described difference in ease of leaving the catalyst surface can also serve to explain why the saturation of monoenes depends more strongly on the hydrogen concentration than the reduction of dienes or polyenes in general. Because the monoene adduct MH* leaves the catalyst surface so readily and takes part in a dynamic equilibrium, its concentration can be assumed to be proportional to both the substrate concentration ([*c*-M] or [*t*-M]) and the hydrogen atom concentration [H*]. In German this is called a "vorgelagertes Gleichgewicht". Since saturation requires another hydrogen atom, the rate of saturation is likely to be proportional to $[H^*]^2$. In the case of the linoleic acid, the concentration of the half-hydrogenated intermediate can be assumed to be proportional to the concentration of the linoleic acid in the bulk of the oil and to the hydrogen atom concentration [H*]. However, since this adduct is not as likely to leave the catalyst surface as the half-hydrogenated monoene, it will have ample time to react with another hydrogen atom so that the concentration of the latter does not affect the rate of saturation. Accordingly, the rate of reduction of linoleic acid is proportional to [H*] whereas for oleic acid, this rate is proportional to $[H^*]^2$. Accordingly, stearic acid starts being produced when the hydrogen concentration in the oil increases.

Discussion

The partial hydrogenation of triglyceride oils leads to saturation and geometrical and positional isomerisation of the double bonds present in the starting material. Because of the current scare about *trans* isomers and labeling requirements, there is interest in reducing the *trans* isomer content of hydrogenation products (Dijkstra, 2006). Accordingly, interesterification of fully hydrogenated products with liquid oil is used to provide semi-solid fats that can be used in margarine and shortening. Palm oil or its fractions are also used as hardstock that does not contain *trans* isomers but processing these *trans*-free fat blends is not that straightforward (Gerstenberg, 2008) and the products tend to lack plasticity.

From the mechanism discussed above, it follows that avoiding the formation of *trans* isomers is at the expense of selectivity. Accordingly, a process that leads to a lower *trans* isomer content (Van Toor et al., 2005) by inducing a high hydrogen concentration by operating at a low temperature and a high pressure thus favors Reactions 10 and 16 in Fig. 6.1, which need hydrogen, while suppressing Reactions 5, 8, and 15, which liberate hydrogen and a *trans* isomer (Beers & Mangnus, 2004; Beers, 2007; Beers et al., 2008). In practice, producing partially hydrogenated oils with a reduced *trans* isomer content implies that these partially hydrogenated oils have a higher stearic acid content and run the risk of causing a sticky mouthfeel. This can of course be remedied by removing the high melting triglycerides by fractionation but it all adds to the cost of the final fat blend.

Chapter 7

Catalysts

H. B. W. Patterson

Necessary Characteristics for Heterogeneous Catalysts

Although, as we saw in Chapter 2, the hydrogenation of fats and oils was practiced on the work scale continuously and batchwise, with fixed-bed and with suspended catalysts, the batch process employing a suspended catalyst remains by far the most popular. This has certain implications. The catalyst must be separable from the hardened oil without particular difficulty; it must offer abundant, easily accessible surface to triglyceride molecules for contact with hydrogen adsorbed there and easy departure afterward from the surface into the main bulk of the oil; finally, if the catalyst is neither too sensitive to chemical attack (poisoning) nor mechanical degradation, this will be advantageous since this twofold durability permits repeated use, and hence, greater economy. Fortunately, all these separate requirements are met to a very useful extent within the same circumstances: that is, when the nickel is distributed throughout a porous material which at one and the same time provides particles large enough to allow easy filtration and a high proportion of pores wide enough for the easy transit of triglyceride molecules. Obviously, such a catalyst will provide even easier access for fatty-acid molecules, yet it may hinder the movement in and out of larger molecules bearing catalyst poisons.

Filterability

Although a cube of nickel weighing one gram has a superficial surface of less than one square centimeter, if the nickel were divided into particles of 50–100 Å in size, the surface area would then grow to 100 m^2/gram or more. However, particles of this small size would be quite impractical for filtration; hence, in the production process (see the "Production" section in this chapter), contrived is that the particles or crystallites of nickel, while still of 50–100 Å in size, are now distributed throughout a porous siliceous material which includes a good proportion of 10 μm (100,000 Å) particles. Now the total particle surface area can have a range of 200–600 m^2)/gram, of which 20–30% is catalytically active. Such supported catalysts filter rapidly, and if the usual precaution is taken (allowing a layer of a few millimeters to accumulate gradually on the filter membrane at the beginning of filtration on a clean filter), the so-called "black run" will probably cease in under 15 minutes; the remainder of filtration can then be completed at the normal rate (see the "Filtration" section in Chapter 4). The mechanical properties of this supported catalyst permit its repeated filtration on 20–40 subsequent occasions, depending on the type of use to which it is subjected. Catalysts which do not already have an inert support because they were directly produced by the thermal decomposition of an organic nickel salt, such as the formate, usually have one added by the catalyst manufacturer.

Activity

The activity of a catalyst expresses how much of the catalytic metal (Ni, Cu, Pd, Rh, etcetera) is needed to promote a fixed drop in iodine value (IV) in a fixed time in very precisely defined conditions of temperature, pressure, hydrogen dispersion (agitation), oil type, etcetera, in comparison to another catalyst, the latter frequently being accepted as the standard. To compare the dose of nickel in catalyst A which is required to drop the IV of neutralized, bleached pilchard oil from 185 to 155 with the dose of nickel in catalyst B required to drop the IV of neutralized, bleached herring oil from 135 to 105 would be useless. The exceptionally plentiful supply of double bonds in the first case helps, by a mass-action effect, to speed the progress of the hydrogenation. Equally misleading would be a comparison of one catalyst's ability to promote the hydrogenation of rapeseed oil with the performance of another in hydrogenating cottonseed oil, since the former is very likely to be richer in catalyst poison. These distinctions are immediately recognizable as valid because the differences are obvious.

Other distinctions need to be made, although the need for them may be less obvious. The further a particular hydrogenation is progressed, the more demanding the task becomes. If, therefore, we take equal weights of two different catalysts—equal, that is, in metal content, hence, equal doses—and find that one promotes a substantially greater drop in IV in otherwise identical conditions, we fail to give adequate credit to this more active catalyst if we relate its superiority merely to the greater IV drop, since the additional drop represents further performance at the most demanding part of the task—toward the end. Rather, we must discover the relative doses of either catalyst which give the same performance; the weights may then be compared: one catalyst may be the chosen standard; in our final comparative test, doses of each catalyst may be so chosen that the respective IV drops or refractive-index drops come quite close to one another, and in a region where a linear relationship exists between the dose and IV drop. We can then apply a correction (which will not be large) to the weight of the catalyst under testing actually used to calculate what the weight ought to have been to achieve precisely the drop of the standard catalyst; then the two doses may be numerically compared. Obviously, if to give the performance of standard catalyst A, only one-half as much of catalyst B is needed; the activity of B is said to be 200%. A closely comparable situation exists in comparing the relative power of two bleaching earths. If the task is to remove a moderate amount of pigment from an oil, earth A may achieve this with an obviously smaller dose than earth B. If we make heavier demands, the performance of A may easily begin to falter, so that for a target final color much lighter than in the first test, actually B achieves this more efficiently because of its greater staying power. Here two quite different results are admissible, entirely depending on how far we wish to take the change in color.

Activity depends on the area of accessible active surface per gram available to the molecules being hydrogenated. Although some doubt was cast on the hypothesis that a nickel catalyst requires relatively wide pores to carry out rapid and selective hydrogenations (e.g., highly active and selective catalysts can be obtained by reducing a nickel and sodium—silicate complex which will not have the pore structure expected in kieselguhr-based catalysts), this still remains the most generally accepted idea. The crucial importance of the qualification "accessible" is well-demonstrated by Coenen (1970, 1978), who first plots the activities of 10 catalysts against their respective total specific surface areas without finding any correlation. Molecular movement into and out of long or deep pores is less than for short ones; evidently, smaller particles will benefit by having fewer long pores. Also, the specific external surface area of the particle is greater when the particle is smaller (i.e., it is inversely proportional to the particular size). The activities of the same 10 catalysts were next plotted against the reciprocal of the average particle size. The particle sizes themselves were distributed over a range of 1–30 μm. Immediately, a distinctly improved correlation could be seen. Finally, knowing that the triglyceride molecules would find difficulty moving within pores less than about 25 Å in width, the surface areas of these 10 catalysts and numerous other examples were corrected by a deduction of that portion of the area due to pores less than 25 Å in width. This was possible from data obtained via nitrogen adsorption. When activity was plotted against corrected surface area divided by average particle size, an excellent correlation was obtained. This even reflected the very small amount of poisoning done by the pure sesame oil used in estimating activity and the falling off in the hydrogenation rate due to the disappearance of linoleate, hence, an apparent diminution of activity. From that point, a catalyst would be working under a comparative handicap, since only monoenes remained. Active nickel surface reached only by pores of 20 Å in width and less is held to contribute nothing to triglyceride hydrogenation, since oil molecules, probably greater than 15 Å in some dimension, are barely able to maneuver inside the pore (Coenen, 1970; Linsen, 1964). To the extent that, during the hydrogenation reaction, some proportion of the active nickel is being destroyed as the reaction progresses, mainly by the acquisition of poisons or an attack by free fatty acids, the final result reflects the specific active surface available and its resistance to attack. The nature of the attack depends on the oil being hardened and the gas used. Quite sensibly, a user of a catalyst could devise his/her own activity test, since this presumably will closely reflect the conditions in which the catalyst is to spend its life. Even so, remaining very desirable is that a catalyst standard which applies across a wide variety of triglyceride oils should exist. The actual material is best preserved as an inorganic compound, subject to little change when kept in sealed containers. A portion of such material may be reduced under standard conditions from time to time; then, as a fresh catalyst, it may be used as the basis of comparison for other catalysts, or it can equally be used to assess the hardening quality of an oil or of a hydrogen supply. The American Oil Chemists' Society (Allen, 1978) supplies a stock of standard catalysts, and one investigation (El Shattory et al., 1980) used this in evaluating four dry- and one wet-reduced catalysts regarding activity and selectivity.

Durability and Poisoning

Activity and durability are not entirely the same, but in practice, especially for supported catalysts, a close relationship exists. If a catalyst is very active, it can adsorb several poisons, and still give good results for a time simply because the level of poison in the oil is not high enough to overcome the plentiful amount of active nickel which the operator makes available in the dose. On the other hand, the construction of the catalyst may be such that at least some poisons present as relatively large molecules find it more difficult to gain access through the pores to the active sites within, and mainly the active nickel on the outer surface suffers. Calculated (Coenen, 1976) was that a milligram of sulfur present in a kilogram of oil poisons 13 m^2 of nickel surface. For solid (nonporous) particles of nickel of 5 μm in size (smaller would give filtration problems), the specific surface is about 0.15 m^2/gram. Therefore, to provide 13 m^2 of nickel surface and to adsorb the sulfur, some 87 grams of nickel would be needed for each 1000 grams of oil, or nearly 9%, a quite impractical dose. However, by spreading nickel crystallites throughout the honeycomb pores of a siliceous support, a nickel surface of 50–100 m^2/gram is achieved in a total surface ranging from about 200 to 600 m^2)/gram. A 10 μm particle of this description would contain about one million nickel crystallites. Because of the vastly increased nickel surface (100 m^2/gram, say, in place of 0.15 m^2/gram), the required weight of nickel per 1000 grams of oil falls from 87 grams to $87 \times 0.15/100 = 0.13$ grams. This gives a practical consumption of nickel as 0.13 kg/ton of oil, which would be normal for many classes of hardening. Similarly, a specific nickel surface of 50 m^2/gram would lead to a nickel consumption of 0.26 kg/ton. This example presumes that all the nickel in the pores as well as the exterior surface is accessible for poisoning by sulfur. The distribution of poisoning can have interesting effects beyond the obvious diminution of activity (Baltes, 1970). If the great majority of the poisoning is at the mouth of the pores and on the outside surface, most hydrogenation will therefore occur inside the pores. This will enhance the chance of more than one contact between an oil molecule and an active surface during the time—presumably now rather longer—that the oil molecule is in the immediate vicinity. Hence, a tendency will exist for a more fully hydrogenated species to desorb from the surface; less isomerization will exist, the change to saturated or nearly saturated groups will accelerate, with a consequential, more rapid elevation of melting point and flattering solid-fat index (SFI) or solid-fat content (SFC) curves. Such a catalyst would be described as having low selectivity. How severe these effects are will also depend on how high the proportion of medium pores is (25–30 Å in width) in which still some restriction of movement exists. When the poisoning is evenly distributed both inside and outside of the pores, we have a reduced chance of more than one contact between an oil molecule and catalyst inside a port during one visit, simply because the number of active sites there was reduced. Now the catalyst is selective, more isomers emerge, and SFC or SFI curves are steeper (Scarpiello, 1979). The contrast in degree of saturation between bulk oil and oil inside the pores at any movement is less marked.

Another situation is one where mainly the inside of the pores was rendered inaccessible to the oil molecules because something else was already adsorbed there, and therefore the easily accessible active sites on the exterior do most of the work. Selectivity improves, and isomerization is also more likely; the change in behavior will be most obvious in catalysts which previously had a considerable proportion of medium-to-narrow pores. Various organic molecules, when adsorbed, displayed this effect, and some patents were granted, but this was not seemingly exploited commercially; this may be because in use the effect diminishes and must be constantly refreshed. The exterior and most accessible nickel can be readily attacked by mineral and organic acids and various substances such as zinc stearate, but this effect is rarely, if ever, sought. Elemental sulfur remains apparently the most useful nickel poison where the aim is to produce even poisoning, and therefore a strongly iso- or *trans*-promoting catalyst. H_2S, CS_2, and SO_2 are reported (Swern, 1964) weight-for-weight of sulfur to be equally effective, but are taken up to an increasing extent in the order shown. When a fresh catalyst was made extremely *trans*-promoting by the addition of 2–3% of sulfur/nickel, its activity may have fallen to around one-tenth that to be expected in its fresh, unpoisoned state. This may be compensated for to some extent by increasing the dose fivefold and accepting a slower-than-normal rate of hydrogenation. The exact balance to be struck depends on how the operator views his program. A poisoned catalyst of this description is likely to permit many reuses. Such catalysts have come to be described in the 1990s as sulfur-promoted catalysts, and are now included in the varieties offered by certain manufacturers (see the "Examples of Commercial Nickel Catalysts" section in this chapter). The mechanism of the hydrogenation reaction using a sulfur-treated catalyst was investigated (Allen & Covey, 1980), with the result that two isomerizing reactions appeared to be proceeding simultaneously, one due to the nickel–sulfur complex and one due to hydrogenation via the remaining active nickel.

The potency of the poisoning action of phosphorous compounds and phospholipids in particular has attracted attention (Ottesen & Jensen, 1980). From a comparison of the hydrogenation of fully refined and deodorized soybean oil and the same oil to which various doses of stearyl palmitoyl lecithin had been added, the calculation was that the equivalent of a nickel dose of 0.0008% of nickel/oil was totally inactivated by 1 ppm of phosphorus (approximately 25 ppm of lecithin). In particular, this class of poison inactivated the exterior and pore mouths of the supported catalyst, which therefore also greatly reduced selectivity, leading to a more rapid elevation of melting point and flatter SFC or SFI curves. Neutralization and adsorptive bleaching prior to hydrogenation naturally reduced the phosphatide level and its ill effects. Additional active catalyst also helped to restore the situation, and in this respect, while keeping the hydrogenation time constant, as further assessed, an additional 0.01% of nickel/oil compensated for 1 ppm of phosphorus (25 ppm of lecithin) (i.e., virtually 12 times as much nickel was required to bring results back to normal as was totally inactivated by the 1 ppm of phosphorus, or, as stated by the investigators, the compensating factor was in this instance 12 times higher than the poisoning factor). The same author (Ottesson, 1975), as mentioned in the "Catalyst Induction, Fatigue, and Poisoning" section in Chapter 1, had assessed the potency of other common catalyst poisons by describing what dose of nickel was inactivated by 1 ppm of the poison: for phosphorus (as lecithin), 0.0008%; for nitrogen (as amino acids), 0.0016%; for sulfur, 0.004%; and for oxy-polymers (as occur, for example, in fish oils), 0.0047%.

Unsupported catalysts (wet-reduced ex-nickel formate) and supported catalysts of different origins (siliceous support) were compared in some detail, regarding their induction period and subsequent hardening rate when contending with sulfur, phosphorus, and soap (Drozdowski & Zajac, 1977) (see also the "Catalyst Induction, Fatigue, and Poisoning" section in Chapter 1). The comparison rightly included the performance of each catalyst against unpoisoned oil and the same oil with different levels of poison. The unsupported catalyst faltered under attack; the best-supported catalyst stood up well. The same investigators (Drozdowski & Gonoj-Moszora, 1980) later explored the potential value of resuscitating partially inactivated catalyst by adding different complete doses of an activated bleaching earth at the beginning of hydrogenation or at differing intervals (15, 30, and 60 minutes) thereafter. The activity of the selected catalysts (unsupported exformate and supported Nysel DM3) was first reduced 50% by the necessary addition of allyl isothiocyanate, phospholipid, or sodium soap. Then, hardening at 170°C and with atmospheric pressure and including one trial hardening with a nil addition of earth as the control, 1.5% of the Czech activated earth used had the optimal effect. This was most noticeable for soap—not unexpectedly—and least so for the sulfur-bearing poison. Results showed that the earth had reacted not only with the poisons in the oil, but also with the poisons on the catalyst, especially if the latter was of the unsupported type. When it did have an effect, earth was best added simultaneously with the catalyst at the beginning. Unfortunately, the improvement when sulfur (allyl isothiocyanate) was the poison was negligible. The benefit to the hydrogenation rate achieved by purifying the feedstock was well-illustrated (Coenen, 1975) in the case of fatty-acid hardening by comparing rates achieved when working on a feedstock which had been (a) previously cleaned, (b) previously split, or (c) previously split and distilled. Not only was the rate vastly improved, but a much lower final IV became readily obtainable.

Stabilization or Passivation

When supported catalyst is dry-reduced in the roaster, it emerges as a pyrophoric powder which must be maintained in an atmosphere of hydrogen until it is dropped into a protective medium, such as melted hardened fat. If, however, the pyrophoric powder falls first into an inert atmosphere and is then gradually exposed to superficial oxidation by the admission of a current of air, the pyrophoric quality is lost and the nonpyrophoric catalyst is then in a passive or stabilized condition. The temperature must be restrained via a water-cooled jacket or the like during the exposure to air; meanwhile, the powder may be gently agitated. When oxygen is no longer taken up from the air, stabilization is complete. This could be regarded as a special case of reversible poisoning. In this form, the catalyst has the advantage that it does not carry with it any organic medium, such as a hardened oil which might be unacceptable in the liquid in which it is to be employed. The superficial oxidation may cause some induction period at the beginning of a subsequent hydrogenation, due presumably to the time taken to penetrate the oxidized layer and possibly for the reduction of the latter back to a catalytically active condition. This need to penetrate or reduce the oxidized layer also renders this type of catalyst generally unsuitable for low-temperature hydrogenations. A catalyst of this type can be obtained with a variety of apparent bulk densities and specific surface areas. It has a variety of applications as a pelleted catalyst, and its use is not confined to the hardening of fats and oils (Harshaw Catalysts, Hoechst Aktiengesellschaft).

Selectivity

One can recognize at least three basic consideration factors as governing what is now understood by selectivity in fat hardening: First, are the questions: Is the accessibility of the active surface to the molecules taking part? How easily and how quickly can they reach it? and How quickly can they escape? For smaller molecules, such as the methyl esters of the fatty acids rather than the triglycerides, movement restrictions imposed by pore widths are much less, so that selectivity can continue to operate easily in their case, when with triglycerides it would be much weaker. This aspect is discussed at length in the "Operation of Selectivity" section in Chapter 1. The second basic consideration is related to the alignment of the unsaturated hydrocarbon chains of the fatty acids concerned when the triglyceride molecule is in direct contact with the array of nickel, copper, palladium, or other atoms at active sites scattered over the catalyst surface, whether inside pores or on the exterior. How effectively they interact governs whether hydrogenation takes place at all and what the chances are that a geometric or positional isomer will desorb from the active surface after this contact. The "Isomerization" section in Chapter 1 considered these possibilities. The third consideration factor is the abundance or scarcity of hydrogen adsorbed at the active sites, and therefore available to be transferred to the opened olefinic links also present there between catalyst and fat. As described in the "Operation of Selectivity" section in Chapter 1, the plentiful supply of hydrogen is what discourages merely partial hydrogenation of the most unsaturated chains, and what encourages a higher proportion of fully hydrogenated chains or even triglycerides to result at an earlier stage in the hardening than would otherwise be the case.

The spectacular rise of soybean oil in the 1960s to become the world leader in terms of annual production among oils and fats stimulated intense efforts to discover a reliable means to reduce its 8% of linolenate, which is typical, to less than 2%, while conserving as much as possible of the 48–58% of linoleate. At the same time, no increase in the amount of saturated fatty acids already present was desirable (mostly palmitic and stearic), and a minimal formation of *trans* isomers was another requirement. The depression of the linolenate assisted flavor stability; the avoidance of even low-melting solid components enhanced the yield of stable liquid able to pass a cold test after the partially hydrogenated oil was fractionated. The United States Department of Agriculture's Northern Regional Research Center in Peoria, Illinois was outstanding in this field, which follows from the fact that the United States easily led the world in increased production. Now other countries are increasing theirs: the advantages to be gained by selective hydrogenation are therefore of widespread interest, and have added considerable momentum to the exploration of other catalysts besides the conventional one of nickel (Mounts, 1981). Numerous contributions to this exploration were briefly reviewed by Gray and Russell (1979). As far as conventional nickel catalysts are concerned, as early as 1964, Evans et al. (1964) hydrogenated soybean oil with 0.2% of nickel/oil and an agitation of 1380 rpm to below 2% of linolenate, using first 170°C and 0.3 atm, and in a second trial, 120°C and 7.0 atm. The selectively hydrogenated oil of the first trial showed no increase in stearate, but understandably, *trans*-isomer content increased as compared with the second. Work continued with conventional nickel catalysts, and sufficient selectivity was obtained to establish the winterization of partially hydrogenated soybean oil as a successful industrial process (*Handbook of Soy Oil Processing and Utilization, 1980*) (see also Chapter 8). The requirement for selectivity, especially S_{II} or linolenate selectivity, has been a major item, as well as activity, when researching the performance both of nickel catalysts made by less conventional means and of other catalytic metals, including the noble metals, and even homogeneous catalysts. Although one can fairly claim that results have increased our understanding of catalytic action in hydrogenating and isomerizing oils, commercial exploitation has not followed, for reasons touched on in the following sections.

Raney Nickel and Other Nickel Catalysts

When a powdered 50/50 nickel–aluminum alloy is added to a caustic-soda solution, the aluminum reacts to form a solution of sodium aluminate, leaving behind a spongy pyrophoric nickel. After thorough washing to remove all alkali, blanketing with an inert liquid, and vacuum drying, the catalyst is ready. Its use is not confined to the hydrogenation of oils and fats. Its ease of preparation has contributed to its popularity, this being more evident in the laboratory rather than on the industrial scale. No particular advantage regarding activity or selectivity in fat hardening is attached to it.

The reduction of nickel salts in solution by alkali-metal borohydrides was investigated (Brown & Brown, 1963; Buisson & Joseph, 1951; Dutton & Koritala, 1966), and found to yield an active catalyst; this also proved to be true of cobalt, palladium, and platinum, but not iron or silver. The activity of nickel in these circumstances was enhanced by the presence of up to 2% of Pd, Cu, Cr, and Pt; Cu and Cr salts when reduced together gave catalysts with an S_{II} of about 7 (i.e., linolenyl groups hydrogenated seven times quicker than linoleyl) and no increase in saturates. No appreciable commercial exploitation of these catalysts appears to have followed, and the same may be said of numerous trials by Russian and East European researchers to exploit the potential of Ni–Cu catalysts (Gray & Russell, 1979), which was even remarked on by Normann (1902 & 1903) and Hilditch & Jones (1932). Mukherjee and co-workers (1975) prepared a variety of Raney-style catalysts from Ni–Cu, Cu–Al, Pd–Al, and Cu–Cr–Al alloys; pelleted $CuCrO_3$; granular Raney Cu–Cr; and pelleted Pd. Widely differing activities and selectivities were noted in the fixed-bed continuous flow system used, but again this has not become popular on the industrial scale.

Copper Catalysts

From the early 1960s, the selective action of a copper catalyst in hydrogenation was vigorously followed up by various groups of workers. The Japanese produced a stable liquid hydrogenated fish oil, the procedure being more practical with marine oils of medium unsaturation such as whale, herring, and a local flatfish oil. Great interest was soon taken in finding a copper catalyst which would offer a practical means of doing what was just described as the target for nickel, that is, of bringing the linolenate in soybean oil below about 2% while leaving a maximum of linoleate and forming no additional saturates and the minimum of *trans* isomers (Unilever, 1963). Okkerse et al. (1967) showed how with copper catalyst, when the linolenate was reduced to 2%, some 49% of linoleate remained, but with nickel only 28% remained. Thus, in hydrogenating to 115 IV with Cu and leaving only 1% of linolenate, this leads, after winterizing, to a liquid oil yield of 86% which would pass a cold test of 18 hours at 5°C without separating solid. Cowan et al. (1970) worked on the same topic, reporting in terms of flavor stability as the first preference for cottonseed oil, then copper-reduced soybean oil, and lastly nickel-reduced soybean oil. These efforts even extended from the more conventional Cu and Ni catalysts to Cu on molecular sieves, co-precipitated catalysts, and supported catalysts, one of the best in terms of activity and selectivity being Cu precipitated on a pure type of silica.

Next to come to public attention was a mixed Ni and Cu chromite which had good selectivity, giving 20% of dienoic and 66% of monoenoic groups; also, citric acid proved useful in removing Cu from hydrogenated oil (Popescu et al., 1968). From the detailed investigations following rapidly after one another in the United States (Koritala & Dutton, 1969), came increasingly strong support for the view that the conjugation of double bonds played a necessary part in the action of copper catalysts. Isolated double bonds would not hydrogenate; copper chromite would only reduce double bonds of monoenes if they were adjacent to the carboxyl group; conjugated double bonds reacted faster than methylene-interrupted unsaturation (skipped double bonds—see the "Fatty Acid Chain Length and Unsaturation" section in Chapter 1). Moulton et al. (1969), to hydrogenate soybean oil below 1% of linolenate without increasing saturates, chose 1% of copper–chromite oil, working at 2 atm of hydrogen pressure, stirring at 1400 rpm, and 170°C (Moulton et al., 1969). Also found feasible was to hydrogenate soybean oil with copper chromite at 0.5–0.1%, 155°C, and 2 atm of pressure, but with linolenate selectivities of only about 5. The presence of 1–20 parts of Ni per thousand Cu appears to decrease slightly the linolenate selectivity as compared with unsaturated copper chromite (Moulton et al., 1973). Johansson and Lundin (1979) describe in detail the conditions they found most favorable for the selective hydrogenation of soybean and rapeseed oils with copper chromite. A flow of hydrogen through the headspace removed water formed, maintained hydrogen partial pressure, and assisted agitation. Coenen (1976) remarks that copper catalysts are in general less active than nickel and more sensitive to poisoning, which lowers their selectivity: since reuse is not feasible, consumption is five to ten times greater than nickel. Koritala (1981) reviews progress in this field, pointing to the established high S_{II} of Cu (de Jonge et al., 1965; Koritala & Dutton, 1966), but its lower activity compared with commercial Ni catalysts (Koritala, 1970). Koritala adds that when Cu is supported on silica of high specific surface, an enhanced activity is obtained. However, a variety of poisons affects Cu on silica catalysts, although copper soaps and carbon monoxide are not among them (Koritala, 1975).

For these reasons, plus the facts that copper catalysts must be isolated from others in the plant and that stringent control must be exercised to remove copper from the finished product, copper has made little headway against nickel in commercial hydrogenation. Another continuing factor is that linolenate-containing oils—which have had to be hydrogenated with a copper catalyst—face competition in price from unhydrogenated vegetable oils whose unsaturation is present only as dienes and monoenes (Patterson, 1974).

Noble Metal and Other Catalysts

As well as the better-known catalytically active metals, Johnson (1972) classified the noble metals and some others which exhibit some degree of catalytic activity, according to how their electron *d* orbital is filled and positioned (Table 7.1).

The expected easier bond formation exists between olefinic links and class III metals rather than for class II; hence, the greater are the exchange and isomerization when using class III metal catalysts and their dominance in the field of industrial hydrogenation. Obviously, considerations of activity, expense, and the need for conservation in handling bear strongly on the choice for full-scale use. Rylander (1970) discussed the industrial application of noble metal catalysts, and concluded that this was curtailed by economic factors. Apart from Pt, the group contains Ru, Rh, Pd, Os, and Ir; their linolenate (S_{II}) selectivity is undistinguished. Although palladium is 15–100 times more active than nickel, it is much more costly; because of the small doses likely to be used, special arrangements would be needed in full-scale plants to control loss.

Unilever Research (Van der Planck et al., 1980) showed in the laboratory that a resin impregnated with $PdCl_2$ can promote the extremely selective hydrogenation of a methylene-interrupted diene, partly by an immediate reduction of one of the double bonds and partly by first forming a conjugated system; geometric but not

TABLE 7.1
Catalytic Metals

Class	Metal							
I	Mo	W						
II	Rh	Ir	Ru	Os	Ti	Re		
III	Fe	Co	Ni	Pd	Pt	Cu	Ag	Au
IV	Zn	Ga	Cd	In	Ge	Sn	Pb	

positional isomerization was very noticeable. The rate of reaction was slower when double bonds were interrupted by -CH_2-CH- rather than just -CH-.

Various palladium metals on carbon support catalysts were extensively tested in hydrogenating soybean oil (Ahmed et al., 1979), the results confirming that the method of preparation and the conditions of use are vitally important if a palladium catalyst is to operate to its greatest advantage and allow a fair comparison with nickel. The results have an interesting confirmation of current beliefs concerning the importance of accessibility of the active surface to a triglyceride molecule in establishing selectivity (see the "Selectivity" and "Operation of Selectivity" sections in Chapter 1). Polymer-bound hydrogenation catalysts (Frankel et al., 1980) are quoted as being interesting because potentially they may provide the desirable selectivity noted for some homogeneous catalysts, along with the ease of separation of heterogeneous catalysts, and to this end, the performance of Pd, Rh, and Ni polymer-bound catalysts was compared when hydrogenating soybean-oil esters. Similarly, homogeneous catalysts were investigated because they are described (Van der Plank et al., 1980), as also suggested above, as being of superior selectivity to heterogeneous catalysts in certain circumstances and also as valuable tools in establishing the mechanism of hydrogenation. Rhodium complexes were investigated in this connection. The use of rhodium coordination catalysts in achieving selective hydrogenation of soybean oil with the minimum of *trans* isomers (Frankel, 1977) was also pursued, since this effect is seen as possibly valuable to health. The description given above of a little of the work done in the field of catalysis by metals other than Ni and Cu, by catalysts deposited on other than siliceous supports, and by homogeneous catalysts, gives only a glimpse of this large area, for which separate detailed accounts are available elsewhere (Frankel, 1977). Research continues in this field either to discover a catalyst of high economic value or to learn more about hardening, or both.

Production

Nickel sulfate is probably the most popular starting point in the manufacture of a nickel catalyst. When one feels the importance of avoiding even a small amount of sulfur in the final product, such as 0.3% of sulfur/nickel, nickel nitrate or chloride is an alternative.

Dry Reduction

The solutions of the nickel salt and an alkali such as sodium carbonate, bicarbonate, or hydroxide are brought together just below the boiling point and with thorough agitation to maintain uniformity throughout the mixing vessel. Kieselguhr is present as a source of silica. A portion of the guhr represents a soluble form of silica, and this plays an important role in helping to form the complex basic nickel silicate which is precipitated. A good catalyst with high specific surface can be made by substituting a sodium–silicate solution for the kieselguhr. In any case, if deliveries of kieselguhr are of variable quality, a worthwhile action is to establish what proportion of guhr-to-nickel sulfate is most effective by preliminary laboratory-scale precipitations.

Batch and continuous co-precipitation are feasible, maintaining a weekly alkaline environment. Next, the hot suspension is filtered, and the filter cake is washed with hot water to remove sulfates as far as possible. Even when the filtrate shows a negative -SO_4 ion test, a little insoluble complex sulfate may be retained by the cake. This may be lessened by including in the washing sequence a mildly alkaline stage before reverting to water alone. The complex basic nickel–silicate precipitate, or so-called green base, almost certainly contains hydrated nickel oxide or carbonate, which also has an important part to play subsequently in obtaining an active catalyst. In fact, Dafler (1977) makes the point that the eventual crystallite size of reduced nickel is probably predetermined by the size of the nickel–oxide clusters in the unreduced green base, and that one of the important roles of an inert support is to limit very stringently the degree of sintering of metallic nickel as it arises during a reduction in the roaster, since this would lead to the loss of nickel surface, and hence, activity. In support of this contention, he quotes an earlier observation by Koestler and Meisel (1966) that in the absence of a support, crystallite growth is extensive, due presumably to sintering. Hence, too little silicate may lead to a high proportion of reduced nickel metal, which then sinters and loses surface area, or too much silicate diminishes the proportion of nickel which is reducible and therefore the yield of surface area is poor, although it may not be exposed to further loss by appreciable sintering. Evidently, the optimum lies between. When washing is complete, the filter cake is likely to contain c. 60% of moisture, and must be dried evenly to under 5% of moisture so that baking to coarse nodules is avoided (Patterson, 1974). Reduction is performed by passing the dried powder on a continuous basis through a roaster where, for a large part of its residence time, it is maintained at 430–500°C while hydrogen flows in the opposite direction. Perhaps 60% of the nickel present is reduced to metal, but that which was is pyrophoric, and carries occluded hydrogen with it. The black powder is therefore dropped into a protective soft or hardened oil while still in the atmosphere of hydrogen, or it may be led continuously into an inert gas and later rendered passive (see the "Stabilization or Passivation" section in this chapter). Although once a common practice was to collect the pyrophoric catalyst in an unhardened oil and retail it as a sludge or soft cake packed in drums, now the common practice is to take the suspension of nickel collected in hardened oil of around 50°C mp and convert it to solidified droplets by a process involving chilling and a passage through a pastillator. These droplets have come to be preferred over the flake form, since appreciably more catalyst can be packed into a drum of the same size (thus, reducing freight and storage costs) and the tendency to cause dust problems virtually disappears.

Of course, the drying of the green base can be performed in a fluid-bed dryer, and the next step of reduction to active nickel can also be achieved by the fluid-bed technique, but the extent to which this procedure is used is kept confidential by manufacturers.

The strong modern trend is to purchase a catalyst from a supplier who specializes in such production and is able to offer a range of supported nickel catalysts tailored to the customers' requirements. These catalysts are very likely to be dry-reduced, and the option of having zirconia or alumina or a proprietary material included as a support may be offered. Outside of the field of the hydrogenation of fats and oils, the use of alternatives to silica as a support is commonplace.

Wet Reduction

Although other organic salts of nickel, such as the oxalate, decompose on heating, the formate is the one which is used in the preparation of a catalyst. Having a low solubility in cold water, nickel formate, $Ni(OOC\text{-}H)_2\text{-}2H_2O$, is easily thrown out of the solution by adding sodium formate to a strong solution of nickel sulfate. When filtered, the crystals are washed with the minimum of cold water to remove sulfate, and are then dried. Take care to avoid the loss of any nickel in filtrate liquors, etcetera. Sometimes the formate is made by the direct addition of formic acid to nickel hydroxide or carbonate.

The plant for the decomposition of nickel formate is uncomplicated, and may be purchased as a package if desired. Basically, it consists of a closed vessel fitted with a generous vent pipe and equipped with a stirrer, heater, and temperature recorder as well as oil feed, a hatch for the addition of crystals, and a bottom outlet for the withdrawal of the final slurry. Nickel–formate crystals are stirred into about twice their own weight of oil in the reaction vessel. A partly-hardened oil is suitable. The temperature is raised steadily while stirring continues. The water of crystallization is driven off at 180°C, which causes a temporary flattening of the ascending curve on the temperature recorder. This step must not be unduly hurried; it may last for an hour, and then the temperature shows a tendency to rise.

Additional heat is now supplied until about 245°C, and then reduced so a steady 250°C results: a temperature above 255°C damages the catalyst. The decomposition of the nickel–formate is proceeding rapidly by now, the total change being shown as:

$$Ni\ (OOC\text{-}H)_2\text{-}2\ H_2O = Ni + 2\ CO_2 + 2\ H_2O + H_2 \qquad \text{[Eq. 7.1]}$$

At this stage, a current of hydrogen passed through the vessel helps sweep away the gaseous products of decomposition. Vacuum was used for this purpose; precautions to cope with the priming of the gaseous liquor up the vent pipe are advisable. Within two hours, the reaction is complete. If the reaction is violent or rapid, a higher proportion of submicron nickel particles may result which, although of high activity, may be less durable and certainly will impose extra strain on the steps taken to promote the subsequent ease of filtration during use, such as the addition of kieselguhr when the mixture is cooled to 90°C. The vessel is then emptied. The suspension can be used as it stands, probably in that case by the hardening plant in which it was produced on the small batch scale: it may be filtered and the cake marketed; the cake may be reconstituted with some fresh hardened oil to displace that used as a reaction medium. A wet-reduced catalyst with an added filter aid has been available for purchase, although this method of preparation apparently was particularly popular in central and eastern Europe, where at one time an emphasis was on hardeners, each producing catalysts on a small scale currently with their hardening programs.

Storage

Pelletized fresh catalyst is supplied by manufacturers in lined metal drums. If a container of fresh catalyst was opened, it should not be left standing in that state after the required amount of catalyst was taken, but closed again if feasible. Likely little change will occur in the activity of a fresh catalyst in an unopened container for up to one year, and possibly double or more that length of time. Black pellets of fresh catalyst may develop a grey appearance, owing to the migration of fat to the surface rather than oxidation, but a good policy is not to leave fresh catalysts exposed to the air unnecessarily.

Superficial oxidation may have occurred with some loss of activity. Pelletized catalyst is easy to dispense in the plant and can readily be incorporated in warm oil. This is true whether the active nickel is produced by a wet or dry reduction. During repeated use in the plant, a catalyst filter cake has to be cleaned from the filter into soft oil, kept warm there, and stirred while portions are taken at intervals to harden subsequent batches. Desirable is that a slurry held in these conditions should all be used within two to three days, as we now have circumstances where not only is the nickel being mildly exposed to oxidation, but also, more so, the oil. One can almost certainly arrange that the filter is cleaned as soon as it contains the amount of nickel which the production plant will take back into circulation during this limited period. When finally discarded from use, exhausted catalyst should be repacked into closed containers, since if left in a draught or the hot sun, it may eventually commence to smolder and then burn, as will residues from the filter of a bleaching-earth process. If the catalyst was purchased originally in drums, these will provide the needed repacking containers.

Recovery

Exhausted or spent catalyst contains about 55% of hardened fat, depending on the conditions of the final filtration. If circumstances allow, this fat is best removed by solvent extraction before one attempts to dissolve the nickel. The value of the fat thus recovered may cover the cost of its extraction or yield a little profit, depending on the local-market demand. If solvent extraction is not feasible, other answers were to burn the fat, to saponify it, or simply to separate the melted fatty layer after thorough boiling with diluted sulfuric acid. The nickel-containing portion is further boiled with dilute sulfuric acid, probably for several hours, and also in the presence of an oxidizing agent which not only assists the attack of the acid on the nickel, but converts the iron present to the ferric state. By these means, only about 3% of the nickel fails to be converted to soluble nickel sulfate. The acid solution is then rendered only feebly acid by the addition of a cheap alkali, and is filtered. The more cautious addition of alkali to the filtrate achieves a pH at which hydrated ferric oxide is thrown out of the solution, but nickel hydroxide is not formed. A final filtration gets rid of the iron which would have a bad effect on the quality of any catalyst made from the recovered nickel.

Most hardeners do not recover nickel from their exhausted catalyst, and are interested to discover if returning it to the original supplier for the credit then allowed is worthwhile, or if the local market provides a better outlet. For example, another manufacturer may have the facilities for extracting the fat, and a second may wish to dissolve the nickel from the defatted residue with a view to using it in, say, the nickel-plating industry. In general, the less fat and the less non-nickel solids which are associated with the nickel, the better price it is likely to command when the exhausted catalyst comes to be sold. To obtain an idea as to the cost of the recovery of 1 kg of nickel, the difference between the cost of 1 kg of nickel as nickel–sulfate crystals and the credit allowed for 1 kg of nickel as exhausted catalyst should be noted (Patterson, 1974).

Examples of Commercial Nickel Catalysts

The catalysts listed below are supplied for the hydrogenation of fats, oils, and in some instances are offered especially for fatty-acid hydrogenation. The list does not include catalysts intended for use in hydrogen manufacture, other gaseous processes, or the general field of organic chemicals. Special features advised by the manufacturer are noted, and these will be seen as relevant to particular hydrogenation methods when those catalysts are mentioned at various points throughout Chapter 8.

Engelhard Catalyst Suppliers:

- Engelhard Corporation, Houston, Texas
- Engelhard Corporation, Iselin, New Jersey
- Engelhard Corporation, Elyria, Ohio
- Engelhard De Meern BV, De Meern, The Netherlands
- Engelhard Sales, Ltd., Sutton, Surrey, England

Available Engelhard Catalysts

Catalyst—Nysosel 325

This is a 22% of nickel catalyst on an alumina support. It is highly active, very porous, and shows marked linoleate selectivity. This catalyst performs at its best on vegetable oils with low levels of catalyst poison.

Catalysts—Nysosel 222; Nyosel 111

Nysosel 222 is a 22% of nickel catalyst on a silica support with good all-around performance; a large specific nickel surface confers good poison resistance; a favorite chosen for use on marine, maize, and rapeseed oils. Nysosel 111 is an alternative catalyst containing 19% of nickel on silica, and used where hydrogen diffusion is less efficient.

Catalyst—Nysel Sp-7

This 18% of a nickel catalyst on a silica/alumina support was treated with sulfur to promote high *trans*-isomer contents, and hence, steep melting-curve fats.

Catalysts—Nysel Sp-10; Nysel DM3; Nysel DM3-22
Nysel Sp-10 is 21% of a nickel sulfur-promoted catalyst on an alumina support, rather more active and quicker filtering than Nysel Sp-7, but equally *trans*-promoting.

The long-established Nysel DM3 (25% of nickel on silica/alumina) and Nysel DM3-22 (22% of nickel on silica/alumina) for general-purpose fats-and-oils hardening are still available.

Catalysts—Nysofact 101-1Q; Nysofact 120; Nysofact 99
Nysofact 101-1Q is 22% of a nickel catalyst, silica-supported and designed to give a rugged, highly active performance when hydrogenating fatty acids under mild conditions.

Nysofact 120 is another 22% of a nickel catalyst, also silica-supported, highly poison-resistant, and specially designed to resist deactivation by the formation of nickel soaps during hydrogenation under severe conditions. Nysofact 99, 20% of nickel on silica, is, in fact, an alternative to Nysofact 101-1Q for optimal performance when conditions of hydrogen diffusion are less effective.

Other Specialized Engelhard Catalysts

Besides those listed above, Engelhard provides, on request, a variety of catalysts whose application extends either into very specialized minor applications of oil hydrogenation or beyond it into general industrial organic chemistry.

Catalysts—Ni5256 P; Tall Cat; Cu-1985 P
Ni5256 P is a nonpyrophoric nickel catalyst devoid of fat. Tall Cat (22% of nickel on silica) is suitable for hydrogenation or monomer acids derived from tall oil. Cu-1985 P is a Cu/Cr catalyst recommended for the selective hydrogenation of more highly unsaturated fatty-acid radicals where this is economically feasible. For those who need them, platinum and palladium catalysts are also available.

Hoechst AG, D6230 Frankfurt am Main 80, West Germany—A Catalyst Supplier

Catalyst—Hoecat 882 OF
This is 22% of a nickel dry, reduced catalyst on 13% of kieselguhr support supplied as pellets containing approximately 63% of fully hydrogenated fat. The catalyst is suitable for the partial or complete hydrogenation of a wide range of fats and oils. Its durability permits several reuses.

Catalyst—Hoecat 882 FA
This is, again, a dry, reduced 22% of nickel catalyst on a 13% of kieselguhr support whose pellets contain approximately 63% of fully hydrogenated fat. It has good resistance to poisons found in crude feedstock, and a mechanical strength allowing it to endure the most rigorous stirring and pumping without an appreciable effect on its filterability. This catalyst is recommended for the complete or partial hydrogenation of free fatty acid.

Hoechst has long-time experience in producing supported nickel, cobalt, and copper catalysts for a wide variety of industrial processes.

Other Catalyst Suppliers

- Mallinckrodt Speciality Chemicals Europe GmbH, D5202 Hennef/Sieg 1, Germany
- D5202 Hennef/Sieg 1, Germany
- Calsicat Catalysts, Erie, Pennsylvania

Catalyst—E472D
This is a dry, reduced 25% of nickel catalyst on 12% of proprietary support held in 63% of hardened soybean droplets. It is particularly recommended for the selective partial hydrogenation of vegetable oils such as soybean, maize, canola, rapeseed, palm, and others.

Catalyst—E428D
This is a dry, reduced 22% of nickel catalyst on a 23% of kieselguhr support held in 55% of hardened soybean droplets. It has a very wide range of applications covering the hydrogenation of lauric, marine, and castor oils as well as greases and fatty acids of different origins.

Catalysts—E230P; E480P
E230P is a 60% of nickel catalyst on an alumina support. It has a 50% degree of dry reduction, and is supplied as a stabilized (fat-free) powder. This catalyst rapidly achieves the partial hydrogenation of soybean oil, as also does E480P, which is a 65% of nickel on a proprietary support.

Catalysts—Cu/Cr; Pt and Pd
Calsicat produces a range of stabilized (nonpyrophoric) catalysts which have numerous applications in industrial organic chemistry. Cu/Cr powdered catalyst is available, and also Pt and Pd catalysts on carbon or alumina supports.

Another Catalyst Supplier

Süd-Chemie AG, 8000 München 2, Germany—Süd-Chemie offers a range of nickel catalysts for fat and fatty-acid hydrogenation under the titles of Girdler and KE Catalysts.

Catalyst—G531
This is a 25% of nickel dry, reduced catalyst supported on 13% of kieselguhr and clothed in approximately 63% of saturated fat as pastilles. The recommendation for its use is for partial and total hydrogenation of most vegetable, animal, and marine oils as well as fatty acids.

Catalyst—G53K
This catalyst is similar to G53, but especially selective for partial hydrogenations of vegetable and marine oils for edible purposes. It is available as pastilles or flakes.

Catalyst—G53L
Again, this is similar to G53, but is recommended for the total hydrogenation of vegetable and animal fats as well as fatty acids. Flakes or pastilles are available.

Catalyst—G94; G53
This is a dry, reduced 21% of nickel catalyst supported on 12% of kieselguhr and clothed in approximately 65% of saturated fat. Available as pastilles, it has a lower nickel content than G53 but may be economical for many of the same uses.

Catalyst—G70
This is a dry, reduced 23% of nickel catalyst where lower temperature activity is enhanced by the inclusion of 1% of zirconium without the loss of selectivity or durability. The catalyst is supported on 13% of kieselguhr, and the pastilles are clothed in saturated fat. At normal temperatures, it may also be used with fats and fatty acids, including those normally difficult to hydrogenate.

Catalyst—G95D
This is a dry, reduced 22% of nickel on 11% of proprietary support, the flakes being clothed in edible-grade saturated fat. It has high-linoleic and oleic selectivity when used in light or touch hydrogenations, and filters particularly well.

Catalyst—G111
A dry, reduced 18% of nickel catalyst on 5% of support, this was treated with sulfur to promote *trans* isomerization, and hence, steep melting curves. It is clothed in approximately 73% of a vehicle melting at 60°C, and is supplied as pastilles.

Catalyst—KE-KTR
A dry, reduced 21% of nickel catalyst on 11% of proprietary support, the pastilles are clothed in edible-grade saturated fat. The special feature is high selectivity, in which connection a thorough preliminary refining of the feedstock is very beneficial.

Catalyst—KE-NF20
A patented, continuous, dry reduction is used to produce the 22% of nickel catalyst on 11% of proprietary support. The pastilles are clothed in hard fat of 59°C mp. This is a good general-purpose catalyst for glyceride oils, and a variety of it with appreciably greater activity was recently developed (T4460).

Catalyst—KE-FS40
This catalyst, produced in a manner similar to KE-NF20, is however, specially designed for the hydrogenation of fatty acids or oils which cause difficulty during hydrogenation. The nickel content is approximately 22% on an 11% of proprietary support held in the particles of the hard-fat mp—59°C. T4357 is a further development of this type.

More Catalyst Suppliers

- Unichema International, D4240 Emmerich, Germany
- PRICAT Catalysts

Catalyst—9900
This is a dry, reduced 22% catalyst on a silicate support in pellets of hardened vegetable oil, mp. 52–62°C. A reliable general purpose catalyst with a long established use for hardening vegetable and animal oils.

Catalyst—9906
This is a catalyst similar to 9900, but with an enhanced selectivity in hydrogenating polyunsaturates.

Catalyst—9908
This dry, reduced 22% of nickel catalyst on a silicate base was treated with sulfur so as to promote *trans*-isomer formation, and hence, hardened fats with particularly steep melting curves. It is supplied as pellets in hardened vegetable oil, at a 52–62°C mp.

Catalyst—9910
This catalyst is a development from 9900 and 9906 with high activity and selectivity in vegetable and animal edible oils. It is available in the same form.

Catalyst—9912
This dry, reduced 22% of nickel catalyst on silicate support is especially produced to give it a rugged poison-resistant performance in hydrogenating industrial oils and fats destined for splitting to fatty acids.

Catalyst—9920
This supported nickel catalyst has the best low-temperature (110–120°C) activity and selectivity of the PRICAT range, and is specially designed for use with well-refined vegetable oils such as soybean. It also shows a very high activity in the normal conditions of 150–200°C.

Catalyst—PRICAT 9932
This catalyst is specially recommended for the hydrogenation of fatty acids after they are split from their parent triglycerides, and preferably distilled. Its resistance to the formation of nickel soaps is especially high. Moderately low temperatures of hydrogenation are feasible, thus minimizing the risk of polymerization of polyunsaturates. Hydrogenation at a 10-atm pressure becomes feasible if desired. The catalyst contains approximately 22% of nickel on a silicate support, and is supplied as droplets in hard fat, at a 52–62°C mp. PRICAT 9932 used with purified feedstock, such as distilled fatty acids, maximizes the chance to obtain very low IV (i.e., under 1.0).

Chapter 8

Hydrogenation Methods

H. B. W. Patterson

Variability in Natural Fats and Oils

One can say the simplest—and indeed, superficial—view of the hydrogenation of oils is the progressive addition of hydrogen at double bonds pursued far enough to produce more and more saturated groups within the triglycerides, until the texture at ambient temperature becomes firmer, ultimately hard, and even brittle. The disappearance of original color, odor, and flavor provides a bonus; an increase of resistance to oxidation is natural and expected. A closer view shows the following: isomerization, to a greater or lesser degree, is taking place at the same time; we can influence this by the conditions we choose; and isomerization can also play an important role in deciding the texture of partly hydrogenated oils. Lastly, we are able to influence the progress of hydrogenation in the direction of adding hydrogen for preference to those groups which have the biggest appetite for it: we can be selective to a degree.

These possibilities are classified and explained in Chapter 2; in this chapter is described, oil by oil, how appropriate use was made of them; for several oils, the range in characteristics of the hardened products is large. First of all, we must say that for many fats and oils commonly regarded as single entities, an appreciable variation exists in the degree of unsaturation as related to the location in the organism, age, season, and climate. This affects their use in the unhydrogenated state; for the hardener, this represents a variation in the starting point, which is helpful to consider. Some examples show what wide possibilities for variation exist. With lard and beef tallow, the firmer fat, therefore of slightly lower iodine value (IV), is found toward the center of the animal; subcutaneous fat is softer and more unsaturated. Obliging a pig to wear a sheepskin coat for a time depresses the IV of the subcutaneous fat slightly.

Diet is another factor in the softness of body fat. As the temperature of Lake Balaton falls, the IV of the oil in the fish there rises and vice versa. When herring are about to migrate, they eat voraciously. The additional body oil (thus stored as fuel for the journey) contains more highly unsaturated material; therefore, the IV reaches a maximum. At the end of the journey and after spawning, the oil content of the fish and the IV of the oil drop noticeably.

Nigerian groundnut oil shows an IV of 87–95, but Argentinean oil from the cooler climate shows an IV of 103–105. This allows the Argentinean oil to contain nearly 40% of linoleate, or twice as much as the Nigerian oil, which is correspondingly richer in oleates. A striking recent example (Morrison & Robertson, 1978) is that sunflower grown in the northern United States contains (typically) 55–70% of linoleate in its oil, but in the South, only 30–50%. Hence, the respective oils are suitable for salad oil or deep-fat frying. This report raises the interesting possibility of the oleic and linoleic groups being tailored for end use by the cultivation of the crop in the appropriate area.

Whatever differences in IV may exist between different parcels of oil from the same species, these will be considerably less by the time the respective oils are hardened to the same slip melting point. What is most useful to remember is that hydrogenation is usually accompanied by the formation of *trans* isomers, and that the greater the drop in IV, the greater is the opportunity for these to reach their usual equilibrium level in the hardened oil of about two-thirds of the remaining double bonds. This means in the partly hardened range, corresponding to slip melting points 32–38°C, the solid-fat content (SFC) or solid-fat index (SFI) at 20°C is already substantial, and a steeper melting curve results. This may be very acceptable for some products but not so for others. When we consider oils from different species of fish, the starting point for anchovy (Peruvian) and pilchard lies at c. 190–210 IV, but for herring, c. 135 IV, and capelin, c. 105 IV. When all these are hydrogenated to 78–80 IV, the possibility exists to reach endpoint 78–80 IV with the last two oils, and maintain an SFI at 20°C of 10-4, but for the first two, double this SFI result. This example is one of the most extreme of its kind.

In the years since this book was first published (1983), substantial changes in oils, plants, and catalysts have continued to grow in importance. At least for some crude vegetable oils, conventional plant breeding established a variety most suitable, technically and economically, for commercial use. Canola oil (see the "Rapeseed (Coza Oil)" section in this chapter), with its vastly reduced content of erucic acid and sulfur compounds, replaced old-style rapeseed oil for edible use in much of the world. Cloning to produce an oil palm with a consistently high yield of good oil is well-established (see the "Palm-Kernel Oil" section in this chapter). Except in the Philippines, the presence of polyaromatic hydrocarbons (PAHs) in crude coconut oil and some seed oil is almost eliminated by more sensible methods of drying. A variety of soybean oil with only 2.2% of linoleic acid in place of the usual 6.4% was produced by selective breeding (Mounts et al., 1992). Overfishing has caused a local scarcity of some fish oils.

Process Control

The control of hydrogenation may depend on the human observation of a test or an automatic instrumental response. In an even wider sense, it also depends on the quality of the oil, hydrogen, and catalyst, as well as the effectiveness of the agitation in bringing these three together. As suggested above, oil quality, whether crude or refined, is being steadily improved. Hydrogen quality has been quite adequate for years; the use of very pure hydrogen may now be more widespread. The nickel catalyst is more sophisticated in its preparation to more readily meet specific requirements such as selectivity, *trans*-isomer promotion, poison resistance, etc. The enhanced activity of most catalysts today as compared to those in use in the early 1980s is such that the recommended dose may well be only about one-third of what used to be the case. The wide selection of catalysts offered in the "Examples of Commercial Nickel Catalysts" section in Chapter 7 is evidence of such development. Lastly, as mentioned several times in Chapter 4, more intensive mixing resulting from improved impeller design is leading either to shorter reaction times or lower catalyst doses, or some combination of these two. An old and inaccurate method of measuring how much hydrogen was passed to an autoclave was to note the change in the hydrogen store–possibly a pressure drop, if this happened to be feasible. Today, the hydrogen is measured by a temperature- and pressure-corrected meter. This is so accurate that in modern hydrogenation plants, it is a usual feature of computerized pre-set control. All such systems infer that hydrogen leaving the store is taken up by the oil, even if some correction factor is applied. The other approach is to measure a change in the oil: this is commonly regarded as more positive. Fortunately, to perform a series of rapid IV tests throughout the hardening is not necessary, although in some instances, one is used at the end. The refractive index (RI) of the oil falls with decreasing unsaturation in an acceptably linear manner, so that when some elementary precautions are observed, this simple and rapid physical check is used to monitor the progress of hydrogenation, indicating when the end point is due to be attained and when it is reached. RI can now be measured quickly and automatically to the fifth decimal place. Since a one-unit fall in the fourth place corresponds fairly closely to a drop of one unit in IV, this technique affords an extremely sensitive monitor. This final result may then be confirmed by whatever test the hardener feels necessary. Often this is a melting point, sometimes an IV, and in some cases, an SFC test. Truly, when an attempt is made to relate RI to molecular structure, and this can include IV, several conditions and restrictions arise (*Handbook of Soy Oil Processing and Utilization,* 1980) but these are avoided in hardening-plant practice by keeping the test procedure close to a set routine (see the "Refractive Index (n_D^t)" section in Chapter 12), which works at a fixed temperature (±0.1°C), and generally compares the series of test results with what was obtained for the same kind of oil during previous operations: an overshoot is then unlikely. Since *cis* and *trans* double bonds contribute to RI and IV, neither of these tests can be taken as a guide in isolation as to precisely what texture was achieved by hydrogenation. What can be done most usefully is to relate the RI of one parcel of hydrogenated oil to the RI and associated texture of another parcel of similar oil hardened under the same conditions. Just because this relationship is easily seen to be empirical does not detract from its usefulness.

In such improved circumstances regarding oil, catalyst, and agitation which may be enjoyed in different degrees by different hardeners, to state a rigid catalyst dose in every instance is not sensible. Some 20 different oils are now about to be considered. Recommendations of the nickel dose must therefore be regarded as a trustworthy guide forming the basis of a subsequent adjustment in the light of results. Also appreciate that the fields of nickel catalyst and activated earth manufacture are in a state of steady development. A succession of products appears which is more exactly tailored to particular tasks. This leads several large manufacturers to offer a range of catalyst products with applications overlapping one another for technical or commercial reasons, or both. Recommendations here take this into account, and may sometimes mention more than one product from a supplier.

The final choice must then rest between the customer and the supplier. Of course, numerous examples are quoted from literature produced by suppliers of catalysts and hydrogenation plants in which specific doses of catalyst and precise results are quoted. All of these naturally refer to the specific circumstances in which the test hydrogenation was performed. This situation is very similar to that situation when dosages of activated bleaching earths are being prescribed for a range of fats and oils (Patterson, 1993).

Cleaning of Oils Prior to Hydrogenation

As part of rising quality standards, the content of sulfur, phosphorus, chlorophyll, trace metals, and oxidized fat in oils offered for hydrogenation is steadily improving by more searching pretreatment. This was described in detail (Patterson, 1993).

"Cleaning of an oil" means the removal of catalyst poisons which would prevent the hydrogenation from attaining its purpose to an adequate degree (Mounts, 1981). Not all poisons are completely removed on all occasions. Sometimes a proportion of a catalyst such as nickel is allowed to neutralize remaining poisons as a matter of course before the bulk of the catalyst dose comes into action. Ultimately, the trouble taken to clean an oil prior to hydrogenation becomes a question of economics: very often, the decision to take the trouble pays good dividends in allowing the end-product specification to be met regularly and in acceptable cycle times. When working with edible-class oils, the following standards are usual, and are attainable without particular difficulty. When centrifugal neutralization and washing are employed, better standards are assured; for some oils, preliminary centrifugal degumming before neutralization is also a means of ensuring a good neutralizing efficiency and a minimal subsequent poisoning of a catalyst.

One should bring free fatty acid (FFA) content to a maximum of 0.1%, although as far as hydrogenation is concerned, two or three times this level will not seriously impede the rate of reaction. Another consideration exists, however; for many vegetable oils, one can be quite practical by bringing the FFA content to 0.05% before hardening, and especially if the hydrogenation is of the low- or medium-temperature class (115–155°C), this probably does not increase to more than about 0.07% by the end. In these circumstances, a popular action, especially with American operators, is merely to give the filtered crude hardened oil a light earth bleach, and then to filter and deodorize, whereupon the FFA is stripped to a lower level. Soap content prior to hydrogenation should be brought under 0.05%; it is often much less in practice. Dry oil to 0.05% of a H_2O maximum, not only to avoid charging the headspace with water vapor, but also to oppose any tendency of the triglycerides to split and increase FFA content during hardening. This risk is greater at 180°C and with liberal doses of old catalyst. Expectedly, oxidized and polymerized contaminants will be substantially decreased during pretreatment, which probably includes adsorptive bleaching. Again, the necessity for this latter step must be judged on the particular merits of the case, bearing in mind that as the efficiency of neutralizing and washing has increased, the value of adsorptive bleaching may easily have fallen.

Many vegetable oils contain very little sulfur, the exceptions being those derived from the Brassica family, such as rapeseed and mustard seed, where 50 ppm of sulfur/crude oil is common. Conventional neutralization–bleaching will bring this down to c. 10 ppm. As suggested, the sulfur-bearing compounds (isothiocyanates, etc.) should be stripped from the oil by deodorization prior to hydrogenation; this will increase the processing cost, which must therefore be compared with the cost increase due to nickel activity destroyed by poisoning. If the hardening task happened to be the production of a steep SFI or SFC curve with the aid of a sulfur-poisoned catalyst, the removal of 10–40 ppm of sulfur/oil would seem pointless; if the aim were to produce a flatter dilatation curve with minimum *trans* isomers in the hardened oil and using a fresh catalyst, the presence of sulfur would then be very unwelcome. In any case, canola oil now usually contains no more than 5 ppm of sulfur after degumming and less again after refining (Patterson, 1989, 1993). Fish oils commonly contain 12–18 ppm of sulfur, which is about one-half that contained in average quality grades of whale oil.

With marine animals such as the whale, appreciable quality differences were expected between top grades produced by the simple rendering of blubber and the inferior oil resulting from the prolonged boiling of entrails and bones. The scale of this trade is now greatly curtailed. Similar quality differences apply in the case of land animals. Because of the marked unsaturation of several fish oils, the likelihood always exists that the crude oil will contain triglycerides, some of whose fatty-acid groups were oxidized, as well as oxidized FFA. This provides an argument for alkali neutralization, washing, and earth bleaching prior to hydrogenation. The dark washes withdrawn from the oil during refining and, even more so, the tarry matter depositing from the soapstock when this comes to be split provide ample evidence of the merits of cleaning such crude oil if fresh nickel catalyst is subsequently to be employed to produce a stable, partly hardened product with a minimum of *trans* isomers (lower-melting solids). In a limited number of locations, crude marine oils were cleaned merely by a generous (say, 2%) earth bleach, hardened, and then neutralized. This produces a hardened-oil soapstock in place of the usual soft one: this is done only where a clear economic advantage is evident when the time comes to disposing of the hard acid oil. The quality of the hardened marine oil itself also has to be kept in view when assessing overall economics. In other cases—probably better grades of vegetable oil—practicality may guide one to simply neutralize, wash, and dry without earth bleaching prior to hydrogenation. A post-hardening earth bleach is almost certain to follow in any case, since nickel has to be removed, and should then not exceed 0.2 ppm of nickel/post-treated hardened oil.

Phosphorus is present in some vegetable oils—the most important example being soybean oil—in the form of phosphatides such as lecithin. Although crude soybean oil may contain 1–3% of lecithin, a degummed crude oil is marketed at no more than 0.5%, and sometimes a first-class degummed crude oil will contain only one-half this amount. If the pretreatment reduces this to no more than 0.01% of lecithin (c. 4 ppm of phosphorus/oil), evidently from the figures quoted in the "Durability and Poisoning" section in Chapter 5 no serious burden will then be imposed on fresh nickel in promoting a low-temperature, noniso-promoting class hydrogenation. If a less rigorous pretreatment coincides with an oil whose initial degumming has brought it to just below the 0.5% of lecithin quoted above, a good chance exists that the oil offered for hydrogenation will then contain c. 8 ppm of phosphorus, and this the fresh nickel will have to take up. The degumming of soybean oil is reviewed at length (Braae, 1976; Carr, 1976; *Handbook of Soy Oil Processing and Utilization*, 1980; Patterson, 1993).

In the notes which follow concerning hydrogenation procedures for various fats and oils, mention is made of the cleaning precautions to be adopted, especially if these relate to special features of particular oils.

Lard

Lard is produced by the dry- or, more commonly, wet-rendering of selected tissues of clean healthy pigs. Pork fat also produced by rendering allows the use of more of the animal carcass, but still excludes certain parts. In the *Codex Alimentarius*, lard is defined in detail by the Recommended International Standard CODEX STAN 28-1981 Suppl. 1 (1983) and rendered pork fat by CODEX STAN 29-1981 Suppl. 1 (1983). The characteristics of the lard depend somewhat on the part of the animal from which it was derived, but also very much on the diet of the animal. Notably, United States lard is frequently softer (c. 33°C slip melting point) than European lard (c. 35°C). Probably rather less than the random proportion of trisaturated triglycerides is in soft American lards, whereas with harder European lards, the proportion could be greater than random. This leads to the interesting result that upon interesterification, the slip point of the former may easily rise a little, whereas that of the latter falls. Table 8.1 (Patterson, 1975, 1976) indicates the standards to be sought in crude lards if they are to be capable of reaching good edible quality, and is the most helpful practical guide in a complex area where standards are primarily empirical.

As far as the characteristics of the inferior grades are concerned, these will depend very much on how much processing effort the manufacturer is able to expend in relation to the standard which he is obliged to meet under local regulations. Expectedly, for normal lard, the phosphatides will be less than 0.05%, and if the fat is given any refining before hydrogenation, the normal standard of 0.1% of FFA maximum, 0.05% of H_2O maximum, and soap nil on the neutralized, and lightly earth-bleached fat will apply. Kaufmann et al. reported (1956a,b) on the means of distinguishing between lard which was simply rendered and lard which subsequently was processed in a way which included earth bleaching. Typical basic fatty-acid composition was often reported (Patterson, 1975, 1976). Since the 1960s, with sensitive methods of analysis, about 200 different fatty acids (Swern, 1979) were identified, including traces of odd-numbered and branched-chain fatty acids.

TABLE 8.1
Characteristics of Unprocessed Edible Grades of Lard

	Superior	Normal	Inferior
FFA (% max.) (m.w. 282)	0.5	1.0	1.5
Unsaponifiable (%)	←	under 1.0	→
Moisture plus impurity (%)	0.1		1.0
Color max. (Y + R 1 in)	10 + 1	12 + 1.2	15 + 1.5
Iodine value	—	63–71 usual	
	—	46–77 possible	
Saponification value		190–203	
Peroxide value (meq.)	5 max. ←	up to 10	→
Bömer (acetone)	74 min.	71	
m.p. (°C)		31–37	
Titer (°C)		32–43	

TABLE 8.2
Typical Fatty-Acid Composition of Lard (Simplified)

Acid Type	%
C14:0	2
:1	0.5
C16:0	24
:1	4
C18:0	14
:1	43
:2	9
:3	
C20:0	0.5
:1	1
:2	1

For a lard of normal unsaturation (62 IV), the typical makeup would be as shown in Table 8.2. Typical triglyceride composition and corresponding dilatation curves are also given by Patterson (1975, 1976).

Hydrogenation

Ultralight or Touch Hydrogenation (see the "Suspended Catalyst" section in Chapter 2)

The feedstock may be a neutralized, lightly earth-bleached lard whose FFA content was reduced to a 0.1% maximum. The aim is to limit hydrogenation to a maximal drop of 2 IV working at 150°C and a pressure between 1 and 3 atm. If a partly-used catalyst is utilized, the dose may be as high as 0.1% of nickel/oil, but if fresh, 0.02% of nickel/oil should be ample, leaving the possibility of reuse. A normal quality of catalyst should suffice for this easy task, and Nysosel 222, Hoecat 8820F, Calsicat E472D, Süd-Chemie KE-NF20, and Unichema 9900 or 9910 would serve. A more expensive catalyst with special selectivity would be effective, but not essential. The moderate operating temperature of 150°C discourages *trans*-isomer formation; hence, realistically, attempt to limit the increase in the percentage of SFC at 20°C to 4, or the SFI increase to 4. This treatment is designed to improve flavor stability. Possibly, the improvement owes as much to the adsorption of minute amounts of off-flavor precursor on the nickel surface as it does to the reduction in polyunsaturated fatty acid groups. If a greater increase in SFC value is acceptable, a moderate increase in hydrogenation would probably enhance stability still further. The product would obviously be a processed or stabilized lard. An RI at 60°C of perhaps 1.4529 would fall to c. 1.4527, but with such a small change, the accuracy of the refractometer may not be reliable. Confirmation by means of normal IV testing would be advisable.

Normal Hydrogenation (see the section with the same title in Chapter 2)
Feedstock similar to that used in the "Hydrogenation" section in this chapter may be hardened to 50 ± 1°C slip mp. End-point IV is 41 ± 2; SFI 20/30/50 is 70 ± 4/67 ± 4/48 ± 4. The drop in IV (or RI) will depend on the starting point, but a fall of 20 units is feasible. An end-point RI (60°C) between 1.4496 and 1.4501 on the first attempt should be checked by an IV test. Hardening temperature is controlled at 180°C and pressure at 3 atm with a maximum of 0.07% of fresh nickel/oil of the catalyst qualities given above.

Fully-Hardened (see the "Higher-Melting and Fully-Saturated Hardened Oils" section in Chapter 2)
When fully-hardened, 58°C slip mp is obtained for 1 IV by using a maximum of 0.15% of fresh nickel/oil of the above catalyst qualities at 180°C and 3 atm. The end point should be checked by an IV test.

Beef Tallow

The *Codex Alimentarius* recognizes a standard for Edible Tallow [CODEX STAN 31-1981 Suppl. 1 (1983)] (synonym: Dripping) as rendered from certain organs of healthy bovine animals and a standard for Premier Jus [CODEX STAN 30-1981 Suppl. 1 (1983)] (synonym: Oleo Stock) which is more restrictive in demanding the low-heat rendering of a limited list of organs, and narrows slightly the limits for liter, IV, etc. The edible end use for these materials is obviously primarily in mind, with the possibility of refining them for the same purpose. The fat found toward the center of the animal is firmer and of slightly lower IV than subcutaneous fat. Diet can also markedly influence the unsaturation and other characteristics of the fat. Even within edible grades, exists an appreciable spread in characteristics, but this is greatly extended when different grades used simply for technical purposes are included. As far as hydrogenation is concerned, parcels of tallow are chosen for their capacity to be hydrogenated fully and give—after neutralization, earth bleaching, and deodorization—a very light-colored product; this, in turn, may be converted to monoglyceride, which is also light-colored and stable; a final purification step may include its distillation. Table 8.3 (Patterson, 1975, 1976) classifies top-grade edible beef tallows into "superior," "normal," and "inferior"; for the manufacture of light-colored hydrogenated tallow, the superior grade would be chosen. Worth noting is that if, instead of giving the selected crude tallow merely a heavy earth bleach prior to hardening, a full alkali neutralization, washing, and bleaching are provided first, a great improvement in the color of the final fully hardened tallow is obtained, especially where the original starting material is at the lower limit of normal acceptability.

Swern (1979) points out that minute amounts of a wide range of fatty acids are now detected in tallows, as is also the case for lard. In Patterson (1975, 1976), a much simplified group of fatty-acid components is given along with an indication of triglyceride types. Table 8.4 shows the main classes of fatty acid in Premier Jus grade that are acceptable for the *Codex Alimentarius.*

When properly rendered, both the phosphatide and protein content of animal fats are already very low, and in refining, these are still further diminished so that they do not present any special hazard of catalyst poisoning. One may use phosphoric or citric acid to assist the removal of gums if necessary (Patterson, 1975, 1976). When neutralized and bleached, tallow should comply with FFA 0.1% maximum, soap nil, H_2O 0.05% maximum. No more than representative color figures, in addition to those quoted in Table 8.3, can be given. Taking, in each case, the yellow

TABLE 8.3
Characteristics of Unprocessed Beef Tallow—Top-Grade

	Superior		Normal	Inferior
FFA (% max.) (m.w. 282)	1.0		1.5	2.0
Unsaponifiable (% max.)	←	0.8	→	1.5
Moisture plus impurities (%)	0.1			1.0
Color max. (Y + R 1 in)	10 + 1		18 + 1.8	25 + 2.5
Iodine value			33–58	
Saponification value			190–203	
Peroxide value (meq.)			2–3	
m.p. (°C)			40–48	
Titer (°C)			40–47	
Swift life[a] (minimum hours to give a peroxide value of 10)	20		14	7

[a]Swift life is the number of hours required to reach a stated peroxide value (P.V.) when a sample is aerated at c. 150 cm^3 air/min at 98 ± 0.5°C.

TABLE 8.4
Advisory Ranges of Fatty-Acid Composition of Premier Jus [CODEX STAN 30-1981 Suppl. 1 (1983)]

Acid-type	%	Acid-type	%
C14	<2.5	C17:0	0.5–2.0
C14:0	1.4–7.8	C17:1	<1.0
C14:ISO	<0.3	C17:ISO	<1.5
C14:1	0.5–1.5	C17:ANTI ISO	<1.5
C15:0	0.5–1.0	C18:0	6.0–40
C15:ISO	<1.5	C18:1	26–50
C15:ANTI ISO	<1.5	C18:2	0.5–5.0
C16:0	17–37	C18:3	<2.5
C15:1	0.7–8.8	C20:0	<0.5
C16:2	<1.0	C20:1	<0.5
C16:ISO	<0.5	C20:4	<0.5

reading to be 10 times the red, a feasible progression in color reduction would be neutral-washed 5 R-18 R (5 ¼"), earth-bleached 2 R-4 R (5 ¼"), fully-hardened, and post-treated 0.2 R (1"). From these results, a really first-class crude tallow may already be as light as some parcels which had to be rigorously processed.

Hydrogenation

Ultra-Light or Touch Hydrogenation (see the "Suspended Catalyst" section in Chapter 2)
Virtually the same technique as described in the "Lard" section in this chapter for the improvement of lard was applied to beef tallow, and for the same reasons, although it is much less commonly used. The neutralized, bleached edible tallow is hydrogenated at up to 150°C max., at 1–3 atm with either 0.02% of fresh nickel/oil or about 0.1% of nickel/oil partly used catalyst; IV drop should be limited to 2 units. The better the quality of the tallow, the milder the hydrogenation conditions may be. The same qualities of catalyst as given for lard hydrogenation may be used in this case.

Fully-Hardened (see "Higher-Melting and Fully-Saturated Hardened Oils" section in Chapter 2)
Taking first the simplest case of neutralized bleached tallow as mentioned at the beginning of this section, the cleaning of the crude fat makes subsequent hydrogenation more efficacious. A dose of 0.15% of fresh nickel/oil, controlled at 180°C and about a 3-atm pressure, should ensure an end point IV of 2 maximum within 3 hours of gassing time. The slip melting point will be c. 58°C. Crude tallow and crude tallow which was substantially cleaned by treatment with activated bleaching earth are important raw materials for the production of fully hardened tallow, which is then split and a range of fatty acids produced by fractional distillation. Of course,

one can use the very important alternative of splitting the tallow first, distilling the fatty acids, and then hydrogenating them.

Some catalysts are now designed to give a rugged, highly poison-resistant performance on impure triglycerides, while others are outstanding in their resistance to attack by fatty acids (Lok & Joosten, 1991). The effect of an additional catalyst in one example (Table 8.5) and a combination of rigorous cleaning and sufficient nickel in another (Table 8.6) illustrate how the removal of catalyst poisons can have a dramatic effect on the hardening rate.

The Süd-Chemie Technical Information Sheets (Nickel Catalyst KE-NF 20 and KE-FS40, on catalyst KE-NF 20 give the results shown in Tables 8.5 and 8.6.

In contrast, the improvement in the hydrogenation rate following bleaching alone is recorded by Mexican operators in Table 8.7 (Goebel, 1981).

In selecting parcels of tallow for hydrogenation, often advantageous, if the situation allows, is the submission of a preliminary sample to a laboratory hydrogenation test—which may include a pretreatment—before purchase. Unichema recommends catalyst 9912 for its poison resistance when hydrogenating triglyceride oils, but 9932 is outstanding in its resistance to fatty acid, whether crude or distilled (Nickel Catalyst for Fatty Acid Hydrogenation, Fatty Acid Hydrogenation, 1990). Süd-Chemie offers KE-FS40 or a more recent T4357 for the hydrogenation of crude fats as well as both crude and distilled fatty acids. Hoechst offers Hoecat 882FA for this class of work, and Calsicat their E4228D. Among the Engelhard catalysts, Nysosel

TABLE 8.5
Hydrogenation of Crude Tallow (IV 42.2; mp 45.0°C; FFA 7.5%) (Nickel Catalyst KE-NF 20 and KE-FS40) (Nickel Dose 0.03%, Nickel Dose 0.08%, Hydrogen Pressure 20 atm, Hydrogen Pressure 3 atm)

	Nickel Dose 0.03%, Hydrogen Pressure 20 atm		Nickel Dose 0.08%, Hydrogen Pressure 3 atm	
Reaction time (min)	Temperature (°C)	IV	Temperature (°C)	IV
0	140	42.3	160	42.3
5	145	39.7	165	35.0
10	150	37.3	170	27.8
30	160	27.1	180	8.4
60	180	14.5	200	1.7
90	190	5.4	200	0.3
120	200	1.9	200	0.25
180	200	0.3	200	0.1

Final m.p. was 58.6°C in each case. The larger dose of Ni has not only got the reaction off to a much more rapid start, but the rate was well maintained, so that in the second case, under 2 IV was reached in 60 min as against 120 min in the first case.

TABLE 8.6
Hydrogenation of Refined and Bleached Tallow (IV 44.1; mp 44.7°C; FFA 0.06%; Soap Nil) (Nickel Catlayst KE-NF 20 and KE-FS40)

	Nickel Dose 0.04%, Hydrogen Pressure 25 atm	
Reaction Time (min)	Temperature (°C)	IV
0	140	44.1
15	150	34.5
30	160	26.6
60	180	13.4
90	180	5.6
120	180	0.5

A final m.p. of 58.7°C was reached, but it appears the time taken to fall below 2 IV would have been much less had only a little more nickel been employed.

TABLE 8.7
Hydrogenation of Crude and Bleached Tallow (IV 49: FFA 12.1%)

Bleaching with Tonsil Optimum Earth for 30 min at 110°C				
	% Tonsil Optimum			
Color (Lov. 5 ¼")	0	0.5	2	3
Red	21	23	10.4	8.7
Yellow	70	70	70	70
Blue	6.2	4.8	0.2	0.1
White	1.0	1.0	1.0	1.0
Hydrogenation 0.1 % Ni (Girdler catalyst G53); 6 atm pressure at 180°C				
Reaction Time (min)	IV			
0	49	49	49	49
10	45	39.5	30.5	22
20	44	38	24.5	10
30	43	37.5	18.0	5
40	43	37	14.0	1
50	43	36.5	10.0	
60		36.5	7.5	
70		4		
80		2		

222 has a high poison resistance with triglyceride oils, and Nysofact 120 has a resistance to fatty acids even under severe conditions.

Coconut Oil

More oil per acre can be derived from the cultivation of the coconut than any other oil seed. It flourishes within 20° of latitude north and south of the equator in coastal regions. The dried coconut flesh or meat, copra, typically contains 66% of oil. In areas of subsistence agriculture, the natural tendency is to consume the fresh coconut meat as such or employ it directly in cooking, but in areas of more sophisticated cultivation, such as the Philippines, it is an important agro-industry. A use was found for every part of the plant (Coconut Processing Technology, 1980). The oil melts at 25°C not because of its content of unsaturated fatty acids, but because of the high proportion of shorter chain fatty acids as compared with most vegetable oils. A typical fatty-acid composition would be as shown in Table 8.8.

The large content of laurates gives to this oil and others of similar composition (palm kernel, babassu, tucum, murumuru, ouricuri, and cohune) the general classification of "lauric" oils. The domination by lauric and myristic acids is held to be the explanation of the abrupt melting behavior and the steep SFI curves before and after hydrogenation. Obviously, with an IV range of 7–11, little room is available for maneuvering in hydrogenation so as to affect texture, and a fully hardened coconut oil melts at 33–35°C. A quite small contamination of 1–2% of nonlauric oil prior to hardening will serve to elevate the mp of fully hardened coconut and other lauric oils by a few degrees; keep this in mind when adding catalyst. The low degree of unsaturation contributes to the oxidative stability of even unhardened coconut oil, but if a little hydrolytic splitting of the triglyceride occurs, the short-chain fatty acids and monoglycerides produced have a strong "soapy" off flavor, due presumably to their greater volatility and solubility. If sulfur contamination of the copra occurs during drying, this results in a marked "rubbery" off flavor in the deodorized oil within seconds of its exposure to daylight. This unwelcome phenomenon is now rare; one can completely eliminate it by exposure to a very brief contact with a nickel catalyst and hydrogen before refining.

TABLE 8.8.
Typical Fatty-Acid Composition of Coconut Oil[a]

Fatty Acid	Type	%	Fatty Acid	Type	%
Caproic	C6:0	T	Palmitic	C16:0	9
Caprylic	C8:0	8	Stearic	C18:0	2.5
Capric	C10:0	6	Oleic	C18:1	7.0
Lauric	C12:0	47	Linoleic	C18:2	2.5
Myristic	C14:0	18			

See also CODEX STAN 124-1981.

Smoke or hot-air-dried copra yields oil which in some cases may be contaminated with up to 3 mg/kg of PAHs (Biernoth & Rost, 1968; Sagredos et al., 1988; Wendt, 1980). The control of drying methods greatly diminishes such occurrences. Only tricyclic and tetracyclic hydrocarbons are partially removed by the usual neutralization–earth bleaching–deodorization. If certain activated carbons are added to the earth at bleaching, PAHs, including the heavier penta- and hexacyclic molecules, are adsorbed (Wendt, 1980). About 0.4% of Norit FND carbon was useful for this purpose. The saponification-value range is 254–264; unsaponifiable (%) is usually less than 0.5, but BS 628 allows up to 0.8%; FFA of 3% is typical of first-class crude oil, but 5–6% is met; BS 628 gives 5% maximum. A typical n_D^{60} is 1.4418, which being low as for other lauric oils, matches the preponderance of short-chain fatty acids. Phosphatides are usually absent; sulfur may be met up to 4 ppm. The color of good crude oil may be as low as 1.5 Y 0.8 R (1"), but double this is common; after earth bleaching, 1.3 R (5 ¼") is usual; more highly colored oils would be acceptable for hydrogenation.

Hydrogenation

Ultra-light Hydrogenation (see the "Ultralight, Touch, Brush, or Flash Hydrogenation" section in Chapter 2)

This ultralight hydrogenation need not even achieve a fall of one unit in IV. The reuse of some old catalyst (ideally from lauric-oil hardening) at ca 0.05% of nickel/oil with only a few minutes of agitation with hydrogen appears to adsorb the traces of what are probably sulfur compounds, which occasionally have given parcels of coconut oil a persistent rubbery flavor after deodorization. It is not intended to modify texture, and most coconut oil does not require this treatment.

Normal Hydrogenation and Full Saturation (see the "Normal Hydrogenation" and "Higher-Melting and Fully-Saturated Hardened Oils" sections in Chapter 2)

Pure coconut oil melts rapidly at about 25°C, and is valued for the steep descent of its SFI or SFC curves. Variations in genuine crude oil may be aggravated by contamination by other oils. The commonest hydrogenation is carried out to at least 2 IV maximum, but may be prolonged even if the RI appears not to change until an IV test shows a value of 0.5 or less. Now no involvement occurs with polymer or isomer formation, so to help the temperature to rise from the commencement at, say, 140°C to reach 180°C is advantageous. A dose of 0.05% of fresh nickel working at up to 3 atm pressure should complete the reaction within 2 hours. For a slip melting point of 33°C, the percentage of SFC range at 10/20/30/35/40°C should then be c. 78/50/5/2/0. Süd-Chemie, for catalyst KE-NF20, has issued data (Nickel Catalyst KE-NF 20 and KE-FS40) for hydrogenation at 2 atm pressure by using: (a) 0.05% of nickel at 180°C; (b) 0.02% of nickel at 180°C; (c) 0.02% of nickel at 160°C; (d) 0.02% of nickel at 120/160°C. As might be expected, conditions (a) and (b), being the most rigorous, achieve under 1 IV in about 25 and 80 minutes, respectively. The characteristics of the hardened oil toward the end of the reaction are described in Table 8.9 and in more detail for conditions (a) above, by using 0.05% of nickel, 180°C, and 2 atm.

TABLE 8.9
Characteristics of Hardened Coconut Oil toward End of Hardening (Nickel Catalyst KE-NF 20 and KE-FS40)

	Solids by NMR (%)			
Temperature (°C)	IV = 2, m.p. = 28°C	IV = 1, m.p. = 32°C	IV = 0.5, m.p. = 34°C	IV = 0.2, m.p. = 35°C
10	77	78	79	80
20	48	50	52	53
30	4	5	5	6
35	1	2	3	3
40	0	0	1	1

Reaction Time (min)	*IV*	*Melting Point (°C)*
15	2.0	28
30	0.95	32
45	0.29	33.5
60	0.25	33.8
90	0.20	34.6
120	0.18	35.0
150	0.18	35.2
180	0.18	35.3
240	0.18	35.4

Other catalysts which would easily hydrogenate coconut oil are Engelhard's N325 or N222, Calsicat E428D, Hoecat 882 OF, and Unichem 9910.

Cottonseed Oil

Cottonseed-oil production in North and South America, Africa, and Asia has exceeded 4 m tons/year; it is one of the more important edible oils. Since the unsaturation is present almost exclusively as oleic and linoleic acids, with only very small percentages of linolenic occasionally reported, naturally, the refined oil keeps well, and is used in cooking as such. Winterized oil is used as a salad oil; hydrogenation produces a range of very suitable components for vanaspati, shortenings, and margarines, but has to face economic competition on one side from the unmodified refined oil itself in areas which have adapted to its use, and on the other side from hydrogenated products which commence from less costly oils. The oil is, of course, a by-product of cotton growing, and the cake, after oil extraction, is an important animal-feed component. Fairly wide variation is present in the oil content of seed from different species and areas; 18–23% (dry basis) is common.

Nonfat components of the crude oil can account for around 2% and include, notably, phospholipids and the pigment gossypol. The crude oil is dark-red; common practice is to give this alkali neutralization and washing before sale, at which point it is then known as "washed cotton oil," and in this form most hydrogenators acquire the oil. Fortunately, efficient refining reduces the level of both phospholipids and gossypol each to about 0.01% (Swern, 1979). Cottonseed oil also contains up to 1% of cyclopropenoid fatty acids, of which sterculic acid is typical, but these are reduced by hydrogenation.

$$\begin{array}{c} CH_2 \\ / \quad \backslash \\ CH_3(CH_2)_7 - C = C(CH_2)_7COOH \end{array}$$

Sterculic acid

These acids are the basis of the qualitative Halphen color test by which cottonseed oil is detected; their disappearance explains why the Halphen test does not apply to hydrogenated oil. Crude oil and especially "washed" oil purchases are regulated by various commercial standards, two important items being FFA and color. The FFA on washed or semi-refined oil is often expected to be not greater than 0.25% and the Lovibond color (5 ¼") to fall within some maximum, such as 2.5 R or 7.6 R, according to grade. Average washed oil of, say, 3 R (5 ¼") can easily be earth-bleached to 3 R (5 ¼") if it is to be used as such; a somewhat higher bleached-oil color would probably be acceptable for hydrogenation. In Table 8.10 are the limits noted as possible by FAO/WHO for use in the *Codex Alimentarius* (see the "Cottonseed Oil" section in the *Codex* publication).

The IV is usually taken as 108; however, Texan oil at 103 and Indian at 116 were noted. An RI n_D^{60} of 1.4573 ± 0.0004 would cover most varieties commonly met; saponification values are 189–198; unsaponifiable (%) is 1.5 maximum; sulfur is absent; and, as stated earlier, the removal of phosphatides in preliminary refining is so effective that if 0.01% remains, this is only equivalent to about 4 ppm of phosphorus, and therefore not seriously damaging to a catalyst.

Hydrogenation

Iso- or trans-*Suppressive Hydrogenation (also see a section with the same title in Chapter 2)*

Severe low-temperature hardening (see the "Low-Temperature Hydrogenation" section in Chapter 2) to remove linolenic groups while minimizing *trans* formation is not needed with this oil, since the linolenic level is naturally so low. However, in considering shortenings and vanaspati (Patterson, 1974), hydrogenation at 140°C, 3 atm, and 0.05% of fresh nickel/oil is advantageous. A drop in IV to 74, corresponding with a fall in RI of n_D^{60} 1.4573–1.4536 approximately, brings a soft solid of 32–34°C mp, and the SFI 20/30 is 17/6.7 (% SFC at 20/30°C = 18/6). The continuation of hardening in the same conditions to 70 IV (c. n_D^{60} 1.4530) brings

TABLE 8.10
Advisory Fatty-Acid Composition of Cottonseed Oil [CODEX STAN 22-1981 Suppl. 1 (1983)]

Fatty Acid Type	%
C < 14	<0.1
C14:0	0.4–2.0
C16:0	17–31
C16:1	0.5–2.0
C18:0	1.0–4.0
C18:1	13–44
C18:2	33–59
C18:3	0.1–2.1
C20:0	<0.7
C20:1	<0.5
C22:0	<0.5
C21:1	<0.5
C24:0	<0.5

the melting point nearer to 36°C, and increases the SFI 20/30 to 22/8.2 (% SFC at 20/30°C = 25/7). Precise results will vary with the efficiency of the hydrogen distribution in the autoclaves concerned. To the material of the 74 IV type, in particular, customarily one adds a small percentage of fully hardened vegetable oil to elevate the SFI 30/35 and thus obtain a still flatter SFI curve. For this easy hydrogenation, Engelhard D325, Calsicat E428D, Hoecat 882 OF, Süd-Chemie G53, KE-NF20, and Unichema 9910 would be very suitable. Catalysts with a specially enhanced selectivity to linolenic and linoleic acids would certainly work well, but are not strictly necessary in this application.

Normal Hydrogenation (see the section with the same title in Chapter 2.)
This class of hardening is commonly performed when components are required for conventional packet table margarine, and when a fairly steep descent occurs in a SFI to a slip melting point of 34–36°C; it is also used when higher melting points are needed. An old catalyst from other vegetable-oil hardening which has lost two-thirds of its original activity may be employed at up to 0.4% of nickel/oil or even higher, at 3 atm pressure, commencing hardening at 140–150°C, and allowing an immediate rise to 180–190°C. Conditions are now *trans*-promoting, so that by 74 IV and n_D^{60} 1.4536, a melting point of c. 34°C has an SFI 20/30 of 38/14.5 (% SFC at 20/30°C = 45/14), markedly higher than was the case in the "Hydrogenation" section above. The continuation to 70 IV (c. n_D^{60} 1.4530) brings the melting point to 36 ± 1°C with an increase in the SFI 20/30 to 50/27 (% SFC at 20/30°C = 61.5/28.5). A RI itself is not an adequate indication of the amount of *trans* isomers present at these end points,

so more than usually important is to check slip melting points. If the conditions of oil quality, catalyst, etc remain substantially the same, however, the processor can then place increasing reliance on an RI reading as a result of experience in associating it with the correct end point. If fresh catalyst is being used, the dose of 0.03% of nickel/oil is likely to give results in an acceptable gassing time. Such general-purpose catalysts are Unichema 9910, Süd-Chemie C.53 and KE-NF20, Hoecat 8820F, Calsicat E472D, and Engelhard N325. All these catalysts have potential for reuse at a somewhat increased dosage.

If melting points above 40°C are sought, one may need to commence with about 0.06% of fresh nickel/oil to keep the reaction time within acceptable limits. In any case, selectivity and *trans* promotion now count for less, as double bonds are disappearing to form saturates. Thus, in producing a hard component for 20% of use in a margarine blend, a final IV of 55 (c. n_D^{60} 1.4513) for a 44°C melting point is quoted (Swern, 1964). From the actual SFI quoted, this would approximate to SFI 10/20/30/35/40 = 66/61/50/42/16 or % SFC at 10/20/30/35/40°C = 95/76/54/43/13.

Iso- or trans-*Promoting Hydrogenation (see this title section also in Chapter 2)*
The use of specially sulfur-poisoned catalyst at up to 0.2% of nickel/oil, at up to 5 atm pressure, and 180–190°C enables the virtually maximal content of *trans* isomers to be achieved as the melting point reaches the mid-thirties, so that an SFI 20/30 of 52/26 is within range. As hardening progresses to a higher melting point, SFI 30 is gaining more rapidly than SFI 20; hence, the steepness in the SFI curve is gradually lost. Several fresh sulfur-poisoned catalysts are commercially available. In these, some activity remains, and the sulfur is so distributed to obtain a maximal *trans*-promoting effect. Sections "Iso- or *trans*-Promoting Hydrogenation" in Chapter 2 and "Durability and Poisoning" in Chapter 7 discuss the use of these catalysts and the possibility of refreshing their useful *trans*-promoting life. The term "sulfur-promoted" is now in use (Okonek, 1986), and such catalysts are Engelhard Nysel Sp-7 and Nysel Sp-10, Süd-Chemie G111, and Unichema 9908. Okonek (1986) concludes that with a sulfur-promoted catalyst, temperature and pressure have virtually no effect on the rate of *trans* formation. The nature of the catalyst is by far the dominating factor; hence, to increase pressure is feasible even up to 7 atm to quicken the hydrogenation reaction. Reuse is feasible.

Higher-Melting and Fully-Saturated Hardened Oils (see the section with the same title in Chapter 2)
The hardened cottonseed oils of over 44°C mp are obtained by the continuation of the same procedure as in the "Normal Hydrogenation" section above. Probably the most widely used item in this class is fully-hardened cottonseed oil of 2 IV maximum and 62°C mp. This is added to the hydrogenated-shortening base (see the "Iso- or *trans*-Suppressive Hydrogenation" section above) to produce a blend with a sufficiently flat SFI curve. Although sometimes referred to, loosely and incorrectly, as a "stearin," this fully-hardened cottonseed oil in fact depends on the 25% of palmitic acid present to make impossible the formation of a very high proportion of tristearates. This has a favorable influence on the crystal form obtained on cooling,

and hence, on the shortening texture in which it was incorporated (Swern, 1964). A good-quality "washed" cotton oil, dry and below 0.3% of FFA, will hydrogenate to 2 IV without particular difficulty.

Grapeseed Oil

In the crushing of grapes, the solid residue (as with olives and in the separation of tomato seeds) is described as pomace. Now the recommendation (Bernardini, 1973) is that the pomace be dried and the grape seed separated from it. The oil content is about 15%; the cooked seed is best flaked and solvent-extracted. Grapeseed oil is clearly of the oleic–linoleic class (Swern, 1979), with a fatty-acid composition generally within the limits shown in Table 8.11.Typically, oleic content would be c. 20% and linoleic c. 67%. The green crude oil is degummed, neutralized, washed, and earth-bleached before hydrogenation (see the "Cleaning of Oils Prior to Hydrogenation" section above). Being of the oleic–linoleic class, the refined unhardened oil has the potential as an ingredient of salad or cooking oils. The generally accepted limits for other characteristics are: IV 124–143; saponification value 178–196; unsaponifiable (%) maximum 1.6; n_{D}^{60} 1.4615 ± 0.0004.

Hydrogenation

The oil has a significant economic interest only in those countries where grapes are processed on a large scale. Separated seed from dried pomace could provide part of the program for an extraction refining plant. If it is not used as a table or cooking oil, it can be hardened like other oils of its class (e.g., cottonseed oil), although some green color may persist.

TABLE 8.11
Advisory Ranges of Fatty-Acid Composition of Grapeseed Oil [CODEX STAN 127–1981 and Suppl. 1 (1983)]

Fatty Acid Type	%
C12:0	<0.5
C14:0	<0.3
C16:0	5.5–11
C16:1	<1.2
C18:0	3.0–6.0
C18:1	12–28
C18:2	58–78
C18:3	<1.0
C20:0	<1.0
C22:0	<0.3
C24:0	<0.1

Normal and Iso- or trans-*Promoting Hydrogenation (see "Normal Hydrogenation" and "Iso- or* trans-*Promoting Hydrogenation" sections in Chapter 2)*
To obtain a quick-melting fat at c. 33°C mp, the hydrogenation should commence as usual at 140°C and receive whatever steam heating is available, being finally controlled at 180–190°C, a pressure of 1–3 atm, and up to 0.4% of nickel/oil either as a partly exhausted catalyst or a specially sulfur-poisoned catalyst. A fall of n_D^{60} 1.4615–1.4549, corresponding to c. 84 IV, should yield for a 33°C mp an SFI 20/30 of 40/10. Continuing hydrogenation to 36–38°C mp, while it elevates the whole SFI curve, will bring about a gradual flattening of it.

Since the proportion of C18 acids could be over 90%, inevitably, the fully-hardened oil must contain, unlike fully-hardened cottonseed oil, a very large amount of tristearate, in which case, because of crystallization behavior, it would be less suitable for addition in small percentages to shortening, etc.

Groundnut (*Arachis*, Peanut) Oil

Groundnut oil is an obvious member of the oleic–linoleic group, since its unsaturates—c. 80%—are almost entirely derived from these two acids. The kernel not only contains nearly 50% of oil, but also is the source of valuable protein; in some countries, a natural tendency is to consume much of the crop of nuts without further processing. As linolenic acid or other fatty acids of equal unsaturation are virtually absent, the oil is stable as well as nutritious. The cultivation of the groundnut has spread from South America, its original home, to many parts of the world, notably West Africa, Sudan, Tanzania, India, and China. The oil is used in cooking, frying, and in margarine blends; it also provides a range of useful hydrogenated fats with either flat or steep melting ranges useful in making vegetable ghee (vanaspati), shortening, and margarines. Its usefulness maintains its price among the more expensive edible oils (Achaya, 1994; Langstraat, 1976; *Oils and Oilseeds*, 1971; van Stuyvrengerg, 1969).

As mentioned in the "Variability in Natural Fats and Oils" section in this chapter, the makeup of groundnut oil is influenced by both genetic factors and environmental factors such as temperature, while the proportion of saturates remains in the 17–20% range, the 80–83% unsaturates may be either mainly oleic or almost equally oleic and linoleic. More temperate climates tend to yield oil of higher IV. Nigerian oil (87–95 IV) and Argentinean oil (103–105 IV) typify opposite extremes.

The significant content of the longer chain fatty acids (C20 and upward) (Table 8.12) distinguishes groundnut oil. Also, oils which have a notably higher proportion of linoleic acid may possess a slightly greater proportion of saturated fatty acid. This seems to explain the apparent anomaly that some Nigerian oil of n_D^{60} 1.4558 was recorded as having a slip melting point of 0.2°C, whereas more unsaturated Argentinean oil of n_D^{60} 1.4568 showed a slip melting point of 20°C. The presence of some high-melting glycerides in the oil also explains the fact that it assumes a jelly-like texture around 5°C, and any crystalline fat separating at low temperatures is difficult to remove by filtration.

TABLE 8.12
Advisory Ranges of Fatty-Acid Composition of Edible *Arachis* (Groundnut) Oil (CODEX STAN 21-1981 Suppl. 1 1983/1987) (Mounts, 1987) and Examples of Nigerian and Argentinean Oils

Fatty Acid Type	CODEX STAN 21 %	Nigerian %	Argentinean %
C<14	<0.4		
C14:0	<0.6		
C16:0	6.0–16	10	11
C16:1	<1.0		
C17:0	<0.1		
C18:0	1.3–6.5	3.5	3
C18:1	35–72	59	39
C18:2	13–45	20	38
C18:3	<0.3	0.5	0.5
C20:0	1.0–3.0	1.5	1.5
C20:1	0.5–2.1	1.5	1.5
C22:0	1.0–5.0	2.5	3.5
C22:1	<0.3		
C24:0	0.5–3.0	1.5	2.0

A common IV for groundnut oil is 91, corresponding to an n_D^{60} of 1.4560. The *Codex Alimentarius* quotes IV limits as 80–106, relative density (20°C) 0.914–0.917, saponification value 187–196, unsaponifiable (%) 1 maximum (usually under 0.8%), and content of arachidic and higher fatty acids 0.48 minutes. The PV should not exceed 10 mequiv. of O_2/kg of oil. The FFA standard on crude oil varied widely in the past, depending very much on how it was extracted and stored; under 2% of FFA is a usual standard; parcels well above this may be on offer. A fair color would be 15 Y 1.0 R (1"), but BS 629 quoted a maximum of 35 Y 4 R (1"). After neutralizing and earth bleaching, 0.5 R-2.0 R (5 ¼") would be typical. The settling of the crude oil in tanks is well worthwhile; as in that case, the lecithin is usually under 0.2% as against a possible 0.5%: at still higher lecithin contents, the efficiency of batch refining especially begins to suffer. After neutralizing and bleaching, the phosphatide derived from such a well-settled crude falls to 0.01% (4 ppm P). If the double bonds in a skipped distribution, as in a linoleic or a linolenic group, become involved in oxidation, conjugated dienes and trienes can result. These are detected by the absorption in the ultraviolet at 232 nm and 268 nm, respectively, and they change according to the manner in which the oil is processed. Thus, possibly begin the assessment of the oxidative state of the crude oil and what was done to it by the spectrophotometric measurement of these characteristics. Some countries use these extinction coefficients in setting quality specifications for crude and deodorized soft oil. Obviously, hydrogenation will destroy the numerical relevance of the degree of oxidation and conjugation, etc, but a well-oxidized crude oil is likely to lead to a poor flavor in a partly-hardened product. An E1% 1 cm 232 nm (sometimes abbreviated as E.1. 1.232) above a value of 5 for crude groundnut oil is an indication that oxidative deterioration is progressing. The "Alkali Isomerization" section in Chapter 8 describes the significance and limitations of such tests in greater detail. The sulfur content is negligible. The pretreatment of the oil before hardening (see the "Cleaning of Oils Prior to Hydrogenation" section in this chapter) should achieve an FFA maximum of 0.1%, soap 0–0.05%, and a H_2O maximum of 0.05% without difficulty.

Hydrogenation

As in the case of cottonseed oil, no necessity exists in the case of groundnut hydrogenation to bring down the content of any highly unsaturated fatty acids such as linolenic to around 2% to attain stability. We are therefore concerned from the outset primarily with texture, although obviously as the proportion of remaining linoleic acid falls, a greater resistance to atmospheric oxidation will arise.

Iso- or trans-*Suppressive Hydrogenation (see same title section also in Chapter 2)*
A fresh catalyst at about 0.05% of nickel/oil with hardening controlled at a 140°C maximum and at a 2–3 atm pressure should succeed in dropping the IV to 70–72 in about three hours, corresponding with a fall in n_D^{60} from 1.4560 to 1.4540. A slip melting point of c. 33°C with an SFI 20/30 of 16.7/5.6 (% SFC at 20/30°C = 18/5) results. If hydrogenation continues to 37°C mp, the texture changes approximately to an SFI 20/30 of 19.4/9.3 (% SFC at 20/30°C = 21.5/8.5). By the time an IV of 46 is attained, the slip mp has risen to c. 46°C and the SFI 20/30 of 65/55.5 (%SFC at 20/30°C = 82/61.5) is still somewhat flatter than when hardening at a higher temperature such as 180–200 °C. Any active general-purpose catalyst will serve here. Engelhard Nysosel 22 and Nysosel 325; Unichema 9900, 9906, and 9910; Süd-Chemie G53, G53K, KE-NF20, and KE-KTR; Hoecat 882 OF; and Calsicat E472 D should all perform well and permit some reuse. Worth noting is that Unichema 9920 and Süd-Chemie G70 have good low-temperature (140°C) activity and selectivity.

Vegetable ghee (vanaspati). Natural (cow) ghee typically has a slip melting point of about 37°C, although some varieties melt as high as 41°C. The texture quoted (Patterson, 1974) as typical for 37°C mp is given in Table 8.13. As one can see, this texture (Table 8.13) is not far removed from that of the groundnut oil hardened at

TABLE 8.13
Natural-Ghee Texture

°C	15	20	25	30	35	39
SFI	32	23.6	17.4	12.4	6.8	2.4
SFC	38	24.8	16.2	10.3	5.1	1.4

140°C to 37°C, and on the figures quoted, had that hydrogenation continued a little further to 38°C mp at most, an even closer fit at 20°C and 30°C could be attained. This procedure fits with a technique of producing the desired result from a single hydrogenation of one vegetable oil or a mixture of vegetable oils. At the same time, one must recognize that many manufacturers find it economical to blend some components of higher melting point with more lightly hydrogenated vegetable oil, or even unhydrogenated oil. The medium and higher melting components allow the repeated use of the same catalyst so that the overall expenditure of nickel is modest. A hardened oil of a medium melting point may have presented a substantial amount of *trans* isomers arising from the conditions of hardening, including the use of a partly-exhausted catalyst. The melting point and solids content at 20°C will then depend on these *trans* isomers, and the SFI or SFC curve may be relatively steep. Liquid-oil components will obviously lower the solid component at 20°C and soften the blend, while a small amount of high-melting component will lift the solid component at 30°C and 35°C, thus establishing a flat melting curve.

Palm oil melts at 35–39°C; therefore, when it is a major (c. 50%) component in the production of vanaspati, the scope for maneuvering in hydrogenation becomes limited; soybean oil, on the other hand, with an IV of 135 and c. 8% of linolenic acid present, not only allows but also requires more hydrogenation. Cottonseed oil and groundnut oil, traditionally popular for vanaspati production in Asia and Africa, are intermediate in this respect. In some countries, local regulations require about 5% of sesame oil to be included in the vanaspati. The sesamol present thereafter gives a positive reaction in the Baudouin color test, and enables the vanaspati to be distinguished from natural ghee in which it would be a profitable adulterant. Other regulations may control the use of artificial coloring and flavoring, and limit the production of vegetable ghee to below a fixed maximal melting point (Gander, 1976; Haighton, 1916; Patterson, 1974; Swem, 1964).

Normal Hydrogenation (also see same title in Chapter 2)

The remarks in the "Normal Hydrogenation" section of Chapter 2 apply very directly to groundnut oil, where the chances of unwanted side reactions such as polymerization during hydrogenation do not exist. Hence, having introduced hydrogen at 140–160°C, the control in this class of hardening is usually 180°C, with 200°C as a possible maximum.

Common practice is to employ a generous dose of partly-exhausted normal catalyst—say, about 0.4% of nickel/oil—depending naturally on how much activity remains in the catalyst and how desirable is completing the hardening within three hours. Alternatively, a fresh catalyst of 0.05% of nickel/oil would probably suffice. A pressure of 3 atm is normal. When obtaining a fairly high SFC at 20°C when using a normal catalyst is desirable, a lower pressure of 1 atm and extended gassing time are likely to be necessary. When a high SFC at 20°C and a minimal gassing time are sought, a sulfur-promoted catalyst could be used and the pressure lifted to 5 atm or higher. The latter approach amounts to moving from what is here called "normal hydrogenation" toward iso- or *trans*-promoting hydrogenation (see the "Iso- or

TABLE 8.14
Hydrogenation of Groundnut Oil in Normal (see "Normal Hydrogenation" Section in Chapter 2) Conditions

Slip	m.p. (°C)	IV	n_D^{60}	SFI 20/30	SFC % at 20/30°C
(i)	28	78	1.4545	28/4	32.5/3.5
(ii)	31	73	1.4532	31.5/9.3	36.5/8.5
(iii)	33	72	1.4529	42.6/13.9	46/13
(iv)	34	70	1.4528	46.3/16.7	56/16
(v)	35	68	1.4527	49/19.4	60/19
(vi)	37	64	1.4524	52/22.2	64/22
(vii)	41	58	1.4518	61/43.7	76/46

trans-Promoting Hydrogenation" section below). The general-purpose catalysts suitable for hardening groundnut oil are: Hoecat 882 OF, Calsicat E472, Süd-Chemie F53 and KE-NF2, Unichema 9910, and Engelhard N325. Bearing in mind that the initial IV may easily lie between 87 and 103, regard the following as no more than typical when starting from an oil of 91 IV corresponding with an n_D^{60} of 1.4560. As hydrogenation progresses into the region of 40°C mp and beyond, differences in IV or RI associated with a particular slip melting point will naturally grow much less in spite of the original groundnut oils having been 15 or 20 units of IV apart. This, as with other oils, is simply due to the fact that fewer double bonds then remain, so that whether they are of a *cis* or *trans* configuration is progressively less important than the presence of increasing amounts of saturates (see Table 8.14).

Iso- or trans-*Promoting Hydrogenation (see same section title in Chapter 2)*
The conditions described in the "Normal Hydrogenation" section in this chapter using a partly-exhausted or "spent" catalyst at 180°C are themselves substantially *trans*-promoting. If this tendency is to be pursued to the maximal extent, probably affording some advantage will be to use a specially sulfur-poisoned catalyst at about 0.2% of nickel/oil concentration in the first batches, and allow this to double itself as the same catalyst is used repeatedly. Also no reason exists why re-sulfuring should not be done at intervals as described in detail in the "Iso- or *trans*-Promoting Hydrogenation" section in Chapter 2. The practical consequence of this approach is to obtain the maximal difference between solids contents at 20°C and 30°C for any one item while following the IV or RI figures quoted in Table 8.14 (i)–(vii). Suitable catalysts would be Engelhard Nysel Sp-7 and Nysel Sp-10, Süd-Chemie G111, and Unichema 9908 working at up to a 5-atm pressure.

High-Melting and Fully-Saturated Hardened Oil
For acceptable gassing times, a dose of about 0.1% of nickel/oil of fresh catalyst is used; the temperature is controlled at the usual 180°C, and the pressure at 3 atm, although an increase of the latter to 5 atm will probably secure a useful shortening

of cycle time, especially for a 48°C mp and above. Hardening to an n_D^{60} of 1.4514 is the equivalent of c. a 56 IV and 43°C mp, for which an SFI 20 of 59 (SFC at 20°C = 74) is the likely minimum. Continued hardening to an n_D^{60} of 1.4505 (50 IV) gives a 46°C mp and an SFI 20 of 65 (SFC at 20°C c. 88). At an n_D^{60} of 1.4486, the IV has fallen to around 18, and the mp has risen to 56°C. The melting curve is now very flat with an SFI of 75 (SFC at 20°C = 95). Finally, at around 2 IV, the fully hardened groundnut oil of 62°C mp results. Suitable catalysts are Süd-Chemie G53 and KE-NF2, Engelhard Nysosel 325, Hoecat 882 OF, Unichema 9910, and Calsicat E472 D.

Linseed Oil

Linseed oil is obtained from the seed of the linseed–flax plant, one variety of which produces a high yield of seed and another a high yield of fiber (flax). Dual-purpose varieties are said (*Oils and Oilseeds,* 1971) to give inferior seed and fiber. Linseed is primarily a warm temperate zone or subtropical plant, major producers being Argentina, the United States, Canada, Russia, and India. Both genetic and seasonal (temperature, rainfall) variations lead to a wide spread in possible IVs of different samples from 130 to 205, but 180–185 may be taken as typical. Dry seed contains 35–45% of oil. Because the oil may easily contain around 60% of linolenic acid, its principal employment is as a drying oil in the paint and linoleum industry, and even here petrochemical products are taking some of the market. The use of hydrogenated linseed oil for edible purposes was known, and it also has limited use for saponification, but these may be regarded as exceptional and of diminishing importance.

Phosphatides, gums, and waxes are present in the crude oil, a proportion of which separates on standing in the store tanks (up to 1% of crude oil). Heating the raw settled oil rapidly to 110°C induces a "break," which means up to 0.5% of a gummy-impurities deposit. Refining follows similar procedures as those used for soybean oil, and is concerned to minimize the level of phosphatides. Chilling may cause some wax to separate from refined linseed oil, and likely to obtain a satisfactory color, up to 3% of the more highly acid-activated earths must be used in bleaching. If the oil is to be hydrogenated, possibly catalyst poisons will be sufficiently taken up by only 1% of activated earth, and the color will then be further lightened during hydrogenation. For crude oil BS 242/243/259/632: 1969, quote a 70 Y 6 R (1") maximum. Commonly, 40 Y 4 R (1") will be met. Refined linseed oil is quoted as a 20 Y 1.5 R (1") maximum. Again, considerably lighter may be met in practice, and for a hydrogenated oil after conventional posttreatment, 1 Y 0.2 R (1") is attainable. The same British Standards give saponification values 188–195 and maximum. FFA in crude oil is 2%. Unsaponifiable (%) should not exceed 1.5. An oil with a fairly high IV of 192 is likely to have an n_D^{60} of 1.4658, but examples are known in which the range is n_D^{60} 1.4647–1.4697. A typical fatty-acid composition would be as shown in Table 8.15.

Prior to hydrogenation, the oil should be refined to the usual FFA of 0.1% of maximum, soap of 0.05% of maximum, and dried to H_2O at 0.05% of maximum.

TABLE 8.15
Fatty-Acid Composition of Linseed Oil

Fatty Acid	%	Fatty Acid	%
C16:0	6	C18:0	4
:1	trace	:1	22
C20:0	trace	:2	16
:1	trace	:3	52[a]

[a]Variation in linolenic (%), 35–62.

Hydrogenation

A potent, somewhat nauseous sweet off flavor and odor are found to be features of partly-hydrogenated linseed oil. To avoid them developing after deodorization, customarily, one hydrogenates to over 40°C mp. Linseed oil was not unique in this respect. Oils containing appreciable amounts of linolenic acid afforded a similar if less severe risk. Keppler et al. (1965) isolated 6 nonenal as the main off flavor, the *trans* 6 nonenal being the most potent individual item arising from, first, the partial hydrogenation of linolenates, then isomerization and oxidation, followed by the final degradation of iso-linoleates. Following this identification, the Vlaardingen laboratory (Keppler et al., 1967) demonstrated how the 9,15 and 8,15 iso-linoleates arising during hydrogenation were specifically the isomers responsible for the subsequent formation of the *trans* 6 nonenal whose threshold of odor perception in paraffin oil is 0.005 ppm, and taste perception is 0.0003 ppm (Meijboom & Stroink, 1972).

$$\begin{array}{l} \qquad\qquad\qquad CH-CH_2-CH_2-CH_2-CH=CHO \\ \qquad\qquad\qquad \| \\ CH_3-CH_2-CH \end{array}$$

trans 6 nonenal

Higher-Melting Hardened Oil (also see similar section title in Chapter 2—"Higher-Melting and Fully-Saturated Hardened Oils")

At 40°C mp, the iso-linoleates have virtually ceased to be present; hence, the so-called "linolenic hardening flavor" cannot form. Conditions for this class of hardening are described in the similar section title in the above title heading to this section. The question presents itself how best to attempt the production of lower-melting hardened linseed oils with acceptable oxidative stability (bringing down the triple unsaturates to c. 2%) and also keep *trans* isomers to a minimum. These requirements are best met by a two-stage hardening as described in the "Two-Stage Hydrogenation" section of Chapter 2. For hardening according to the "Higher-Melting and Fully-Saturated Hardened Oils" section in Chapter 2, the catalysts to be chosen are: Süd-Chemie G53 and KE-NF 2, Engelhard Nysosel 325, Hoecat 882 OF, Unichema 9910, and Calsicat E472 D.

Two-Stage Hydrogenation (see also the same-titled section in Chapter 2)
The well-refined linseed oil is given a dose of 0.10–0.20% of nickel/oil (fresh catalyst), and hydrogenated within the 120–140°C range at 2–3 atm. Catalysts which should perform well in this task are Unichema 9920 and Süd-Chemie T4460 with G70, G53 K, and G95 D as possible alternatives. Among other catalysts, Engelhard Nysosel 325, Hoecat 882 OF, and Calsicat E472 D also provide alternatives. If the initial IV is 192, which as quoted above is fairly high and suggests around 62% of linolenic acid in the composition, hydrogenation continues at the restricted temperature to 130 IV, then is allowed to rise as rapidly as possible to 180°C so that the fall of 120–105 IV takes place in the 170–180°C range, the final IV being 103–105. From the time the temperature is increased from 140°C, the pressure is reduced to c. 1 atm. Note that this regime is at the limit of the rule to restrict hardening temperature until any danger of cyclization, etc has passed (see the "Cyclization and Polymerization" section in Chapter 2). However, in this example, a high IV was chosen, and significant hydrogenation continues as the temperature moves steadily upward from 140°C so that the risk is now very small. In these circumstances, the linolenic percentage drops below 2% as 105 IV is approached, acceptable oxidative stability becomes very probable, but complete freedom from linolenic hardening flavor is less certain, owing to the extremely low concentration at which it can be detected by the consumer. The above, however, is a useful example of how to deal with this type of problem, and with other oils (e.g., soybean), it may be much less severe. As hydrogenation progresses below 105 IV, the slip melting point rises through the low thirties, while from 95 IV onward the oxidative stability reaches a steady and satisfactory level.

The inducement to employ hydrogenated linseed oil for edible purposes was heavily diminished by the wider cultivation of soybean in particular; also sunflower oil and rapeseed oil containing a low level of erucic acid are increasingly available. In conclusion, the shorter the time between the conclusion of pretreatment and the commencement of hydrogenation, the better, since in such an interval, oxidation of the oil is taking place.

Maize (Corn Oil)

Maize or corn oil is present in the germ of the maize grain to the extent of around 54% (*Oils and Oilseeds,* 1971), but as the germ represents only a small percentage of the whole grain, the fact that c. 200,000 tons of oil are produced annually depends on the large production of maize grain itself, mostly in the United States. Even so, much maize grain is consumed as such, without the separation of the germ by either of the so-called wet or dry processes.

The oil is another member of the oleic–linoleic class; 80–85% of all fatty acids present consist of these, other unsaturated fatty acids being present in little more than trace amounts. The inherent stability of such an oil is further helped by the antioxidant quality of its tocopherol content. In these circumstances, much of the oil is used as such as a table oil or source of polyunsaturated fatty acid in a domestic margarine.

Partly-hydrogenated oil is also used in some margarine compositions. IV is normally c. 125, although a variation of 103–133 is quoted and *Codex Alimentarius* (CODEX STAN 25-1981) suggests the range 103–128. The *Codex Alimentarius* also gives saponification values of 187–195, with an unsaponifiable (%) not above 2.80. The range in RI corresponds approximately to n_D^{60} 1.4580–1.4610. Phosphatides may amount to 2% at most, so that degumming, neutralizing, and earth-bleaching procedures applied to soybean oil will be effective here. Both copper and iron contamination at a fraction of 1 ppm are not hazards; color (BS651:1967) may reach 70 Y 7 R (1"), but much less is quite normal; and FFA content is likely to be between 1.5 and 4%. A feasible average fatty-acid composition for maize oil would be as shown in Table 8.16. Degumming, neutralization, and washing will reduce FFA content to a 0.1% maximum, soap to a 0.05% maximum, and, after active earth bleaching, phosphorus under 10 ppm and soap nil are likely. At this stage, the color will probably have dropped to under 5 R (54").

Hydrogenation

The oil is valued principally for its quality as a soft oil, but if hydrogenation is economically attractive, one would have no difficulty producing a range of low- and medium-melting-point fats, as was done in the case of groundnut and cottonseed oils.

Normal and Iso- or trans-*Promoting Hydrogenation (also see "Normal Hydrogenation" and "Iso- or* trans-*Promoting Hydrogenation" sections of similar titles in Chapter 2)*
Partly-exhausted catalyst at c. 0.3% of nickel/oil is suitable, with the proviso that if the maximal gradient in the melting curve is sought, this is most likely to be obtained

TABLE 8.16
Advisory Fatty-Acid Composition for Edible Maize Oil [CODEX STAN 25-1981 Suppl. 1 (1983) and Amended in 1987] (Mounts, 1987)

Fatty Acid Type	%
C12:0	<0.3
C14:0	<0.3
C16:0	9–14
C16:1	<0.5
C18:0	0.5–4.0
C18:1	24–42
C18:2	36–62
C18:3	<2.0
C20:0	<1.0
C20:1	<0.5
C22:0	<0.5
C24:0	<0.5

with the specially sulfur-poisoned catalyst produced for this purpose. Since the starting point is at a higher (Frankel et al., 1980) IV than groundnut and cottonseed oils, particular values of SFI 20/30 and their corresponding slip melting points are attainable some 10–12 IV units earlier. Hence, for a drop n_D^{60} of 1.460–1.4545, we may be already in the mid-thirties for slip melting point and SFI 20 of 46–50. Specifically for the hydrogenation of maize oil, Unichema recommends Pricat 9920, Engelhard Nysosel 222, and Calsicat, their own E472. Süd-Chemie's G 95 D and Hoecat's 882 OF would appear to be suitable in this application. If fresh catalyst is to be used in the 180°C and 3 atm pressure range, 0.05% of nickel/oil would be a sensible dosage level at which to commence trials with refined corn oil of normal quality.

Normal versus Low-Temperature Hydrogenation

In 1990, Unichema published (*Selective Hydrogenation of Edible Oils, Edible Oil Hydrogenation,* 1990) a detailed study of the hydrogenation of corn oil by using Pricat 9910 and Pricat 9920. Refined oil (sulfur 9.9 ppm and phosphorus 0.8 ppm) of 117.5 IV and 0.16% of FFA was used. A constant pressure of 3 atm (absolute) (i.e., 30 psig) was employed. To favor polyene selectivity and *trans* formation, a temperature range of 150–200°C was used in one series of tests and 120°C in a second to suppress *trans* formation. At both temperatures, 9910 was the more active, but at 120°C, only moderately so; the activities of both were good. At 150–200°C, 9910 produced the lowest *trans* and 9920 the lowest solids at 30°C and 35°C. At 80 IV, 9920 had generated the least C18:0 and the most C18:1. A dose of 0.02% of nickel/oil was used. For the 120°C series, 0.04% of nickel/oil was used. Pricat 9920 again demonstrated its higher selectivity for polyenes, and produced marginally higher *trans.* The report emphasizes that the quality of oil, hydrogen, and plant performance, as well as product specification, influences the type and dosage of the catalyst required.

Olive Oil

The ripe olive may contain 10–60% of oil (*Oils and Oilseeds,* 1971) or 35–70% (dry basis) (Swern, 1964, 1979), and this is obtained by as many as three successive pressings, the first of which is described as virgin olive oil. The residue from the last pressing, known as olive pomace, contains up to 10% of oil, and this has commonly been extracted by solvents. When carbon disulfide was used as the solvent, this gave rise to the clumsy title, "sulfur olive oil." Other solvents are now used to extract what CODEX STAN 33-1981 describes as "olive residues." This oil may be made edible by "refining methods which do not lead to the alteration in the initial glyceridic structure." For the full recommended international standard for olive oil, virgin and refined, and for refined olive-residue oil, one must consult the *Codex Alimentarius* CODEX STAN 33-1981 Suppl. 1 (1983). Abundantly evident is that in most circumstances, the hydrogenation of olive oil affords no economic advantage. Should a temporary local excess of poor-quality oil exist, an attempt might then be made to convert this to a stable and more presentable fat.

TABLE 8.17
Fatty-Acid Composition of Olive Oil [CODEX STAN 33-1981 and Suppl. 1 (1983)]

	Fatty Acid	%
Lauric	C12:0	Not detectable
Myristic	C14:0	0–0.05
Palmitic	C16:0	7.5–20.0
Palmitoleic	C16:1	0.3–3.5
Stearic	C18:0	0.5–3.5
Oleic	C18:1	56.0–83.0
Linoleic	C18:2	3.5–20.0
Linolenic	C18:3	0.0–1.5
Arachidic	C20:0	Minute only
Gadoleic	C20:1	Minute only
Behenic	C22:0	Minute only
Erucic	C22:1	Not detectable
Lignoceric	C24:0	Minute only

The *Codex* puts forward this possible range in fatty-acid composition for olive oil (Table 8.17),bearing in mind that, like other vegetable oils, its makeup is strongly influenced by its environment; oils from cooler climates have higher IVs, but even when the IVs are very similar, this result perhaps was achieved by different combinations of various fatty acids.

The refined olive-residue oil would have: IV 75–92, unsaponifiable (%) 2.5 maximum, saponification value 182–193, n_D^{20} 1.4680–1.4707 (say, n_D^{60} 1.4540–1.4567), FFA 0.3% maximum, PV 20 mequiv. of O_2/kg maximum, E. 1.1.232 nm 6.0 maximum, E. 1.1.270 nm 0.2, soap nil, insoluble impurities 0.05% maximum, volatile at 105°C 0.1% maximum. The most notable nonoil constituent of olive oil is the hydrocarbon squalene (0.1–0.7).

Hydrogenation

With linoleic acid present often at less than 10%, it is evident that saturates soon begin to form in any hydrogenation; hence, a flat melting curve for medium melting point would not be difficult to secure with a fresh catalyst (0.04% of nickel/oil) and temperature restricted to 150°C maximum; possibly, 120°C could be chosen and 3–5 atm pressure. More probably, a partly-exhausted catalyst at 0.25% of nickel/oil, 180–200°C, and 1 atm pressure would be chosen to promote as high a proportion of low melting-point solids (*trans* isomers) as possible within an IV drop of 10–15 units. Very probably those solvent extraction plants working at times on olive residues or pomace and in association with a refinery will have no difficulty in disposing of their product in the market for edible oils and soapery fats without having to resort to hardening. Any of the commercial catalysts listed as suitable for vegetable-oil hydrogenation would serve here.

Palm Oil

The oil palm which originated in West Africa is also extensively cultivated in the Congo, Southeast Asia, and Central and South America. An old palm now growing in the Botanic Gardens at Bogor some miles inland from Jakarta proclaims it was introduced there in 1848 and is the mother plant of the Southeast Asian plantations. Evidently, four trees were sent from the Botanic Gardens, Amsterdam, to Java via Mauritius about this same time for purely ornamental purposes (Langstraat, 1976). Huge plantations in the Congo were later organized by the first Lord Leverhulme, demonstrating the feasibility of improved oil yield and quality. In the decade 1970–1980, the scientific cultivation of the oil palm advanced Malaya to the preeminent position in this trade (Bek-Nielson, 1974). The estimated world production of palm oil in 1992 is almost 12 million metric tons—second only to soybean oil. Hundreds of varieties of palm exist; the selection and breeding of types most suitable to a particular environment has long been an important scientific study in the world's vegetable-oil business (Achaya, 1994).

Although oil from the fruit (c. 56% of content) is the larger part and more commercially important, oil from the kernel (c. 47% of content) is also very valuable; it is one of the most important lauric oils (along with coconut oil, which it resembles in several respects), being therefore markedly different from palm oil. Palmitic and oleic acids in nearly equal proportions account for about 80% of the fatty-acid composition of palm oil; which predominates helps to decide the physical character of the particular variety in which they occur, but another factor of outstanding importance, especially in the case of palm oil, is the way in which the fatty acids are grouped to form the triglycerides (Swern, 1979). This was clearly explained in some detail (Loncin et al., 1970; Taylor, 1973), but is not the immediate concern of this book. Suffice it to say that the proportion of trisaturates (6–9%) is appreciably less than would result from an overall strictly random distribution; oleo dipalmitin and dioleo palmitin account for about one-half of the triglycerides present; linolenic acid is absent from mature fruit. Potentially, then, palm oil has a composition which is resistant to atmospheric oxidation. Some typical fatty-acid compositions are shown in Table 8.18 with the possible variation which was observed over the years for the three principal fatty acids: palmitic, oleic, and linoleic. The composition of fatty acids in oil from hybrid varieties and its possible nutritional value were also the subjects of investigation (Clegg, 1973).

Reportedly, the β-carotene content of palm oil acts as a pro-oxidant in opposition to the antioxidant influence of the tocopherols also present, and when the β-carotene is much diminished by refining, bleaching, etc, this lifts the oxidative stability very markedly (Meara & Weir, 1976; Swern, 1979). As much as 0.2% of carotene is found in some varieties of palm oil, giving the oil its characteristic orange–red coloration. Top-quality palm oil has a color of c. 26 Y 26 R (1"). This

TABLE 8.18
Fatty-Acid Composition of Palm Oil

Fatty Acid Type	CODEX STAN 125 1981 %	Nigerian %	Congo %	Sumatra %
C12:0	<1.2		<0.05	<0.05
C14:0	0.5–5.9	1	1	1
C16:0	32–59	40	44	46
C16:1	<0.6	0.5	<0.05	<0.05
C18:0	1.5–8.0	5	5	4
C18:1	27–52	40.5	40	39
C18:2	5.0–14	12	10	10
C18:3	<1.5	0.5		
C20:0	<1.0	0.5	0.5	<0.05

pigment is markedly thermolabile and decreases rapidly in the 180–220°C range; even higher (250–270°C) temperatures are employed very briefly in some processes; it can be decreased by activated earths and by hydrogenation. Once oxidized, it is most resistant to further removal by any means, and is then responsible for a persistent dull brown color. A bleachability test for palm oil, such as the Bernam test (Bek-Nielson, 1974; Krishnan, 1975), allows the oxidative state of a palm oil to be assessed because the test is especially sensitive to the amount of undamaged carotene present; hence, it shows a considerable lightening of color when this amount is high (i.e., a greater improvement), whereas some other tests looking possibly to a technical rather than an edible employment of the oil are designed more to indicate the limit of color removal by more drastic methods. Interestingly but not surprisingly, note in this connection that the palm oil of the most intense red coloration—such as Sumatra oil—gives the lightest color when hydrogenated; also, if crude hardened palm oil is allowed to stand for some hours after the catalyst is filtered, one may observe a gradual darkening of color. Johansson (1975) discusses the correlation of primary and secondary oxidation with ultraviolet absorption E.1.1. 235 and 270 results; he recommends that, to assess the quality of palm oil, measure—a useful move—the carotene content photometrically at 450 nm as described by Cocks and van Rede (1966). This approach is not new, and some users monitored Nigerian palm-oil shipments by this kind of method, expecting to find the peak value of E. 1.1.458 (in chloroform solution) between 2.5 and 3 for oil which had sustained little damage, whereas if the figure fell to 1.5, this suggested a loss of carotene and presumably some damage to the oil. Palm oils of different origin might possess a peak less pronounced than the example quoted and yet be of good quality. For hexane solutions, the peak is to be found at 450 nm, and where the oil has sustained little oxidation, multiplying the adsorption figure by the quite empiric factor of 427 was used to obtain the carotene content in ppm. The heating of the oil to above 100°C, even without oxidation, may displace the peak absorption to slightly below 450 nm.

Nigerian palm oil very often was found to have c. 35°C mp as compared to 37°C mp for Malaysian oil; the corresponding percentage of SFC at a 20°C variation would be approximately 18–22; variations in Malaysian oil of 35–38°C mp were noted (*Malaysian Palm Oil,* 1974, 1975), with SFI 20 ranging from 21 to 24. This type of difference is a variable feature, and must be considered in processing (hydrogenation, fractionation, etc) as it occurs. If the intention is to derive a product with a particularly steep SFI or SFC curve when hydrogenated, obviously the advantage is first to remove the higher melting triglycerides by fractionation and then to hydrogenate the more unsaturated (58 IV upward) fraction, which will usually account for about three-quarters of the whole. This softer or "olein" portion may also be used as a frying oil because of its basic stability; fractionation to produce the first olein will occur quite sharply in the 28–32°C range; if required, the first olein may be fractionated again after the removal of the stearin by filtration and then cooling to 20–22°C according to what is required in the way of the final texture.

Palm oil which is handled or stored in a warm temperature (i.e., above melting point) and a moist environment easily splits to give an extremely high FFA (5–50%). So-called "soft" palm oils were at the lower end of the FFA scale in the 5–10% FFA region, and are still normally well-fitted for edible use after appropriate refining. At the other end of the scale, the palm oil would be destined for technical use. For intermediate situations where the FFA has risen to over 12% and therefore where double that percentage of the removal of fatty matter has to be envisaged by the refiner, the situation will be strongly influenced by whether the conversion of palm oil to soap is almost as profitable as its use in edible products. Fortunately, the quality of crude palm oil has shown a steady improvement for years by the application of a more rational method of handling (Jacobsberg & Jacqmain, 1973). Although a contractual basis of 5% of FFA as norm may apply to many purchases of palm oil, far superior qualities are available from Malaya and Zaire with limits covering FFA at 2%, the metal content and degree of oxidation. Naturally, these superior grades primarily envisage the use of the palm oil as being for edible purposes. In any event, the level of phosphatides is expected to be less than 0.1%, while the sulfur content, so far as hydrogenation is concerned, is negligible. Prooxidants such as Cu and Fe have attracted much attention. For the best-quality crude oil, not more than 0.2 ppm of Cu and 5 ppm of Fe are to be expected, at most (*Malaysian Palm Oil,* 1974, 1975). Commercially, the accepted maximum for moisture and impurities is normally 0.5%, of which only a little, c. 0.02%, is likely to be a solid impurity. IV ranges from 52 to 58, RI n_D^{60} 1.4495–1.44515, unsaponifiable (%) up to 0.8, and crude oil PV 4–16 mequiv. O_2/kg. A well-sustained effort is to improve the quality of crude palm oil as delivered to users. This is well-summarized by Johansson (1975) in his description of expectations for current (1975) palm-oil quality (Table 8.19) compared with feasible improvements in the future.

TABLE 8.19
Current (1975) and Future Quality Standards for Crude Palm Oil (1975)

Characteristics	Present Specification	Future Specification
Water	max. 0.5%	max. 0.5%
Insoluble impurities	max. 0.5%	max. 0.5%
Free fatty acids	max. 3%	max. 2.5%
Anisidine value	max. 6	max. 4
Peroxide value	max. 5 meq. v/kg	max. 4 meq. v/kg
Oxidation value	max. 15	max. 10
Iron	max. 5 ppm	max. 3.5 ppm
Copper	max. 0.08 ppm	max. 0.05 ppm
AOM-stability	min. 53 h	min. 60 h

The question was raised (Loncin, 1975) as to whether the thermal bleaching and deodorization of palm oil at temperatures which reach 270°C for a few minutes, or 220–240°C for somewhat longer, give rise to PAHs from the carotenoid pigments present. After a full investigation (Biernothl & Rost, 1968), this proved to be a quite negligible effect, and no more noticeable than with soy, corn, and groundnut oils (Biernothl & Rost, 1968; Rost, 1976). The temperatures discussed here for hydrogenation do not exceed 200°C.

Hydrogenation

A slight elevation of the melting point to 41°C by dropping the IV about 5 units with the aid of a largely exhausted catalyst is feasible, but the resulting, very lightly hydrogenated palm oil may show signs of flavor reversion. Generally speaking, moderate hydrogenation is carried a little further. When palm oil is blended in large proportion with other, more unsaturated vegetable oils to produce a feedstock for an all-hydrogenated vanaspati or shortening, obviously—since the final melting point is limited by law or performance considerations—the bulk of the hydrogen in this particular case will be taken up in reducing the more unsaturated components (linoleate, linolenate) of those oils present other than palm oil (e.g., cottonseed or soybean oil). A large amount of palm oil is hardened to a mid-forties melting point in normal or *trans*-promoting conditions; hence, weight for weight it contributes substantially to the solid component of margarine blends; a much smaller tonnage is hardened further to melting points ranging from 52 to 58°C, the latter corresponding with almost complete saturation.

Normal Hydrogenation (see section with the same title in Chapter 2)

This class of hardening provides an opportunity for the repeated use of partly exhausted catalyst at 0.06–0.20% of a nickel/oil level, 1–3 atm, and 180°C. Gassing may commence at 160°C, although for those hardened oils where only modest IV drops of up to 15 units are involved, obviously considerable dependence is upon the steam heat applied to reach the control temperature of 180–200°C. Temperatures above 160°C offset catalyst poisoning due to any remaining carbon monoxide in the hydrogen being used (nowadays a much smaller problem), promote *trans* isomers, help shorten the cycle time, and assist in the removal of carotenoid pigment. In dropping an initial n_D^{60} 1.4513–1.4502, a fall of about 5 IV is obtained to correspond with 42°C mp and an SFI 20/30 of 52/38.5 (% SFC at 20/30°C = 64/40). A further fall to n_D^{60} 1.4500 lifts the slip melting point to c. 43°C and the SFI 20/30 to 57.4/40.4 (% SFC at 20/30°C = 71.5/42); at c. n_D^{60} 1.4498 and IV c. 42. The slip melting point is c. 45°C and SFI 20/30 is 61/50 (% SFC at 20/30°C = 83/54). At c. n_D^{60} 1.4494 and IV 37, the slip melting point is 46–48°C, and the SFI 20/30 is 66.7/55.6 (% SFC at 20/30°C = 91/61), and at IV 25, the slip melting point reaches c. 50°C, and the SFI 20 is 74 (% SFC at 20°C = 95).

Obviously, the original IV slip melting point of the palm oil being used, whether 1 or 3 atm pressure is being employed, and how large the dose of catalyst is, all have some bearing on the figures just quoted, and these can merely be regarded as typical. If the reaction time for the higher-melting items proved unacceptably long, a practical answer would be to change to a lower dose of fresh catalyst, or even to enliven the old catalyst with a modest addition of fresh catalyst, the latter being of a reasonably selective type. Süd-Chemie (Nickel Catalyst KE-NF 20 and KE-FS40) gives an extensive range of information on the hardening of a palm oil (52 IV, 36.4°C mp) at 1 atm, 180°C, and 0.03% of nickel/oil fresh KE-NF20 catalyst, and also gives results when hardening at 2 atm, 180°C and 0.01%, 0.015%, and 0.03% of nickel/oil, respectively. Between 44 and 42 IV, the proportion of linoleic acid falls to a low level (c. 2%), and stearic acid rises steadily. In this situation of a decreasing number of double bonds and a falling proportion of *trans* isomers as against a rise in saturates, the SFI curve begins to grow less steep. Thus, in the test run, employing 0.01% of nickel/oil, 180°C, and 1 atm at 44.4 IV, the difference in solid-content figures [nuclear magnetic resonance (NMR)] at 20–30°C is 23.3%, but by 36.5 IV, this difference has fallen to 15.6%. Other catalysts suited for use with palm oil would be Engelhard Nysosel 222, Calsicat E472 D, Hoecat 882 FA, and Unichema 9910.

Iso- or trans-*Promoting Hydrogenation (see section by the same title in Chapter 2)*

For the lower-melting items in the "Higher-Melting Hardened Oil" section of this chapter, one could substitute the specially sulfur-poisoned catalyst available commercially (Engelhard De Meern BV, The Netherlands; The Versatile Catalyst for Vegetable Oil Hydrogenation. Influence of Catalyst Structure on Hardening of Edible Oils and Fatty Acids, Unichema International, Germany) (see also the "Examples of Commercial Nickel Catalysts" section in Chapter 7) for the old catalyst which arises from other operations in the hardening plant. This would maximize the SFI 20/30 difference when a steep curve is being sought. This advantage is substantially enhanced if a start is made not from palm oil itself, but with the soft (olein) fraction, which amounts to around 75% when the parent oil has the higher-melting components removed by simple chilling and filtering or a selective solvent-wetting action in support of separation. Such olein will have an IV of 58 or some units higher. If a temperature of 180–190°C and an up to 5 atm pressure are used in conjunction with c. 0.2% of nickel (sulfur-promoted)/oil, the 36–38°C mp products show a difference in SFI 20/30 of about 23 units (SFC c. 21%) corresponding with a drop in IV of 5–10 units, depending on the IV at the starting point. See also the "Iso- or *trans*-Promoting Hydrogenation" section of this chapter.

Higher-Melting and Fully-Saturated Hardened Oils (see section with the same title in Chapter 2)

For this class of hardening, a 3–5 atm pressure and 180°C are best, combined with 0.2% of fresh nickel/oil. The nickel dose may be decreased or increased according to how convenient the gassing time is found to be. At n_D^{60} 1.4480, the IV has fallen to c. 19, and the slip melting point reaches 52–54°C; at 15 IV, the slip melting point is 54–56°C; and finally, at 3 IV and below, a slip melting point of 58°C is obtained. Just as the olein fraction is a suitable starting point for medium-melting products, so the hard or "stearin" fraction is a convenient feedstock for the production of high-melting palm products. The catalysts suitable for this type of hardening are: Süd-Chemie KE-NF2, Engelhard Nysosel 222, Calsicat E472 D, Hoecat 882 FA, and Unichema 9910.

Palm-Kernel Oil

The kernel of the oil-palm fruit contains up to 53% of oil (dry basis), but 46–48% is a fair average (*Oils and Oilseeds,* 1971). As another lauric oil, it so closely resembles coconut oil (see "Coconut Oil" section in this chapter) that the two were, in large measure, interchangeable in both the edible-products and soap-making fields. Worldwide production of palm-kernel oil amounts to about 1.5 m tons p.a. as against 3 million tons p.a. coconut oil. In general, palm-kernel oil has a rather higher IV (13–23, average c. 18) than coconut oil (7–11 IV) because its content of oleic acid is higher; its content of C6, C8, and C10 fatty acids is lower, which tends to give it slightly the higher melting point of the two—i.e., palm-kernel oil typically 26°C, coconut oil 25°C—but the complete ranges overlap considerably. Typical fatty-acid composition of a palm-kernel oil is given in Table 8.20.

The low degree of unsaturation of palm-kernel oil gives it a marked resistance to atmospheric oxidation, but, like coconut oil, if short-chain FFAs or monoglycerides do form by the hydrolytic splitting of the fat, these are potent soapy off flavors. BS 652 envisages no higher than 6% of FFA (as lauric acid) in a good crude oil. Sulfur and lecithin are virtually absent; copper under 0.1 ppm and iron c. 2 ppm are satisfactory; PVs over 8 indicate oxidative damage, and this may lead to color reversion after the deodorization of the refined oil. The rigorous removal of pro-oxidant metals at the bleaching stage and hydrogenation both assist color stability. The usual unsaponifiable (%) is 0.4, although up to 1.2% may be met: typical saponification

TABLE 8.20
Fatty-Acid Composition of Palm-Kernel Oil (CODEX STAN 126–1981)

	Fatty Acid	%
Caproic	C6:0	<0.5
Caprylic	C8:0	2.4–6.2
Capric	C10:0	2.6–7.0
Lauric	C12:0	41–55
Myristic	C14:0	14–20
Palmitic	C16:0	6.5–11
Stearic	C18:0	1.3–3.5
Oleic	C18:1	10–23
Linoleic	C18:2	0.7–5.4

values ranging from 230 to 254 RIs of n_D^{60} 1.4430–1.4439 are known, and, like other lauric oils, are appreciably lower than for most other classes of oils. BS 652 suggests a 20 Y 1.5 R (1") maximum for crude oil. The color depends on how the oil was obtained, shipped, and stored. Poor color—35 Y 3.7 R (5 ¼"), 10 Y 1.0 R (1")—occurs, and excellent crude material at c. 22 Y 1.3 R (5 ¼"), 4 Y 0.4 R (1") is available. After full refining, 10 Y 1.0 R (5 ¼") at most is expected of a food product, and half this color is obtainable. No problem arises in pretreating the crude oil to the standard suggested (see the "Cleaning of Oils Prior to Hydrogenation" section in this chapter) prior to hydrogenation.

Hydrogenation

Note: To obtain the expected results consistently in hydrogenation, a catalyst made up in a lauric oil (coconut was commonly used for many years) or a nonpyrophoric (stabilized) catalyst free from organic material must be used: Nysel N-14 (Harshaw Catalysts) is an example, but other catalyst manufacturers should not find any difficulty in preparing a fresh catalyst incorporated in coconut or palm-kernel oil. If other oils are present during the hydrogenation of a lauric oil, either the required slip melting point is obtained earlier at a rather higher IV, or if the oil is hardened to a very low IV, a slip melting point is obtained which is above that normally obtained from uncontaminated oil. This, of course, may well suit the purposes for which the product is intended: this is mentioned here and later simply so the manufacturer may recognize the fact. Since palm-kernel oil has a higher IV than coconut oil, this gives greater room for maneuvering in hydrogenation. As with coconut oil, however, the major fatty acids in palm-kernel oil differ among themselves in melting point by less than 20°C (Swern, 1979), hence, the quick-melting behavior of the triglycerides of the natural fat itself. Hydrogenation does sharpen this melting behavior, and especially when carried definitely below 1 IV, confers an even greater stability on an already stable fat. Like coconut oil and palm oil, palm-kernel oil was fractionated by conventional dry and wet methods. The "stearin" and "olein" fractions, perhaps better referred to as "solid" and "liquid" fractions, found their separate uses in confectionery coatings, bakery, and soap-making applications. They also were hydrogenated as separate items, in many instances like the oils themselves, with the use of a catalyst clothed in a nonlauric oil such as a well-hardened vegetable oil of another class. The level of nonlauric oils introduced in this way affords the chance of elevating the slip melting point modestly, possibly by 2°C, so that a fully hydrogenated product shows a 41°C mp rather than 39°C. The extent to which a group of triglycerides containing short-chain saturated fatty acids and longer chain unsaturated fatty acids can be separated from the remainder is illustrated by Thomas and Paulika (1975) by using 2-nitromethane as a solvent, chilling, and filtration.

An average palm-kernel oil of 18 IV and c. 26°C slip mp would probably have an SFI 20/30/40 of 36–41/1.9–2.8/0.7–1.9 (% SFC at 20/30/40°C = 43–45/2–4/0.5–1). In Table 8.21,the increase in the liquid fraction of oleic and linoleic acids (21.4% and 4.0%) affords an obvious clue as to why when such a fraction, or blend containing a considerable proportion of such a fraction, is hardened, the attainable slip melting point is higher than would otherwise be obtainable. The increased proportion of C18 acids makes possible in the first place a larger proportion of C18 *trans* isomers (low-melting solids), and as these are hardened in turn, progressively more stearic-acid groups appear in the triglycerides.

Normal Hydrogenation (see section with the same title in Chapter 2)

Hydrogenation at 180°C and c. 3 atm pressure is normal; obviously, since IV drops are small, the attainment of 180°C is dependent on heat supplied via heating coils. A dose of 0.05–0.15% of fresh nickel/oil suffices for most hydrogenations, and a considerable reuse of catalyst is feasible since catalyst poisons in the oil are likely to be at a low level. The following sequence of results is to be expected, the catalyst possibly being incorporated in hardened oil other than a lauric oil. Commencing at 18 IV, n_D^{60} 1.4437 and a 26°C slip melting point, a drop to n_D^{60} 1.4426 (9 IV) gives a slip melting point of c. 31°C, SFI 20/30/35 = 56/13/1 (% SFC at 20/30/35°C = 70/12/1); n_D^{60} 1.4423 (8 IV), 33°C mp, SFI 20/30/ 35– 58.5/17/1.9 (% SFC at 20/30/35°C = 73/12.5/1.5); n_D^{60} 1.4122 (7 IV), 35°C mp, SFI 20/30/35 = 59/20.5/3.7 (% SFC at 20/30/35°C = 73.5/20/3); n_D^{60} 1.4420 (3.0 IV), 38°C mp, SFI 20/30/35 = 60.3/23/8.1 (% SFC at 20/30/35°C = 75/22.5/6.7); n_D^{60} 1.4417 (0.8 IV), 39 °C+ mp, SFI 20/30/35 = 65/28/10.7 (% SFC at 20/30/35°C = 88/26/9).

The slip melting point may be further increased if moderate proportions of other vegetable oils containing C16 and C18 fatty acids are added to the hardening feedstock. Conversely, if the solid fraction alone is hydrogenated, the D20 increases very rapidly, so that in falling to a final IV of 1, a 36°C mp and SFI 20/35 of 69/6.5 (% SFC at 20/35°C = 94/5.3) are attainable, and hence, a very quick-melting fat results. Reference to Table 8.21 again provides the answer to this phenomenon, as evidently both lauric (C12) and myristic acids (C14) have substantially increased as against the decrease in the liquid fraction.

TABLE 8.21
Fractionation of Palm-Kernel Oil (Thomas & Paulika, 1975)[a]

		Palm kernel Oil	Solid Fraction	Liquid Fraction
Wiley melting point (°F)		83.6	87.7	72.4
Wiley melting point (°C)		28.6	30.9	22.4
Iodine value		15.3	5.2	25.6
Yield (wt.%)		—	50	50
Fatty acid composition (wt.%)				
Caproic	C6:0	0.3	0.1	0.6
Caprylic	C8:0	4.3	2.1	6.9
Capric	C10:0	3.9	3.1	4.6
Lauric	C12:0	49.5	57.5	42.5
Myristic	C14:0	16.1	21.5	11.0
Palmitic	C16:0	7.8	8.3	7.0
Stearic	C18:0	2.3	1.9	1.9
Oleic	C18:1	13.7	4.8	21.4
Linoleic	C18:2	1.9	0.6	4.0
Linolenic	C18:3	0.1	T	0.1
Arachidic	C20:0	0.1	0.1	T

T = trace (<0.05%).

Finally, when very low final IV results are being sought (e.g., less than 1 IV), the hydrogenation is continued past the point where an obvious change in RI ceases, the minimal value for which is c. n_D^{60} 1.4417. The IV should be checked in the final stages of a rigorous hydrogenation of this type, and this may continue to fall by fractions of one unit for an hour or longer after 1 IV has been attained. Similarly, Süd-Chemie's results with coconut oil hardening (Nickel Catalyst KE-NF 20 and KE-FS40) illustrate that even after there is little detectable change in IV, the melting point may continue to rise fractionally for about another two hours hydrogenation (see the "Normal Hydrogenation and Full Saturation" section in this chapter). The same catalysts as mentioned for use with coconut oil (see the "Coconut Oil" section in this chapter) will serve here.

Rapeseed (Colza) Oil

New Varieties of Canola Oil, Canbra Oil, etc.

Members of the Brassica family, such as rapeseed and mustardseed, have been cultivated in Asia for up to 4000 years (Achaya, 1994) for the production of edible oil whose spicy flavor was highly appreciated and derived from the presence of sulfur-containing compounds–the glucosinolates. Several varieties of Brassica have been found to thrive in northern and eastern Europe; France, Sweden, Germany, and Poland have produced large tonnages for years; (Thomas, 1982). Canada commenced large-scale production during the 1940s; Britain commenced production in the 1970s, and by 1982 became Europe's second largest producer (Ward, 1985). Characteristically, classic rapeseed oil contained sometimes as much as 50% erucic acid *(*C22*:*1), which in the 1960s become suspect as the principal reason why rats fed liberal amounts of the oil developed heart disease and other disorders. Some distinction in this regard has been made between the unhydrogenated oil and the oil after hydrogenation when erucic has been converted to behenic acid (Nolen, 1981). Steffansson and Downey, working in Canada, succeeded by 1969 in developing a strain of rapeseed in which the erucic acid content of the oil was under 5%. This low-erucic oil became known for a period as "Canbra" (Canadian Brassica), but the title was dropped when it was adopted as a corporate title by a processing concern.

The switch to low-erucic oil for human consumption appeared both prudent and commercially attractive (Hudalle, 1977; Vles et al., 1976; Ilsemann et al., 1976; Anon., 1978). Rapeseed meal or cake remaining after extraction of the oil could be fed to livestock in limited amounts; otherwise, growth was retarded and milk yield reduced. These disadvantages were linked in the 1960s with the presence of glucosinolates, which hydrolyze enzymatically to toxic isothiocyanates, oxazolidinethione, and nitriles (Thomke, 1981). In 1970, the Canadian government advised the production of low-erucic low-glucosinolate rapeseed, the so-called "double low." Hence, by 1974, the old-style high-erucic (HEAR) rapeseed had been replaced by a low-erucic (5% max.) (LEAR) variety in Canada, and then by 1980, some 80% of the Canadian crop was "double low," in which the erucic acid had fallen to under 3% and the glucosinolates were successfully limited to a maximum of 3 mg glucosinolate/g dry meal (Langstraat, 1976; Anon., 1981; Daun, 1986) traces the improvement up to 1985. Thomas (1982) states the name Canola is usually taken to mean a low erucic acid rapeseed oil and a rapeseed meal with a maximum of 2 mg–or 18 μmol–glucosinolate, expressed as 3-buterylisothiocyanate per g of dry oil free meal. Teasdale gives 3–5 ppm as the sulfur content of water degummed Canola, and this falls to 2 ppm after alkali refining. The old style crude rapeseed oil used to contain c. 60 ppm sulfur, which fell to c. 10 ppm after alkali neutralization and active earth bleaching. Obviously, this modem low sulfur content reduces hydrogenation times and nickel dosages (Teasdale, 1975; Devinat, 1980; Embong, 1980; Ahmad & Ali, 1980; Daun & Hougen, 1977). Also, improved flavor may result and the risk of color deterioration during deodorization, because traces of iron and nickel are present, is much less (Mag, 1983). It must be noted that some old-style HEAR oil continues to be produced, since it is obviously a valuable source of long-chain fatty acids such as erucic (C22:1) and behenic (C22:0) which continue to be needed by industry, especially for inclusion in special lubricants. Thus, in the United States as elsewhere, there is a demand for this class of oil (Swern, 1979, pg. 414ff; Mag, 1983). In Asia, there is less concern to obtain low-erucic oil as such, but the advantage of being able to feed much higher amounts of meal to livestock because its glucosinolate content is lower has an important appeal. The 1992 world production of rapeseed oils was over 9 m metric tons next to palm (almost 12 m) and soybean (16 m). There is a continuing move to double low type (Anon., 1981). At this point, some indication may conveniently be given of the fatty acid composition of the classic and new varieties of rapeseed oils (Table 8.22).

From a very large range of possibilities, the three compositions in Table 8.22 have been chosen as fitting in well with hydrogenation data provided later in this section. The "classic" composition (X) accords well with Asiatic and former European production. The "reduced erucic" (Y) represents an advanced degree of transition towards the new so-called "zero erucic" types, and indeed would qualify as LEAR if only the C22:1 did not exceed 5%; this is the oil on which Engelhard provides hydrogenation data using their Nysel DM3 catalyst (Harshaw Catalysts), and these would obviously be little affected if the erucic content was two or three percentage points lower. The IV of the neutralized bleached oil was 113-6 and the sulfur content 4.5 ppm. The "low to zero" (Z) composition is derived from those recorded in 1975 (Teasdale, 1975) as typical of "Tower Brand" Canbra oil, and the latest variety of this oil after prerefining shows a sulfur content of only 2 ppm and a typical IV of c. 118; this would now fit with Canola; hydrogenation

TABLE 8.22
Fatty Acid Composition of Rapeseed Oils

Fatty Acid		X Classic %	Y Reduced Erucic %	Z Low to Zero Erucic %	CODEX STAN 123-1981/7 (Mounts, 1987) %
	C<14:0				<0.1
Myristic	C14:0	0.5	0.1		<0.2
Palmitic	C16:0	3.5	4.3	4.9	2.5–6.0
Palmitoleic	C16:1	—	0.3		<0.6
Stearic	C18:0	1.0	1.8	1.9	0.9–2.1
Oleic	C18:1	13.0	52.7	57.0	50–66
Linoleic	C18:2	14.0	20.7	24.0	11–23
Linolenic	C18:3	9.0	10.4	10.4	5–13
Arachidic	C20:0	1.0	0.6		0.1–1.2
Eicosenoic	C20:1	7.5	3.4		0.1–4.3
Eicosadienoic	C20:2	1.0	—		
Behenic	C22:0	0.5	—		<0.5
Erucic	C22:1	47.5	5.7	c. 1.5	<0.5
Docosadienoic	C22:2	1.0	—		
Lignoceric	C24:0	0.5	—		0.2

data for this class of oil are given because of the importanee it has already established in Europe and Canada, which will increase as cultivation of "double low" types extends to other parts of the world. This seed contains 41% oil, of which 58% of the fatty acids may be oleic. The polyunsaturates (linoleic and linolenic) at c. 32% are lower than in soybean, sunflower, maize, and cottonseed oils, but higher than in groundnut and palm oils, while the saturates (5% palmitic, 2% stearic) are notably lower than in the great majority of other important vegetable oils (Anon., 1981). It is unfortunate that the proportion of linolenic to Iinoleic is as high as 10% to 22% in Canola as against 9% to 54% in soybean oil, but breeding programs may diminish linolenic and increase linoleic. Even so, the unhydrogenated Canola has found use as such in the composition of domestic cooking salad oils.

Work is also progressing to obtain a strain of HEAR for industrial use with a much-reduced glucosinolate content, so that the meal from such seed may be readily acceptable for cattle feeds. Another difficulty arising from the greatly increased proportion of C18 fatty acids in the new oils is that when hardened, their crystallization habit is such that if used at a high level in shortening and margarine compositions, grittiness or sandiness has very obviously developed within two months. This is largely avoided by incorporating about 30% of other oils, such as partly hydrogenated soybean oil in the blend (Teasdale, 1975; Anon., 1981). This same difficulty (and answer) would apply to any oil containing a very high proportion of fatty acids of the same chain length; C18 fatty acids in sunflower oil is another example; the difficulty becomes more noticeable the greater the degree of hardening. The progressive elimination of the long-chain monounsaturate erueic acid (fall in molecular weight) and the increase in linoleic acid (rise in unsaturation) in the new oils lifts the former upper IV and saponification value limits. A typical classic European rapeseed oil would have an IV of 103 (limits 97–120), whereas Canola would have an IV of c. 118 (limits 110–126). Conversely, high-erucic rapeseed oils tend to the lower IV and saponification values. The saponification value for classic rapeseed oil falls in the range 170–180, but for the low-erucic oil, 188–193 is typical. In relation to its iodine values, the classic rapeseed oil (high molecular weight) has a high refractive index of n_D^{60} 1.4580–1.4620; the lower-erucic oils, being influenced by higher IV but lower molecular weights, appear to fall in the range n_D^{60} 1.4580–1.4600.

Phosphatides in non-degummed oils have been noted (Teasdale, 1975) as high as 3.6%, but after conventional degumming, 0.6% is the likely maximum. The high sulfur content of crude Brassica oils of up to 60 ppm is well known; conventional alkali neutralization and active earth bleaching could be relied upon to bring this down to 10 ppm or lower, and deodorization removes a very large part of the sulfur compounds. Hence, high glucosinolate rapeseed oils when deodorized do not consume so much nickel catalyst, but whether the cost of the extra steam used would be compensated by the nickel saved is a matter for calculation. As stated above, Canola after degumming shows only about 5 ppm sulfur, which falls to about 2 ppm after alkali refining. This welcome benefit to the hardener arises, of course, from the greatly diminished glucosinolate content; hence, less isothiocyanate and other sul-

TABLE 8.23
Grade Requirements for Rapeseed Applied in Connection with Grading According to Quality on Arrival at Extraction Plant (Dahlean, 1973)

Analysis	Seed Grade 1	Seed Grade 2	Unacceptable
Chlorophyll (ppm)	30 max.	31–70	over 70
Free fatty acids (%)	1 max.	1.1–3.0	over 3
Moisture (%)	8 max.	8 max.	over 8
Seed purity (%)	9–100	90–100	under 90

fur compounds remain to poison the catalyst. Detailed examinations of the sulfur compounds in rapeseed oils and how these diminish during conventional processing have been made (Biasini et al., 1980; Embong, 1980; Ahmad & Ali, 1980; Daun & Hougen, 1977). Free fatty acids in crude oil do not usually exceed 1.5% (expressed as oleic acid), and half this is very acceptable.

From 1966, Swedish Oil Extraction Ltd. (EXAB) and the Swedish Oilseed Association monitored production of rapeseed in Sweden to establish quality grades. Chlorophyll content was found to be a reliable indicator of seed maturity, and hence, an augury of the flavor, stability, and color of the oil to be obtained (Dahlean, 1973). Table 8.23 shows some of the conclusions.

Freshly harvested seed should yield an oil of under 1.0 P.V. Unsaponifiable (%) should be less than 2. A rather poor crude oil color of 5 R (1") after alkali neutralization and a generous activated earth bleach (1%) drops to 5 R (5 ¼") and further after deodorization to c. 2.5 R (5 ¼"); a very good crude oil would give half of these color readings. If the oil is to be hydrogenated, only about half the dose of activated earth will be justified. Color removal from damaged or immature seed is reported as difficult (Teasdale, 1975). The oxidative state to be expected of good crude and processed oils is discussed in Johansson (1975), Teasdale (1975), and Dahlean (1973). A good crude oil with P.V. 1.0 and anisidine value (An.V.) 0.8, giving a TOTOX (2 P.V. + An.V.) of 2.8, would on neutralization and bleaching lose all P.V. and show only a slightly increased An.V. Crismer values for various low-erucic rapeseed oils (containing up to 4.1% C22:1) were all below 70 for high-erucic oil (20–45% C22:1), the values were from 76 to 82: hence, it is concluded that rapeseed oils falling within the category of maximum erucic acid content of 5% will all show Crismer values under 70 (Sahasrabudhe, 1977). With OLC tests becoming commonplace, the Crismer test appears likely to fall into disuse. The increase in specific gravity with declining proportion of long-chain fatty acid in the new Brassica oils has been well illustrated by Ackman and Eaton (Table 8.24) (Ackman & Eaton,1977).

Degumming, alkali neutralization, and bleaching are performed much as for soybean oil, phosphorus being then reduced to quite manageable levels (see "Cleaning of Oils Prior to Hydrogenation"). Similarly, trace metals in the crude oil (Fe under 2 ppm, Cu under 0.5 ppm) are adequately controlled by the pretreatment.

TABLE 8.24
Specific Gravities of Rapeseed and Canbra Oils[a] (Ackman & Eaton, 1977)

Oil Type		Specific Gravity[b]
45–55%	C22:1	0.9064
23.1%	C22:1	0.9123
20.1%	C20:1	0.9123
4.1%	C22:1	0.9159
3.2%	C20:1	0.9159
0.7%	C22:1	0.9171
1.6%	C20:1	0.9171
Refined soybean oil		0.9187

[a]Brassica oil of less than 5% C22:1.
[b]At 25°C the weight of a volume of oil in air compared with the weight of an equal volume of water.

As already indicated, if the sulfur content in the pretreated oil remains up at the 10 ppm level or more, this causes difficulty in hardening, reaction times are prolonged, consumption of nickel goes up, and it may prove impractical to attempt to produce partly hydrogenated oil of minimum *trans* isomer content by low temperature hydrogenation. Canola oil after degumming contains less than 5 ppm S, so this kind of restriction is disappearing.

Hydrogenation

Brassica oils now in use in different parts of the world present different possibilities to the oil hardener, according to whether they are of the old-style, high-erucic high-glucosinolate type, or the new type in which these are very greatly diminished. It is necessary to consider the likely success of each kind of hydrogenation procedure in relation to these contrasting types of oil. Published data are by no means complete, but where there are gaps, much can be inferred from what is already known in order to decide how best to proceed to obtain a useful result most readily: this is the approach used in the following sections. Today, principal catalyst manufacturers offer at least one catalyst which performs well with Canola. Also, such a catalyst, because of its activity and poison resistance, can reasonably be expected to perform with the old style rapeseed oil substantially better than those in use up to the early 1980s. Suitable catalysts are: Unichema 9920, Süd-Chemie T4460, Engelhard Nysosel 222, Hoecat 882 OF; Calsicat E472 D and E428 D both have application in rapeseed oil hardening.

Low-Temperature Hydrogenation (see Chapter 2, "Low-Temperature Hydrogenation")
As explained in Chapter 2, "*Low-Temperature Hydrogenation*" section, the object here is to bring linolenic acid down to c. 2% with a minimum formation of solids, the latter induding the low-melting *trans* isomers. To achieve this, the disappearance

of some of the desirable linoleic acid has to be accepted as well as a very modest increase in saturates such as stearic acid.

Canola (Z, Table 8.22). Conditions such as 0.05–0.10% fresh Ni/oil, 110–120°C, and 3–5 atm pressure, assuming first-class dispersion of high quality hydrogen, all favor this kind of hydrogenation and even allow a limited re-use of catalyst. Teasdale (1975) using 0.05% Ni/oil, 135°C and 3.8 atm pressure, reduced linolenic to 0.4% by the time the original IV of 118.5 had fallen to 88.0, and the *trans* isomers merely amounted to 24.6% of the double bonds at this point. Admittedly, 135°C is appreciably above what has been classed as "low-temperature hydrogenation" (Chapter 2, "*Low-Temperature Hydrogenation*"), but it may reasonably be inferred that a 90–95 IV as an end-point would afford an excellent chance of obtaining a linolenic content of c. 2% and an even lower *trans* isomer percentage-probably under 20%. Similarly, Engelhard (Harshaw Catalysts) on reduced erucic oil (Y, Table 8.22) with (0.05% Ni/oil, 140°C and 3 atm pressure, obtained 2% linolenic at 90 IV with 23.6% *trans* isomers and a rise from 2 to 8.5% stearic acid. The melting points of both of the above would be well below 20°C and poorly defined.

Classic rapeseed oil (X, Table 8.22). Here, 0.1–0.15% Ni/oil would be advisable at 110–120°C and 3–5 atm. The slip melting point is likely to be about 25°C at 86 IV, but again, such slip melting points are notoriously imprecise. It is very unlikely that anyone will be attempting this type of hardening today with old style rapeseed oil.

Iso- or trans-*Suppressive Hydrogenation (see Chapter 2, "Iso- or* trans-*Supressive Hydrogenation")*

These non-selective conditions are often favored when an all-hydrogenated vegetable shortening is being produced. To flatten the melting curve even further, up to 5% fully hardened cottonseed oil is added. Alternatively, the main component vegetable oil may be lightly but selectively hardened to achieve oxidative stability, and then 25–30% of the composition made up by hardened vegetable oil(s) of higher melting point, again to obtain the typical flat melting curve (Teasdale, 1975).

Canola (oil Z, Table 8.22) hardened with 0.05% Ni/oil at 135°C and 3.8 atm (Teasdale, 1975) to 71.6 IV (a continuation of "Low-Temperature Hydrogenation" section above) still showed only 31.7% *trans* isomers with a rise in stearic acid to 18.3%. The SFI 10/21.1/28.7/33.3/40 was *24.5/13.4/8.2/4.5/0.2.* This would correspond with a slip melting point of about 35°C. Engelhard (Harshaw Catalysts) with 0.05% Ni/oil, 140°C and 3 atm (oil Y, continuing as in "Low-Temperature Hydrogenation" section above) at 71.2 IV, obtained 39.8°C melting point, 35.4% *trans* isomers and 18.2% stearic acid. The percent solids by NMR *10/20/30/35/40/45* was 36/20.5/11/8.5/2.5/0.5.

Normal Hydrogenation (see Chapter 2, "Iso- or trans-*Supressive Hydrogenation")*

This has been taken to mean control at 180°C, although gassing may start at 140°C. The linolenic content of Brassica oils (old or new types) is only about 10%, so that by the time the temperature has reached 180°C, the risk of cyclization (Chapter 2, "Cyclization and Polymerization" section) is slight.

Englehard (Harshaw Catalysts) gave full details of hydrogenation with 0.05% Ni/oil at 170°C and 3 atm (oil Y, Table 8.22). This combination of temperature and pressure has noticeably accelerated the reaction so that the IV now drops to around 50 in a little over half an hour. At 82.0 IV (13 min gassing), the linolenic acid has fallen to 0.5% and the stearic acid has risen to only 7.5%. The *trans* isomers have just reached 31.9%, so it is not surprising that the melting point is no more than 20.5°C. At 68.2 IV (22 min gassing), the melting point is 42.8°C, the *trans* isomers having now risen to 41.4% and the stearic acid to 17.3%. The percent solids by NMR 10/20/30/35/40/45 is 48.5/31/20/10.5/5.5/0.5. For oil Z, Teasdale (Mukherjee et al., 1975) reports for 0.05% Ni/oil at 0.38 atm and 205°C, an experimental hydrogenation where pressure and temperature are taken to a condition of very high selectivity as against the usual 180–200°C temperature limit on the full plant scale. In these conditions, the linolenic has disappeared to a trace at 88.2 IV, as might be expected; the stearic has risen to only 4.3%, but the *trans* is already 34%. By the time the IV drops to 79.5, the slip melting point of a soft solid with a steep melting curve is evidently about 24°C (SFI 21.1/26 7 = 7.8/0.1). By 72.8 IV, the melting point appears close to 35°C. The *trans* form amounts to 51.9% of the double bonds, and the stearic acid has jumped to 12.9%-still a moderately steep melting curve, but from then on, further hardening produces, as usual, progressively less steep melting curves. By 60 IV, the slip melting point is about 48°C.

Classic rapeseed oil (oil X, Table 8.22) has been hardened in fairly selective conditions of 180°C, c. 3 atm pressure, and generous doses of partly used catalyst. Depending upon how much activity remains, about 0.5% Ni/oil might be used, but frequently half this dose is acceptable. Some hardeners enliven the performance of partly used catalyst by the addition of a smaller proportion of fresh nickel. As with all Brassica oil hardenings, if gassing commences at 140°C, much of the 10% linolenic acid will have disappeared as the temperature approaches 180°C. Particularly steep melting curves are a feature of hydrogenated high erucic oils, this behavior being associated with the disparity in chain length between the large proportion of C22 and other (mainly C18) fatty acids, and hence the lack of durable crystal structure. In these circumstances, as the original n_D^{60} 1.4581 drops to n_D^{60} 1.4549, a slip m.p. of 30°C results with a steep SFI 20/30 of 37.8 ± 2.2/2.8 ± 1.7 (% SFC at 20/30°C = 45 ± 3.0/2 ± 1.5); at n_D^{60} 1.4542, a 34°C m.p. for an SFI 20/30 of 42.4 ± 3.1/13 ± 3.0 (% SFC at 20/30°C = 51 ± 3.5/12 ± 3); at n_D^{60} 1.4540, a 36°C m.p. for an SFI 20/30 of 49.1 ± 3.1/23 ± 3.0 (% SFC at 20/30°C = 60 ± 4/21 ± 3). By the time the IV has dropped to 65 and the slip m.p. risen to c. 41°C, the melting curve becomes less steep, the SFI 20/30 is 59.3/40 (% SFC at 20/30°C = 74/40). At 43°C m.p., the SFI 20/30 is 63/48 (% SFC at 20/30°C = 79/52), and at 45°C m.p. (56–59 IV), the SFI 20/30/40 = 65/57/21.5 (% SFC at 20/30/40°C = 82/66/18.5). The pressure may be increased to 5 atm, and some fresh nickel added to the partly used nickel dose when batches above 40°C m.p. are sought, since this will help to keep the gassing time within 3–4 h.

Two-Stage Hydrogenation (see Chapter 2, "Two-Stage Hydrogenation")
For Canola starting at 118 IV, this kind of hardening would suggest dropping the IV to 92 at 120°C with up to 0.1% fresh Ni/oil at 3–5 atm. This would entail a fall in refractive index of n_D^{60} 1.4581 to 1.4556 (the correlation of IV drop and change in refractive index should be verified); then the temperature is given every help from the heating coils to rise to 180°C, and at the same time, the pressure is dropping to 1 atm or even lower if the reaction rate is not too slow. An end-point of 70–73 IV will establish what is to be expected in terms of texture from the parcel of crude oil concerned. This procedure is modelled on that applied to soybean oil (see "Soybean Oil, Hydrogenation, Two-Stage Hydrogenation" section) with the intention of producing a stable medium-melting fat suitable as a component in domestic shortening, vanaspati, or soft (tub) margarine. As indicated in Chapter 2, "Two-Stage Hydrogenation" section, the initial hydrogenation temperature and the location of the switch point before the temperature is helped to rise are both flexible in the light of results obtained. With classic rapeseed oil using around 0.1% fresh Ni/oil when the refractive index is dropped merely 15 points to around n_D^{60} 1.4566 at 110–120°C and 3 atm pressure and then brought up to 180°C for an n_D^{60} 1.4548 (say, 72 IV) end-point, a fat of c. 35–37°C m.p., and an SFI 20/30 of 20.7/7.4 (% SFC at 20/30°C = 23/7 is the product, which will, however, still contain appreciable amounts of C22:1. This texture should be contrasted with that obtained in "Normal Hydrogenation" of this section for normal hydrogenation of oil X to 30°C m.p., but whose final refractive index and IV are quite similar.

Iso-or trans-*Promoting Hydrogenation (see Chapter 2, "Iso- or* trans-*Promoting Hydrogenation")*
For Canola (oil Z, Table 8.22), it is suggested that to obtain a melting curve of maximum steepness, 0.1–0.3% Ni/oil as sulfur-poisoned catalyst should be employed with a final controlled temperature of 180–200°C and a pressure of 1-3 atm, depending upon results. An SFI 20/30 difference of more than 20 units (% SFC difference c. 22) in the 69–73 IV region (32–37°C m.p.) seems assured, as this is obtainable with ordinary active catalyst in good selective hydrogenation conditions. For high-erucic rapeseed oil, an SFI 20/30 difference of 30 (% SFC at 20/30°C = 35) is feasible in the same conditions and for the same final IV.

High-Melting and Fully Saturated Hardened Oil (see Chapter 2, "Higher Melting and Fully Saturated Hardened Oils")
High-erucic rapeseed oil can reach a slip melting point of 68–70°C when hardened below 2 IV. Fresh catalyst at up to 0.15% Ni/oil, 180–200°C and 3–5 atm should be used to obtain a gassing time under 4 h, especially if the hydrogenation is commenced at 140°C so as almost to eliminate linolenic acid by the time 180°C is reached.

Rice Bran Oil

The brown coating of the unmilled rice grain constitutes about 6% of the whole and is known as rice bran. It contains up to 23% oil, although 10–16% is more common, whereas the remainder of the grain contains less than 1%. Husk and broken rice grain must first be separated from the bran as part of the milling process; otherwise, its oil content is much reduced. With minimum delay, the bran must next be briefly steamed (90–95°C) to inactivate enzymatic lipases, then dry heated to 3–4% moisture content and cooled. If this is not done, the action of the enzymes is so rapid that the FFA of the oil in the bran can increase at a rate of 5% per day, so that it soon ceases to be of edible interest, and at best can only serve some technical purpose. The economic production of rice bran oil depends upon having a sufficiently large quantity of bran easily accessible within the time necessary to inactivate it, and this is achieved best in large rice growing areas with good communications, such as Japan. Here it has been estimated (Yokochi, 1972) that some 90,000 tons p.a. are produced and 60,000 tons p.a. are refined. Much greater quantities are lost to edible use elsewhere because of the hazards indicated. The crude oil must be degummed, and neutralization-washing-drying followed by chilling-filtration to separate waxes and other valuable unsaponifiable items must be carried out. These matters are given more detailed explanation in recent publications (Patterson, 1989; Patterson, 1993). Rice bran oil is very obviously a member of the oleic-linoleic group, as the composition range in Table 8.25 shows (Bernardini, 1973; Swern, 1970, p. 408).

Iodine values range from 92 to 115 and saponification values from 175 to 190. The unsaponifiable (%) is high at typical refractive index range is n_D^{60} 1.4577–1.4607. The degummed oil is neutralized and bleached to the standard indicated in the "Cleaning of Oils Prior to Hydrogenation" section in this chapter before hydrogenation, if this is intended. From the above composition, it is evident that C18:3 is so low that no serious problem of oxidative stability exists, and therefore the incentive to hydrogenate appears minimal in normal circumstances. In addition, because of a variety of natural antioxidants present in the oil, it is one of the most stable of vegetable oils and finds domestic use as such in Japan in particular. If hydrogenation is performed, a substantial amount of linoleic acid is present, and this gives the opportunity for selective and trans-promoting hardening reactions. In such an application, a specially sulfur-poisoned nickel catalyst would give the steepest melting

TABLE 8.25
Fatty Acid Composition of Rice Bran Oil

Fatty Acid	Range (%)
C14:0	<1.0
C16:0	12–18
C16:1	<0.4
C18:0	<3.0
C18:1	40–50
C18:2	29–42
C18:3	<1.0
C20:0	<1.0
C22:0	<1.0

curves in the 32–36°C m.p region. Enough C16 fatty acids appear to be present to render any problem of graininess or sandiness in the hydrogenated oil, which has been stored for some weeks, quite slight; blending with other oils to diversify the fatty acid chain length composition should get rid of any difficulty completely.

Safflower Oil (Cartamo, Kusum)

The safflower is tolerant of a wide range of climatic conditions, from moderately dry tropical to temperate. The oil has been used for centuries in various parts of the world, and cultivation is pursued principally in the United States, Mexico, and India, but also to a smaller extent in Russia, the Middle East, and parts of Africa. Seed yields vary from dry conditions at around 350 lb/acre to 4000 lb/acre in irrigated areas. The oil content varies from 25 to 43% and is extremely high in linoleic acid; linolenic acid is virtually absent, and saturated fatty acids are generally lower than 10%. Because the level of natural antioxidants in the crude oil is low, it does not have very strong resistance to oxidation, but varieties of safflower have been produced wherein the content of oleic acid has been increased to a dominant position at the expense of the linoleic acid, and in these circumstances, as might be expected, the stability is enhanced (Bernardini, 1973; *Oils and Oilseeds*, 1971; Swern, 1970, p. 400). The oil is used both as a table oil (to which synthetic antioxidant may be added) and as a drying oil in paints and in alkyd resins which show high color stability. Phosphatides are satisfactorily removed in degumming, neutralization, and earth bleaching; sulfur is not a problem as a catalyst poison. The saponification value ranges from 186 to 197, unsaponifiable (%) does not exceed 1.5; the IV ranges from 140 to 150, n_D^{60} 1.4597–1.4627. The common "highlinoleic" safflower oil has a fatty acid composition as shown in Table 8.26; to this has been added an indication of what is possible in a high-oleic variety.

TABLE 8.26
Advisory Fatty Acid Composition of Safflower Oil [CODEX STAN 27-1981, Supp. 1 (1983)]

	Fatty Acid	High Linoleic (%)	High Oleic (%)
Shorter than	C14	<0.1	
	C14:0	<1.0	
	C16:0	2–10	4–8
	C16:1	<0.5	
	C18:0	1–10	4–8
	C18:1	7–42	74–79
	C18:2	55–81	11–19
	C18:3	<1.0	
	C20:0	<0.5	
	C20:1	<0.5	
	C22:0	<0.5	

The unsaturated fatty acid composition of high-linoleic-type safflower oil lends itself to the production of steep melting point curves in the 32–36°C region using either partly exhausted catalyst (up to 0.6% Nil/oil) or specially sulfur-poisoned catalyst (up to 0.3% Ni/oil) at 180–200°C and 1–3 atm pressure. If, on storage, graininess develops in the hardened oil owing to the crystallization habit arising from the great preponderance of fatty acids of the one (C18) chain length, a blend with some other oil(s) richer, for example, in C14 and Cl6 fatty acids, should reduce or eliminate the problem. It seems likely that as the amount of safflower oil produced continues to increase, the opportunities for the economic use of hardened safflower oil will also grow; price will, however, continue to be a major factor in its employment.

Sesame Oil (Gingili, Sim-Sim, Til)

For many centuries, this oil, has been highly prized as a very stable odorless table and cooking oil. In its long history, it has found diverse employments in soap manufacture, perfumery (fixative), pharmaceuticals, cosmetics, and insecticides, and with and without hydrogenation in shortening, vanaspati, and margarines; the seed is also used in confectionery. The seed occurs in so-called white, red, or brown varieties. Local preferences may exist, such as for oil from the white variety, and in the Sudan, the white seed is preferred for sweetmeats. The red variety may have a lower percentage of hulls, and hence, rather more oil. Press cake is used as a component of poultry and livestock feed and contains about 35% protein. Principal regions of cultivation are China, India, Mexico, and the Sudan (Bernardini, 1973; *Oils and Oilseeds*, 1971; Langstraat, 1976; Swern, 1979, p. 387; Moore 1959 & 1969). The yield of seed can vary from 250 lb/acre in primitive conditions to 2000 lb/acre (Mississippi, United States). The oil content of the seed is about 50%, and a unique characteristic is that besides its content of tocopherols (up to 500 ppm), there is present a phenolic-type antioxidant, sesamol, in a combined form, sesamolin (up to 0.6%), which breaks down in a variety of circumstances and releases the sesamol, which then considerably enhances the stability of the oil. The conditions present during treatment with acid-activated bleaching earth and during hydrogenation, and even sometimes during storage, are among those which progressively liberate more sesamol from combination in sesamolin; on the other hand, deodorization and possibly some types of bleaching earth remove sesamol. The structure and chemical relationship of sesamol, sesamolin, and closely similar compounds such as sesamin have been investigated and summarized in detail by Budowski (Budowski, 1964). In the presence of hydrochloric acid and furfural, these substances give an intense carmine red color (Villavecchia and Baudouin tests) which can be employed as a semi-quantitative test (Cocks et al., 1966; Section 2.30 of *Methods of Analysis of Oils and Fats*, 1958). This feature has led to the compulsory inclusion of sesame oil in vanaspati and margarines in various countries, so that its use as an adulterant in ghee or butter can more readily be detected. The level of inclusion is usually about 5%, and this also helps to increase the essential fatty acids content (*cis*, *cis*, 9,12 linoleic acid). The phosphatide content of sesame oil is very low at under 0.1%; also, sulfur content is negligible. It is not surprising, therefore, that neutralized bleached sesame oil has been popular for many years as a medium for the assessment of nickel catalyst activity, the iodine value drop procured by a set dose of catalyst in standard time, pressure, temperature, and agitation conditions being a measure of activity. As the test is not a severe one, it does not adequately reflect the durability of nickel catalysts which may be called upon to work regularly in more arduous conditions, such as the hardening of marine and other oils which contain considerably more catalyst poison. In these circumstances, oil hardeners may well adopt a performance standard based on an oil which more accurately represents their everyday experience. Ultimately, comparison of one catalyst with another, especially where gaps in time may amount to years and distance is global, depends upon holding an inorganic substance, such as the base from which the dry-reduced catalyst may be prepared (in set conditions) as required, in the laboratory. This is not to say that a particular catalyst activity test which is recognized to apply to a particular field of hydrogenation must be condemned because it is empirical. The key word is "recognized." Iodine values range from 103 to 120 and generally occur at 109–113; commonly, the refractive index is n_D^{60} 1.4580-1.4620 with n_D^{60} 1.4582–1.4585 most usual; saponification values are 187–195, and unsaponifiable (%) less than 2.0. FFA content is expected to be less than 2.5%. Sesame oil is neutralized and earth bleached in much the same way as washed (semi-refined) cottonseed oil or groundnut oil. The color of the crude oil in BS 656:1967 is given as 30 Y 25 R (1") max.; a typical color would be 35 Y 4.3 R (5 ¼"), falling to about 35 Y 1.5 R (5 ¼") after neutralization and under 3 Y 0 6 R (5 ¼") after moderate earth bleaching. The possible spread in fatty acid composition, together with a typical simplified composition corresponding to 113 IV, is shown in Table 8.27.

TABLE 8.27
Advisory Fatty Acid Composition of Sesame Oil [CODEX STAN 26–1981, Supp. 1 (1983)]

Fatty Acid	Possible Range (%)	Typical Composition at 113 IV (%)
<C14	<0.1	
C14:0	<0.5	
C16:0	7–12	8.5
C16:1	<0.5	
C18:0	3.5–6.0	4.5
C18:1	35–50	42.0
C18:2	35–50	44.5
C18:3	<1.0	
C20:0	<1.0	0.5
C20:1	<0.5	
C22:0	<0.5	

There is no difficulty in pretreating sesame oil to the standard given in the "Cleaning of Oils Prior to Hydrogenation" section in this chapter prior to hydrogenation; indeed, if washing and drying after alkali neutralization have been performed efficiently, the oil could easily provide one of the examples where earth bleaching can be omitted prior to hydrogenation without appreciable detriment. If hydrogenation of the 113 IV oil were to be carried out with perfect selectivity, the linoleic acid would be converted to oleic by a fall of 39 units to 74 IV. Since perfect selectivity is not attained, we can expect that even in selective conditions of 180–200°C, 1–3 atm and a dose (0.2–0.4% Ni/oil) of partly exhausted catalyst, selectivity will begin to break down by 80 IV, and it is between there and 70 IV that high proportions of *trans* isomers and steep melting point curves are obtainable for fats below 40°C slip melting point. As with other oleic-linoleic class oils, the use of sulfur-poisoned catalyst will enhance the steepness of the SFI or SFC curve. If, because of the very high proportion of fatty acids all of the same chain length (C18 in this case), there is a crystallization-graininess problem on storage, the same answer as with other oils will apply, and fats containing some different fatty acid chain lengths ought to be introduced into the composition blend (see "*Rapeseed (Colza) Oil; New Varieties of Canola Oil, Canbra Oil, etc.*" and "*Sunflower Oil (Tournesol, Girasol)*").

Soybean Oil

Native to Manchuria and Japan, the cultivation of the soybean has spread to many parts of the world (United States, Brazil, China, and Argentina, pre-eminently) to reach a production of around 241.7 million metric tons during 2010, far outstripping any other oilseed. At 20% (dry basis), its oil content is well below that of many other oilseeds, but the rest is a rich source of protein for animal feeds: the growth of vegetable protein use in human foods is slow by comparison as yet. Its high yield of edible protein per acre and the promise of a good return to the farmer have made a large contribution to its success. There is, as might be expected, a reasonably large spread in certain characteristics, such as degree of unsaturation of soybean oil, according to variety of bean and climate. In following the detailed directions for hydrogenation given below, the operator should have some regard as to whether the oil being processed is typical or lies more nearly towards an extreme. Apart from the obvious fact that oil of particularly high IV must show a larger drop in IV and refractive index to reach a set final IV, it is well to remember that the opportunity for isomerization is increased in a longer hydrogenation, and the extent to which this is effective will depend on whether the conditions of the hydrogenation themselves favor isomerization or have been chosen to suppress it. This effect is much more marked when hardening oil from different species of fish. Again, a parcel of oil which has sustained substantial oxidation may easily be a less satisfactory candidate for lighter degrees of hydrogenation than an equivalent fresh oil. Even so, the hydrogenation, when performed, is likely to improve its oxidative stability. Variations in soybean oil are reviewed in detail in *Bailey's Industrial Oil and Fat Products*, (1979) and *Handbook of Soy Oil Processing and Utilization*, (1980). The composition given

TABLE 8.28
Advisory Fatty Acid Composition of Soybean Oil [CODEX STAN 20-1981, Supp. 1-1983, Amended 1987 (Mounts, 1987)]

Fatty Acid	Range (%)
C<14	<0.1
C14:0	<0.5
C16:0	7.0–14
C16:1	<0.5
C18:0	3.0–5.5
C18:1	18–26
C18:2	50–57
C18:3	5.5–10
C20:0	<0.6
C20:1	<0.5
C22:0	<0.5
C24:0	<0.5

in Table 8.28 could be considered typical. BS 653 and *Codex Alimentarius* quote an IV range of 120–143; typical is 133. An unsaponifiable (%) of 1–5 max, is noted, but well under 1% is usual. Density (20°C) is 0.919–0.925; refractive index (n_D^{60}) 1.4660–1.4700; for n_D^{60}, a temperature correction of –0.00035 (60–40) = –0.0070, is applied (see Chapter 8, "Unsaturation, Refractive Index" section) and then gives a range n_D^{60} *1.4590–1.4630.* Saponification values range from 185 to 195, but 191–193 is usual. Tocopherols range from 0.15 to 0.21%. Free fatty acids may reach 1%, but 0.3–0.7% is to be expected. Phosphatides in crude oil range from 1–5 to 3.0%; in commercial de gummed oil, 0.5% is poor, but 0.3% represents good degumming, and this corresponds with 0.012% phosphorus (120 ppm P). After refining, it is suggested (*Handbook of Soy Oil Processing and Utilization*, 1980) that lecithin should range from 0.003 to 0.045% corresponding with 1–15 ppm P. If 4 ppm P is obtained, this is good, but starting from some crude oils, it may prove necessary to accept 10 ppm P in oil about to be hydrogenated. Crude soybean oil may have a color of 10 R (5 ¼"), which after degumming, neutralizing, and earth bleaching is expected to drop to 3.5 R (5 ¼") with the best quality crude, but nearly double this may be found at times (*Handbook of Soy Oil Processing and Utilization*, 1980). Hydrogenation itself diminishes color, as does deodorization of this oil. The processor will therefore balance the rigor of the earth bleach against what is expected to be achieved in the way of color removal by subsequent hydrogenation, deodorization, etc., if these are to be performed. The National Soybean Processors Association, United States (NSPA) envisages (*Handbook of Soy Oil Processing and Utilization*, 1980) for refined, bleached, partially hydrogenated, winterized, and deodorized pure soybean oil a maximum Lovibond color of 20 Y 2.0 R (5 ¼") and various other requirements relating to oxidative stability. Various methods of purification of the crude oil have recently been described (Patterson, 1993). For a final edible product, refiners are now looking for chlorophyll contents of under 100 ppb, and sometimes under 50 ppb (Hastert, 1988; Leysen, 1989).

Hydrogenation

A huge amount of data has been published on the hydrogenation of soybean oil, not simply on account of its commercial importance, but because these studies have helped establish the pattern of the hydrogenation reaction, especially for oils containing a proportion of polyunsaturates. Figs. 1.9 and 1.10 illustrate the limitations on texture imposed by the selectivity of the catalyst then in use c. 1978. Although various catalysts with superior performance have been produced since then, the illustration of the principle remains valid. Commercial brochures (Unichema; *Katalysatoren der Süd-Chemie AG;* Engelhard De Meern BV; Hoechst Aktiengesellschaft; Mallinckrodt) give with explicit detail the dosages of nickel recommended, either to obtain flat melting curves with adequate selectivity (lower temperatures), or steep curves depending on *trans* promotion where selectivity is more easily obtained (higher temperatures). Mounts (1981) and Garibay (1981) discuss the effects of processing soybean oil. The simplest way to treat this wealth of material is to adopt the same approach as used throughout this chapter and describe specific conditions as applied to neutralized and bleached soybean oil (see "*Cleaning of Oils Prior to Hydrogenation*") which relate to the different hydrogenation techniques described in general terms in Chapter 2.

Low-Temperature Hydrogenation (see Chapter 2, "Low-Temperature Hydrogenation")
Oil is dosed with c. 0.07% fresh Ni/oil and hydrogenated at up to 3 atm and 110–125°C to 105 ± 5 IV. A typical R.I. drop would be n_D^{60} 1.4600–1.4570; approximate % SFC at 20/30°C = c. 5/2; slip melting point is notoriously inaccurate at this level, but should be c. 28°C. If the linolenate present has been reduced to under 3%, a very soft fat of enhanced oxidative stability and improved color will result. Isomerization is so suppressed by these conditions that only a small proportion of double bonds, probably well under a quarter, have attained the *trans* form. Depending partly on how large was the original catalyst dose, it may be possible to obtain at least one re-use with virtually the same results. For such low-temperature hydrogenation, Unichema 9920 is an obvious choice; others which deserve consideration are Engelhard Nysosel 325 and Süd-Chemie KE-NF2 or T4460. Doses in the range 0.007–0.03% fresh Ni/oil are mentioned in brochures.

Two-Stage Hydrogenation (see Chapter 2, "Two-Stage Hydrogenation")
A medium-melting fat (slip melting point c. 35°C) is sought in which *trans* isomers have been rigorously suppressed by control at a low hardening temperature, followed by rapid elevation to 180°C towards the end of hardening to create selective conditions in which the less stable linolenates will be reduced to quite negligible amounts. This produces a hardened oil base from which to blend a "compound

shortening," but it may easily feature as a component in a soft-textured margarine (*Bailey's Industrial Oil and Fat Products*, 1964). Contrast with hardened winterized soybean oil (HWSB) in this chapter under "Soybean Oil, Hydrogenation" section.

Catalyst dose c. 0.07% fresh Ni/oil; harden at 110°C and 2–3 atm until there is a fall in IV of 35–40 units, corresponding to an n_D^{60}, of 1.4600–1.4560; drop pressure to 0.5–1.0 atm, and with full assistance from steam coils, raise temperature towards 180°C until an end-point of 75 IV, corresponding to approximately n_D^{60}, 1.4540. Ideally, the last few units fall in IV will occur at 180°C, but in any case, temperatures above 150°C become increasingly favorable to selectivity. Slip melting point 33-35°C; SFI 20/30/40 is 19/7/1, or % SFC at 20/30°C is 21/6/1. The risk of cyclization with soybean oil is small; the extent of the IV drop at 110°C more than defeats this possibility. All the catalysts recommended for use with well refined soybean oil, such as Hoecat 882 OF, Calsicat 472D, Engelhard Nysosel 222, Unichema 9920, and Süd-Chemie T4460, should perform this task without difficulty. The exact nickel dose required and the possibility of re-use should be checked with the supplier.

Hardened Winterized Soybean Oil (HWSB)

Although the rigorous full application of the restriction to 150°C for an IV drop of 0.002 (original IV) (British Patent 1515 (1903)) given in Chapter 2, "Cyclization and Polymerization" section for the early part of the hardening cycle is most desirable for the more unsaturated fish oils, if hydrogenation of soybean oil commences at c. 140°C and rises steadily towards 180°C, practical tests on oil from the autoclaves concerned may show their hydrogen dispersion is powerful enough to defeat the formation of cyclic monomers in such a low-risk example as soybean oil hydrogenation. For some time, it has been important in the United States to hydrogenate selectively good quality soybean oil to under 3% linolenate prior to winterization to obtain the maximum yield of liquid fraction (salad oil). Typical conditions to secure this effect are reported (*Handbook of Soy Oil Processing and Utilization*, 1980) as 0.2% fresh Ni/oil, 170°C and 0.3 atm. to a final IV of 110–115. Above 115 IV, insufficient oxidative stability is achieved, and below 105 IV, the yield of liquid fraction falls markedly. Early detailed experiments are given in Okkerse et al., (1967) and Evans et al., (1964); see also Goebel (1981), List (1980), and Mounts (1980). All the catalysts just mentioned in "Two-Stage Hydrogenation" section above would serve here, but improvements in catalysts since 1980 (the date of ref. 61) suggest a dose of 0.07% fresh Ni/oil should be adequate.

Iso- or trans-*Suppressive Hydrogenation (see Chapter 2, "Iso- or* trans-*Suppresive Hydrogenation")*

Englehard (Harshaw Catalysts) has issued a study of the hydrogenation of soybean oil under the conditions given in Table 8.29.

Samples taken at intervals from the start down to nearly 55 IV show speed of reaction, development of *trans* isomers, solid content and disappearance of unsaturates, etc. It is then possible to contrast the results of varying temperature and

TABLE 8.29
Hydrogenation of Soybean Oil (Harshaw Catalysts)

	DM3 Catalyst (% Ni)	Agitation (rpm)	Temperature (°C)	Pressure (atm)
(a)	0.02	950	140	0.5
(b)	0.02	950	140	3.0
(c)	0.02	950	170	0.5
(d)	0.02	950	170	3.0
(e)	0.02	950	200	0.5
(f)	0.02	950	200	3.0

constant pressure, and vice versa. Under conditions (b) in Table 8.29, linolenates have fallen well below 2% at 100 IV, and *trans* isomers have risen only to 27%. A slip melting point of c. 28°C is attained at 81 IV, but by then the percentage of *trans* isomers has risen to 44. Since the results as shown in Table 8.29 were obtained, catalyst DM 3 has been improved substantially by the activity performance of Nysosel 222, which required only about two-thirds of the DM 3 dosage. Later still, Nysosel 325 has been shown to require only about half of the dosage of DM 3, and this is combined with an enhanced selectivity towards polyunsaturates with marginally slower filtration. Strikingly lower *trans* formation has been shown for Unichema 9920 operating 120°C as against 150→200°C on refined soybean oil.

Normal Hydrogenation (see Chapter 2, "Normal Hydrogenation")

For many oils, this kind of hardening commences at c. 140°C and continues at 180°C for a large part of its course. In the case of soybean oil, some restraint at 150°C, as mentioned earlier, will discourage cyclization if the risk is present with a dose of c. 0.07% fresh Ni/oil (or several times that amount of partly used catalyst) and up to 3 atm pressure, hardening commencing at 140°C will soon show a drop from, say, n_D^{60} 1.4600 to 1.4565, even if the temperature is kept to 150 ± 5°C. Hardening continues at 180 ± 5°C to c. 60 IV (c. n_D^{60} 1.4521). SFI 20/30/40 is 58/41/16 or the approximate (% SFC at 20/30/40°C is 74/45/12). Slip melting point is 43 ± 1.0°C. Table 8.29 shows for conditions (c) at 62 IV, m.p. 43°C and solids content % 20/30/40 is 61.5/44.5/13.5 (by wide line NMR). For conditions (d) in Table 8.29 at 60 IV, the m.p. is 45.3°C; for conditions (e) at 59.8 IV, the m.p. is 44.8°C; and for conditions (f) at 60 IV, the m.p. is 45°C.

The full graphical presentation of the detail from Table 8.29 bears out that higher temperature and lower pressure favor selectivity (Sr) and *trans* promotion, while increase in temperature, like increase in pressure, accelerates the reaction rate.

Under conditions (f) (Table 8.29) at 57 IV, an m.p. of 47.1°C is reached with solids content % 20/30/40 of 71/58/25.5 (by wide line NMR). Süd-Chemie (Nickel Catalyst KE-NF 20 and KE-FS40) for KE-NF 20 catalyst has issued similar extensive data for the hydrogenation of soybean oil (Table 8.30).

TABLE 8.30
Hydrogenation of Soybean Oil (Nickel Catalyst KE-NF 20 and KE-FS40)

	KE-NF20 (% Ni)	Pressure (atm)	Temperature (°C)
(a)	0.015	0.5	160
(b)	0.015	0.5	180
(c)	0.015	0.5	200
(d)	0.Q15	1.0	160
(e)	0.015	1.0	180
(f)	0.015	1.0	200

TABLE 8.31
Hydrogenation of *nbd* Soybean Oil (132 IV) with G 95 E (*Katalysatoren der Süd-Chemie AG*)

Catalyst conc.	0.006% Ni
Pressure	2 bar (abs)
Temperature	140 →185°C
Agitation	1000 rpm
IV	74
m.p.°C	35
SFC % 20°C	40
SFC % 30°C	14
SFC % 35°C	4

Average rate of hydrogenation = 1.17 IV/min
Other catalysts offered for production of margarine stock are Calsicat E472 D, Hoecat 882 OF, and Unichema 9910 or 9920.

The conditions in Table 8.30 range from the least selective least *trans* promoting to the opposite. In (a), the linolenates have fallen to 2.9% at 105 IV, but in (c), they are down to 2.0% at the same IV. Whereas the *trans* reach 38.7% in (a) at 76 IV, already at 861 IV in (d) they are 38.6%. For (d) at 74.5 IV, the m.p. is 30.8°C and the *trans* 37.5%. In (f) at 76I IV, the m.p. is already 32.9°C and the *trans* up to 44.3%. Finally, in (f) at 61.5 IV, a m.p. of 44.1°C is obtained.

More active catalysts than KE-NF20 are being developed as seen by the laboratory performance of one of them given in Table 8.31. Other catalysts offered for production of margarine stock are Calsicat E 472 D, Hoecat 882 OF, and Unichema 9910 or 9920.

Iso- or trans-*Promoting Hydrogenation (see Chapter 2, "Iso- or* trans-*Promoting Hydrogenation")*

The object of this type of hardening is to obtain a steep SFI or SFC curve; this implies the maximum difference between SFI at 20 and 30°C. The lower edge of the shaded area in Fig. 1.10 traces a rapidly increasing SCI 20 (Whether solid content index [SCI] as in original paper or solid fat index [SFI], the reasoning is the same.) with small increase in SCI 30 up to a value of SCI 20 of about 43 (say, % SFC of 51–54). After that point, the curve ascends rapidly because of the much larger SCI 30 values. It has long been recognized (*Bailey's Industrial Oil and Fat Products*, 1964) that 74.5 corresponds with the iodine value of a typical soybean oil which has been hydrogenated with complete selectivity (13.4% saturates, 88.6% monounsaturates); hence, as hydrogenation proceeds towards this point with the assistance of a correctly sulfur-poisoned nickel catalyst, chances of maximum difference between SFI 20 and 30 exist. Low pressure (0.5 atm) and a temperature of 180–200°C assist the *trans* promotion even if there has been some pause at 150°C to defeat any tendency to polymerization. Polymers tend to form most easily in conditions of hydrogen scarcity on the nickel surface, and would in any case, if present in sufficient amount, disrupt the normal SFI 20/30 relationship. Fortunately, the specially made sulfur-poisoned catalyst, in which the sulfur is distributed evenly inside pores as well as outside, may be less dependent on high temperature for its *trans* promoting efficiency than a catalyst poisoned through repeated use. Because the activity of the special catalyst will be low by normal standards, doses of up to 0.2% Ni/oil may be found helpful. It is not to be expected that even in the conditions just described, complete selectivity corresponding to the 74.5 IV mentioned above will be obtained; hence, it will be necessary to take samples from about 82 IV onwards at intervals of 2 IV to discover the optimum SFI 20/30 difference, after which these conditions may be adopted as a standard by the processor for similar oil (Harshaw Catalysts). A typical result might be at 80 IV, an SCI 20/30 of 44/14 with a slip melting point of 34°C (Okonek et al., 1980; Okonek, 1986).

High-Melting and Fully Saturated Hardened Oil (see Chapter 2, "Higher Melting and Fully Saturated Hardened Oils")

When hydrogenated to 2 IV, soybean oil attains 68°C slip m.p. Depending upon the desired speed of reaction, up to 0.1% fresh Ni/oil may be used at up to 3 atm and 180–200°C, with some interval at 150°C if necessary, as discussed for other hardened soybean oils above. This pattern would also be classed as normal hydrogenation.

Sunflower Oil (Tournesol, Girasol)

A native of Central America, the sunflower was brought to Europe in the sixteenth century and began to be cultivated as an edible-oilbearing crop in the nineteenth century, notably in Russia, where a use had been developed for all parts of the plant (*Oils and Oilseeds*, 1971; Langstraat, 1976; Moore 1959 & 1969; Sullivan, 1980; Achaya, 1994). It is very adaptable, thriving best in temperate zones, but cultivation in the tropics is possible. In 1992, world production of sunflower oil amounted to nearly 8 m tons; the republics of the former Soviet Union easily make the largest single contribution but Eastern Europe and Argentina are large producers: cultivation is rising in many other countries, especially the United States, since the latest varieties of seed can contain 50% oil and the cake is a source of high-grade protein.

TABLE 8.32
Advisory Fatty Acid Composition of Sunflower Oil [CODEX STAN 23-1981, Supp. 1 (1983), Amended 1987 (Mounts, 1987)]

Fatty Acid	Possible Range (%)	Typical Composition at 132 IV (%)
C<14	<0.4	
C14:0	<0.5	
C16:0	3.0–10	6.5
C16:1	<1.0	0.5
C18:0	1.0–10	4.5
C18:1	14–35	23.0
C18:2	55–75	63.5
C18:3	<0.3	<0.3
C20:0	<1.5	0.5
C20:1	<0.5	0.5
C22:0	<1.0	0.5
C22:1	<0.5	
C24:0	<0.5	
C24:1	<0.5	

A yield of 1000 lb seed/acre in reasonably good conditions is common, and 1800 lb/acre has been known. Very little linolenic acid is found in the oil, and this, no doubt, assists its resistance to atmospheric oxidation. There is great variation in the relative proportions of oleic and linoleic acid depending upon climate and plant variety, the cooler climates favoring the higher proportion of linoleic acid, as is usual. Table 8.32 shows this variation and a simplified make-up corresponding to 1321 IV has been added; however, examples are found, as with other oils, where some limits are exceeded, especially in cases of experimental production.

If oxidative degradation of the refined oil does occur, the off flavors are not as potent as those arising from oils containing more linolenic acid, such as soybean, i.e., the threshold of their detectability is not so low. Crude oil may reach 40 P.V., but 30 is more usual, and some supplies do not exceed 10. For good crude oil, the anisidine value (An. V.) is unlikely to exceed 5.0; above 12 indicates an inferior oxidative state. The size of the extinction coefficients E.1.1.232 and E.1.1.268 and their proportion to one another are also used for quality control, sometimes by national legislation. The data are empirical and are to be interpreted in the light of past experience with the kind of oil concerned; i.e., sunflower, groundnut, rapeseed, etc. Iodine values are in the range 110–143; saponification values 187–194 and a refractive index n_D^{60}, 1.4588–1.4620 appear possible, but a typical value is n_D^{60}, 1.4612. *Codex Alimentarius* gives n_D^{60} 1.467–1.469 (NB 40°C). The unsaponifiable (%) should not exceed 1.5, and generally is not more than 0.8. Tocopherols are quoted at

700 ppm or higher; phosphatides are low at 0.2%; sulfur compounds are negligible and waxes are a nuisance, more especially if the product being marketed is a table oil required to pass some form of cold test. Various wax contents (0.02–0.35%) have been quoted for crude sunflower oil (Ostric-Matijasevic, 1973), and these may have been influenced by the fact that waxes, like phosphatides, will settle out in store tanks, particularly when some moisture is also present. Certainly 0.04% wax (400 ppm) may be present in fresh, virtually unsettled crude oil, but the lower layers of a sunflower oil store tank may eventually reach 0.2% wax. The bulk of the supernatant oil drawn from such a tank will yield an oil which, with the usual degumming-refining treatment, is quite suitable for hydrogenation and, after deodorization, for use in margarine shortenings. Where table oils are concerned, chilling followed by a filter-aid-assisted filtration still at a reduced temperature was an obvious early choice (Sullivan, 1980).

Since phosphatides act as crystallization inhibitors, the importance of first degumming an oil before chilling and separating wax crystals has come to be recognized. In the absence of significant amounts of phosphatide, the particular waxes concerned have been found to be sufficiently hydrophilic to cause them to concentrate at the oil-aqueous interface and thus lend themselves to centrifugal separation, which is made even more effective in the presence of a wetting agent (Sullivan, 1980; Ostric-Matijasevic & Turkulov, 1973; Rac, 1972; Patterson, 1993). The processing details are not of direct concern here but are discussed in the references given.

Up to 5% FFA in crude sunflower oil may be met, but usually FFA content is not above 1.5%. A crude oil of 50 Y 4 R (1") on neutralization and washing shows 45 Y 5 R (5 ¼"), although not such a pale color as the latter is required if the oil is going forward for hydrogenation. For shipments of oil which have been derived from seeds dried in hot flue gases, a dose of 0.2% activated carbon should be included with the bleaching earth so as to adsorb polycyclic aromatic hydrocarbons (Biernoth & Rost, 1968; Wendt, 1980; Patterson, 1993) (see "Coconut Oil" section in this chapter). Since over 90% C18 fatty acids are usually present in sunflower oil, the hardened sunflower oils are likely to cause a problem of grainy texture on storage of products. This difficulty is related to the high level of saturates and *trans* isomers all with the same chain length. Of the various ways of reducing or eliminating the problem, the inclusion of some oil or fat with appreciable proportions of fatty acid other than Cl8 acids may prove the simplest.

There is no special difficulty in bringing the oil to the standard suggested in the "Cleaning of Oils Prior to Hydrogenation" section of this chapter prior to hydrogenation.

Hydrogenation

Low-Temperature Hydrogenation (see Chapter 2, "Low-Temperature Hydrogenation")

As just explained, the very low content of linolenic acid in sunflower oil already contributes to its oxidative stability; hence, there is little need to engage in the low temperature hydrogenation to eliminate linolenic acid with a minimum formation of *trans* isomers, as may be required with other oils, such as soybean and rapeseed. If, however, sunflower oil is hydrogenated in such conditions (120°C, 3–5 atm, 0.05–0.10% fresh Ni/oil), by the time the IV falls from 132 (n_D^{60} 1.4612, say) to 100 (n_D^{60} 1.4575, approximately), the slip melting point rises to c. 20°C. This may be contrasted with Süd-Chemie's hydrogenation of sunflower oil (Nickel Catalyst KE-NF 20 and KE-FS40) of 132 IV in rather more selective conditions (130–170°C, 1 atm, 0.02% fresh Ni/oil), where by 91.2 IV, the slip melting point has only risen to 118°C and a further hydrogenation to 82.6 IV (n_D^{60} 1.4564, approximately) is needed to raise the slip melting point to 22.4°C. Morrison and Robertson hydrogenated samples of two distinctly different sunflower oils with a view to determining the effect on stability of each when employed on the more vigorous duty of frying (Morrison & Robertson, 1978). The oil produced in the cooler northern climate (Dakota) was notably higher in linoleic acid than that produced in the south (Minnesota), where most cultivation of the oil seed type sunflower takes place. Table 8.33 shows the change in relative proportions of oleic and linoleic acids in each oil; the other recorded changes, being individually very small, have been omitted here. The more unsaturated (northern) oil has been given a larger IV drop; hydrogenation in all cases was at 125°C 20 psig (1.3 atm) using 0.05% Nysel catalyst (Engelhard De Meern BV).

The oxidative stability as seen by the initial active oxygen method (AOM) values improved for both oils after hydrogenation, the improvement being more obvious in the northern oil; both southern oil AOM results were better than those for northern oil; the rate of loss of oxidative stability during several hours heating appeared much the same for all samples; both oils were very suitable as frying oils.

Catalysts well suited for this hydrogenation of sunflower oil are Unichema 9920, Engelhard Nysosel 325, Süd-Chemie KE-NF 20 or G70, Hoecat 882 OF, and Calsicat E472 D.

Iso- or trans-*Suppressive Hydrogenation (see Chapter 2, "Iso- or* trans-*Supressive Hydrogenation")*

Sunflower oil provides a good starting point for the production of hydrogenated fats with a relatively flat melting curve in the 32–36°C m.p. range by dropping the IV

TABLE 8.33
Changes in Oleic-Linoleic Acids During Hydrogenation of Sunflower Seed Oil

	Northern Oil		Southern Oil	
Fatty Acid	Unhydrogenated (%)	Hydrogenated (%)	Unhydrogenated (%)	Hydrogenated (%)
C18:1	26.0	32.1	55.5	56.9
C18:2	64.4	53.3	33.4	30.8
IV	133.9	119.9	105.6	102.3
% *trans*	0	6.4	0	3.3

to 76–72 (n_D^{60} 1.4543) with 0.05–0.08% fresh Ni/oil at 140–150°C and 3 atm. The addition of a proportion of some other vegetable oil containing other than C18 fatty acids may be desirable to avoid graininess; up to 5% fully hardened cottonseed oil helps flatten the melting curve and contributes some C16 fatty acids. The same catalysts as recommend in this section under "Hydrogenation, Low-Temperature Hydrogenation" would be suitable here.

Normal Hydrogenation (see Chapter 2, "Normal Hydrogenation")
Since linolenic acid is virtually absent, the normal hardening control temperature of 180–200°C may be approached without imposed delay from the usual beginning of gassing at 140°C. A fresh catalyst dose of up to 0.1% Ni/oil will carry hardening a long way at a satisfactory rate, but if it is desired to use older catalyst and at the same time approach a high proportion of *trans* isomers, then up to 0.4% Ni/oil would serve well, and on occasion, an even higher concentration would be used. Pressures of 1-5 atm are normal here, according to plant convenience. Dropping from 132 IV (n_D^{60} 1.4612) to n_D^{60} 1.4543 with the old catalyst brings a 32–34°C m.p. at about 80 IV and an SFI 20/30 of 38.9/12 (% SFC at 20/30°C = 46/11); further hydrogenation under the same conditions to n_D^{60} 1.4537 (c. 71 IV) shows a 35–37°C m.p. with an SFI 20/30 of 50/23.3 (% SFC at 20/30° C = 61/21); and by the time the m.p. is in the 39–41°C range, the SFI 20/30 is 5704/37 (% SFC at 20/30°C = 72/37). Above 38°C m.p., the SFI 20/30 curve loses its steepness like most other hardened oils.

Süd-Chemie (Nickel Catalyst KE-NF 20 and KE-FS40) gives considerable detail for the hydrogenation of sunflower oil (132 IV) with 0.02% fresh Ni/oil as the KE-NF 20 catalyst (see Chapter 7, "*Examples of Commercial Nickel Catalysts*") and 1 atm pressure. The temperature ranges used are 130–170°C, 160–170°C, and at 180°C. The effect of the higher (180°C) temperature in promoting additional *trans* isomers (low-melting solids) in the medium-melting range (c. 32°C) is seen by the larger SCI 20/30 differences at that point in hardening, again, as might be expected. As far as fresh catalysts are concerned, the same ones as mentioned in this section under "Hydrogenation, Low-Temperature Hydrogenation" will serve here.

Two-Stage Hydrogenation (see Chapter 2, "Iso- or trans-Promoting Hydrogenation")
This technique is less relevant to sunflower oil than to those which contain appreciable quantities of triunsaturates and even more unsaturated fatty acids. Nonetheless, if the equivalent procedure for soybean oil is initiated, a hardened fat of useful characteristics for shortening, vanaspati, and domestic margarines, including the tub variety, results. Using 0.05–0.10% fresh Ni/oil at 2–3 atm, hydrogenation is first performed at 110°C from n_D^{60} 1.4612 (132 IV) to c. n_D^{60} 1.4568 (92 IV); then, encouraging the temperature to rise towards 180°C, hydrogenation is concluded at n_D^{60} 1.4549 (75 IV), the pressure being allowed to drop to 1 atm max. during this latter period. The IV and refractive indices quoted will depend upon the degree of unsaturation of the original oil, and must be adjusted so that the texture of the final product best suits the purpose in mind. As described here, a slip melting point of c. 35.0°C results, with an SFI 20/30 of 20 ± 1.5/8.6 ± 1.1 (% SFC at 20/30°C

22 + 2/6 ± 1). The important step of 110°C hydrogenation has restrained the development of low-melting solids content at 20°C to half that which could result with old catalyst and 180°C hardening. Repeated use of the same catalyst on the same task, possibly three to four times, is worth exploring. When using fresh catalysts, those already mentioned in this section under "Hydrogenation, Low-Temperature Hydrogenation" would be most suitable.

Iso- or trans-*Promoting Hydrogenation (see Chapter 2," Iso- or* trans-*Promoting Hydrogenation")*
It is to be expected that melting curves of maximum steepness will result when the procedures of normal hardening in this section under "Hydrogenation, Normal Hydrogenation" are employed using specially sulfur-poisoned catalyst at c. 0.2% Ni/oil and controlling hardening at 180-200°C. Re-use of the catalyst is feasible. Engelhard Nysel Sp-7 and Sp-10, Unichema 9908, and Süd-Chemie G111 are suitable.

High-Melting and Fully Saturated Sunflower Oil (see Chapter 2, "Consistent Quality in Hydrogenated Oil Deliveries")
A slip melting point of 68°C is obtained when sunflower oil is hardened at 180°C and 3 atm, using up to 0.1% fresh Ni/oil to a maximum 2 IV. Like fully hardened soybean, maize, and sesame oils, the slip melting point is some 6°C higher than the fully hardened cottonseed oil (62°C) popular for addition in small percentages (up to 5%) to shortenings to prolong their plastic range; i.e., to flatten the melting curve. It appears that, because of its higher melting point, an even smaller percentage of fully hardened sunflower oil would be needed to cause the same prolongation while not elevating the solids content at 20°C to quite the same extent. Against this, it has been pointed out (*Bailey's Industrial Oil and Fat Products*, 1979) that where C18 fatty acids predominate, so will tristearin on solidification, and this can give rise to texture difficulties in blended shortening. Where the level of addition of fully hardened sunflower oil is as low as 3%, compared with the 10–15% of component low in C18 fatty acids, however, it appears a legitimate matter of trial as to whether advantages may be gained for little or no disadvantage. All general purpose catalysts recommended for use with vegetable oils will serve here.

Teaseed, Tomato Seed, and Other Oleic-Linoleic Class Oils

A variety of oils of this class have a merely local importance, frequently as a byproduct of a much bigger agro-industry. Generally speaking, the content of linolenic acid is negligible, saturated fatty acids are commonly less than 20%, and the remainder is made up from variable proportions of oleic and linoleic aeids. If the question of hydrogenation arises at all, it is probably as a means of improving color and oxidative stability so as to provide an acceptable component of a margarine or cooking fat type composition. In this event, the same general approach as that given for grapeseed oil (see "Grapeseed Oil" section in this chapter) will apply.

Marine Oils-General Considerations

Whereas for many years, a substantial number of vegetable oils have each enjoyed a distinct identity on account of differences in behavior in use as such, and when hydrogenated, marine oils, commercially at least, have been regarded more as a group. Marine animals such as the whale were early recognized as providing an oil which could be hardened for edible and soap-making purposes just like the much less important seal and sea elephant. The sperm whale is exceptional in that about three-quarters of its oil content comprises not triglycerides, but long-chain (mainly C14, C16, C18, and C20) fatty alcohols, saturated and monounsaturated, in combination with long-chain fatty acids, that is, waxes. Fish oils could vary from 99 to 211 IV, but all were low-melting oils looking and smelling very similar, having origin in the sea and, unless used as drying oils in paints and varnishes, requiring hydrogenation to convert them to edible fats (or, much more rarely, polymerization to produce, like whale oil, an oil of enhanced viscosity and stability).

With steadily falling supplies of whale oil and indeed the species facing extinction, the potential replacement by fish oil came to be explored more thoroughly and proved successful from the early 1960s. From about the same time, more sophisticated methods of analysis became available; hence, finer and more accurate distinctions could be made, especially in recognizing differences in polyunsaturated fatty acid compositions. Steadily, it came to be recognized that not only did differences in composition occur between species of fish, but that within the same species, fatty acid make-up varied with maturity, temperature of the environment, diet, and location in the body. The marine fish oils are distinguished by their high content of long-chain (C20, C22) fatty acids which may contain from zero to six double bonds. The triglycerides of fresh water fish, like fresh water plants, contain very little of such material, but contain fatty acids which predominately are in the C16 and C18 bracket. It is instructive to note that young salmon, in migrating from fresh to salt water, change the type of their body fat; young fresh water fish reared on a diet including cottonseed oil will develop unsaturated fatty acids in their body oil typical of fresh water fish, but if menhaden oil is substituted for the cottonseed oil, their body oil comes to resemble that of the menhaden. Herring about to migrate and spawn eat voraciously, greatly increasing their oil content, and (diet being a big factor) the degree of unsaturation also increases. The oil stored as energy for the migratory journey is consumed by the time the herring leave the spawning grounds; thus, trawlers may recognize from the lower IV and lower oil content of a particular catch whether they have come upon a shoal before or after spawning. An early investigation showed how the IV of oil from fish caught in Lake Balaton rose as the temperature of the lake fell and vice versa in a seasonal cycle (Hilditch & Williams, 1964; Stansby, 1967; *Bailey's Industrial Oil and Fat Products*, 1964, Reed, 1964; De Witt, 1963).

Evidently, fish have some capacity for building up the unsaturated fatty acids they need, but diet enters into the process. The minute marine and fresh water phytoplankton living near the surface are plants which contain mainly C16 and

C18 unsaturated fatty acids. As the temperature of the water falls, the plankton produce oils of greater unsaturation, and vice versa. Plankton are eaten by small animal organisms which prolong these chain lengths to C20 and C22 unsaturated fatty acids. Then the small animal organisms in their turn are consumed by fish; hence, the pattern for fish oil fatty acid composition is being influenced by the diet immediately available, even though the fish metabolism may continue the process (Masson & Mella, 1980). Furthermore, from different parts of the same fish, there may occur marked differences in the relative proportions of some of the major fatty acids present such as, for example, C16:0, C18:0, C18:1, C20:1, C20:5, C22:1, and C22:6 in dorsal and ventral samples from Pacific herring (*Bailey's Industrial Oil and Fat Products*, 1964). Stansby (1980 & 1981) has drawn attention to the fact that, just as inadequate sampling from different fish of the same species at first gave misleading results as to truly typical oil content, so now, for the same reason, erroneous beliefs as to typical fatty acid composition may be spread. To this complication was added another, at least temporarily; because both chain length and number of double bonds present influence the sequence in which fatty acid methyl esters move through a GLC column, some confusion between these has arisen. Fortunately, other analytical approaches, such as chromatography, are now helping to clear up these difficulties as well as the changes in both positional and geometric isomerism which are continually occurring during hydrogenation. An altogether more accurate picture is therefore emerging; the detailed investigations now extend into marine plant life.

Following the discussion on the possible unwanted effects of erucic acid (C22:1 Δ13) as a component of oils in the human diet (see "Rapeseed (Colza) Oil" section) and the restrictions placed upon it in the European Economic Community (*Official Journal of the European Communities*, 1976), it became inevitable that the effects of erucic acid in certain fish oils or their partially hardened products would come under close scrutiny. Barlow and Duthie (Barlow & Duthie, 1980; Duthie, 1980) failed to find any substantial link between partially hardened marine oils and coronary and ischemic heart diseases in humans, pointing out that Greenland eskimos with marine C22:1 fatty acids in their serum triglycerides are nonetheless noted for their low susceptibility to cardiovascular disorders. The investigations were pursued in considerable detail for a few years and confirmed the first findings (Duthie, 1988).

It has been noted by Ackman et al., (1981) that in the partial hydrogenation of herring oil, although there was a slight increase in the total C22:1 isomers present, the amount of erucic acid (C22:1 Δ13 isomer) fell slightly. In spite of the several factors which give rise to appreciable variation in the degree of unsaturation in the overall body oil content of one species of fish-and degree of maturity is one of the most important of these-it remains possible to classify marine animal and fish oils into two broad groups (A and B, see Table 8.34) which are of particular interest to the hardener (see "Marine Oils-General Considerations, Hydrogenation" section).

A common feature of marine oils is that, by reason of the process of separation from the body tissue (wanted for fish meal) and the fact that highly unsaturated oil tends to oxidize when exposed during initial processing and shipment, considerable amounts of mucilage pass readily into the soapstock upon neutralization and washing: it is very beneficial to give dilute caustic soda-water washes to free the oil from these impurities since they are active catalyst poisons. It is upon splitting the soap stock with mineral acid that the mucilage precipitates as a tarry gel which must be regularly cleaned from the plant handling it. Oxidized fatty acids will precipitate from oils and acid oils when a sample is poured into petroleum ether. It is, naturally, the most highly unsaturated fish oils which generally occasion this nuisance, although some of these, such as Peruvian anchovy oil, are often neutralized-washed to about 0.3% FFA prior to shipment as a service to buyers. Alkali neutralization-washing of fish oils, whether by batch or continuous processes, is well established and easily achieves 0.1% FFA max. and 0.05% soap max., desirable for active earth bleaching prior to hydrogenation (see "Cleaning of Oils Prior to Hydrogenation" section). With continuous centrifugal neutralization-washing, the soap content may be no more than one-tenth of the limit indicated. With some fish oils, both yield and quality are likely to be improved if steps are taken to allow hydration of gummy material immediately prior to caustic soda addition, and in the case of centrifugal neutralization, this may take the form of up to 0.1% phosphoric acid addition. The dark color of washes subsequent to the removal of the soapstock is an indication of what is being achieved in the way of removal of catalyst poisons. A 1% dose of moderately acid activated bleaching earth operated at 95°C will normally be adequate to complete the pre-hydrogenation treatment, but if very strenuous efforts are considered advantageous to allow fresh active nickel the maximum opportunity subsequently to work unpoisoned, this earth dose may be doubled. It will also be beneficial to keep the contact between the highly unsaturated warm bleached fish oil and air to a minimum as regards situation and time prior to hydrogenation. Some recent comments on the benefit of various levels of pre-bleaching marine and land animal fats are available (see "Beef Tallow, Hydrogenation, Fully Hardened" section) (Goebel & Romero, 1980; Patterson, 1993). In connection with earth bleaching, it would be prudent for those using both metal gauze filter elements in automatic self-cleaning filter units and siliceous earths which have been activated by hydrochloric acid to check whether any corrosion problem they have experienced in the interstices of the gauze or at its periphery can be related to the Cl- ion. Bleaching earth manufacturers have available the relevant data concerning chloride content, if any, of their products, and can be consulted without difficulty.

Hydrogenation

Characteristics of individual marine oils and the procedures for their hydrogenation are given in the sections which now follow. Certain comments which apply in varying degree to all are given first.

Hydrogenation of marine oils is pursued to promote a stable bland flavor and to obtain the texture-flat or steep dilatation curve-most suited to the performance of the product in which the fat is to be included; disappearance of fishy odor (Meijboom & Stroink, 1972) and dull, yellow-green color is taken for granted; post-treatment may include further mild neutralization-washing-drying earth bleaching and deodorization, or in favorable circumstances, it may amount simply to adsorbent bleaching (to make sure nickel content is no more than 0.2 ppm), followed by deodorization.

Just as in the case of vegetable oil hydrogenation, a principal concern has been to reduce linolenic acid to c. 2% or less to avoid oxidative instability; so with marine oils, it is necessary to get rid of the more elaborate unsaturated groupings in which not merely three but four, five, and six double bonds occur as a potentially very reactive arrangement in which they are each separated by one reactive methylene ($-CH_2-$) group. It is important to make clear immediately that double bonds which are separated by more than one $-CH_2-$group, even though there may be several of them, are isolated and without the same vulnerability. This opens up the probability that, whereas in getting rid of a small or modest component of highly reactive unsaturated fatty acids we may drop from an original 120 IV to 78 IV, if we commence at 210 IV, the drop will be much greater but could cease earlier at 85 IV. or more; by that time, the double bonds, both by hydrogenation and migration, have assumed a more stable general system. It is evident that if, by migration, a double bond were to assume a position in which it and its neighbors became more susceptible to reduction, this is just what would occur; so the whole process, because of its mobility, not only enhances stability by elimination of double bonds *per se,* but also by maneuvering those which remain into the less reactive arrangement. The selectivity of the catalyst and the other conditions of the reaction help to decide how expeditiously the maximum number of remaining double bonds in a system of minimum reactivity is achieved. This may be sought by the hardener, along with the least possible formation of low-melting solids (*trans* isomers). This is asking for the benefits of positional isomerism (migration) with the least hindrance from geometrical isomerism *(cis → trans).* A modern siliceous supported catalyst with under 0.3% S/Ni will be an advantage here. A compromise must be made, and the price paid by the formation of some saturates if and when the hydrogenation reaches the point of minimum reactivity for the remaining double bonds. If, on the other hand, creation of low-melting solids (*trans* isomers) is not merely acceptable but welcome, then the conditions making for selective hydrogenation can be enjoyed, minimum reactivity attained comparatively early, and the melting curve will descend steeply. The more powerful analytical tools now available enabled Ackman et al., (Ackman & Sebedio, 1980) to trace the course of double bond disappearance and formation of *trans* isomers during the selective hydrogenation, using 0.2% Ni of menhaden oil from 159 to 84.5 IV. *Trans* isomers rose progressively, 3.4% (150 IV). 24.9% (120.5 IV), 45.1% (84.5 IV): tetraenes fell from an initial 5.6% to a final 0.4%; pentaenes plus hexaenes fell from 8.8% to a trace. Monoenes rose throughout from 23.9 to 34.2%, dienes from 3.9 to 13.2% and trienes from 4.2 to 8.3%, there being in all this very little increase in saturates-merely to 41.6% at 150 IV and a final 43.8% (84.5 IV). It was also noted that C20:1 rose 1.3% to 4.5% and C22:1 rose 0.2% to 1.7%, *trans* isomers predominating in the newly formed monoenes. Oxidative stability, as might reasonably be expected, rose during the course of the hydrogenation, so that a sample of the 150 IV oil during two months storage increased its P.V. from 4.2 to

7.6 while the P.V. of the 84.5 IV sample increased from 0.4 to 2.0. Questions still remain as to the disposition of double bonds in newly formed systems which would certainly be described as polyunsaturated. This has also been explored by Ackman et al., (Ackman & Sebedio, 1980; Ackman, et al., 1981) in considerable detail during the hydrogenation of refined herring oil (119 IV) with experimental conditions of high selectivity; i.e., 0.2% fresh Ni, 190–225°C, and c. 70 kPa (c 0.64 atm) hydrogen pressure. Samples were taken at the start (119 IV) and at 107, 101, 88, and finally, 74 IV. Several important conclusions are drawn from the results and the sequence in which they occurred. Down to 88 IV, there is little change in saturates and monoenes have been increasing. This fits well with Coenen's deduction (Boerma & Coenen, 1968; Coenen, 1970; Coenen, 1976; Coenen, 1978) covering the general situation that *cis* polyunsaturates virtually monopolize active catalytic sites up to the point where the concentration of the poly unsaturates has been greatly diminished. The methylene-interrupted four-, five, and six-double-bond polyunsaturates disappear first, their places being taken by the less active non-methylene-interrupted trienes (NMITs) and nonmethylene-interrupted dienes (NMIDs). Later, NMITs are hydrogenated to NMIDs. Concerning the C18 polyunsaturates, it is noted that between 118 and 101 IV, for example, C18:3 Δ9,12,15 almost vanishes and C18:2 Δ9,12 falls by half. It is after 88 IV that NMITs fall, and by this IV, the natural *cis* polyethylenic four-, five-, and six-double-bond material has gone completely. For C20 and C22 monoenes, it is concluded that the transit through tetraene and triene stages brings about such randomization in location and geometric form that the original position of remaining double bonds cannot be identified. The final distribution of *cis* and *trans* monoene isomers, it is suggested, is only adequately accounted for if both the Horiuti-Polanyi mechanism (see "Isomerization" section in Chapter 1) and the formation of the following allyl complex

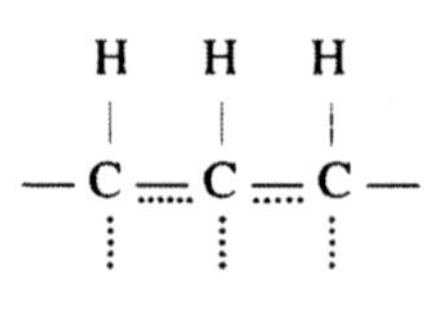

from a double bond and an adjacent-CH_2-group (Coenen, 1970; Gault, 1960) are accepted as taking place during hydrogenation, a general concept which had been put forward earlier by Van der Plank and other Vlaardingen workers (Van der Plank & Van Oosten, 1975; Heertje & Wosten, 1974; Van der Plank, 1972; den Boer & Wosten, 1968). The question as to how far C20 and C22 polyunsaturates are affected in their isomerization and rate of hydrogenation by reason of their location in the triglycerides at the l, 3 or 2 position is also discussed in the light of the results obtained. (Note remark at end of Chapter 1 "Selectivity" section.)

This clarification of what is happening as the composition of hydrogenated fish oils approaches a state of acceptable oxidative stability will be very welcome to those who ask themselves at how high an IV it is acceptable to terminate such a hydrogenation, always with regard to conditions used and texture required. The conditions of the experiments are not necessarily exactly those which hardeners on the plant scale would feel able to employ. A practical point is raised by the authors who early pointed out that discrepancies between IV as calculated from GLC and determined by Wijs may reflect some polymerization of polyunsaturates; Ackman and Eaton mentioned the same phenomenon in a previous paper (Ackerman & Eaton, 1966). This danger has been recognized for over 20 years, and the precautions to be taken have been explained in Chapter 2 "Cyclization and Polymerization" section. Not only is it prudent to avoid side-effects which look biologically undesirable, but the presence of polymers causes erratic results to appear in plant performance when, for example, refractive index is used to terminate hardening at what are supposed will be appropriate SFI or SFC results. Superior mechanical dispersion of hydrogen appears likely to reduce the risk.

Lastly, a distinction must be made between oxidative stability as seen from controlled storage, and accelerated oxidation chemical analyses and organoleptic acceptability, for the simple reason that a poor flavor is not only a question of how much oxidation is taking place, but what is oxidizing. The "Hydrogenation" section in this chapter describes this kind of difficulty in relation to linseed oil. Marine oils can cause similar problems, the off flavor frequently being more noticeable if the taster goes immediately into the open air. There is also the question of whether some particular off flavor precursor derives from the unsaponifiable rather than the fatty components. An understanding of the changes taking place as the more reactive polyunsaturates disappear during hydrogenation helps decide at how high an IV we may confidently terminate hardening a fish oil so as to obtain a low-melting (c. 30°C) edible fat. The task appears much easier in starting at 130 IV or less rather than 210 IV, because in the latter case, the greater hydrogenation increases the opportunity for *trans* isomers to form. As a partial compensation, the replacement of *cis* methylene-interrupted bonds by NMIDs and NMITs mentioned earlier may lift the acceptable final IV several points. Conditions for hydrogenation of marine oils in the two groups given in Table 8.34 follow and include some information on the individual oils.

Group A-Herring Oil, Capelin Oil, Liver Oils of Cod, Halibut, and Haddock; Whale, Seal, and Sea Elephant Oils

Herring Oil

The North Atlantic, including the North Sea area, has been easily the richest source of herring for many years, but over-fishing has greatly reduced stocks, as it has for mackerel (Vedeler, 1982). The North Pacific is a less important source. Iodine values in winter range from 115 to 130; in summer, up to 160; 130 IV is regarded as typical. Saponification values range from 180–192, and unsaponifiable (%), between 0.5 and 2.5, although 1.0 to 1.6 is common. Proper handling of the catch, processing

TABLE 8.34
Fatty Acids of Marine Oil Groups

	IV	Fatty Acid Type	(%)
Group A			
		Saturated	19
Capelin[a]	c. 100	Monoene (very high)	76
		Polyunsaturated (very low)	5
Herring	c. 130		
Cod (liver)			
Halibut (liver)	c. 160	Saturated (moderate)	under 30
Haddock (liver)		Monoene (high)	c. 50
		Polyunsaturated	15–30
Whale			
Seal, etc.	c. 120		
Group B			
Anchovy	c. 200		
Pilchard	c. 200	Saturated	c. 35
Sardine	c. 170	Monoene (low)	under 30
Menhaden	c. 165	Polyunsaturated (moderately high)	35–45

[a]A small member of the salmon family found in Norwegian waters and the Barents Sea; exports (1981) from Iceland and Norway c. 160,000 tons.

and storing should give an oil of less than 3.0% FFA. The use of synthetic antioxidants to retard deterioration of the crude oil has received some attention (Vedeler, 1982). Usually, the refractive index is n_D^{60} 1.4591–1.4623. The sulfur content normally lies at the 12–18 ppm level in average crude oil; phosphorus compounds are adequately removed by the pretreatment; oxidized fatty matter also present in crude fish oils is a possible catalyst poison (see Chapter 1 "Catalysts Induction, Fatigue, and Poisoning" and Chapter 7 "Durability and Poisoning" sections). Ignoring a large number of very small amounts of unsaturated fatty acids, the principal fatty acid composition is given in Table 8.35.

Hydrogenation

Regarding the fatty acid composition given in Table 8.35, chief interest attaches to reducing the 28% of fatty acids with more than two double bonds to acceptable oxidative stability, and this, in general, appears to entail a fall in IV from 130–140 to 80–83.

TABLE 8.35
Principal Fatty Acid Composition of Herring Oil

Fatty Acid Type	%	Fatty Acid Type	%
C14:0	6	C18:3	2
C16:0	11	C18:4	3
C16:1	7	C20:1	13
C16:3	7	C20:5	7.5
C16:4	1.3	C22:1	21
C18:0	1.4	C22:6	7
C18:1	10		
C18:2	1	*Total*	98.2

Two-Stage Hydrogenation (see Chapter 2, "Two-Stage Hydrogenation")
Following the procedure (see Chapter 2, "Cyclization and Polymerization" section) to avoid cyclization of polyunsaturates, a drop of some 34–40 IV units is to be achieved before the temperature exceeds 150°C. If the end-point is to be 80–83 IV, this leaves a further drop in IV of only c. 17–20 units in which to rise to c. 180°C. A dose of 0.05–0.10% fresh Ni/oil is adequate with 3 atm hydrogen pressure until 150°C, and 1 atm thereafter. How low the temperature is for the first stage depends on how important the need is to minimize the SFI or SFC value at 20°C and how easily the temperature can be assisted after the first stage to rise above 160°C, for a short time at least, in the second. Commencing with n_D^{60} 1.4618, the above entails a fall to c. n_D^{60} 1.4574 at some level between 120–150°C, then a fall to c. n_D^{60} 1.4544. In these circumstances, a slip m.p. of 30–32°C should result with an SFI 20/30 of 14.4 ± 3.0/3.7 ± 1.9 (% SFC at 20/30°C 15 ± 3.5/3 ± 1.5).

When a hydrogenated herring oil of SFI 12 (% SFC at 20°C = 12) is combined with an addition of up to 5% fully hardened vegetable oil (hardened cottonseed or sunflower oil, maximum 2 IV), the texture of a useful shortening fat at low cost becomes available. A hardened herring oil with such comparatively low solid content at 10/20/30°C provides a possible very major component for a soft (tub) margarine. In some countries, any remaining C22:1 fatty acid content may need to be checked. Major catalyst suppliers give examples of fish oil hydrogenation under various conditions. The activity of catalysts in the 1990s means the fresh nickel concentration range quoted above of 0.05–0.10% is adequate for nearly all operations. Catalysts offered for fish oil hydrogenation include Engelhard Nysosel 222, Unichema 9910, Süd-Chemie KE-NF20 or T4460, Calsicat E428 D, and Hoecat 882 OF.

Normal Hydrogenation (see Chapter 2, "Normal Hydrogenation")
In this case, the first stage of the hydrogenation as described in preceding section may in effect take place at 145–150°C. Fresh catalyst may be replaced by partly exhausted catalyst at up to 0.5% Ni/oil; otherwise, 0.05% fresh Ni/oil is used. It is probably acceptable to allow pressure to remain at 3 atm, but this can easily be

lowered if the reaction is moving rather more quickly than desired. Following this pattern allows more *trans* isomers to form, and hence, the possibility of steeper melting curves until slip m.p.s in excess of 38°C are reached. Now at the same end-point, c. n_D^{60} 1.4544, the slip m.p. moves up to 32–34°C and the SFI 20/30 is 22.2 ± 1.3/5.6 ± 1 (% SFC at 20/30°C = 25 ± 1.5/5 ± 1). A further slight hardening, especially when performed a little more slowly, brings us at n_D^{60} 1.4542 to 34–36°C m.p., and an SFI 20/30 of c. 26.8/8.5 (% SFC at 20/30°C = 30/8). By 41°C m.p., the tendency to a flatter melting curve becomes noticeable, and an SFI 20/30 of c. 52/37 (% SFC at 20/30°C of c. 64/38); and at 48°C m.p., the SFI 20/30 is c. 68.5/64.8 (% SFC at 20/30°C of 86/74). The same catalysts as listed in the previous section will serve here.

High-Melting and Fully Saturated Hardened Oils (see Chapter 2, "Higher Melting and Fully Saturated Hardened Oils")
When fully hardened to below 2 IV, marine oils in general reach a 58 ± 1°C m.p., which is appreciably lower than for most vegetable oils, presumably because the greater variety in chain length of fatty acids causes a less stable crystal mix.

Iso- or trans-*Promoting Hydrogenation (see Chapter 2, "Iso- or* trans-*Promoting Hydrogenation")*
Since there is the wish to restrict the chances of cyclization in the early part of the hydrogenation by not allowing the temperature to rise above 150°C, this fits in very well with minimizing *trans* isomers: indeed, this latter effect, as explained in Chapter 2, "Two-Stage Hydrogenation" section, may be sought by working for a time at 20 or 30°C below 150°C. However, when it comes to encouraging *trans* isomers to form, there is little opportunity in the IV drop which remains for herring oil to achieve more with specially sulfur-poisoned catalyst than appears likely to result from the use of a liberal dose of catalyst partly exhausted through use.

Capelin Oil-Hydrogenation

Capelin oil presents the hardener with the easiest task of any marine oil, since there is such a small amount of polyunsaturated material to be reduced. Table 8.36 shows

TABLE 8.36
Principal Fatty Acid Composition of Capelin Oil

Fatty Acid Type	%	Fatty Acid Type	%
C14:0	8	C20:0	2
C16:0	11	C20:1	23
C16:1	10	C22:0	1
C18:0	2.5	C22:1	24
C18:1	13.5	C20 and C22	
C18:2	0.7	Polyunsaturated	1.7
		Total	97.4

a slightly simplified typical composition, the IV often being close to 100. As with other marine oils, this no doubt undergoes some seasonal variation. Much of the balance is made up from very small amounts of saturated fatty acids below C18. It seems that the C20 and C22 polyunsaturates would be reduced to dienes during an initial fall in IV of about 6 units. Thereafter, the opportunities are at their best to produce a low-melting fat for soft margarine, compound shortening base, and, conceivably, by one or more fractionations, a useful "olein" or liquid fraction. Hydrogenation conditions are the same as those for herring oil for the corresponding hydrogenated fat, except fresh catalyst doses may be reduced in view of the easier hydrogenation task involved. Again, where legal restrictions are in force, it will be necessary to take into account how much fatty acid group of the C22:1 variety is present in the product offered.

Fish Liver Oils (Cod, Halibut, etc.)

Liver oils from these fish have long been valued for their vitamin A and D content, but parcels not up to the required potency have been made available for edible and technical use. In contrast with capelin oil, these liver oils have an IV appreciably greater than the average for herring, and 160 IV may be taken as typical in a total range of 120–190. Yet again, seasonal influences are at work in determining the exact characteristics of the oil. Saponification values vary from 182 to 191, usually for cod 185-187, and for a 160 IV average, there is a refractive index range of n_D^{60} 1.4633 to n_D^{60} 1.4639. There is typically an unsaponifiable (%) of up to 1.5 for cod, but other liver oils may be much higher (Van der Plank, 1972). Refining to the standards of "Cleaning of Oils Prior to Hydrogenation" section in this chapter is adequate. Table 8.37 indicates the high level of monoenes which is typical of group A.

TABLE 8.37
Fatty Acid Composition of Atlantic Cod Liver Oil (*Bailey's Industrial Oil & Fat Products*, 1979; Ackman et al., 1964)

Fatty Acid Type	%	Fatty Acid Type	%
C14:0	3.5	C20:1	14.6
C15:0	0.5	C20:4	1.7
C16:0	10.4	C20:5	5.0
C16:1	12.2	C22:1	13.3
C17:0	0.1	C22:5	2.0
C18:0	1.2	C22:6	10.5
C18:1	19.6		
C18:2	0.8	*Total*	96.3
C18:3	0.2		
C18:4	0.7		

Hydrogenation

According to the "Cyclization and Polymerization" section of Chapter 2, hydrogenation for a marine oil of 160 IV requires a first stage of 51 IV units drop before exceeding a temperature of 150°C in order to avoid cyclization. If hydrogenation is continued at a low temperature, there is the virtual certainty of a sweet, nauseous hardened oil flavor developing in the hardened product in spite of deodorization, but if a period at 180°C is allowed, this effect is greatly diminished and at c. 38°C m.p. is often virtually absent. Where it is important to keep as soft a texture as possible, then early hydrogenation with fresh catalyst at 0.1% Ni/oil 120–140°C and 3 atm will help, the temperature afterwards being immediately assisted to rise (the pressure may be dropped here to 1 atm) so that by 85 IV, there have already been several minutes hardening near the higher (180°C) control level. Hardened fats with slip melting points of c. 34°C are achieved by 80 IV with an SFI 20/30 of c. 24/9.3 (% SFC at 20/30°C is c. 27/8). Performing the first stage of hydrogenation at 145-150°C may be classed as normal and frequently partly exhausted catalyst (0.2-0.5% Ni/oil) is used here at 3 atm pressure. For the same IV or n_D^{60} end-point, the SFI 20/30 will probably be 4/2 points greater, the melting curve tending to be rather steeper. The same catalyst types as recommended in the "Two-Stage Hydrogenation" section may be used here.

Whale, Seal, and Sea Elephant Oil

Whale oil was for about 50 years (1910-1960) the classic marine oil sought by hardeners to produce an acceptable, modestly priced fat for kitchen and toilet soaps, margarine, and cooking fats. All these aims were achieved, but because of over-catching stocks shrank greatly and whales are now protected, some countries forbidding the use of whale oil and others restricting their catch. The blubber of the sei, fin, blue, and humpback whales produced edible grades; oil from the sperm and bottlenose whales, because of its high content of long-chain fatty alcohols (35%) in combination with fatty acids, was reserved for technical outlets, although in the process much sperm whale oil was fully hydrogenated.

From time to time occasional parcels of seal or sea elephant oil became available and proved to have the same or very similar characteristics for hydrogenation purposes as the familiar whale oil. Most of the basic technique of hardening marine oils was learned from long years spent in hardening whale oil in one or more stages, and at low or fairly high temperatures. The catalysts then used were inferior to those of 1960 and after; worthwhile improvements in the quality of nonelectrolytic hydrogen and the mechanical dispersion of gas into oil date from about the same time. Table 8.38 shows how the composition of a typical whale oil fits it for inclusion in group a typical IV is 120, with many samples being between 110 and 130, but extremes of 107–162 are known (Stansby, 1967). The unsaponifiable (%) should not exceed 2, and it is here that at least a semi-quantitative estimation may be made as to whether any contamination of marine oil by mineral fuel oil in ship or land tank has taken place.

TABLE 8.38

Fatty Acid Type	%	Fatty Acid Type	%
C14:0	5	C20:1	15
C16:0	12	C20:4 and :5	5
C16:1	7	C22:1	11
C18:0	3	C22:5 and :6	8
C18:1	28		
C18:2	2	*Total*	97
C18:3 and :4	1		

Hydrogenation

In view of the effort being made to conserve whale stocks no advice is offered.

Group B-Anchovy, Pilchard, Sardine, and Menhaden Oils

Anchovy and Pilchard Oils

Vast quantities of anchovy migrate in the Humbold current along the Pacific coast of South America, and as supplies of whale oil dwindled in the late 1950s, "Peruvian Fish Oil" entered world markets on a spectacular scale. At about the same time, fish oil was exported from South and Southwest Africa, also in large amounts, and was known, in the United Kingdom at least, as "South African Pilchard Oil." Both oils had an IV range of 180-210; their conversion to a stable hardened oil for use in margarine and cooking fats presented a technical challenge and the opportunity for considerable economic gain. Prejudice against their use declined rapidly, and within a year or two in some countries hardened fish oils (including herring) became major constituents of popular margarines and cooking fats.

A new factor has entered into the situation since 1979, and that is the requirement to restrict the amount of C22:1 fatty acid groups in products sold for human consumption. This was enacted by the EEC. This tells more heavily against group A than group B marine oils.

A simplified fatty acid composition of what is believed to be a typical sample of each oil is given in Table 8.39. It appears that the fish concerned have been feeding on identical diets, and from the standpoint of the commercial hardener, they are interchangeable. Small percentages of saturated fatty acids with odd numbers of carbon atoms, branched chains, etc., are not shown.

The FFA in the crude oils is normally less than 3.0%, and Peruvian oil may have been refined to 0.3%. Unsaponifiable (%) is c. 1, but when the oil content of the fish is at its lowest, this figure may be much higher. Saponification values are covered by the range 191–196. If the oil has been semi-refined prior to shipment, it will hydrogenate after being dried and earth bleached; results will depend in part on the condition of the original crude oil; a neutralization-washing-earth

TABLE 8.39
Principal Fatty Acid Compositions of Anchovy and Pilchard Oils

Fatty Acid Type	Anchovy Oil (Peruvian) (%)	Pilchard Oil (South African) (%)
C14:0	7.5	7.3
C16:0	17.5	16
C16:1	9	9
C16:2	1	2
C16:3	2	2
C16:4	2	3
C18:0	4	4
C18:1	12	10
C18:2	2	2
C18:3	1	1
C18:4	3	—
C20:1	2	1
C20:2	1	—
C20:3,4,5	19	21
C21:5	—	1
C22:1	1	2
C22:3,4,5,6	11	12
C24:1	0.5	0.7
Totals	95.5	94.0

bleaching prior to hardening is the safest procedure and likely to save some catalyst and give a rather better hardened oil, especially in point of color (see "Cleaning of Oils Prior to Hydrogenation" section in this chapter). No more than 1% mildly activated earth should be needed; half this dose may suffice. Rarely, a green color in the oil, presumably derived from the food the fish has consumed, persists stubbornly even through hydrogenation. Heavier earth bleaching is only partly effective. If this problem is severe, it is worth noting that a number of activated earths were developed in the 1990s. These have particular capacity for removal of chlorophyll. These are Engelhard Nevergreen, Süd-Chemie Tonsil Optimum FF, Laporte Fulmont XMP 2, and Mizusawa Galleon DC. A simple program of laboratory tests will show how effective anyone of these may be. In the normal course of events, after hardened fish oil has been post-refined, the color will be less than 20 Y 2 R (5 ¼"). Crude fish oils contain about 18 ppm S as against 30 ppm for good whale oil. The mucilage present passes easily into soapstock, but as mentioned in the section on "Marine Oils-General Considerations" in this chapter, it separates as a tarry mass from acid oil.

It has already been emphasized (in the "Hydrogenation" section) that a long drop in IV during hydrogenation affords the opportunity for the ever-diminishing number of double bonds to assume positions where they are less ready to react than when previously the immediate neighbors of a reactive methylene group. This clearly applies to anchovy and pilchard oil; in discovering the acceptable highest level at which to terminate hydrogenation, the character of the catalyst used, the efficiency of the dispersal of high quality hydrogen, and the temperature-pressure regime employed all have to be taken into account.

Two-Stage Hydrogenation (see Chapter 2, "Two-Stage Hydrogenation")
To minimize *trans* formation, a fresh selective catalyst of under 0.3% S/Ni is preferable. A first use at 0.05–0.10% Ni/oil may permit a few re-uses (with increasing concentration) before it becomes transporting in turn. Commencing at 200 IV (c. n_D^{60} 1.4668–1.4672), there is a requirement in any case to keep below 150°C for the first 80 IV drop (see the "Cyclization and Polymerization" section of Chapter 2, to avoid cyclization), but if stronger *trans*-suppressive conditions are decided upon, the first-stage temperature may be 120–130°C and prolonged to a fall of 90 units IV. This presumes the end-point will be at 85 IV or a little less, so that the interval from 110 to 85 IV is taken up by the temperature rising (with assistance) and the hardening being concluded under control at 180 ± 5°C. The first-stage pressure is likely to be 3 atm and the second stage 1 atm. Success means obtaining a hardened oil of acceptable oxidative and flavor stability. If more selectivity is desirable, the temperature of the first stage may be lifted nearer to 150°C. This will have the further effect of giving a little more time above 150°C in the second stage. The best balance for the particular catalyst and autoclave system in use has to be found by trial. For an 85 IV end-point, an n_D^{60} 1.4544–1.4548 is probable and easily checked. Slip m.p. is in the 32-33°C range, SFI 20/30 is c. 22/5.7 (% SFC at 20/30°C is c. 24.5/5). The same catalyst types as recommended in the "Two-Stage Hydrogenation" section in this chapter may be used here.

Normal Hydrogenation (see Chapter 2, "Norma Hydrogenation")
Here, the hydrogenation may well commence at 140°C and be controlled at 150°C until 120 IV. Fresh (0.07–0.10% Ni) or partly exhausted catalyst (0.2-0.5% Ni according to convenience) may be used. From 120 IV (c. n_D^{60} 1.4588), the temperature is helped to rise to conclude under control at 180 ± 5°C. A pressure of 1–3 atm should be satisfactory, lower pressure favoring a higher SFI 20. For approximately the same end-point as in the "Two-Stage Hydrogenation" section in this chapter, the texture is now nearer to an SFI 20/30 of 26.3/7.8 (% SFC at 20/30°C = 30/7) and a slip m.p. of 34°C. Continuing at 180°C to c. 76 IV, the slip m.p. reaches 37°C and the SFI 20/30 is 35.2/14.8 (% SFC at 20/30°C = 41.5/14); at c. 50 IV, the slip m.p reaches 42°C and the SFI 20/30 is 55.6/41.7 (% SFC at 20/30°C = 69/43.5). Below 40 IV (c. n_D^{60} 1.4496), the slip m.p. reaches 48°C and the SFI 20/30/40 is 68.5/64.8/37 (% SFC at 20/30°C = 87/73). The same catalyst types as recommended in the "Two-Stage Hydrogenation" section in this chapter may be used here.

High-Melting and Fully Saturated Hardened Oils (see Chapter 2, "Higher Melting and Fully Saturated Hardened Oils" section*)*

The first stage of the hydrogenation is controlled not to exceed 150°C (see Chapter 2, "Cyclization and Polymerization" section), then allowed to rise to 180°C. A pressure of 3-5 atm would be sufficient; a fresh catalyst dose of c. 0.1% Ni/oil would be usual in seeking to obtain less than 2 IV in a few hours. The slip m.p. is 58 ± 1°C. Failure to reach 2 IV is an indication (see "Cyclization and Polymerization" section) that some polymerization has taken place.

Zschau (1981) reported in detail upon the results obtained in hydrogenating Peruvian fish oil (197 IV, FFA 0.08%, P.V. 37.3, S under 1 ppm, phosphatide 1.4 ppm) with 0.05% Ni/oil of four different catalysts, all experiments employing 3 atm pressure. Each hydrogenation was performed according to two temperature conditions. In one experimental condition, the hardening temperature was fixed at 180°C, as this had the convenient effect of ensuring a drop to 60 IV (40–46°C m.p., approximately) within 3 h when a standard catalyst was in use. In the second condition, the temperature was programmed to rise steadily at 1°C/min from 130 to 190°C, although this appears to have fallen well short of ensuring an IV drop of c. 80 being obtained before passing the 150°C level. One test catalyst is shown as having maintained an equal or slightly better hardening rate in the 130–190°C regime, two were little changed, and one slowed markedly. The catalysts displayed different powers of S_I and *trans* promotion, which had clear and expected effects upon the steepness or flatness of the solid content curves for samples taken at 2°C intervals between 30 and 40°C m.p. Unichema (Unichema, 1987–1992; Balemans & Osinga) has published extensive details on the economic re-use of dry-reduced nickel catalyst in fish oil hardening (see Chapter 4, the "Catalyst Handling and Economy" section).

It will be evident that when polyunsaturated systems in which *cis* methylene-interrupted double bonds have been reduced to a very small percentage (for partially hardened vegetable oils, 2% linolenic acid is about the level), the softer components can then be separated by fractionation and still retain acceptable oxidative stability. This may be performed, again as with vegetable oils, in one or two steps, the temperatures chosen being related to the stringency of the cold test applied to the product in which the hardened fish oil final liquid fraction is to be included. Such a liquid fraction, having few saturates and also being unlikely to cyclize, might provide a steep melting curve if partially hydrogenated to a slip m.p. in the mid-thirties centigrade with the aid of a specially sulfur-poisoned catalyst. In performing the sequence: hydrogenation, fractionation, hydrogenation, the costs involved–assuming technical success--must bear comparison with what is available from vegetable oil sources such as the hardened lauric oils and vegetable oils hardened with sulfured catalyst.

Sardine and Menhaden Oils

What has been known as sardine oil in the United States is known in Canada as pilchard oil. This oil is not derived from the same species of fish as Japanese sardine

TABLE 8.40
Principal Fatty Acid Composition of Sardine and Menhaden Oils

Fatty Acid Type	Sardine Oil (%)	Menhaden Oil (%)
C14:0	8	7
C16:0	16	19
C16:1	9	9
C16:2		2
C16:3,4		4
C18:0	4	4
C18:1	11	13
C18:2	1	2
C18:3,4	3	5
C20:1	3	2
C20:2		1
C20:3,4,5	19	13
C22:0		
C22:1	4	1
C22:2		
C22:3,4,5,6	15	12
Totals	93	94

oil; "Portuguese fish oil" would be regarded in the United Kingdom as "sardine oil." These fish oils are treated here as being members of group B, but with a typical IV of 165, rather than 200, which is commonly the value for the anchovy and South African pilchard oils previously described. For practical hydrogenation purposes, "sardine" oil and menhaden oil, which also is frequently found with an IV of 165, may be processed in the same way. Table 8.40 gives fatty acid compositions which in their simplified form clearly place these oils in group B on account of their modest content of monoenes (not above 30%) and considerable (c. 40%) polyunsaturates.

Sardine oil has been reported with an IV up to 190, but 165 is taken here as more representative; the IV bracket 150–170 for menhaden oil fits well enough, with a value of 165 IV also being taken as typical. The unsaponifiable (%) for both oils is normally close to 1.0. The FFA content and oxidative state of these oils depends, as with other marine oils, on how carefully the fish have been handled and processed; both are refined in either batch or continuous equipment to the standards suggested in the "Cleaning of Oils Prior to Hydrogenation" section of this chapter prior to hardening. The more rigorous the pretreatment, the lighter the poisoning if fresh catalyst is to be used. Dark washes and the separation of tarry mucilage when soapstock is acidified testify to the cleansing which the crude oil receives (see

"Marine Oils—General Considerations" section earlier in this chapter). These oils have been used extensively for technical purposes such as drying oils, for sulfonation, and after hydrogenation for saponification.

Note that the figures given are to be interpreted as a guide for the hardening of what is taken as a typical starting IV. If the initial IV is higher, the drop will be appreciably greater, if not proportionally so. Similarly, the refractive indices in relation to IV are approximate. Conversion of dilatation results to percent solid fat content were made in accordance with the published table (van den Enden et al., 1986). In practice, the exact sequence of operating conditions on a particular autoclave leading to the texture required as tested by the plant concerned will soon become apparent.

Two-Stage Hydrogenation (see Chapter 2, "Two-Stage Hydrogenation")

It may be desired to keep the % SFC at 20°C of a partially hydrogenated oil to a minimum. A catalyst of established selectivity should be used and a dose of 0.07–0.15% fresh Ni/oil employed. If successful, a smaller nickel dose may be tried subsequently. If the dose is reduced beyond a certain point (say, under 0.04% Ni/oil) the inevitable poisoning of the catalyst will lift the % SFC at 20°C and prolong hardening times. Conversely, a larger initial dose of fresh nickel will improve the chances of a satisfactory re-use, if not in this *trans* suppressive type of hardening, but in what here is classed as normal hydrogenation. Gassing commences at 120°C; the IV drop required before 150°C is achieved (see Chapter 2, "Cyclization and Polymerization" section) and will be 54 for an original IV of 165, bringing the value down to 111. Since the end IV to be expected is no greater than 84, this allows a further fall of 27 IV from IV 111 to 84, which should permit the attainment of 180°C for the last 10 min of hydrogenation, especially as the temperature is assisted to rise in the second stage. In case of difficulty, the temperature may be allowed to come up slowly from 120°C in the first stage to attain 150°C at the time the IV has fallen to 111. Experience will indicate where the most satisfactory balance lies. At least 45 psig pressure is likely to be convenient for the first stage; I5 psig may be sufficient for the second. For an initial n_D^{60} 1.4646, the first stage will end at c. n_D^{60} 1.4586; for an 82 IV end-point, an n_D^{60} 1.4545 is to be anticipated. In these circumstances, a 33°C slip m.p. is attainable with an SFI 20/30 of 18.5/8.3 (% SFC at 20/30°C =20/5.5). Prolongation of the first stage to lower than 111 IV in the above example will tend to flatten the melting curve, but this decrease in the remaining hydrogenation at the higher temperature may so lower selectivity that oxidative-and hence, flavor-stability are adversely affected. Unichema (Unichema International, 1992) describes the hydrogenation of fish oil (175 IV), using c. 0.11% fresh Ni/oil (PRICAT 9910) at 120°C throughout and 4 bar (44 psig) hydrogen pressure. A final IV of 90 was obtained in 50 min gassing. This is the class of hydrogenation when subsequent fractionation of the partially hydrogenated oil is intended to produce a stable (liquid) olein. Chapter 2, "Low-Temperature Hydrogenation" and "Iso- or *trans*-Suppressive Hydrogenation" sections should be read in this connection. Engelhard offers Nysosel 222 for low temperature (140°C) hydrogenation of fish oil (Engelhard De Meern BV), but as Nysosel 325 has still higher selectivity, it might be worthwhile exploring its use in this application. The Süd-Chemie catalysts C70 and T4460 should be tried for the same task.

Normal Hydrogenation (see Chapter 2, "Two-Stage Hydrogenation")

It is sufficient in this case to perform the first stage of the hydrogenation in the previous section just below 150°C, then to complete at 180°C. With the same quality of catalyst as recommended in this chapter in "Herring Oil, Hydrogenation, Two-Stage Hydrogenation" section, the result of slightly better *trans*-promoting overall conditions is likely to be an SFI 20/30 of 22.2/7.4 (% SFC at 20/30°C = 25/6.7) with a marginal increase in slip m.p. above 33°C. If, as is common practice for reasons of catalyst economy, an increased dose of partly exhausted catalyst (0.2-0.5% Ni/oil) is used, the SFI 20/30 will approach 26/7.4 (% SFC at 20/30°C = 30/6.7). From then on, continued hydrogenation results in a *37°C* slip m.p. with an SFI 20/30 of 37/20.4 (% SFC at 20/30°C = 44/19.7), after which the SFI 20:30 ratio begins to lose its steepness, and any rapid descent of the melting curve will be more obvious in the SFI 35/40 region, with very similar values as for other hardened fish oils such as anchovy and South African pilchard oils.

Süd-Chemie has described in detail (Nickel Catalyst KE-NF 20 and KE-FS40) the laboratory hydrogenation of a 152 IV fish oil, which suggests it is of the group B type. A dose of 0.06% Ni/oil of KE-NF 20 catalyst is used at 2 atm pressure throughout with moderately fast agitation. The first example commences gassing at 130°C; the temperature is then programmed to rise at 1°C/min to 190°C and continues at that level thereafter. By the time the IV falls to 113 (a drop of 39 units), the temperature has risen to 165°C, so that this approaches the first stage described earlier in this section. Evidently, the catalyst is selective, since at 85 IV the open tube m.p. is 26°C and SC% (by NMR) 20/25/30 is 17/6/2; at 75 IV, the m.p. is 32°C and SC% 20/25/30/35 is 29/20/1¼. For comparison, a hydrogenation uniformly at 180°C (ignoring the kind of 150°C restriction described here for plant-scale practice) shows the reaction rate is faster at the start but falls later, so that by the end at 54 IV, the overall reaction time is a few minutes longer. As regards texture, it is noticeable from the graphs provided that when comparing the SC% at 80 IV, the 130-190°C programmed hardening has the higher figures across the 10–40°C range, but by the time the IV drops to 70 and beyond, the differences in SC% are negligible. Unichema (Unichema International, 1992) also describes hardening of fish oil (160 IV) with c. 0.07% fresh Ni/oil (PRICAT 9910) at 44 psig and a temperature rising from 150°C to 180°C.

High-Melting and Fully Saturated Hardened Oils (see Chapter 2, "Higher Melting and Fully Saturated Hardened Oils")

The situation here is similar to other hardened marine oils, and a final slip m.p. of 58 ± 1°C for 2 IV max. is to be expected.

Castor Oil

The castor oil plant grows in tropical, sub-tropical, and warm temperate regions, preferring a dry environment. It is cultivated both as an annual and a perennial, the size of plant and yield of beans per acre showing very large variation according to cultivation. In difficult conditions in India, as little as 160 lb/acre of beans may be obtained, though in Brazil 2200 lb/acre, and in California even 3000 lb/acre may be obtained. Oil content of the beans averages 45–55%. Brazil, India, and the USSR are the most important producers (*Oils and Oilseeds*, 1971). Oil is obtained by pressing, or pressing followed by solvent extraction. Pharmaceutical grade oil is high quality oil derived solely from the first pressing of beans, and accounts for only a very small percentage of total sales. Technical grade castor oil obtained by pressing, repeated pressing, and solvent extraction has an extremely wide range of uses because of its very exceptional chemical and physical characteristics, which relate to its uniquely high content of an unsaturated hydroxy fatty acid, *cis* 9,12 hydroxy octadecenoic acid, commonly known as ricinoleic acid:

$$CH_3-(CH_2)5-\overset{12}{CHOH}-CH_2-\overset{9}{CH}=CH-(CH_2)_7COOH$$

A typical fatty acid composition (Janson, 1974) of castor oil is given in Table 8.41.

Various sources have reported a range in ricinoleic acid content of 80–95%. It is therefore evident that heredity, environment, and the limitations of sampling and analysis operate here as elsewhere. There is a spread in IV of 81–91, saponification value of 176–187, density of 0.945–0.965 (25°C/25°C), FFA up to 2% (as oleic) for first class quality, and color up to 2.2 Y 0.3 R (1") for top quality. The refractive index at 20°C may be n_D^{20} 1.4748–1.4810; for an average good quality oil at 60°C n_D^{60}, 1.4648 is a feasible starting point in hydrogenation control. Hydroxyl values of 155–165 are quoted; 162–165 appears usual. Castor oil has a very high density for an oil, and is miscible with ethyl alcohol, but only to a limited extent with

TABLE 8.41
Fatty Acid Composition of Castor Oil (Janson, 1974)

Fatty Acid Type	%
C16:0	0.9–1.2
C18:0	0.7–1.2
C18:1	3.2–3.3
C18:2	3.4–3.7
C18:3	0.2
C18:1Δ9,–OH12	89–89.4 (ricinoleic)
C18:-OH9,10	1.3–1.4 (dihydroxy stearic)

aliphatic hydrocarbons. It is somewhat hygroscopic; its viscosity withstands temperature change very markedly, and if it burns, little residue remains; hence, it is extremely useful in lubricants. Under the influence of the polar hydroxyl group, it is compatible with resins, polymers, and waxes. Castor oil may be dehydrated in such a way that the hydroxyl group, along with a neighboring hydrogen atom, is eliminated, leaving two double bonds. This product is used in drying oils and resins to improve gloss, flexibility, and color. Sulphonated castor oil is a very long-established wetting agent for dispersal of dyes. When hydrogenated to below 3 IV in conditions that leave the hydroxyl group virtually undisturbed, a brittle hard odorless fat of 84–88°C slip m.p. results, which is compatible with both natural and petrochemical waxes to give a product whose melting point is both higher and more sharply defined; it increases the effectiveness of lubricants over a wider temperature range, it assists the dispersal of pigments in paint as well as their brushing quality, and enhances the resistance of enamels to yellowing. If hydrogenation is allowed to proceed so that hydroxyl groups split off and double bonds are all hydrogenated, the fat will then obviously consist mostly of stearates.

A good quality crude castor oil, especially from first pressings, may be given merely an adsorptive cleansing or bleaching with activated earth-carbon, under vacuum and steady stirring. Seed temperature is kept at 60°C during crushing to safeguard oil quality, especially as regards color, and it is recommended (Janson, 1974) that this oil be dried immediately in the refinery to stabilize both FFA and color, which will mount steadily if the oil remains moist. Castor oil of up to 1% FFA hydrogenates without particular difficulty. If some deacidification of the crude oil is needed, this works satisfactorily in batch equipment, the alkali being sprayed onto the warm oil with gentle stirring and the temperature being kept up to a level which also opposes the formation of an emulsion (90–100°C). The oil is next settled, after which soapstock is withdrawn from the base. Brine and hot water connections to the neutralizer are available so that the oil may be washed soap-free, then dried and earth bleached. A so-called "stillwash" (sprayed with no agitation) may be best. It is important to keep the filters warm so as to maintain a satisfactory rate of filtration, which really means it is desirable to keep them operating until full and ready for cleaning (Janson, 1974).

Hydrogenation

Dry oil is dosed with up to 0.2% Ni/oil; the catalyst may be of the stabilized, i.e., non-pyrophoric type if wished (Harshaw Catalysts; Hoechst Aktiengesellschaft) or another dry-reduced catalyst made up in an acceptable hardened fat. A pressure of 5 atm should be satisfactory; a rather higher pressure may reduce hardening times, but not greatly. Hydrogen should also be of normal dryness, i.e., not more than 0.1% v/v (see Chapter 5, "Quality" section). A hardening temperature of 110°C is a safeguard against loss of hydroxyl value and decreased melting point. Within 8 h, an IV of under 3 results, with a slip m.p. of c. 86°C and a drop in hydroxyl value of only 2 units, e.g., 162–160.

It will be very helpful if the non-return valve on the hydrogen line immediately adjacent to the autoclave is so weighted (or otherwise controlled) that it closes firmly as soon as the pressure in the line ceases to be at least 0.3 atm greater than the pressure in the autoclave. This safeguards against pressure fluctuations in the hydrogen distribution system, allowing the gradual seepage back from the autoclave of c. 86°C m.p. fat. After the autoclave, the temperature of filter lines, pumps, and filter presses must be maintained comfortably above the melting point of the hardened castor oil. Specially designed steamheated filter presses are available. Examples of castor oil hydrogenation are given by some catalyst manufacturers.

Harshaw (Harshaw Catalysts) quotes 0.06% Ni/oil of their Resan-22 catalyst to hydrogenate castor oil from 85.3 IV, zero moisture, 162 hydroxyl value, and 0.8% FFA to 2.6 IV, zero moisture, 159.9 hydroxyl value, 85°C open tube m.p., and 0.6% FFA using 6 atm in 300 min, as shown by a graphical plot at 135°C. The plot for 120°C shows a longer time (over 420 min) to approach 3.3 IV, but an 85°C m.p. and a hydroxyl value of 161.4. Drying the oil below 100°C prior to hydrogenation and under vacuum, followed by addition of 1-2% activated earth before the temperature was raised to 120-125°C for 15-30 min, was employed as a purifying pretreatment. Süd-Chemie (Nickel Catalyst KE-NF 20 and KE-FS40) illustrates the effects of hydrogenating castor oil (85.6 IV, FFA 1.0% c. hydroxyl value 164.6) under three different ,conditions using KE-NF20 catalyst in each:

(a) 0.06% Ni/oil, 6 atm, 120°C.

(b) 0.04% Ni/oil, 6 atm, 100–120°C.

(c) 0.04% Ni/oil, 10 atm, 80–120°C.

All melting points were 86°C, FFA contents were all satisfactory at 1.1, 0.9, and 1.1%, respectively, and the hydroxyl values were also satisfactory at 162.9, 162.7, and 162.5. However, whereas (a) reached an end-point of 2.6 IV in 4 h, (b) only reached 3 IV after 6 h and (c) 2.3 IV after 6 h. The higher nickel dosage in (a) appears to have been justified. Because of its good low temperature activity, Unichema PRICAT 9920 would be suitable with PRICAT 9910 as a possible alternative; Calsicat offers E428D, and Hoechst recommends Hoecat 882 OF.

The usual aim of castor oil hydrogenation is to obtain a high-melting saturated fat with almost undiminished content of hydroxyl group so that the special physical advantages of this combination can be employed in the varied products already mentioned. To that extent, such a procedure might be described as "normal" hardening, but such a term has not been used here because the operating conditions are markedly different from those described as "normal" for the general run of edible oils (see Chapter 2, "Normal Hydrogenation" section). Other, less usual, aims may be pursued in the hydrogenation of castor oil, and an example of these is provided by vant Hof, Linsen, and Pastoor (1968) who describe four approaches to reducing castor oil in such a way that the hydroxyl groups are removed while to the maximum possible extent, the number of double bonds originally present remains - i.e., dienes resulting from elimination of water from an existing monoene (castor oil) are hydrogenated. Thus, the dienes persisting into the final product are kept down. Fats are obtained with slip melting points in the 21–32°C range, and are seen as having technical applications such as soap making.

(a) Hydrogenation under dehydroxylating conditions.
0.75% Ni/oil of sulfurized catalyst was used at 240°C. Hydrogen flow removed the water liberated.

(b) Dehydration followed by selective hydrogenation.
Dehydration with 0.5% $NaHSO_4$ at 240°C, followed by hardening at 180°C using 0.5% Ni/oil of sulfurized catalyst.

(c) Low temperature hydrogenation followed by dehydration.
Hydrogenation at 100°C (120 min), then at 180°C (10 min) with 0.5% fresh Ni/oil; followed by dehydration under vacuum at 240°C with 0.5% $NaHSO_4$ for 3 h.

(d) Dehydrating hydrogenation.
3% Ni as a sulfurized catalyst plus 2% $NaHSO_4$ were employed at 240°C; also, 2% Ni as a sulfurized catalyst plus 1% $NaHSO_4$ at 180°C were found to give acceptable results at the lower temperature. An excess of hydrogen was here passed through the oil to remove water formed, since in a closed system, it would halt the reaction.

In none of the above hydrogenations did hydrogen pressure exceed 1 atm. Dienes reported in the products were (a) 5%, (b) 9%, (c) 2.5%, (d) 8.5% in both. Finally, results on (b), (c), and (d) were considered acceptable, it being noted that practical difficulties might arise on the large scale in venting moisture from an autoclave if using route (d); (b) and (c), although effective, are two-stage operations. Detailed analyses of the products are given in the report (Linsen, et al., 1968). These experiments again show how hydrogenation may be combined with other reactions in addition to the usual ones of fractionation and interesterification.

Fatty Acids

While the basic factors governing the hydrogenation of fatty acids remain the same as those for neutral oils, some obvious special features present themselves. Fatty acids attack both the autoclave and the nickel catalyst. The first hazard is adequately covered by the use of stainless steel (see Chapter 4, "Material of Construction" section) (Rice, 1979), and equipment manufacturers take this into account. Selectivity is much less commonly required in hardening of fatty acids than neutral oils. It is frequently a case of direct hydrogenation to a saturated condition, and this affords the opportunities for comparatively long uninterrupted runs for the same purpose, a circumstance which maximizes the utilization of a continuous unit. For a set production program, a continuously operating reactor is considerably smaller than a batch one, and this harmonizes with the need to employ materials of construction which, because they are more chemically resistant, may also be more expensive.

Other advantages of a continuous unit in space, services, labor, and control operate here. Batch autoclaves do exist and work well.

Most suppliers of systems for hardening neutral oils will also supply a system for hardening fatty acids (*Fatty Acid Technology*, 1991; Leuteritz, 1971; *Flow Sheet SFS-164*; *Semi-continuous Hydrogenation System*; *Hydrogenation Plant with Post Treatment*; Lightnin Mixers Ltd.; Wurster and Sanger Div.; Rosedowns & SA De Smet NV; Fried, Krupp GmbH, Technical Handout; Fratelli Gianazza S.p.A., Technical Handout; Hastert, 1990; Lally, 1987; Oldshue, 1987; Costruzioni Meccaniche Bernardini). The nickel catalyst employed is of the dry-reduced siliceous supported type. Since the molecules of a fatty acid are smaller than those of a triglyceride, they can penetrate the pores of a catalyst more easily; this may also mean they can move into and out from catalyst pores more easily than many other poisons. There is, however, a further consideration, and this has been described with commendable clarity by Lok and Joosten (Joosten & Lok, 1991). Fatty acids attack nickel crystallites to form nickel soap. The molecule of soap comprising one atom of nickel and two molecules of fatty acid is much more bulky, and is able to clog and block pores more easily than a fatty acid molecule alone. On the other hand, hydrogenated fatty acid molecules can, of course, escape, and the nickel in the pore then remains to continue the work. Hence, a pore width of 40 Å (4 nm) suits fatty acid hydrogenation, but double this and more is important for selective hydrogenation of triglycerides. Strategically, a high proportion of active nickel held in narrow pores is thus better situated to survive and continue the hydrogenation to a low IV. Secondly, a high hydrogen pressure opposes the reaction, nickel plus fatty acid yields nickel soap plus hydrogen.

Thus, a higher pressure—say 20-25 bar—means lower nickel soap concentration. As remarked by the authors, "Indeed, it is the stability of the catalyst towards fatty acids which is enhanced by narrowing pores, rather than intrinsic activity." Even so, it is generally accepted that in fatty acid hardening, the catalyst will sustain only one use. It will help greatly if the feedstock is cleaned well; it must also be dried, firstly to hinder the attack of fatty acids on catalyst, and secondly to provide less water vapor to obstruct the gas space. Poison resistance is probably linked in most instances to a generous proportion of active nickel surface per unit weight of catalyst, which in this context is only another way of saying that there is available a reserve of nickel to combine with and immobilize poison, while the remainder suffices to promote the hydrogenation to the required end-point. Under attack from fatty acids, some nickel is likely to form nickel soaps which can escape into the body of the fatty acid charge, promoting an obvious pale-green tinge if they reach 10 ppm Ni. Obviously, such soap must be removed at a subsequent stage via dilute acid washing, bleaching earth adsorption and filtration, distillation of the fatty acid, or some combination of these. This end-point of near-complete saturation is difficult enough to attain even with some neutral oils for reasons of mass action. It is not surprising, therefore, that fatty acid hardening is commonly performed at 20–30 atm.

Hydrogenation is intended, along with other effects, to improve color, stability, and odor. Operating temperatures are therefore curtailed, according to sensitivity of the feedstock, to 150°C or 180°C, but in some circumstances (such as fish fatty acid hardening) allowed to reach 210°C. A lower temperature also reduces the rate of attack on nickel, but it may prolong the reaction. Taking these different considerations into account, the high pressure hardening of many reasonably clean distilled fatty acids can be completed with a dose of 0.05% fresh Ni/oil at 180°C; it will conserve the catalyst activity if the fatty acid is preheated to 140–160°C under hydrogen before the catalyst is introduced (Unichema International, 1992). Batchwise, the reaction could be complete in very few hours, but if working at lower pressures (up to 5 atm), much more time would be needed; and even then, the nickel consumption is multiplied several times.

In a batch plant, one or more purges of the gas space to atmosphere in order to dissipate moisture may speed the reaction. This will be advantageous when only one-quarter or one-third of the anticipated cycle time has elapsed, because by then, more than half of the hydrogen has been used. In a continuous plant, the design (*Fatty Acid Technology*, 1991) may well allow for the concurrent flow of an excess of hydrogen to separate after the autoclave and then be returned to the input, being dried in the process. Hydrogen of maximum moisture content 0.1% v/v is best (see Chapter 5, "Quality" section); moist hydrogen is reported (Nickel Catalyst KE-NF 20 and KE-FS40) to increase the amount of nickel dispersed in the hardened product. As with wet feedstock, additional moisture facilitates the attack of fatty acid on the exposed active nickel. Again, it is said (Nickel Catalyst KE-NF 20 and KE-FS40) that where a satisfactorily rapid rate of hydrogenation can be initiated by commencing gassing at 120–140°C, this should be done; the attack on the nickel is presumably slower at this temperature level. Exothermic heat of reaction will raise the temperature steadily to 180°C and then the reaction is driven on to reach a low IV if that is required. Dropping working pressure cautiously is a preferred means of slowing a reaction which is becoming too rapid. If cooling water is used suddenly, this may have so drastic an effect that re-establishing a good rate of reaction may be difficult. Both facilities will exist on a normal autoclave, however. Where the fatty acid has a very high IV (170-210), initiating the hydrogenation at temperatures as low as 75°C will probably be feasible, and if so will be beneficial; in cases where temperatures of c. 200°C have to be used in order to make progress and reach the required low IV, the nickel content of the hardened product must be expected to rise appreciably.

A number of the above effects have their parallel in hydrogenation of neutral oils, even if in that case they are less striking. Hastert (Hastert, 1979) has drawn attention to several of these items, and Coenen (Coenen, 1975) emphasizes that, although the rate of hydrogenation rises with rise in temperature, there comes a point, usually in the 180–210°C region, where the maximum is reached, and beyond this the reaction rate drops much more steeply than it had previously risen. So sunflower fatty acids hardening at 14 atm with 0.04% Ni/oil reach a maximum hardening rate at 195°C, after which, by 220°C, the rate has fallen to what was recorded for c. 160°C. In general, as might be expected, increases in catalyst concentration, pressure, agitation, and temperature are all subject to the rule of diminishing returns in fatty acid hardening, as in neutral oil hardening.

Producers of hardened fatty acids have a choice between two basic routes; first, to hydrogenate a triglyceride-which presumably has been cleaned to some degree beforehand-and then to split the hardened fat and distill the fatty acids; alternatively, the triglyceride may be split, the fatty acids distilled, hydrogenated, and then given a post treatment which may include a second distillation. The second route appears to be preferred-cost permitting-when a very low IV of around one is being sought. In this case, a catalyst with an outstanding resistance to fatty acid poisoning and nickel soap production is the choice. If high pressure hydrogenation is not an option, then the first route may be advantageous. In that case, a catalyst with good resistance to the usual range of catalyst poisons may be preferred. Unichema offers PRICAT 9932 especially for fatty acid hardening; PRICAT 9912 has a high resistance to poisons in general, and also performs well with fatty acids (Unichema International, 1992). In the case of fatty acids which are polymers (dimers and trimers), these must be thoroughly dried before hydrogenation. A hardening temperature of 240°C is possible, and the catalyst supplier should be consulted as to which of the latest products is recommended. The higher temperature will also facilitate dispersion of hydrogen into what might otherwise be a particularly viscous medium. If distillation has not featured in the post-hardening treatment of fatty acids, it is important to remember that the post-treatment adopted must secure the removal of nickel and other pro-oxidant trace metals. If present, these traces will hasten the discoloration of the product.

Süd-Chemie (Nickel Catalyst KE-NF 20 and KE-FS40) gives extensive information on the use of their long established nickel catalyst KE-FS40 in the hydrogenation of fatty acids and the conditions of temperature and pressure best suited to individual cases. Although several examples of hydrogenation at 5 bar are quoted, the usual range is 10–20 bar. Hydrogenation is commenced at the lowest feasible temperature, generally in the 100–140°C range, except that for marine oil fatty acids with an IV above 180, a start is made at 75°C. The idea is to use the heat of reaction as the driving force, adding to it when necessary but avoiding cooling, which tends to inhibit the reaction. As might be expected, for the distilled fatty acids of soybean, cottonseed, sunflower, and palm to be hydrogenated to 1 IV nickel doses vary from 0.05% to 0.2%, but if crude vegetable oil fatty acids are being processed, the final IV is taken as 5 and the fresh nickel dose lifted to 0.2-0.4%. For distilled tallow fatty acids, up to 0.1% Ni/oil may be necessary, and for distilled fish fatty acids, up to 0.3%. Again, the corresponding crude fatty acids require substantially bigger doses. Since higher temperatures mean greater risk of forming nickel soaps, the maximum to which the reactions are allowed to rise are for, vegetable fatty acids, 180°C; for animal fatty acids, 190°C; and for fish fatty acids, 200°C. Süd-Chemie's catalyst, T3457, has proved more active and poison resistant than KE-FS40. This opens up the usual choice of using less nickel and accepting the same gassing time, or using the same amount of nickel and reducing the gassing time. The performance of both catalysts is appreciably improved by pre-cleaning of the feedstock.

Engelhard's standard catalyst for hydrogenation of fatty acids is Nysofact 101 IQ and data covering reaction times and doses of nickel at 20 bar and 180°C for fish, tallow, and palm fatty acids are published (Engelhard De Meern BV). The catalyst N 120 is also used in hydrogenating fatty acids when conditions of high temperature, relatively low pressure, and poor hydrogen dispersion encourage the production of nickel soap. Of the non-pyrophoric, non-fatty catalysts, of which Engelhard supplies a number, the Ni 5256 P would be the one of most interest to fatty acid hardeners who seek to avoid any triglyceride in their product. Other catalysts designed for fatty acid hydrogenation are Hoecat 882 FA (Hoechst Aktiengesellschaft) and Calsicat E 428 D (Mallinckrodt). Calsicat also offers a range of powdered, non-pyrophoric nickel catalyst free from any fatty vehicle.

It is evident that distilled dry fatty acids are very much the preferred feedstock for hydrogenation. This may entail cleansing prior to complete splitting of an acid oil, or as a first step in processing of a dirty neutral oil. The most elementary cleaning simply consists of boiling the oil with about 7–10% of its own weight of 10% brine for several minutes, settling, withdrawing the discolored brine, then following with one or more water washes. Where more impurities, such as mucilage and soaps (especially calcium soaps), are present, the cleaning substitutes up to 2% of a 10% sulfuric acid solution for the brine and a contact of up to an hour. This acid digestion may be increased to 3% of a 15% sulfuric acid. Unichema (Unichema International, 1992) also refers to digestion for one hour at 90–100°C with 2% of a 78% sulfuric acid. With these treatments by acid of increased strength, some classes of impurity will flocculate and can be drawn off with the washes. A very dilute alkali wash may be needed to get rid of the last mineral acid. If it is an acid oil which has reached this stage, it will now be ready for hydrolytic splitting, after which the resultant fatty acids may be distilled and then hydrogenated, or the fatty acids may be given a cleansing, such as contact with up to 2% of activated bleaching earth (95-120°C), filtration, and then hydrogenation. If, on the other hand, it is a neutral oil which has just been cleaned, it may now be bleached, hydrogenated, and then split; or, alternatively, it can be hydrolytically split and then, after further cleaning (distillation or bleaching), hydrogenated as before. Acid washing and earth bleaching can go some way towards replacing the expensive step of distillation, for which the proper facilities may not be easily available in any case. Lastly, the hydrogenated fatty acids must be freed from trace metals because of their pro-oxidant influence. This takes the form of a light (0.5%) adsorptive bleach. If cost allows, hydrogenated fatty acids may be purified by distillation. Efforts made to cleanse feedstock will certainly improve ease of hydrogenation, saving nickel and time. The cleansing routine depends upon the amount and type of impurities in the feedstock and the specification to be met by the product. The most economical way of bridging this gap decides the cleansing sequence to be adopted. In very difficult cases where heavily oxidized and dark material is being considered, it may prove advantageous in the long run to saponify it completely and then grain out the normal fatty acid soaps with brine, leaving impurities, including soaps of oxidized fatty acids, to remain in the separated lye as in conventional soap-boiling practice. Purified soap curd is then split to fatty acids, and the way is opened to hydrogenation in easier circumstances.

Technical Oils (Soap-Making)

As regards texture, we are no longer concerned with the position of the fatty acids in the triglyceride, since they are all split off by the act of saponification. Nonetheless, the quality of the final soap is very dependent on the proportions and types of fatty acids present in the original fat charge, since short-chain and unsaturated fatty acids tend to promote ease of solubility of the soap, whereas longer-chain fatty acids, especially if saturated, are more hydrophobic, mix more easily with oils and grease, and promote a more substantial and stable lather. The best compromise as regards chain length would be C14, of which not very large amounts occur in the fats easily available to soap makers. However, by having a balanced range of fatty acids as regards chain length and degree of unsaturation, the necessary standards of solubility, wetting action, stability of lather, firmness in the tablet form, and resistance to atmospheric oxidative attack can all be achieved. Branched-chain fatty acids are not present to any significant extent; *trans* isomers in the unsaturated fatty acids are not favored by soap makers. All of this amounts to performing a nonselective hydrogenation with minimum *trans* formation when hardening an oil which is intended to become a major component in the fat charge of a tablet soap product. This technique has been discussed in detail in Chapter 2, "Iso- or *trans*- Suppressive Hydrogenation" section. Suffice it to say here that 135–155°C would be a suitable range for hardening in this application; 3 atm pressure should prove adequate; if the oil has been reasonably well cleaned from catalyst poisons, then a fresh nickel dose of 0.05–0.10% Ni/oil will produce a fat in the low thirties centigrade melting point range, with the titer which is being sought. How often the catalyst may be reused and what other minor fat component is best included in the charge to secure necessary quick lathering is then a matter of trial. There is a wide choice of catalysts suitable for this purpose, especially if the feedstock has been reasonably well cleaned by an adsorptive bleaching with an activated earth. Catalysts which will be quite effective are: Nysosel 222, Unichema 9910, Süd-Chemie KE-NF20, Calsicat E 428 D, and Hoecat 882 OF.

A possible alternative hardening technique is the two-stage hydrogenation described in Chapter 2. Here the advantage would seem to lie with pursuing the low-temperature first step, which might be as low as 120°C, as far as possible, leaving only about *15* units IV drop to be achieved as the temperature rises with assistance above 150°C. In both types of hydrogenation, saturated fatty acids will be produced ahead of the complete disappearance of polyunsaturates; the brief period of more selective hydrogenation towards the end of the hydrogenation is designed to protect the final soap from oxidative instability. A more diverse mixture of fatty acids is produced by these means. It was pointed out earlier in this chapter in the "Hydrogenation" section, when the hydrogenation of marine oils was being considered, that non-methylene-interrupted double bonds were less reactive than the methylene-interrupted double bonds of the skipped arrangement so common in natural oils. This consideration would apply to soap charges equally as to edible oils.

Chapter 9

Electrocatalytic Hydrogenation of Edible Oils

Peter N. Pintauro

Department of Chemical and Biomolecular Engineering, Vanderbilt University

Introduction

The partial hydrogenation of soybean oil is carried out to produce a solid or semi-solid fat and/or to improve the oil's oxidative stability. The standard chemical catalytic hydrogenation of edible oils is carried out in a batch reactor, operating at 150–225°C and 10–60 psig with Raney nickel or supported nickel catalysts (Hastert, 1981; Coenen, 1986). Hydrogenation at a high temperature leads to a number of undesirable side reactions, in particular the production of *trans* fatty acids which have been linked to coronary heart disease (Ascherio & Willett, 1997). Electrochemical reaction methodologies are well suited to edible oil hydrogenation, due to their mild reaction conditions. Prior studies on this subject are limited, although the published work has been promising. This chapter will review the basics of organic electrochemistry, electrocatalytic hydrogenation reaction schemes, and the application of such schemes to the partial hydrogenation of edible oils.

In an organic electrochemical process, heterogeneous electron transfer reactions at the anode and cathode replace chemical oxidizing or reducing agents in solution. Electrons are added to or removed from the organic species of interest at the cathode or anode surface or the electron transfer reactions at the electrodes generate anion or cation radicals, which are transported into solution where they initiate various homogeneous reactions. The electrode potential (a measure of the energy of the electrons involved in the electrode reactions) often controls the nature of the organic product (or the electro-generated intermediate) and the applied current (a measure of electron flow) is related to production rate. The product of current and voltage is the power requirement for the electro-organic synthesis. The use of electrochemical methods for the production of organic compounds is attractive for a number of reasons: (1) high yields and high selectivity can be attained under appropriate electrolysis conditions so it may be possible to synthesize new products at a lower cost, as compared with the use of traditional methods, (2) electrical energy is used instead of chemically reacting agents (e.g., $LiAlH_4$ for organic reductions), thus, electrochemical processes may be environmentally friendly with less pollution by spent reagents (e.g., it may be possible to regenerate electrochemically spent redox agents in-situ, so that large quantities of product can be produced from a small inventory of reagent with minimal environment impact), (3) reactions take place under mild conditions of moderate temperature and pressure, and (4) oxidation and reduction reaction rates can be varied by controlling the electric current or electric potential.

All electrochemical reactors must contain an anode and a cathode, with an ionically conducting electrolytic solution (a solvent and supporting electrolyte salt) between the electrodes. This system is inherently more complex than a chemical catalytic reactor. There are other complicating factors when carrying out an organic reaction in an electrochemical reactor:

1. The rate of the electron transfer step at the anode or cathode during an electro-organic synthesis is fast for a simple electron transfer reaction but is much slower for a complex reorganization of the substrate. The low rate manifests itself as a low reactor operating current density and a low space-time yield. This problem is often circumvented by use of three-dimensional, packed-bed electrodes with a high electrode surface area to volume ratio or by use of mediated reactions (a fast redox reaction at the electrode followed by a bulk solution homogeneous reaction with the organic substrate which consumes the electrogenerated reagent).
2. The high charge requirement of 96,487 coulombs (one faraday) per mole of electrons is a detriment for electro-organic reactions, especially those that involve multiple electron transfer steps. Thus, the synthesis of high value-added products, compounds that cannot be made by traditional catalytic methods, and organic compounds with a high molecular weight (where the number of faradays per pound of product is low even though the required number of faradays per mole is large) are preferred.
3. The difficulty in mixing the organic substrate with a conducting salt (either dissolving the organic substrate in an aqueous electrolyte solution or solubilizing an inorganic supporting electrolyte salt in the organic starting material). This problem has been addressed by use of: (i) alcohol or alcohol/water mixed solvents that can solubilize certain combinations of organic species and inorganic salts, (ii) special supporting electrolyte salts, such as quaternary ammonium and hydrotropic salts, e.g., tetramethylammonium acetate or tetraethylammonium p-toluenesulfonate (Pintauro & Bontha, 1991), (iii) an emulsified reaction medium (where the organic reactant is dispersed in a water/supporting electrolyte solution (Yusem & Pintauro, 1992), and (iv) reactors of special design where the anode and cathode are in close proximity to compensate for the low ionic conductivity of the electrolytic reaction medium.
4. The requirement of having two reactions in an organic electrochemical reactor (an oxidation reaction at the anode and a reduction reaction at the cathode), where one reaction (referred to as the counter electrode reaction) is often unwanted and where the counter electrode product(s) may interfere with the primary organic electrode reaction. This problem is often circumvented by use of a semipermeable membrane or separator that divides the anode and cathode

compartments in semipermeable the reactor. Having only one electrode involved in product formation also wastes electrical power (at least 50% of the reactor's energy demand can be consumed by the counter electrode reaction). This drawback has been addressed by carrying out "paired" electro-organic syntheses where the reactions at both the anode and cathode are used to generate one or more useful organic product(s) (Pintauro et al., 1984).

Electrocatalytic Hydrogenation Reactions

Most organic electrochemical reductions in a proton-containing solvent like water occur by the direct transfer of one or more electrons to the chemical reactant at the cathode surface. Often, there is simultaneous (and unwanted) H_2 gas evolution. Thus the cathode reactions are:

$$\text{Primary Reaction:} \quad \text{Org} + nH^+ + ne^- \rightarrow \text{Org} - H_n \qquad \text{(Eq. 9.1)}$$

$$\text{Secondary (unwanted) Reaction:} \quad 2nH^+ + 2ne^- \rightarrow nH_2 \qquad \text{(Eq. 9.2)}$$

Organic reductions are normally carried out at a high hydrogen overpotential electrode such as lead, zinc, or mercury where the kinetics of H_2 gas production are sluggish (Beck, 1979). For a water-based electrolytic solution, the anode reaction is usually the oxidation of H_2O,

$$2H_2O \rightarrow O_2 + 4H^+ + 4e^- \qquad \text{(Eq. 9.3)}$$

An alternative reduction route is to perform an electrocatalytic hydrogenation at a low hydrogen overpotential cathode (e.g., Raney nickel or Pt-black). Here, atomic hydrogen is generated electrochemically on the catalyst surface by the reduction of protons or water molecules in the electrolytic reaction medium. The electrogenerated hydrogen then reacts chemically with the organic substrate to give the reduced product. The overall reaction sequence for the electrocatalytic hydrogenation of an olefinic organic compound is as follows:

$$2H^+ + 2e^- \rightarrow 2H_{ads} \qquad \text{(Eq. 9.4)}$$

$$2H_{ads} + R-CH=CH-R' \rightarrow R-CH_2-CH_2-R' \qquad \text{(Eq. 9.5)}$$

An unwanted side reaction which consumes current but does not affect the product yield is the formation of H_2 gas, either by the chemical combination of two adsorbed hydrogen atoms or by the electrochemical reduction of adsorbed hydrogen and protons,

$$2H_{ads} \rightarrow H_2(g) \qquad \text{(Eq. 9.6)}$$

$$H_{ads} + H^{+} + e^{-} \rightarrow H_{2}(g) \tag{Eq. 9.7}$$

Since hydrogen is generated directly on the catalyst surface using electrical energy, high operating temperatures and pressures (common for chemical catalytic hydrogenations) are not required. Also, such a reaction route circumvents the need to produce, store, compress, and transport hydrogen gas. There have been numerous studies showing that high surface area electrically conducting catalysts that promote electrochemical hydrogen evolution (so-called low hydrogen overpotential catalysts) can be used to electrocatalytically hydrogenate a variety of organic compounds, including benzene and multi-ring aromatics, phenol, ketones, nitro compounds, dinitriles, and glucose (see, for example, Miller & Christensen, 1978; Chiba et al., 1983; Pintauro & Bontha, 1991; Song & Pintauro, 1991). Reactions were carried out in batch and semi-continuous flow cells at moderate temperatures (less than 100°C) and atmospheric pressure, where the electrocatalytic cathode was Raney nickel, platinum black, palladium black, and Devarda copper. In these studies, the reaction products were similar to those synthesized in a high temperature chemical catalytic reaction scheme.

An electrocatalytic reaction route for the partial hydrogenation of the double bonds in edible oil triglycerides is particularly attractive. The low reaction temperatures will minimize side reactions such as the formation of unwanted *trans* fatty acid isomers, the concentration of hydrogenation on the catalytic cathode surface can be easily controlled by adjusting that applied current which may improve hydrogenation selectivities, and there is little free hydrogen gas present; thus the risk of explosion and fire is lowered significantly.

The performance of an electrocatalytic oil hydrogenation reaction is characterized in terms of the hydrogenation current efficiency, production rate, reactor power consumption, hydrogenation selectivity, and formation of *trans* fatty acid isomers. These terms are defined as follows:

1. *Current Efficiency* - The current efficiency (CE) of an electrochemical reaction is defined as the ratio of the electric charge used in forming the product of interest to the total charge passed through the reactor during an electrolysis. For the electrocatalytic hydrogenation of an edible oil, the current efficiency is computed from the change in total moles of double bonds in the oil and the total charge passed in an electrolysis (the time-integral of the current)

$$\mathrm{CE}(\%) = 100 \times \left[\frac{\Delta \text{ moles of double bonds}}{(2 \text{ equiv / mol})\, Q\, F} \right] \tag{Eq. 9.8}$$

where F is Faraday's constant (96,487 C/equiv) and Q is the total charge passed during a hydrogenation experiment with units of coulombs (for constant current reactor operation, Q is the product of the applied current and time). For soybean oil, containing 18 carbon fatty acids with one, two, and three double bonds, the current efficiency is given by

$$CE(\%) = \frac{F \times M\left[\frac{2\times\Delta_{C18:1}}{MW_{C18:1}} + \frac{4\times\Delta_{C18:2}}{MW_{C18:2}} + \frac{6\times\Delta_{C18:3}}{MW_{C18:3}}\right]}{I\times t}\times 100 \quad \text{(Eq. 9.9)}$$

where I is the applied constant current, t is the reaction time, and $\Delta_{C18:1}$, $\Delta_{C18:2}$, and $\Delta_{C18:3}$ are the change in the weight fraction concentrations of oleic, linoleic, and linolenic fatty acids before and after a hydrogenation experiment (as obtained from GC analysis). The factors 2, 4, and 6 are the electrical equivalents for hydrogen saturation of one, two, and three double bonds in the fatty acids, respectively, M is the total mass (in grams) of soybean oil utilized in an experiment, and MW is the molecular weight of each fatty acid.

2. *Production Rate (PR)* – The size of the reactors (the required electrode area) for an electrocatalytic oil hydrogenation process is determined by the net production rate of hydro-oil per unit area of cathode. The electrochemical production rate (with units of kg/h-m^2) for a partially hydrogenated soybean oil 132 initial IV is directly proportional to the applied current density (current divided by the apparent/geometric electrode area) and current efficiency,

$$PR = i\left(\frac{CE}{100}\right)\left(\frac{1}{\left[\frac{\Delta IV}{IV_{init}}\right]\times 1.054\times 10^{6}}\ kg/C\right) \times\left(1\times 10^{4}\,cm^{2}/m^{2}\right)\left(3600 s/h\right) \quad \text{(Eq. 9.10)}$$

where i is current density (A/cm^2), ΔIV is the change in the oil's iodine value after hydrogenation, IV_{init} is the initial iodine value of the oil, and the constant 1.054×10^6C is the charge required to fully saturate one kilogram of soybean oil (IV 132). For oils with an initial IV different from 132, the factor 1.054×10^6 must be multiplied by $IV_{init}/132$.

3. *Power Consumption (PC)* – The power consumption of an electrochemical reactor per unit weight of oil product is related to the production rate (PR) given by Equation 9.10 and the product of the applied current density and the anode/cathode voltage difference (the cell voltage, V). With units of kW-h/kg, the power consumption is given by

$$PC = \frac{i\,V\left(2.78\times 10^{-7}\ kWh/J\right)\left(1\times 10^{4}\ cm^{2}/m^{2}\right)\left(3600\ s/h\right)}{PR} \quad \text{(Eq. 9.11)}$$

4. *Extent of Hydrogenation and Hydrogenation Selectivities* – During the electrocatalytic hydrogenation of an edible oil, the iodine value (IV) is used as the usual measure of the degree of double bond saturation (Patterson, 1994), as is done in conventional chemical catalytic hydrogenation schemes. Standard fatty acid hydrogenation selectivities are used to quantify the fatty acid distribution in the final electrocatalytically hydrogenated oil product. For soybean oil, the following reaction sequence for the hydrogenation of C_{18} fatty acids can be written

$$linolenate\,(C18:3) \xrightarrow{k_1} linoleate\,(C18:2) \xrightarrow{k_2} oleate\,(C18:1) \xrightarrow{k_3} stearate\,(C18:0) \qquad \text{(Eq. 9.12)}$$

and the reaction selectivities are defined by the following rate constant ratios (Mounts, 1996):

$$SLn = \frac{k_1}{k_2}; \quad SLo = \frac{k_2}{k_3}; \quad So = \frac{k_1}{k_3} \qquad \text{(Eq. 9.13)}$$

5. Trans *Fatty Acid Isomers* – *Trans* fatty acid isomers affect the physical and nutritional properties of edible oils. A *cis* to *trans* fatty acid isomerization reaction occurs simultaneously with double bond hydrogenation in edible oils. *Trans* isomers have high melting points which influence the room temperature liquid/solid characteristics of the oil, e.g., the *cis* isomer of oleic acid (*cis*-9-octadecenoic acid) has a melting point of 5.5°C while elaidic acid (*trans*-9-octadecenoic acid) melts at about 42°C, making it a solid at room temperature. Thus, the texture of the end product is markedly affected by both the proportion of *trans* and *cis* double bonds, as well as the oil's IV (*i.e.*, the total number of double bonds). Also, it has been established that *trans* fatty acids have negative nutritional properties in that they raise the level of harmful low density lipoprotein (LDL) in the blood (which contribute to heart disease) while lowering the protective high density lipoprotein (HDL) level (see, for example, Mensink & Katan, 1990; Willet et al., 1994; Aro et al., 1997; Ascherio & Willet, 1997; Aschiero et al., 1999). In fact, *trans* fatty acids raise blood cholesterol levels in a manner similar to saturated fats. Consequently, the US Food and Drug Administration now requires food labels to list the total *trans* isomer plus saturated fat content. Unfortunately, researchers and commercial oil processors have been unsuccessful in decoupling selective edible oil hydrogenation rates from unwanted *trans* isomer formation using conventional chemical catalytic reactions methods.

Electrocatalytic Hydrogenation of Soybean Oil with a Raney Nickel Cathode

Yusem and Pintauro (1992) were the first to show that soybean oil could be hydrogenated electrocatalytically (electrochemically) at a moderate temperature

(70–75°C) and one atmosphere pressure. They carried out their initial experiments in a flow-through packed bed reactor without an external supply of pressurized H_2 gas. The cathode (where hydrogen was electrochemically generated from water) was composed of Raney nickel power and the anode (where oxygen was generated from water) was an expanded titanium grid coated with IrO_2 (also referred to as a DSA, a Dimensionally Stable Anode). A schematic diagram of the cylindrical glass reactor (4 cm diameter) is shown in Fig. 9.1. The Raney nickel powder was trapped against a fine porosity glass frit by the upward flow of the electrolytic solution, forming a thin (0.9 cm) packed bed. The reactor was inserted into a bench-top batch recycle reaction loop (Fig. 9.2), consisting of a holding tank immersed in a constant temperature water bath, a pump, and associated flow meters and electrochemical instrumentation. The recycle loop apparatus simulated a single pass of oil reactant through multiple reactors, to achieve the desired product IV. Thus, oil was continuously circulating through the reactor for a specified time, to synthesize an oil with a given degree of hydrogenation. The reaction medium (typically 100 mL in

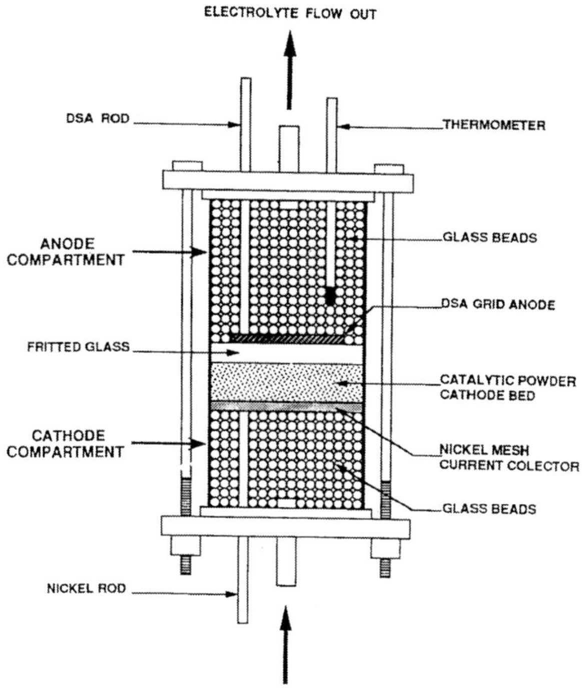

Fig. 9.1. Schematic diagram of the flow-through Raney nickel powder cathode reactor for the electrocatalytic hydrogenation of soybean oil. The glass walled cylindrical reactor was 4.0 cm in diameter. Glass beads were placed upstream and downstream from the electrodes to decrease the dead volume in the reactor and to ensure a uniform electrolyte flow through the electrodes. Source: Yusem & Pintauro, 1992.

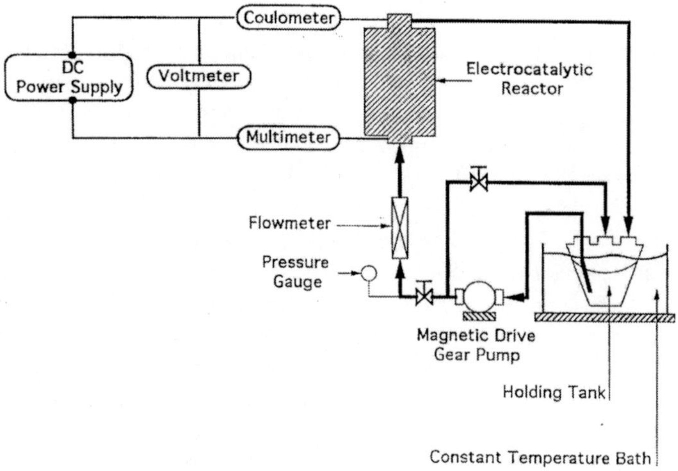

Fig. 9.2. The batch recycle reaction loop for electrocatalytic oil hydrogenation.

volume) was a two-phase mixture of soybean oil in a water/t-butanol solvent, where the vol% t-butanol was varied from 0 to 75% and the soybean oil content ranged from 10–48 wt/vol%. The supporting electrolyte salt was tetraethylammonium p-toluenesulfonate (hereafter denoted as TEATS) at a concentration of 0.5–2.0 M. The reactor was operated at 70°C, 1 atm. pressure, and a constant DC (direct current) current density between 10 and 20 mA/cm^2 (0.1 and 0.2 kA/m^2).

The effect of current density and electrolyte oil content on the current efficiency for soybean oil hydrogenation is shown in Fig. 9.3. The current efficiency decreased with increasing current density, due to an imbalance between the rate of hydrogen species production and consumption, where the current losses were due to unwanted hydrogen gas evolution (cf. Equations 9.6 and 9.7). Oil hydrogenation current efficiencies also decreased slightly with increasing oil content in the reaction medium; but, overall, the efficiency of hydrogen addition to the double bonds in soybean oil was high (50–80%).

The fatty acid profiles and *trans* isomer content of soybean oil products from the experiments in Fig. 9.3 are listed in Table 9.1. The initial fatty acid profile is listed, as well as the profile from a chemical catalytic hydrogenation process (Hastert, 1981). The electrocatalytic process produced an oil with a slightly higher concentration of stearic acid (C18:0) and approximately the same triene (linolenic acid, C18:3) content as that synthesized by chemical catalysis. The total *trans* isomer content in the electro-hydrogenated oil product was significantly less than the 20% *trans* reported in the literature for a high temperature chemical catalytic hydrogenation process. This important result was attributed to the lower reactor operating temperature of the electrocatalytic process. Additionally, the free fatty acid content of the electro-hydrogenated oil products was low (typically 0.05–4.0%) and the dissolved nickel concentration in the oil was very low (0–4.6 ppm). The latter result was attributed to the cathodic protection of the Raney Ni electrode

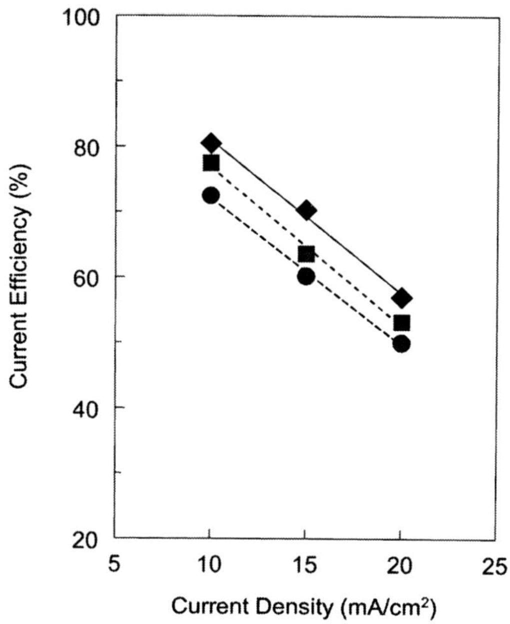

Fig. 9.3. The effect of current density and electrolyte oil content on the current efficiency for soybean oil hydrogenation in a Raney nickel cathode flow-through reactor. Reactor temperature: 70°C; reactor pressure: 1 atm.; solvent composition: 50/50 vol% t-butanol/water containing 1.0 M TEATS. Electrolyte oil concentration (wt/vol%): ◆ 24%; ■ 36%; ● 48%. Source: Yusem & Pintauro, 1992.

during reactor operation, where electrons were pumped into the cathode to stop all Ni dissolution (corrosion).

In a follow-on paper, Yusem et al., (1996) reported on the electrocatalytic hydrogenation of soybean oil in a radial flow-through Raney nickel powder reactor with a high anode/cathode interfacial area, where the design could be easily scaled up for industrial applications. The unique features of the reactor are: (1) a tubular element anode/cathode design where Raney nickel catalyst powder is packed in the annual space between two concentric porous ceramic tubes and (2) a radial flow direction for the electrolyte, to minimize flow resistance and to ensure a uniform flow distribution through the catalyst power bed cathode for a tube of arbitrary length. The total size (capacity) of the reactor can be increased without changing the relative anode/cathode position and the electrolyte flow pattern by increasing the length and number of tubular anode/cathode elements (for multiple tubular elements, the reactor configuration would be similar to that of a shell-and-tube heat exchanger). The current efficiency for soybean oil hydrogenation at 75°C in a single tubular anode/cathode element reactor was dependent on the applied current density and the oil content of the liquid reaction mixture (where the solvent was 50/50 vol% water/t-butanol with 0.5 M TEATS). For soybean products with an IV in the range of 87–110, current efficiencies ranged from a low of 25% (at

TABLE 9.1
Effects of Electrolyte Composition and Applied Current Density on the Chemical Properties of Soybean Oil that was Electrocatalytically Hydrogenated at a Raney Nickel Cathode at 70°C. 2.5 V anode/cathode voltage drop; 1 atm. pressure; 50/50 vol% t-butanol/water solvent containing 1.0 M TEATS. Source: Yusem & Pintauro, 1992

Oil Content (wt/vol %);	Fatty Acid Profile (%)					
Current Density (mA/cm2)	C18:0	C18:1	C18:2	C18:3	IV Value	% Total *Trans*
Initial Oil	4.1	24.2	53.4	7.3	132	0
24%; 10 mA/cm^2	5.6	46.5	34.8	2.2	106	9.9
24%; 15 mA/cm^2	5.9	42.6	38.4	2.1	109	12.1
24%; 20 mA/cm^2	6.4	37.2	41.3	4.0	114	8.3
36%; 10 mA/cm^2	5.5	46.2	35.2	2.1	106	11.7
36%; 15 mA/cm^2	7.1	38.3	40.4	3.1	111	8.3
36%; 20 mA/cm^2	5.2	39.8	40.6	3.5	114	7.7
48%; 10 mA/cm^2	5.1	44.6	37.2	2.1	108	9.9
48%; 15 mA/cm^2	6.5	39.7	39.3	3.5	111	8.8
48%; 20 mA/cm^2	5.7	36.6	42.8	3.8	116	7.9
Chemical Catalytic Hydrogenation[1]	4.8	43.4	38.0	2.9	110	20.0

[1] at 140°C and 3 atm. pressure with a supported Ni catalyst (Hastert 1981).

45 mA/cm^2 with an electrolyte oil content of 45 wt/vol%) to 100% (at 10 mA/cm^2 and an electrolyte oil content of 10 wt/vol%). As compared to a traditional high temperature chemical catalytic oil hydrogenation route with hydrogen gas, the electro-hydrogenated soybean oil products were characterized by a somewhat high C18:0 stearic acid content (4.5–5.8%), a comparably (low) concentration of C18:3 linolenic acid (generally below 3.0%), and a very low concentration of *trans* fatty acid isomers (4.7–16.2%).

Pintauro and coworkers identified and sought to correct a number of drawbacks with the electrochemical hydrogenation of edible oils in a Raney Ni cathode flow-through reactor. Those shortcomings included: (1) the use of a nonfood-grade co-solvent and supporting electrolyte salt (t-butanol and TEATS, respectively), which were needed to stabilize the emulsified oil/water reaction medium and to achieve a reasonable solution-phase ionic conductivity, (2) the low operating current density (0.010–0.045 A/cm^2), which necessitates the use of large area electrodes and large volume reactors; and (4) the high anode-cathode operating voltage drop (which translates into a high reactor power requirement) due to the low ionic conductivity of the two phase reaction medium.

Problems with the Raney Ni flow-through reactor design were corrected by carrying out edible oil hydrogenations in a solid polymer electrolyte (SPE) reactor, where a hydrated thin polymeric cation-exchange membrane separates the anode and cathode and carries out the function of the solvent/supporting electrolyte (An, 1998; An et al., 1998a; An et al., 1998b; An et al., 1999). Thin film electrodes composed of precious metal catalyst powders (one anode and one cathode) are fixed to opposing surfaces of the membrane, forming a "Membrane-Electrode-Assembly" (MEA), similar to that employed in proton exchange membrane (PEM) hydrogen/oxygen fuel cells. Water or humidified hydrogen gas is circulated past the back-side of the anode, where either water molecules are oxidized to O_2 gas and protons, according to Equation 9.3 or H_2 gas is oxidized to two protons and two electrons (the reverse of the reaction in Equation 9.4). Protons migrate through the ion-exchange membrane (under the influence of the applied electric field) to the cathode catalyst component of the MEA where they are reduced to atomic and molecular hydrogen (Equations 9.4 and 9.6). The electrogenerated hydrogen reacts with an organic substrate which is circulated through the cathode chamber and flows past the back-side of the cathode. Ion (proton) conductivity between the anode and cathode occurs through the wetted cation-exchange membrane so that neat organic reactant and distilled water (or humidified H_2 gas) are circulated in the cathode and anode chambers, respectively. In this regard, the SPE reactor represents a significant advancement in electrochemical reactor design since inorganic supporting electrolyte salts are not needed (such salts might contaminate the oil and would have to be removed from the reaction medium downstream from the reactor, often with considerable effort). The close proximity of the anode and cathode on a MEA (the electrodes are separated by the ion-exchange membrane, which is at most 200 μm in thickness) and the high ion-exchange capacity of the cation-exchange membrane ensure facile H^+ transport between the anode and cathode.

Solid polymer electrolyte reactors have been examined previously for a number of different organic electrochemical syntheses (both oxidation and reduction reactions). The first applications of the SPE process for electro-organic synthesis were published by Ogumi et al., (1981), Sarrazin and Tallec (1982) and Grinberg et al., (1984). There have since been numerous publications on the SPE reactor technique (see, for example, Inaba, 1993; Chen & Chou, 1994; and Jorissen, 1996). Typically, the cathode material was Pt, Au, or Au-Pt deposited onto the surface of a Nafion® cation-exchange membrane.

Electrocatalytic Hydrogenation of Edible Oil in a Solid Polymer Electrolyte (SPE) Reactor

A schematic diagram of the SPE oil hydrogenation reactor that was used by An and co-workers (An et al., 1998a; An et al., 1998b; An et al., 1999; Pintauro & An, 2001) is shown in Fig. 9.4, along with the principle reactions associated with electrochemical H_2 generation and oil hydrogenation. The key component of the reactor is a membrane-electrode-assembly composed of either RuO_2 or Pt/C powder (for the anode) and either Pt-black or Pd-black powder (for the cathode). The powders are mixed with a polymeric binder in an alcohol/water solvent to make an

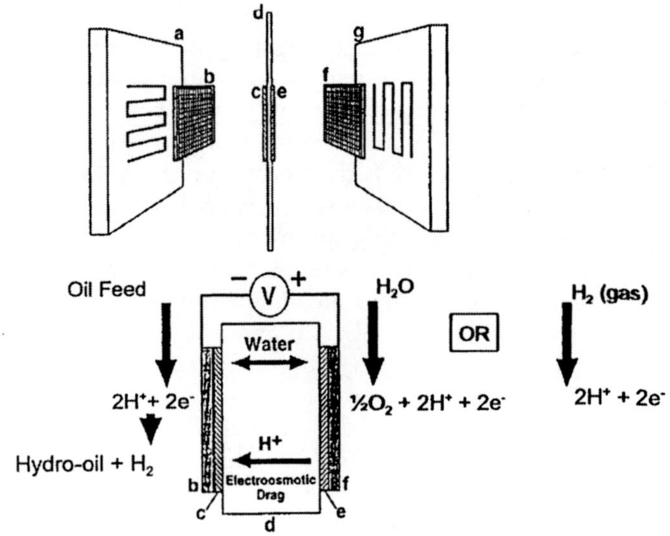

(a and g) Graphite (stainless steel) blocks with cross-patterned flow field.
(b) Gas /oil permeable backing, cathode side.
(c) Cathode thin film catalyst layer.
(d) Nafion cation-exchange membrane.
(e) Anode thin film catalyst layer.
(f) Gas /water permeable backing, anode side.

Fig. 9.4. Schematic diagram of a SPE reactor for edible oil hydrogenation. Source: An et al., 1998a.

ink, which is then coated onto a carbon paper or carbon cloth current collector. The catalyst coated carbon sheets are hot-pressed onto the opposing surfaces of a cation-exchange membrane where the catalyst forms a thin film at the membrane carbon paper/cloth interface. During reactor operation at a constant applied current, water is back-fed to the RuO_2 anode (or H_2 gas is fed to a Pt/C anode) where it is oxidized electrochemically to produce H^+. Protons migrate through the membrane under the influence of the applied electric field and contact the Pt or Pd catalytic cathode where they are reduced to atomic and molecular hydrogen. Neat oil is circulated past the back-side of the cathode and unsaturated triglycerides react with the electro-generated hydrogen species.

Initial studies on the SPE oil hydrogenation reactor utilized water as the anode feed and source for H. The cathode feed was undiluted soybean oil, canola oil, or cottonseed oil (mixtures of fatty acids and fatty acid methyl esters were also examined). The SPE reactor was operated at a constant applied current density of between 0.05 and 0.5 A/cm^2 (0.5 and 5.0 kA/m^2) and a temperature between 50–80°C. Reaction products with IV in the range 60–105 were characterized by a higher stearic acid content and a lower percentage of *trans* isomers, as compared to hydro-oils produced from a conventional chemical catalytic hydrogenation process.

The variation of current efficiency and cell voltage with current density for the electro-hydrogenation of soybean oil at 60°C in a SPE reactor with either a Pd-black or Pt-black cathode is shown in Figs. 9.5a and 9.5b. The Pd cathode is more active for oil hydrogenation, so the current efficiencies are higher than those with Pt at a given current density. Current losses (current efficiencies less than 100%) were due to unwanted hydrogen gas evolution at the cathode (i.e., two H_{ads} atoms reacted with themselves according to Equation 9.6 before they could add to an oil double bond). The range of current densities and current efficiencies for the SPE reactor are considerably higher than those for the Raney nickel flow reactor discussed previously in this chapter, due primarily to better contact of oil with the catalyst surface (a consequence of using neat oil as the cathode feed). Also, the anode/cathode voltage drop in the SPE reactor was lower than that in the Raney Ni reactor even though the applied current densities were 5–25 times higher. The linear dependence of cell voltage drop on current density in Fig. 9.5b is associated with IR losses in the cation-exchange membrane, due to the finite ionic conductivity of membrane.

Representative fatty acid profile data for soybean oil hydrogenation with either a Pt-black or Pd-black powder cathode are listed in Table 9.2. The results show: (i) hydrogenated oil products contain a high concentration of stearic and linolenic acid, as compared to products from a chemical catalytic hydrogenation with Ni catalyst and (ii) oil products from the Pd-black cathode electrolyses contained slightly less C18:0 and C18:3, as compared to similar IV oils that were hydrogenated at a Pt cathode. Fig. 9.6 compares the variation in *trans* isomer content with soybean oil product IV, from the SPE reactor and from a conventional chemical catalytic high temperature hydrogenation with H_2 and Ni catalyst (Hastert, 1981). There was significantly less *trans* isomer content in the electrocatalytically hydrogenated oil due to the combined effects of a lower reactor operating temperature and the use of a precious metal cathode.

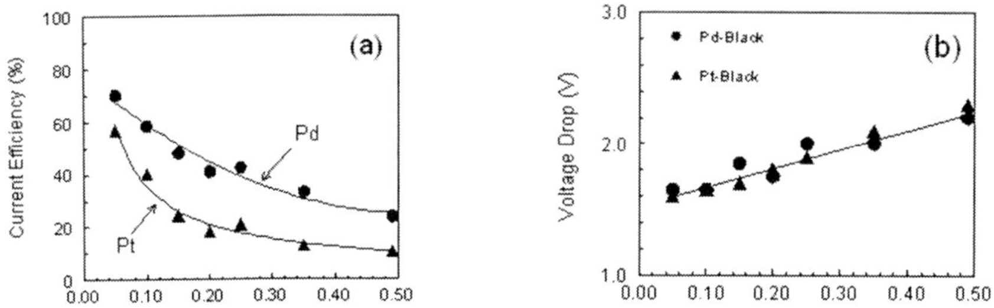

Fig. 9.5. Variation in current efficiency (a) and anode/cathode voltage drop (b) on current density for the electrocatalytic hydrogenation of soybean oil in a SPE reactor with water as the source of hydrogen (RuO_2 anode catalyst loading = 2.5 mg/cm^2, cathode catalyst loading = 2 mg/cm^2, T=60°C, P=1 atm). With kind permission from Springer Science+Business Media: Journal of Applied Electrochemistry, Current efficiency for soybean oil hydrogenation in a solid polymer electrolyte reactor, 28, 1998, 947–954, An et al., Figure 6 & 7.

TABLE 9.2
Electro-hydrogenated Soybean Oils Made in a Solid Polymer Electrolyte (SPE) Electrochemical Reactor. T=60°C, 1 atm. Reactor Pressure, 0.10 A/cm² Constant Applied Current Density, Cathode Catalyst Loading = 2.0 mg/cm²

	Fatty Acid Profile (%)						Anode/
	C18:0	C18:1	C18:2	C18:3	IV	*Trans* Isomers (%)	Cathode Voltage (V)
Initial Oil	4.0	22.5	54.6	7.7	134		0
Pd-black Cathode	23.8	28.9	32.5	3.6	91	6.5	1.65
Pd-black Cathode	28.1	31.7	26.2	2.7	80	**	1.55
Pd-black Cathode	37.4	26.9	22.2	1.7	66	7.4	1.65
Pt-black Cathode	28.4	21.3	34.8	4.2	90	2.8	1.85
Pt-black Cathode	32.7	23.5	29.5	2.7	78	**	1.55
Pt-black Cathode	38.7	22.4	24.9	2.2	68	3.5	1.55
Chemical Catalytic Hydrogenation[1]	6.0	64.1	14.9	1.2	90	35.0	
	12.0	71.0	5.5	0.0	75	39.0	
	23.9	61.3	0.5	0.0	60	41.0	

[1]140°C, 3 atm pressure, supported Ni catalyst [Hastert, 1981].

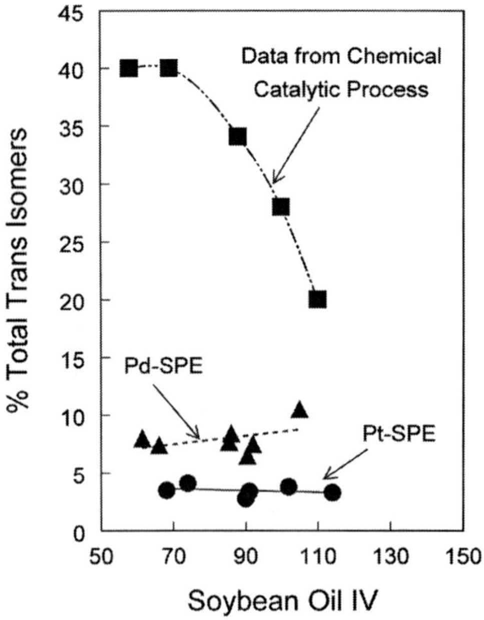

Fig. 9.6. Total *trans* isomer content vs. soybean oil iodine value (IV) for the SPE reactor with water as the source of hydrogen (An et al., 1998). The reactor operated at ambient pressure and a temperature of 60°C. Chemical catalytic process data is from Hastert (1981).

Canola oil and cotton seed oil were also partially hydrogenated in a SPE reactor (An et al., 1998a). The resulting fatty acid profiles, % *trans* isomers in the oil product, cell voltage, and hydrogenation current efficiency for two representative experiments are listed in Table 9.3. The results are consistent with soybean oil SPE reactor results.

The variation in production rate with current density for an IV 90 hydrogenated soybean oil in a SPE reactor operating at 60°C is shown in Fig. 9.7 for a Pd-black and a Pt-black cathode. As expected, the production rate increased with increasing current density, i.e., as the rate of hydrogen generation on the Pt or Pd cathode increased, the rate of addition of hydrogen to the oil's triglycerides increased. A greater amount of IV 90 soybean oil was produced with the Pd cathode as compared to Pt because of the higher current efficiencies with Pd at a given current density (see Fig. 9.5a). The hydro-oil production rate leveled off at a current density of ~0.35 A/cm^2 with a Pd-black cathode, indicating that there was a maximum rate of hydrogen reaction with the oil (i.e., increasing the catalyst surface concentration of H_{ads} by further increasing the current density did not produce more hydrogenated oil). This maximum rate will in all likelihood be a function of temperature, depending on the temperature dependence of the rate constant for oil reaction with H vs. the reaction of 2 H atoms to form H_2 gas. Based on the data in Fig. 9.7, there will be a four-fold decrease in the required size (and cost) of the SPE reactor when the reactor's current density is increased from 0.05 A/cm^2 to 0.35 A/cm^2 (assuming the cost of the reactor scales linearly with the geometric cathode area).

The SPE reactor power consumption per unit weight of IV 90 oil product (kW-h/kg) increased with increasing current density as shown in Fig. 9.8, due to a decrease in current efficiency and an increase in cell voltage with current density (cf. Figs. 9.5a and 9.5b). Thus, for electricity priced at \$0.08/kW-h, the SPE reactor power cost for an IV 90 soybean oil product at an operating current density of 0.150 A/cm^2 (1.5 kA/m^2) is \$0.024/lb when the cathode is composed of Pd-black powder. In general, for an SPE oil hydrogenation reactor, it is expected that reactor capital costs will dominate over electricity (power) costs, in which case a high

TABLE 9.3
Electrocatalytic Hydrogenation of Canola Oil and Cotton Seed Oil in a SPE Reactor with a Pd-black Cathode (2.0 mg/cm^2) and Water as the Source of Hydrogen. T=60°C and Ambient (1 atm)

	Fatty Acid Profile (%)				%	Current	Anode/Cathode
	C18:0	C18:1	C18:2	C18:3	*Trans*	Eff. (%)	Voltage (V)
Initial Canola Oil	4.0	22.5	54.6	7.7	0	**	**
IV 94 Canola Oil	23.8	28.9	32.5	3.6	2.8	41	1.65
Initial Cotton Seed Oil	2.4	18.1	55.4	0	0	**	**
IV 90 Cotton Seed Oil	10.7	27.3	38.5	0	4.3	48	1.65

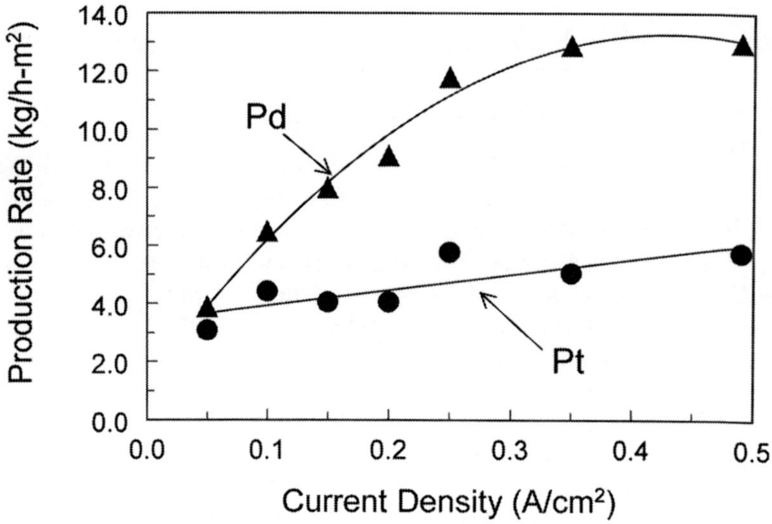

Fig. 9.7. IV 90 soybean oil production rate as a function of SPE reactor current density for Pd-black and Pt-black cathodes. Reactor temperature was 60°C. With kind permission from Springer Science+Business Media: Journal of Applied Electrochemistry, Current efficiency for soybean oil hydrogenation in a solid polymer electrolyte reactor, 28, 1998, 947–954, An et al., Figure 8.

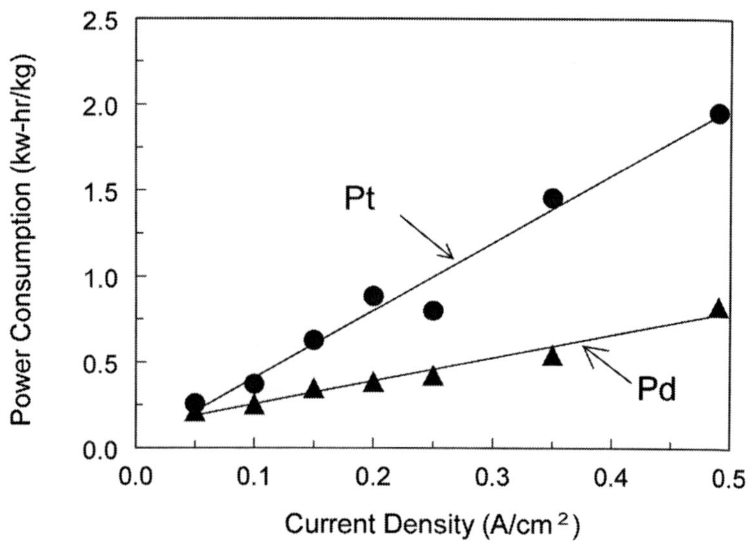

Fig. 9.8. Variation in SPE reactor power consumption per kg of oil product with current density for a Pd-black and Pt-black cathode. IV 90 soybean oil product; 60°C reactor temperature. With kind permission from Springer Science+Business Media: Journal of Applied Electrochemistry, Current efficiency for soybean oil hydrogenation in a solid polymer electrolyte reactor, 28, 1998, 947–954, An et al., Figure 9.

operating current density is favored (to maximize the production rate per unit electrode area thereby minimizing the required MEA area).

In follow-on SPE reactor studies (An et al., 1998b; An et al., 1999), various factors that affect oil hydrogenation current efficiency and reaction selectivity were investigated, including the type of cathode catalyst, catalyst and polymer binder loading in an MEA, and reactant flow rate. The current efficiency ordering of different cathode catalysts powders was found to be Pd > Pt > Rh > Ru > Ir. (This ordering mimics the catalytic activity of the same metals for the chemical reaction of oil and hydrogen). The optimum cathode catalyst loading for both Pd-black and Pt-black was 2.0 mg/cm^2. When the oil feed flow rate was increased from 80 ml/min to 300 ml/min, the oil hydrogenation current efficiency at 0.10 A/cm^2 increased from 60% to 70%. A current efficiency of 70% could also be achieved at a low flow rate (80 ml/min) by inserting a nickel screen turbulence promoter into the oil flow channel located behind the cathode (see Fig. 9.4). There was always a low concentration of *trans* isomer fatty acids in the hydro-oil products, ranging from about 2% with a Pt cathode to about 9% with Pd-black. When a second metal (either Ni, Cd, Zn, Pb, Cr, Fe, Ag, Cu, or Co) was electrodeposited onto a Pd-black powder cathode, a substantial increase in the fatty acid selectivities (SLn, SLo, and So, as defined by Equation 9.13) was observed, due presumably to changes in oil adsorption/desorption on the catalyst surface. For example, a Pd/Co cathode was used to synthesize an IV 113 soybean oil with 9.5% total *trans* isomers, 5.3% stearic acid, 38.5% oleic acid, 42.5% linoleic acid, and 2.3% linolenic acid (where the corresponding selectivities are SLn=2.93, SLo=4.85, and So=14.2). Variations in the reactor's current density had no effect on fatty acid hydrogenation selectivities.

The reactor operating conditions and MEA design for the synthesis of partially hydrogenated soybean oil with low, moderate, and high fatty acid hydrogenation selectivities (i.e., low, medium, and high values of SLn, SLo, and So) were identified. The products from these SPE reactor experiments compared well with literature data for the chemical catalytic hydrogenation of soybean oil with Ni or Pd catalyst in a slurry or fixed bed reactor (see Table 9.4). Of particular importance is the very low concentration of *trans* fatty acid isomers in all of the electrochemically processed oil products.

The Electrocatalytic Hydrogenation of Soybean Oil with H_2 Gas

An SPE electrocatalytic oil hydrogenation reactor was operated using H_2 gas as the anode feed and source of hydrogen, rather than water (Pintauro et al., 2005). In such a reactor, hydrogen is first converted (oxidized) to H^+ at the anode and then the applied electric field between the anode and cathode drives protons to the cathode where they are reduced to H_{ads} and H_2 species. The use of H_2 gas lowers the power requirements of the reactor (the electrochemical production of hydrogen from water is an energy intensive reaction) and it allows for reactor operation at a temperature greater than 100°C at atmospheric pressure (with a water feed to the anode, reaction temperatures > 100°C require pressurizing the reactor to prevent water boiling).

TABLE 9.4
Selectivity Comparison of Chemical Catalytic and Electrochemical/SPE Reactors for Soybean Oil Hydrogenation (SPE reactor operated at 60°C, 1 atm. pressure, 0.10 A/cm² Current Density, and Water as the Source of Hydrogen). From An et al., (1999)

	Fatty Acid Profile								
Reaction Conditions	C18:0	C18:1	C18:2	C18:3	IV	SLn	SLo	So	% *Trans*
SPE Reactor - Low Selectivity 2.0 mg/cm² Pd TCP17 current collector 80 ml/min oil feed flow rate	23.8	28.9	32.5	3.6	91	1.24	0.84	1.04	6.5
SPE Reactor - Moderate Selectivity 2.0 mg/cm² Pd TCP17 current collector 400 ml/min oil feed flow rate with turbulence promoter	12.2	41.8	32.6	2.3	98	1.89	2.65	5.03	9.1
SPE Reactor - High Selectivity 1.0 mg/cm² Pd and 1.0 mg/cm² Co CDC current collector 180 ml/min oil feed flow rate with turbulence promoter	5.3	38.5	42.5	2.3	113	2.90	4.85	14.2	9.5
Chemical Catalytic Hydrogenation Ni catalyst in a slurry reactor 140°C and 44 psig (Hastert, 1981)	5.0	43.6	37.8	2.9	110	1.84	11.3	20.8	20
0.5% Pd/Al_2O_3 in a fixed bed reactor 80°C and 17.8 psig (Mukherjee et al., 1975)	12.2	26.9	43.4	5.4	110	1.62	0.75	1.22	8
1% Pd/C in a fixed bed reactor 170°C/50 psig (Heldal et al., 1989)	10.2	33.2	39.5	5.6	112	0.80	1.40	1.12	17
5% Pd/C in a slurry reactor 60°C/27.5 psig (Ray, 1985)	4.3	34.1	43.9	5.3	118	1.70	8.3	14.1	22.2

The key components of the SPE fuel cell reactor for edible oil hydrogenation with H_2 gas are the same as those shown in Fig. 9.4. Membrane-electrode-assemblies (MEAs) with an apparent electrode area of 25 cm^2 (dimensions of 5 cm × 5 cm), were prepared using a Nafion® 117 cation-exchange membrane with Pd-black as the cathode and anode catalysts (electrodes were hot-pressed to the opposing surfaces of the Nafion® membrane, where the cathode loading was 2.0 mg/cm^2 and the anode loading was 0.5 mg/cm^2). Since Nafion® conducts protons only when hydrated, the hydrogen gas was fully humidified (bubbled through hot water) before it was back-fed to the anode. The various conditions that were examined for soybean oil hydrogenation are listed in Table 9.5. Since no liquid water was present, the reactor temperature was increased above 100°C.

The effects of current density and temperature on soybean oil hydrogenation current efficiency for an IV 90 product are shown in Fig. 9.4. Current efficiency increased with temperature at a given current density because the chemical reaction rate of hydrogen with unsaturated fatty acids in soybean oil increased with temperature; whereas, the hydrogen production rate was constant for a given current density. As was the case with an anode feed of water, the oil hydrogenation current efficiency decreased with increasing current density due to an imbalance in the production and consumption rates of hydrogen. Oil hydrogenation current efficiencies with H_2 were always higher than those with water as the anode feed and source of H (compare, for example, the data in Fig. 9.9 at 60°C with the corresponding current efficiencies with water from Fig. 9.5b, where the CE was 59% at 0.10 A/cm^2, 49% at 0.16 A/cm^2, and 42% at 0.20 A/cm^2). Current (charge) losses at the cathode were due to hydrogen gas generation, i.e., electro-generated H species combined to form H_2 gas before they could react with unsaturated triglycerides. The higher current efficiencies with H_2 gas were attributed to an oil/water/catalyst reaction zone of greater area within the cathode powder layer.

TABLE 9.5
Materials and Operating Conditions for the SPE Reactor for the Electrocatalytic Hydrogenation of Soybean Oil with H_2 Gas

Electrodes:	Pd-black or Pt-black cathode (for oil hydrogenation) Pt-C anode (for H_2 gas oxidation)
Cation-Exchange Membrane:	DuPont's Nafion® 117
Electrode Area:	Single MEA (25 cm^2)
Gas Diffusion Media:	E-Tek, Inc. carbon cloth or Toray carbon paper
Applied Current Density:	0.10–0.20 A/cm^2
Temperature:	50–120°C
Pressure:	1–4 atm.
Hydro-Oil Iodine Value (IV):	60-110 (i.e., 17–55% reduction in the number of double bonds)
Reactor Operation:	Batch recycle mode (small reactor, holding tank, and recirculation pump as per Fig. 9.2) with RBD soybean oil (from C&T Refinery, Charlotte, NC)

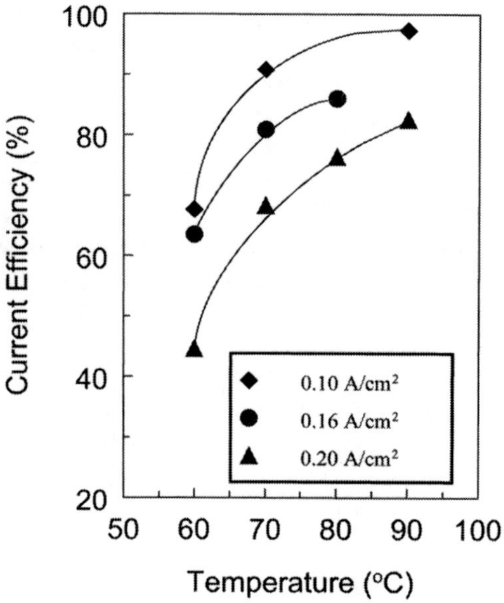

Fig. 9.9. Electrocatalytic hydrogenation of soybean oil with H_2 gas. The effects of temperature and applied current density on the current efficiency in a SPE reactor with a Pd-black cathode, where H_2 gas is the source of H (1 atm H_2). Reprinted with permission from Pintauro et al., 2005. Copyright 2005 American Chemical Society.

The selectivity of soybean oil hydrogenation in the SPE reactor with H_2 gas was similar to that for water as the anode feed (see Table 9.6), with an improvement in hydrogenation selectivity by: (1) operating the reactor at a higher oil flow rate with a turbulence promoter in the SPE reactor's oil feed channel to limit the contact time of oil triglycerides with the cathode catalyst and (2) using a bimetallic cathode catalyst (e.g., Pd/Co or Pd/Fe). When the selectivity was improved (i.e., when there was less saturated fatty acid and less triene in the oil product), the current efficiency decreased.

As expected, the % *trans* in the oil product was very low (< 6% for an IV 90 hydro-oil); but increased with increasing reaction selectivity (i.e., as the C18:1 concentration in the product increased, more *trans* isomers were produced). There was a loss in current efficiency when a turbulence promoter (nickel mesh) was placed in the reactor's oil feed channel, due presumably to an increase in H_2 gas evolution at the cathode when the contact time of oil with the catalyst was shortened. The *trans* isomer and saturated 18-carbon fatty acid content of hydrogenated oils are often lumped together when characterizing a hydro-oil. The C18:0 plus *trans* content for various IV soybean oils that were hydrogenated electrocatalytically with H_2 gas are contrasted to similar IV oils made via a chemical catalytic route (with a Ni catalyst at high temperature) in Fig. 9.10. Although the electrochemical process produced more saturates, the sum of saturates and *trans* isomers is low, thus the oil product is healthier.

The use of hydrogen gas as the anode feed improves the economics of the SPE reactor oil hydrogenation process in two ways: (1) the anode/cathode voltage drop

TABLE 9.6
Soybean Oil Hydrogenation Fatty Acid Profiles with a Modified Pd Catalyst Cathode (0.10 A/cm²; 1.0 atm H_2 pressure, 70°C, 140 ml/min oil feed flow rate). Reprinted with permission from Pintauro et al., 2005. Copyright 2005 American Chemical Society

	Fatty Acid Profile (%)						
Cathode Conditions	C18:0	C18:1	C18:2	C18:3	IV	% *Trans*	Current Efficiency (%)
Base Case:	28.2	26.9	31.1	3.7	87	9.2	81
No cathode	25.1	26.7	33.7	3.8	91	8.0	91
modification,	18.4	28.0	38.8	4.0	102	6.2	80
no TP[a]							
TP[b] and	18.1	37.7	29.3	2.3	89	19.2	69
no cathode	17.0	38.2	30.6	2.7	93	12.0	60
modification	15.8	35.3	34.7	3.4	99	14.7	69
	9.3	32.7	43.2	4.5	115	11.8	59
With TP	15.2	42.9	29.3	1.7	92	20.0	80
and Pb-Co	12.8	40.3	33.7	2.4	99	18.0	65
cathode	10.0	35.2	40.0	3.7	109	16.0	65
	7.5	30.6	45.9	5.1	119	7.0	84
With TP	15.9	45.2	27.5	1.1	90	20.0	72
and Pb-Fe	11.7	40.1	34.9	2.3	101	16.0	76
cathode	9.1	34.1	40.8	3.5	109	12.0	70
	7.2	31.3	46.0	4.8	119	6.0	68

[a]TP denotes turbulence promoter.

is lower than that with a water feed because hydrogen is not produced in-situ from water; the voltage drop was ~0.25 V with H_2 vs. 1.65 V for a water feed (cf. Fig. 9.5b) vs. 2.5 V for an emulsified electrolyte solution with a flow through Raney Ni cathode (data in Table 9.1) and (2) the current efficiencies for oil hydrogenation are higher with an H_2 feed (compare the results in Figs. 9.5a and 9.9). The production rate and power consumption for the PEM reactor for an IV 90 oil with anode feeds of either H_2 or H_2O at 70°C are shown in Figs. 9.11 and 9.12, respectively. The PEM reactor with H_2 gas consumed less electrical power and produced more IV 90 soybean oil per unit area of electrode than a similar reactor that operated with an anode feed of water. The increase in production rate with hydrogen gas, which is attributed to the high current efficiencies, means that fewer reactors will be needed (and lower capital costs will be realized) in an oil hydrogenation plant when the reactors run on H_2. The drop in power consumption is due to the combined effects of

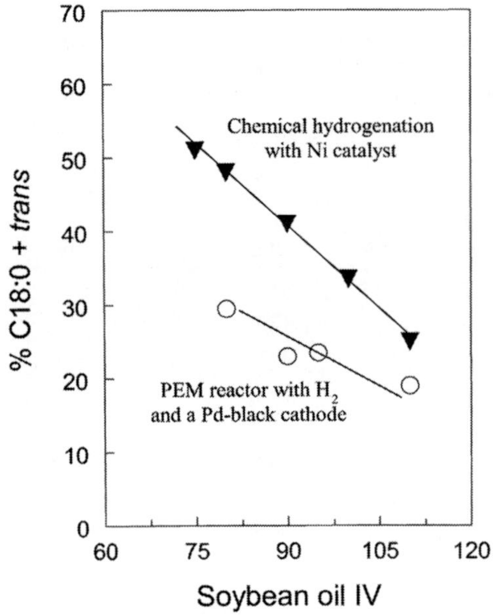

Fig. 9.10. Total amount of 18-carbon saturated fatty acids (stearic acid) plus *trans* isomers in partially hydrogenated soybean oils produced either by an electrocatalytic process (SPE reactor with H_2 gas) or in a conventional chemical catalytic reactor with nickel catalyst (Hastert, 1981). Reprinted with permission from Pintauro et al., 2005. Copyright 2005 American Chemical Society.

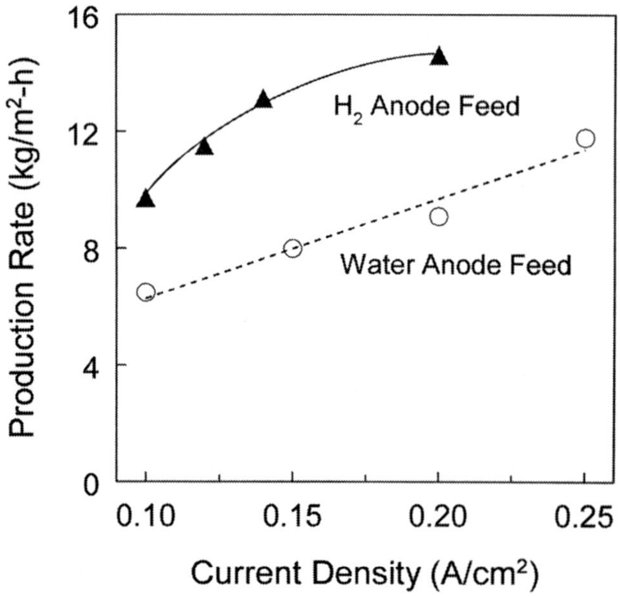

Fig. 9.11. IV 90 soybean oil production rate vs. applied current density for an SPE reactor with a Pd-black cathode and either H_2 gas or water as the anode feed and source of H. Reaction temperature was 70°C. Reprinted with permission from Pintauro et al., 2005. Copyright 2005 American Chemical Society.

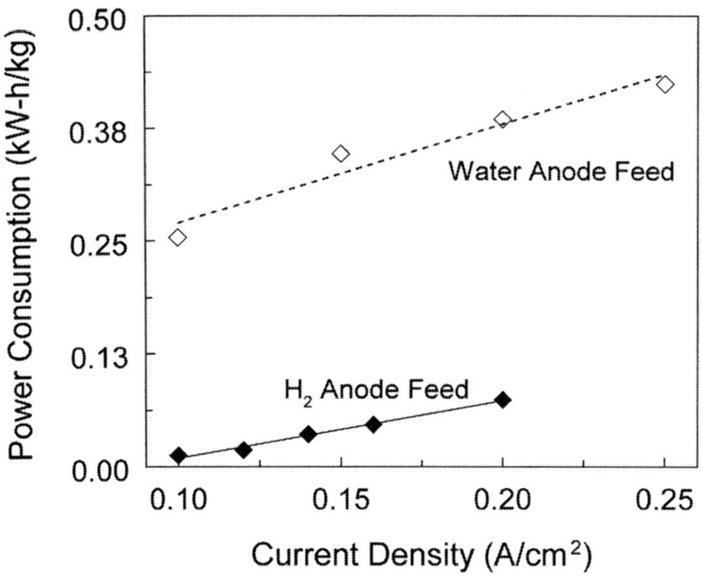

Fig. 9.12. Dependence of the SPE reactor power consumption per kilogram of oil product (IV=90) on the applied current density for an anode feed of either H_2 gas or water. Reactor temperature was 70°C. Reprinted with permission from Pintauro et al., 2005. Copyright 2005 American Chemical Society.

a high hydro-oil production rate and the lower electrochemical energy requirement (lower cell voltage) for the production of H^+ from H_2 as compared to the generation of H^+ from water.

Sensory and Compositional Characteristics and Blending of Electrocatalytically Hydrogenated Soybean Oil

Warner and co-workers (2000) performed a thorough analysis of the compositional and physical property differences between partially hydrogenated soybean oil products from a commercial chemical process with Ni catalyst and an electrocatalytic product of similar IV. The electrochemically hydrogenated soybean oil had significantly more C18:0, C18:2c,c, and C18:3, whereas the chemical catalytic (commercial) oil sample had significantly greater levels of C18:1t and C18:1c (see Table 9.7). The *trans* fatty acid content was 17% for the commercially hydrogenated oil and 9.8% for the electrochemical sample. Free fatty acid and peroxide values were low in the electrocatalytic oil sample, indicating no hydrolytic or oxidative degradation of triglycerides during hydrogenation in the SPE reactor. In room odor evaluations of oils heated at a frying temperature of 190°C, commercial soybean oils (IV 94 and 68) showed strong intensities of an undesirable characteristic hydrogenation aroma and an overall room odor, whereas the electrocatalytically hydrogenated samples (IV 104 and 90) showed only weak odor intensities (see Fig. 9.13).

In addition to the sensory analyses, solid fat values and melting point data were collected by the USDA on various soybean oil samples that were electrocatalytically

TABLE 9.7
Fatty Acid Composition and Chemical/Physical Analyses of Partially Hydrogenated Soybean Oil (Refined, Bleached, and Deodorized Soybean Oil)

	Chemical Catalytic Process (IV=94)	Electrocatalytic Hydrogenation[1] (IV=90)
C16:0	10.9	11.7
C18:0	6.4	20.6
C18:1t	13.1	8.1
C18:1c	40.1	23.5
C18:2 c,t/t,c/t,t	3.9	1.7
C18:2 c,c	23.5	30.8
C18:3	0.9	3.2
Dropping point (°C)	22.3	56.9
Peroxide value	0.1	0.2
Free fatty acid (% oleic)	0.01	0.02

[1]Made in the SPE reactor with Pd-black catalyst and water as the source of H at a temperature of 60°C.

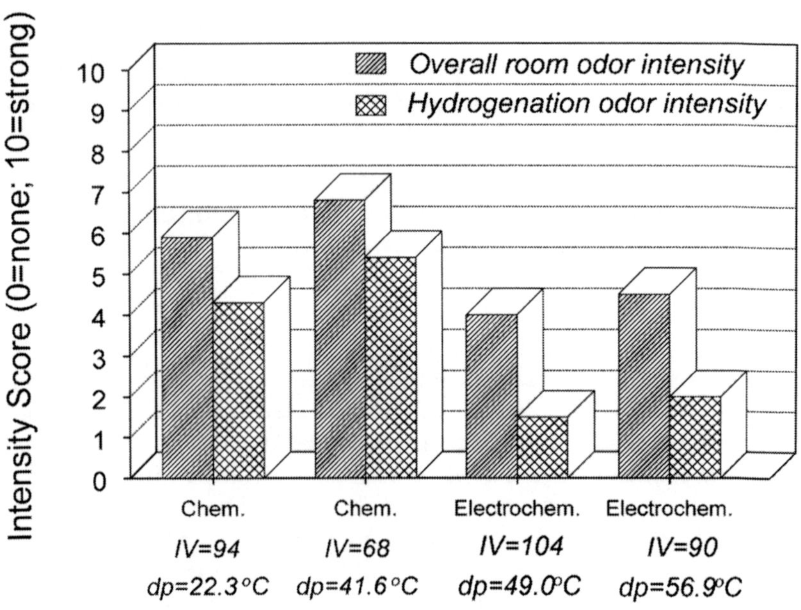

Fig. 9.13. Overall room odor intensity scores and hydrogenation odor intensity scores of partially hydrogenated soybean oils. Chem. refers to chemical catalytic hydrogenation with a Ni catalyst; Electrochem. refers to electrochemically hydrogenated soybean oil in a SPE reactor with a Pd-black cathode at 50–60°C; dp denotes dropping point. Odor analysis at 190°C. Source: Warner et al., 2000.

TABLE 9.8
Chemical and Compositional Properties of Electrocatalytically Hydrogenated Soybean Oils. Source: List et al., 2007

Iodine Value	C18:1 *Trans* (%)	Stearic Acid (%)	NMR Solids at temperature (°C) 10.0	21.1	26.7	33.3	40.0	Dropping Point (°C)
90	13.1	15.3	25.1	17.4	16.6	13.7	9.2	47.4
100	9.9	12.8	17.5	12.3	11.4	9.5	6.4	48.9
110	6.5	9.9	10.5	7.7	7.1	5.8	4.1	44.9

hydrogenated in an SPE reactor with H_2 gas as the source of hydrogen using a Pd-Co and a Pd-Fe cathode, which worked equally well in synthesizing the same products (List et al., 2007). The properties of IV 90, 100, and 110 samples are listed in Table 9.8 (the data are average values from multiple experiments carried out at 90°C). The dropping points are comparable to those of baking shortenings produced via a conventional chemical catalytic hydrogenation process, which are typically 41–48°C. The NMR solid profiles in Table 9.8 compare well with solid fat content melting point curves from baking industry shortening products, as shown in Fig. 9.14. The

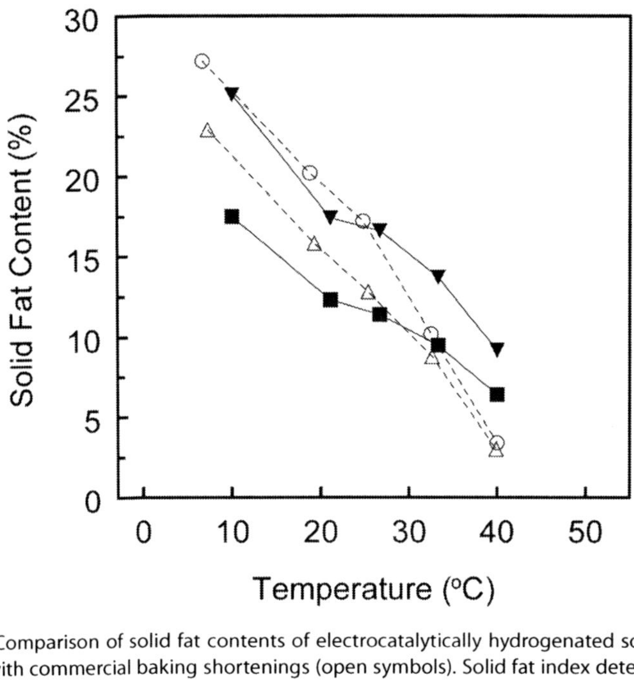

Fig. 9.14. Comparison of solid fat contents of electrocatalytically hydrogenated soybean oil (closed symbols) with commercial baking shortenings (open symbols). Solid fat index determined by pulsed NMR. ▼ IV 90 from Table 9.8; ■ IV 100 from Table 9.8; Commercial shortenings (List and King, 2006): ❍ Pie ; Δ Cake.

TABLE 9.9
Low-*trans* Spread Oils via Blending of Electrocatalytically Hydrogenated Soybean Oil (IV 111) with Liquid Oil. Source: List et al., 2007

Hydrogenation		Base			Solid Fat at Temp (°C)			
Catalyst	Temp (°C)	(111 IV) (%)	Liquid (%)	*Trans* (%)	10.0	21.1	33.3	Dropping Point (°C)
Pd-Co	70	70	30	4.2	6.0	4.3	3.1	37.4
Pd-Co	70	60	40	3.8	4.3	3.5	1.8	35.3

combined *trans* fatty acid and stearic acid content for the products in Table 9.8 is very low (16–28%) as compared to that for typical commercial shortenings, which is in the 39–52% range.

An electrochemically hydrogenated soybean oil was also evaluated as a spread oil or liquid margarine by blending an IV 111 (6.4% C18:1 *trans*, 8.8% stearic acid, and a dropping point of 42.3°C) with liquid soybean oil. The IV 111 sample was synthesized in the SPE reactor at 70°C with a Pd-Co cathode. The blending results, shown in Table 9.9, are promising. The two low-*trans* blends (containing about 4% *trans* fatty acids) exhibited melting points suitable for spread oils and possessed the proper solid fat content for a liquid margarine product (a required solid fat content of 4.0 at 10°C and 2.0 at 33.3°C).

Chapter 10

Low *Trans* Hydrogenation

Gary R. List[1] and Michael A. Jackson[2]

[1]USDA, Retired; [2]NCAUR/ARS/USDA

Introduction

Although hydrogenation has been the technology of choice for fat formulation for many years (List & Jackson 2007, 2009; Latondress, 1980; O'Brien, 2009) recent concerns over the health and nutrition (Ascherio et al., 1999; Eckel et al., 2007) of *trans* fatty acids (TFA) have had a profound effect on the edible oil industry. In 1994, it was estimated that fractionation and interesterification accounted for about 10% of food fats (Haumann, 1994) processed worldwide. Today, this figure is undoubtedly higher because of TFA nutrition labeling enacted by the US Congress in 2003 (Anon., 2003). Since January 1, 2006, TFA has been required on nutrition labels along with saturated and polyunsaturated acids. Canada has enacted even stricter regulations. Whereas, in the US, foods containing less than 0.5 gm TFA/serving (12–14 gm serving) may be declared zero; Canadian regulations set the limit at 0.2 gm TFA/serving. Obviously, the reformulation of foods to meet these requirements has posed a challenge to the food industry (Flöter & Van Duijn, 2006; Tarrago-Traini et al., 2006; Kodali & List, 2005, List et al., 2007). Many companies began this task in about 2003 and, by 2007, over 1,900 foods had been reformulated to meet TFA regulations (www.ers.usda/amberwaves.2007). Recent statistics (Anon., 2009) indicate that TFA reduction has been achieved by using several strategies including:

1. Increased use of naturally stable and trait modified oils (Heintz, 2009).
2. A shift from heavily hydrogenated fats to liquid or lightly hydrogenated oils (Gupta, 2008).
3. Replacement of hydrogenated fats with palm/palm kernel oil and their fractions (Van Dujijn, 2000).
4. Chemically or enzymatically interesterified fats (Anon., 2004).
5. Blends of liquid oils with tropical fats, and fractions thereof (Sundram, 1998).
6. Modified hydrogenation technologies including the use of noble metals, altered nickel catalysts and a switch to non-selective hydrogenation conditions (Berben et al., 1994; Ariaansz & Okonek, 1998; Beers et al., 2008).
7. A few other approaches still under development include the use of electrochemical hydrogenation and preparation of low iodine value oils show promise for reduced TFA food oils (List et al., 2007a, b). Also see Chapter 9.

8. Several processes have been patented for the production of low-*trans* fats via hydrogenation (Higgins, 2007; Van Toor et al., 2005).
9. Hydrogenation in critical fluids has received attention as *trans* isomer reduction technologies. (King et al., 2001; Härröd & Möeller, 2001) A detailed review of hydrogenation in critical fluids is given in Chapter 3. This chapter will review modified hydrogenation technologies with emphasis on *trans* fatty acid reduction. The theoretical aspects of *trans*-isomer formation is discussed in Chapter 6. Since publication of this book, a number of reviews of the hydrogenation process have appeared and should be consulted for further information (Hastert, 1996, 2000; Farr, 2005; List & King, 2006; Koetsier, 1997; Dijkstra, 1997, 2006; Erickson & Erickson, 1996; Farr & List, 2004; Gupta, 2008; O'Brien, 2008; Okonek et al., 1995).

Historically, hydrogenation of edible oils in batch systems have been carried out under selective conditions which favor reduction of polyenoic acids over monoenoic acids (Dutton, 1968). Typically, selective conditions include high temperatures, low pressure, moderate agitation and low catalyst loadings. These conditions favor high *trans* acids formation (Stingley & Wrobel, 1961; Allen & Covey, 1970; Farr & List, 2004). Published figures indicate that soybean oil hydrogenated under selective conditions results in about 0.73% increase in *trans* acids per iodine value drop. (Evans et al., 1964).

On the other hand, non-selective conditions including lower temperatures, higher pressures, increased agitation and higher catalyst loadings tend to suppress *trans* isomer formation (Hasman, 1995; Puri, 1980). Evans (1964) found that soybean oil hydrogenated under non-selective conditions (120°C, 100 psi) results in 0.43% increase in *trans* per iodine value drop and similar results were reported by Stingley & Wrobel (1961).

Effects of Agitation

In stirred batch reactors, agitation may be provided by either turbine or gas dispersion type designs. Early work reported by Beal and Lancaster (1954) showed that gas dispersion agitators promote the selective removal of polyunsaturates from soybean oil with increased *trans* isomer formation compared to turbine agitators. They concluded, however, that agitation cannot be defined in terms of speed. Puri (1980) claims that although increased agitation increases transfer of hydrogen, it has no effect on *trans* isomer formation.

Effects of Temperature

The effects of temperature on *trans* isomer formation has been reported by Stingley & Wrobel (1961), Allen & Covey (1970), and Puri (1980). Allen & Covey (1970) concluded that factors affecting *trans* fatty acid formation in stirred batch reactors increase in the following order: temperature, agitation, pressure, and catalyst concentration.

TABLE 10.1
Effect of Process Variables on Hydrogenation

	Effect on Rate of Hydrogenation		Effect on Selectivity	
Process variable	Transfer	Consumption	S	*Trans* isomer formation
Agitation	+	0	–	–
Design	–/+	0	+/–	+/–
Pressure	+	+	–	–
Temperature	+[1]	+	+	+
Amount of Catalyst	0	+[2]	+	+
Type of Catalyst (nature)	0	+/–	–/+	–/+

+ = increases; – = decreases; 0 = unaffected; [1] = slight increase; [2] = up to a certain limit.
Source: Puri, 1980, J. Am. Oil Chem. *Soc.* 57, 850a–853a.

Effects of Pressure

Pressure can be used to control *trans* fatty acid formation during hydrogenation (List et al., 2000; Eller et al., 2005). Under selective conditions (e.g., high temperatures, low hydrogen pressures, and low nickel catalyst loadings) soybean oil hydrogenated to iodine values of 65–80 show *trans* fatty acid contents in the 32–40% range. At these iodine values, these oils serve as margarine and shortening bases which, after blending with liquid oil or hardstocks, yield *trans* fatty acid contents in the 12–25% range.

On the other hand, hydrogenation carried out at 140–170°C at pressures of 200 psi in the presence of 0.02% nickel catalysts showed *trans* contents ranging from about 16–18% (Table 10.1). At iodine values of 66–70, these oils show about a 60% reduction in *trans* compared to oil hydrogenated to a similar IV level under commercial conditions. Table 10.2 shows the properties of blends prepared from Table 10.1 oils (30%) and 70% liquid soybean oil. Many of these oils meet the physical properties for soft margarine oil and allow preparation of spreads meeting a zero gram *trans* fat/serving (Table 10.3).

It should be pointed out that most commercial hydrogenation converters are designed to operate at pressures in the 1–3 bar (15–50 psi) range. Adoption of higher pressures will require retooling existing equipment.

Low Iodine Value Oils

During hydrogenation of soybean oil under selective conditions, *trans* fatty acids increase from an iodine value of about 130 to about 70 at which *trans* bonds begin to saturate. Cottonseed oil shows a similar pattern at an iodine value of about 60. Fig. 10.1 shows the compositional changes occurring during the selective hydrogenation of soybean oil. Historically, the processing industry has employed an iodine value range of 65–80 for margarine/spread and shortening bases with *trans* fatty acid contents in the 32–40% range. Blending of these basestocks with liquid oil and/or

TABLE 10.2
Properties of Hydrogenated Soybean Oil: Effect of Pressure

				IV			Selectivity			Fatty Acid Composition (%)					Solid Fat Content (°C)			
Oil	Temp (°C)	Pressure (psi)	Nickel (%)		Reaction Time (min)	Trans (%)	Ln	Lo	Melting Pt. (°C)	C16	C18	C18:1	C18:2	C18:3	10	21.1	26.7	33.3
A	221	20	0.02	65	50	39.7	_[a]	-	41.2	11.3	13.6	75.2	0	0	73.7	54.1	44.7	22.3
B	140	200	0.02	66.2	152	16.6	2.4	4.8	46.4	10.7	22.1	55.4	10.5	0.1	46.5	31.2	27	20.9
C	150	200	0.02	63.6	72	17.4	1.6	5.0	48.2	10.7	23.5	55.4	8.8	0.3	51.4	35	31.6	24.3
D	160	200	0.02	70.2	33	16.8	1.6	5.3	46.1	10.7	19.4	56.4	11.8	0.5	43.7	27.6	23.8	17.6
E	170	200	0.02	68.8	21	17.9	1.5	5.5	48.2	10.7	20.1	56.8	10.8	0.5	46.4	30.4	26.7	19.5

[a]Cannot be determined.

TABLE 10.3
Properties of Blended Oils

					Solid Fat Content (°C)					Gms TFA/14 gm Serving			
Oil	%	Liquid SBO (%)	Trans (%)	Drop Pt. (°C)	10	21.1	26.7	33.3	40	80% fat	70% fat	60% fat	50% fat
A	(30)	70	11.9	31.4	17.2	8.6	5.0	1.0	0.0	1.33	1.16	1.00	0.83
B	(30)	70	5.0	37.7	11.1	6.3	5.0	3.1	1.3	0.56	0.49	0.42	0.35
C	(30)	70	5.2	39.4	11.8	7.9	6.1	3.6	1.8	0.58	0.51	0.47	0.36
D	(30)	70	5.8	36.9	9.5	6.0	4.4	2.2	0.6	0.65	0.51	0.44	0.41
E	(30)	70	5.4	37.8	10.3	6.6	5.1	2.6	1.5	0.60	0.53	0.45	0.38
Comm. Spread oils			5.3–14.5	31.6–33.1	10.0–18.0	4.4–8.0	2.8–4.7	0.9–1.7	0				

Source: Eller et al., J. Agric. Food Chem., 2005, 53 (15), pp 5982–5984.

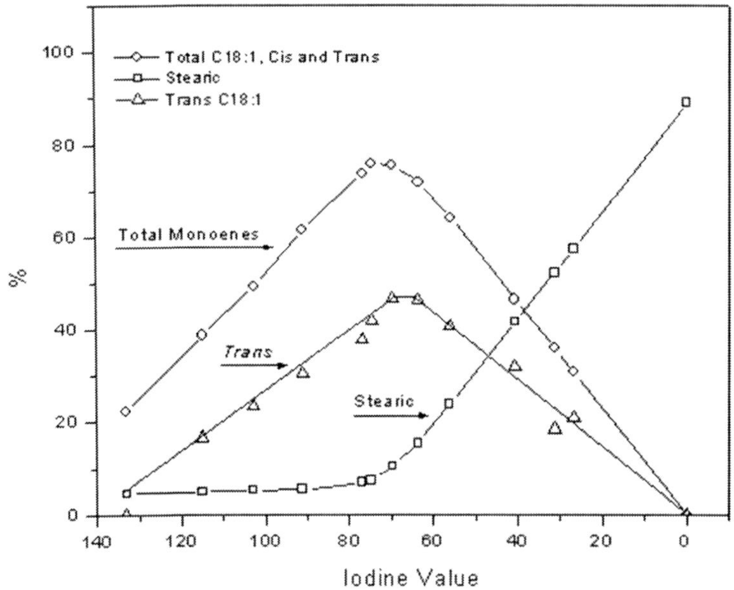

Fig. 10.1. Formulation of *trans* and stearic acids in hydrogenated soybean oil under selective conditions (175°C, 15 psi, 0.02% nickel). Source: List et al., 2007, J. Am. Oil Chem. Soc. *84* 609–612.

fully hydrogenated stearines yield finished products with *trans* contents in the 12–25% range (List & King, 2005). Studies carried out in the authors' laboratory (List et al., 2007b; Jackson et al., 2008) have shown that hydrogenation of soybean and other oils to low iodine values (i.e. 42–45) provide basestocks for the formulation of low *trans* margarine/spread and shortening oils. In effect, functionality including melting point and solid fat results from a combination of reduced *trans* and increased stearic acid (Table 10.4).

Catalysts for Reduced *Trans*

A number of studies have reported that use of noble metal catalysts provide reduced *trans* fatty acids. These include platinum, palladium, and ruthenium (Berben et al., 1994, 2005; Hasman, 1995; Ariaansz & Okonek 1998; Beers & Mangnus, 2004; Beers et al., 2004). Major impediments to commercial acceptance include increased costs compared to nickel or higher catalyst loadings. In some cases, pressure limitations on existing equipment may discourage adaptation of alternative hydrogenation technologies.

Modified Nickel Catalysts for Low *Trans* Hydrogenation

Chemical modification of commercially available nickel catalysts has been employed as a route to low *trans* oils via hydrogenation (Higgins, 2007; Van Toor et al., 2005).

TABLE 10.4
Properties of Low Iodine Value Soybeans

					Solids by MNR (percent @ °C)									
IV	Percentage Low IV oil	Percentage of Soy oil	Percentage of *Trans*	Drop Pt. °C	10	21.1	26.7	33.3	40	45	50	55	60	65
40.7	100	0	32.1	55.3	92.4	82.6	81.9	78.5	59.0	-	18.0	3.6	0.0	0.0
	5	95	1.6	30.5	4.4	2.5	2.2	17.0	0.4	0.0	0.0	0.0	0.0	0.0
	10	90	3.2	36.7	6.9	4.5	4.7	2.8	0.8	0.2	0.0	0.0	0.0	0.0
	15	85	4.8	42.2	12.4	9.5	9.4	6.7	3.6	1.3	0.0	0.0	0.0	0.0
	25	75	8.0	46.7	20.2	17.4	16.6	10.5	7.5	3.8	0.0	0.0	0.0	0.0
31.4	100	0	20.2	60.6	95.5	92.8	93.0	91.7	86.8	-	41.7	21.9	-	-
	5	95	1.0	37.0	5.1	4.4	3.2	3.1	2.0	1.1	0.2	0.0	0.0	0.0
	10	90	2.1	46.3	10.5	8.9	7.7	7.0	5.3	3.4	1.8	0.6	0.0	0.0
	15	85	3.2	49.2	15.0	14.5	13.0	11.8	9.5	6.1	3.9	2.0	0.0	0.0
	20	80	4.2	51.2	19.1	18.1	16.3	15.9	13.0	9.0	5.6	3.0	0.0	0.0
	25	75	5.3	53.1	23.8	22.9	21	19.1	17.5	11.9	7.6	4.3	0.2	0.0
	50	50	10.5	57.9	46.9	46.2	44.9	42.4	39.0	29.5	21.3	15.5	6.7	0.0
26.9	100	0	17.9	61.9	96.2	92.4	95.0	94.2	88.8	-	50.5	26.7	-	-
	5	95	0.9	35.6	3.4	4.8	3.9	3.4	2.8	1.5	0.7	0.2	0.0	0.0
	10	90	1.9	43.8	11.4	10.3	8.8	8.7	7.6	4.4	2.6	0.4	0.0	0.0
	15	85	2.8	48.1	15.0	14.2	12.8	12.0	10.3	7.0	4.2	2.4	0.0	0.0
	20	80	3.7	49.9	19.2	17.9	16.3	15.8	14.3	9.3	5.9	4.0	0.3	0.0
	25	75	4.7	51.4	24.3	23.6	21.5	21.1	19.0	13.6	9.3	6.4	1.9	0.0
	50	50	9.4	56.5	47.4	47.4	44.5	43.5	39.1	33.5	23.8	17.2	8.8	0.7

Source: List et al., 2007, J. Am. Oil Chem. Soc. 84 609–612.

Several example areas follow. A narrow pore nickel catalyst, Nysofact 120 Catalyst (9 lbs, 4.01 kgs.), was placed into a slurry tank along with 1.85 lbs, 0.83 kgs. of phosphate mono- and diglycerides. Slurry development proceeded about 45 minutes. The pre-conditioned catalyst was charged into a converter containing 60,000 lbs, 26,786 kgs. of soybean oil and hydrogenation carried out at 45 psi and 270°F (132°C). Reaction time was 52 minutes. *Trans* fatty acid content was about 4% (Higgins, 2007).

Pricat 9920, from Johnson Matthey, consisting of 22% nickel on an alumina support and coated with hardened vegetable oil was heated at atmospheric pressure to 200°C with a hydrogen flow of approximately 1 liter/min. After 1½ hours, the pressure was increased to 2 bar and hydrogen flow stopped. The catalyst was then cooled to 20°C yielding a solid composition. Canola and soybean oils hydrogenated with this catalyst show *trans* content in the 3–4% range and provide pourable semi-solid products (Van Toor et al., 2005).

Low *Trans* Oils via a Catalyst Switching Strategy

Jackson et al., (2008) expanded the use of low iodine value oils by a catalyst switching strategy. Various oils including canola, high-oleic safflower/sunflower and soybean were hydrogenated under selective conditions using a nickel catalyst where *trans* fatty acids are maximum (e.g. iodine values in the 70–80 range) in the presence of a platinum catalyst at 80°C to iodine values in the 45–49 range. The physical properties of these oils and blends made from them are shown in Table 10.5. The results show that many of these oils offer low or zero *trans* options for margarine/spread and shortenings.

TABLE 10.5
Physical Characteristics of Basestocks Prepared Using Two-step Hydrogenation and Blends

				SFC[b] (°C)										
Basestock	Percentage of Basestock[a]	% *trans*	Dropping Point	0	10	21.1	26.7	33.3	40	45	50	55	60	65
Canola oil														
IV46	100%	13.1	63.8	63.6	57.9	51.3	50.6	47.5	39.5	32.6	26.9	18.7	11.1	0.0
	5%	0.7	30.9	2.9	2.6	1.9	2.1	1.4	1.2	0.2	0.0	0.0	0.0	0.0
	10%	1.3	39.3	5.2	4.7	3.8	4.3	3.0	2.5	1.3	0.8	0.0	0.0	0.0
	15%	2.0	45.5	7.7	6.5	6.5	6.5	6.2	4.2	2.7	1.6	0.0	0.0	0.0
	20%	2.6	48.7	10.7	10.0	8.1	8.2	7.9	5.8	4.6	2.6	0.0	0.0	0.0
	25%	3.3	51.1	14.3	12.7	10.9	10.7	10.0	7.5	5.6	4.2	1.1	0.0	0.0
	50%	6.6	57.5	30.1	26.6	23.2	23.1	20.1	17.7	14.5	10.4	5.7	0.0	0.0
Hi-oleic safflower oil														
IV49	100%	11.2	61.9	53.0	50.1	43.9	43.5	40.0	33.1	26.7	21.7	14.4	4.7	0.0
	5%	0.6	36.8	2.0	2.4	2.0	1.6	1.2	0.6	0.6	0.0	0.0	0.0	0.0
	10%	1.1	37.8	4.9	4.2	3.6	3.0	3.0	2.0	1.1	0.7	0.0	0.0	0.0
	15%	1.7	42.9	6.7	6.2	5.4	4.7	4.2	2.9	1.8	1.1	0.2	0.0	0.0
	20%	2.2	46.9	9.0	9.2	7.3	7.6	6.3	4.2	3.4	2.3	0.5	0.0	0.0
	25%	2.8	51.3	13.2	12.0	9.5	9.2	8.3	6.6	4.3	3.3	0.6	0.0	0.0
	50%	5.6	56.0	27.3	23.6	20.3	19.8	18.1	13.9	11.3	8.8	4.8	0.4	0.0

(*continued*)

TABLE 10.5 *(Continued from previous page)*

Hi-oleic sunflower oil														
IV49	100%	15.5	62.7	70.7	64.5	58.5	58.2	53.9	44.6	37.5	31.1	20.4	9.3	0.1
	5%	0.8	42.2	3.6	3.0	2.8	2.7	2.2	1.3	0.9	0.6	0.0	0.0	0.0
	10%	1.6	46.1	5.5	5.0	4.3	4.4	4.0	2.7	1.5	1.0	0.0	0.0	0.0
	15%	2.3	47.6	9.1	8.1	7.2	6.5	5.7	4.1	3.4	2.4	0.0	0.0	0.0
	20%	3.1	49.9	14.2	11.7	10.1	10.1	9.1	6.9	4.7	3.7	0.4	0.0	0.0
	25%	3.9	52.0	16.9	14.8	12.8	12.1	11.3	8.1	6.1	4.9	0.7	0.0	0.0
	50%	7.8	58.0	34.1	30.2	26.5	26.3	23.5	19.1	16.0	12.3	6.8	0.1	0.0
Soybean oil														
IV49	100%	18.2	60.0	81.0	76.6	58.9	52.0	47.8	40.5	31.2	22.3	13.4	1.4	0.4
	5%	0.9	29.3	3.0	2.9	1.9	2.3	1.5	1.2	0.5	0.0	0.0	0.0	0.0
	10%	1.8	38.8	7.5	6.3	4.5	4.0	3.4	2.2	1.0	0.1	0.0	0.0	0.0
	15%	2.7	44.5	12.2	9.6	7.0	6.4	6.8	2.4	2.2	0.6	0.5	0.5	0.0
	20%	3.6	47.7	16.0	12.6	10.2	9.4	8.3	5.7	3.4	1.7	0.4	0.3	0.0
	25%	4.6	49.3	18.4	15.8	11.9	11.0	9.3	6.6	4.3	2.5	0.6	0.2	0.0
	50%	9.1	55.1	26.3	32.7	26.4	24.2	21.7	17.1	11.6	7.5	3.6	0.0	0.0

[a]Balance soybean oil.

[b]Solid fat content by pulsed NMR .

Source: Jackson et al., 2008, J. Am. Oil Chem. Soc. 85, 481–486.

Chapter 11

Safety

H. B. W. Patterson

Safety, Security, and the Prevention of Error

As with other kinds of plants and indeed a wide range of activities in which some operational risk is present, the action taken to contain the danger must relate to the probability of an event occurring and how serious its effects could be. For example, an important hardening plant located in a target area vulnerable to bombing could have its capacity duplicated elsewhere in a less accessible position; normally, this class of risk does not face management. Probabilities and consequences may not be easy to quantify; nevertheless, this approach is the realistic one.

Risks can also be related to the actions of people or of things such as equipment. Most accidents are found to relate to the actions of people; these include actions which are well-intentioned and those which are foreseeable and preventable. This brings us at once to the cardinal principle of "designing out of trouble." This approach may mean arranging the operation in such a way that, to say the least, a positive action to invite trouble is required before it can occur. For example, valves on an oil pipeline may be left incompletely closed (especially at the change of a shift) or they may leak because of a mechanical fault: the result—good oil is lost or spoiled. A center-point connection is immediately seen to be made or undone; leaks or spills compel attention by pouring onto the floor. A wrong connection at the center-point can cause a wasteful oil mixture, but if male and female unions on a pipe system relate to one particular oil group and no other, the incorrect coupling is prevented. To diminish risk, the same philosophy must then be pursued from the plant to the tank farm and in favorable circumstances, such as when the company controls oil tankers regularly on fixed oil-handling duties, to the oil vehicles themselves. At this stage, the risk of an oil mixture, the resultant loss in value, and the cost of the center-point installation are likely to be in realistic and profitable balance with one another. The next question is who puts what into which tanker; the study then repeats itself elsewhere. The design staff will normally discuss details with those who are destined to operate a plant or its modification, and in any case will have a list of safety checkpoints which have their origin in national legislation and accumulated experience within the industry. This is a class of "know-how" which is readily disclosed. The designers must check against their list, pay particular attention to amendments of the design to be certain these do not frustrate safety precautions already incorporated, and ensure that the operating management is aware of precautions built into the design, particularly if actions within the scope of the production staff could impair these precautions. As already indicated, a good design keeps the extent of these options to a minimum, consistent with an efficient operation.

Safety and Personnel

Part of job training is safety training, but beyond this, adherence to particular security measures should be included in the job description and the point made to the employee that payment is made for compliance with these, like other rules. This is not merely in the employee's interest, but it may also save the employer money otherwise to be paid in compensation. Finally—and this is particularly relevant in a hardening plant—compliance with safety regulations is equally in the interest of fellow employees.

Two practical points relating principally to people emerge here. Firstly, lapses in security are more common at or immediately following a change of shift. This aspect includes the downgrading or loss of material as much as injury to personnel. Difficulty arises from operating a system whereby shift supervision changes at a different time from the operating personnel—perhaps an hour afterward or before. Supervisors, in conjunction with this, may be pleased to operate shifts of unequal lengths—e.g., 9-hour day, 6-hour evening, 9-hour night—in locations where a change of their shifts occurs each week. The underlying thought is to foster a net continuity of the control of operations by personnel throughout time. Secondly, when a break occurs from normal operation, probably to repair or modify an item of the plant, fresh safety risks arise. To keep these risks to a very low level, to operate a "permit-to-work" routine is useful whereby, when a plant is to be taken out of operation and probably opened to the atmosphere, a written authorization that includes the statement that certain precautions were taken is issued by a manager or supervisor nominated for this duty. The dated and signed permit covers the following:

1. Location and nature of work to be done.
2. Precautions to be taken (fuses out, blanks in lines, etcetera, see the "Autoclaves" section in this chapter).
3. Protective equipment required.
4. A statement that precautions were completed.
5. Names of persons covered by permit.
6. Times between which permit operates (as appropriate).
7. Completion of work certified, permission for withdrawal of nominated personnel and equipment, and the return of the item of the plant to normal duty.

Management must obtain legal advice as to whether the proposed wording of such a permit is consistent with any national or local regulations on the subject.

Safety and Equipment

Buildings, plant, and layout are all commonly the subject of laws and codes of practice. The hydrogenation autoclaves are pressure vessels and must be designed as such; this also applies to hydrogen storage vessels according to their duty. Remember that since pressure vessels are to be given a hydraulic test before commissioning, and at intervals thereafter, their foundations must be strong enough to bear the weight of water when the vessel is full as well as the vessel itself. Having the smallest molecule, hydrogen diffuses very rapidly through small gaps. This emphasizes that during erection, connections must not be distorted and therefore become the source of gas leaks. Agreement was reached by many countries concerning the classification of different areas or zones according to the extent of the hazard associated with them (Critchfield, 1976; International Electrotechnical Commission, 1972, 1973; Patterson, 1974). Hydrogen is only one of the materials recognized in this classification, which runs as follows:

- Zone 0. An area or enclosed space in which a hazardous gas, vapor, or volatile fluid is present in sufficient quantity at all times to constitute a flammable concentration.
- Zone 1. An area or enclosed space in which a hazardous gas, vapor, or volatile liquid of flammable concentration may be released from time to time during normal operation.
- Zone 2. An area or enclosed space in which a hazardous gas, vapor, or volatile liquid is so well under control that a flammable concentration can be present only when temperatures and/or pressures are abnormally high. (The hazard is only present in abnormal or fault conditions.)

The experience of hardening plants places them in Zone 1. In the United Kingdom, this is classified as Division 1; in the United States, Class 1, Division 1; and in other countries, Division 1 may be further divided into less or more dangerous parts of the same area. In an attempt to classify a risk, some basic questions are posed—Is the material used flammable? How much is used and what are the other important characteristics? Is the process enclosed? Is the plant indoors or outdoors? How near is another plant or source of ignition?

From these broad considerations, the attention then moves to the detailed safety requirements for individual items, which may be electrical, structural, mechanical, etcetera. A detailed list of sources of safety information given by Churchley (1977) covers several American and European publications.

Safety and Hydrogen

Hydrogen quality—as it influences the hydrogenation reaction—is discussed in detail in the "Quality" section in Chapter 5, and several safety precautions in relation to electrolytic manufacturing appear in the "Security" section of Chapter 5; the solubility of hydrogen in glyceride oils is described in the "Hydrogen Dispersion" section of Chapter 1. The general physical and chemical properties of hydrogen are well-described in a large number of inorganic chemistry textbooks. Therefore, one must draw attention here only to those characteristics of hydrogen which are easily seen to relate immediately to its safe handling as a gas. The storage of hydrogen as a liquid is coming steadily more into prominence for industrial use, especially in the United States; a large amount of information on handling and storage is available, but one should obtain this information from the supplier because of its very specialized nature (Newton, 1967; Reeitff, 1965). Hydrogen is colorless, odorless, tasteless, nontoxic, nonirritant, and does not support respiration. It is the lightest gas, and therefore accumulates at the top of a confined space, displacing other gases downward to the extent that, temporarily, it does not completely mix with them. Its density is 0.0899 g/liter = 0.0056 lb/ft^3 at 0°C and 760 mm of Hg. Compared with air, its specific gravity is 0.069. Hydrogen has one of the highest specific heats of any gas, C_p = 3.4 g cal/g or BTU/lb at 100°C; it conducts heat about seven times as effectively as air. These thermal properties enable its purity to be monitored continuously by physical means, and for alarms and safety action to be activated automatically.

Hydrogen combines with oxygen in the ratio of a two-volume to a one-volume to form water with a speed which varies enormously with circumstances. If both gases are dried well beyond the limits open to usual industrial processes, but obtainable in the laboratory, the dry gases may be heated (Normann, 1902) by an incandescent silver filament, and then they combine only gradually. In the presence of a trace of moisture, without any source of ignition, the combination process proceeds slowly at 180°C or explosively in the temperature range of 550–700°C. At 526°C, the compression of the correct proportions of hydrogen and oxygen ($2\ H_2 + O_2$) leads to ignition; for hydrogen-rich mixtures, the necessary temperature rises (e.g., 544°C), and for oxygen-rich mixtures it falls (e.g., 478°C). At ordinary ambient conditions (15°C, 760 mm of Hg), the lowest concentration of hydrogen which will sustain an explosion when mixed with air is 4.1% by volume (hence, 95.9% of air). Below this concentration, the fuel (hydrogen) is so diluted that the flame front is unable to propagate itself rapidly. At the other end of the scale, when hydrogen exceeds 74.2% by volume, the flame front can then no longer propagate itself because of the lack of oxygen. These two concentrations of hydrogen in air, 4.1% and 74.2%, are spoken of as the "explosive limits." For hydrogen and oxygen mixtures, the values are little different. If the gases are saturated with moisture (as is the case during production by many methods) and the concentration of water vapor rises considerably with a rise in temperature, the lower limit for hydrogen in the mixture as compared with oxygen, if both were corrected back to a dry basis, shows only a slight rise from c. 4 to 5% between 60 and 80°C. The prudent practice adopted, therefore, is to keep to a level of hydrogen which, even when corrected back to a dry basis, is distinctly inside the safe limit. If pressure or temperature is increased for the hydrogen–air mixture, the explosive limits widen slightly in the sense that a mixture slightly poorer or richer in hydrogen will now sustain an explosion. Although for normal atmospheric pressure, the auto-ignition of an explosive mixture is quoted as occurring at 585°C: temperatures above 450°C are regarded as suspect and possibly conducive to auto-ignition; direct contact between the gases (as gases) and an active catalyst surface promotes ignition; an electric spark of a fraction of a millijoule, such as might arise from an electrostatic discharge at over 1000 V, would ignite a mixture of the correct explosive proportions. The most extensive information concerning the flammability of gas–air mixtures and by how much this is depressed by inert diluents is often held by various national bureaus of mines. When pressure is dropped to a small fraction of one atmosphere, even the correctly balanced mixture of 2:1 hydrogen–oxygen combines silently to water when sparked. Although the flame propagates, insufficient energy is liberated to create an explosion. Broadly speaking, when a hydrogen–air mixture is caused to explode, the pressure built up is about 20 times that of the original mixture. Thus, in the experiment familiar to so many chemistry students, hydrogen and oxygen are mixed in a glass eudiometer tube and then the pressure reduced by lowering the flexible leg of the eudiometer, causing the mercury level below the gas to drop. The spark is then passed, and only the appearance of condensate—especially when normal pressure is restored—makes it obvious that a combination has occurred. By contrast, if the pressure is inadvertently first raised instead of being reduced and the spark then passed, the report is loud, the apparatus disintegrates, and mercury is thrown over the bench and observer. Although extracurricular, this experiment, too, is educational.

Common Precautions

Common precautions fall naturally into three groups:

1. Prevent hydrogen from leaking out from the system.
2. Prevent air or oxygen from leaking into the system.
3. Avoid sources of ignition.

Leaks Out

With a correctly designed plant where due care was taken in its original assembly to avoid distortion at joints, etcetera, the avoidance of leaks is primarily a question of sensible operation and continual inspection and maintenance. At sensitive places such as pipe joints, agitator glands, valve spindles, screwed connections, relief valves, and connections for sampling and instrumentation, the application of a soap solution is a useful, simple test for hydrogen leaking out under pressure. The use of hydrogen pressure alone to displace hardened oil from the autoclave through the filter and to clear the filter is not recommended. Hydrogen may indeed be used via the low-pressure or "balance gas" top connection to follow the hardened oil as its level in the autoclave falls, but once the vessel has emptied, the filter pump below, pipe line, and filter should be cleared by nitrogen. Apart from the distinct change in the sound of the filter pump as the last oil passes, one can use various instruments to detect the change and to give warning or initiate action or both (see the "Hydrogen Distribution: Limitation of Uses" section in Chapter 4). A second line of defense which should never be absent in a hardening-plant layout is good ventilation. Small leaks from running glands are a continuing occurrence, but happily the hydrogen is inclined to disperse so rapidly because of its lightness that a natural draft will achieve this if provided. In warm climates, plants may well be located in the open, and have at most an overhead cover. In any case, good practice is to locate hydrogen-handling equipment and connections at as high a level as is practical. This reduces the risk at ground level to a minimum, and then the question of deciding how far a downward-facing gas discharge from an above fault could travel laterally needs to be addressed. The continuous monitoring of any chosen area for the accumulation of a combustible gas or vapor has long been possible with detectors provided by companies who specialize in this field. Lastly, remains the special case where venting unwanted hydrogen into the open is done as a safety measure—the question is how one should handle such vented gas. This is considered in the "Autoclaves" section of this chapter, as it has particular relevance to autoclaves.

Leaks In

Leaks of air or oxygen into a hydrogen system are more dangerous because they may then be confined and may pass unnoticed more easily. The suction side of a hydrogen compressor is the classic vulnerable point for this hazard. Apart from normal maintenance, the action taken is to ensure that the hydrogen pressure in the suction line must always remain slightly above atmospheric pressure (0.5 in WG is usual), and if not, a pressure-sensitive switch cuts out the compressor. Other safety switches affect the operation of the compressor, which transfer hydrogen from the low-pressure (LP) store to the high-pressure (HP) store, but these depend on the amount of hydrogen in the stores and are discussed under hydrogen storage (see the "Hydrogen Storage" and "Security" sections in Chapters 11 and 5, respectively). One may point out here, however, that if the low-level safety switches on the LP store function as intended, the compressor will never be allowed to run on until the 0.5 in WG differential pressure above atmospheric pressure is lost. In the case of autoclaves, the filling with oil commences after excess pressure, if any, of hydrogen or nitrogen, was vented and vacuum was applied. At the end of the filling, the vacuum valve being shut, the closed vacuum in the headspace is then broken steadily with hydrogen. One can arrange that hydrogen is not admitted while the vacuum line is open, or that hydrogen is not admitted until some very LP is attained—say, 50 mm of Hg—and that the subsequent breaking of the headspace vacuum is steady. Other operators may prefer to use a brief nitrogen purge of the headspace before purging it in turn with hydrogen. Precautions against forming an explosive mixture when opening an autoclave are described in the "Autoclaves" section of this chapter.

Ignition

The basic conditions necessary for the ignition of hydrogen were given in the "Safety and Hydrogen" section of this chapter. If precautions against leakage fail, dispersal is the next line of defense. Then comes the third defense line: the exclusion of sources of ignition. One way of achieving this is that, in many important areas of the plant, equipment such as electric motors, switches, lighting, and instruments should be made so that they cannot ignite gas in a way that allows flame to propagate. As far as instruments and some other electrical equipment are concerned, collecting them in a positively pressurized control room approached through a gas lock may be feasible. All such plans need local approval. Where gas may be present, one should use nonsparking tools. Notices stating the existence of a fire–explosion risk must be so disposed that entering the area without seeing them easily is impossible, and they must demand the prohibition of the carrying of matches, lighters, naked lights, or other means of ignition, as well as smoking. To help workers to be in compliance with these rules, a facility can provide a receptacle for depositing forbidden items at least at one point of entry. Do not allow welding at times when gas may be present. The "permit-to-work" procedure covers this situation (see the "Safety and Personnel" section in this chapter). The plant layout should be checked for ducts used by services to see that these do not create a chimney effect in drawing gas from an area where no means of ignition exists to a higher one where it does (e.g., boiler house). This type of risk may be greater with vapors heavier than air, but one must exclude it.

Autoclaves

These are pressure vessels, and so must conform in the material of construction, fabrication, and testing with whatever local regulations exist. This applies to any internal pipework and fittings. Coils and headers should be stress-relieved and annealed before installation. For edible-oil hydrogenation, to allow for working conditions up to 210°C is common, and from full vacuum to 6 atm. Local regulations may then require hydraulic-pressure testing of the autoclave to 50 or even 100% above working pressure (in this example, therefore, to 9–12 atm), although steam coils would be tested to a higher pressure (e.g., 25 atm). These figures are merely examples, and one must follow local regulations. Further, some autoclaves are designed to work at 10 atm, while others used in the hardening of fatty acids are designed for operation at 25–30 atm and 240°C (see the "Fatty Acids" section in Chapter 8). Vacuum was widely used with hydrogenation autoclaves internationally, and from the earliest days of industrial hydrogenation. It is probably a feature of the majority of present plants, and is likely to remain so in the future. The equipment now used to provide the vacuum service is the steam ejector. One can easily fit hydrogen feed lines and vacuum lines with a double set of control valves on a short stretch of pipe carrying between them a small vent with a valve. When the service is to be closed off, both main valves are closed and the vent is opened. As mentioned in the "Common Precautions" section in this chapter, the hydrogen and vacuum services can be interrelated as an additional safeguard. To lead safety-vent pipes to a discharge point outside of the building is important, preferably at an upper level and never directed toward any other construction in the immediate vicinity. In a solvent-extraction plant employing hexane (Critchfield, 1976), an automatic CO_2 extinguisher acted on a false signal, flooding the body of the extractor with CO_2 and driving a cloud of hexane vapor into the work space. A serious explosion followed; some evidence suggested that ignition was caused by the discharge of static electricity on CO_2, "snow" particles.

These remarks apply with much greater force to the vent system—usually about 7.5 cm i.d.—attached to the safety valve situated on the crown of each autoclave, which lifts at—what will here be called—the "design pressure" of the autoclave. This design pressure should be moderately above the normal maximal operating pressure at which the autoclave will be set to work: a difference of 2 atm would be quite adequate. This now means that the maximal operating pressure may be employed without the safety-valve hunting or chattering.

In situations where a bursting disc is also provided on the crown of the autoclave, sensibly this could blow at c. 0.3 atm above the pressure of the safety valve. Remember that the maximal normal hydrogen pressure which is developed in an autoclave is no greater than that in the HP store feeding it, and this itself is likely to be stepped down from the HP-store pressure by a reducing valve to the currently desired working pressure for the autoclaves. For example, an HP store at 7–10 atm might be reduced to 3.5–6 atm. If, however, a coil failure occurs in an autoclave operating at 180°C, this is a new situation with which the vents might have to cope. Also necessary is to fit flame arresters in vents, since a high-velocity hydrogen discharge is well-known to inflame. The flame, if directed into space, is in itself a convenient way of disposing urgently of the hydrogen not required, but as pressure diminishes at its source, the flame front will attempt to travel back (Cude, 1974; Maisey, 1965). The pipe used for returning very moist or impure hydrogen from the headspace to the LP store, or as a direct vent to the atmosphere in the popular dead-end systems, is, of course, separate from the safety vents just described. Advisably, include in this vent line an orifice plate which limits the rate of discharge; a hole of c. 3 mm in diameter in the orifice plate would be adequate for autoclaves of a capacity of 10–20 tons, and permits the refreshment of their gas space within a few minutes. The provision of an easily accessible gas-sampling point on this line is a welcome convenience. In plants which contain several autoclaves, the further possibility exists of what may be described as a convenient option. This simply amounts to locating a pressure-relief valve on the central hydrogen feed line after the reducing valve, and arranging that this relief valve acts above the maximal operating pressure of the autoclaves, but below their design pressure—at which the individual autoclave safety valves are set to lift. This one valve now provides a safeguard for the group. Its existence does imply, of course, that the reducing valve could fail and not bring down the line pressure from the HP store sufficiently. If such a central pressure-relief valve were installed, to locate it in the open would naturally be convenient, with the reducing valve just upstream of it.

The sequence, in ascending order, at which various pressures could be associated with the work and safety of an autoclave is shown together in Table 11.1 as an example. One must emphasize that all schemes of this kind must comply with whatever local regulations exist.

Opening an Autoclave

To carry out inspections and repairs, one must take an autoclave out of service from time to time; hence, an obvious need exists for a safe procedure. Similarly, when the

TABLE 11.1
Example of Relative Pressure Levels Associated with a Hydrogenation Autoclave

Type of Limit	Pressure (atm)	Control
Hydraulic test	9–12	Bursting discs
	8.3	
"Design pressure"	8	Individual safety valves
	6.6	Optional central relief valve
Maximum operating pressure	6	Central reducing valve
Usual operating pressure range	up to 5	

All autoclaves tested to withstand full vacuum.

autoclave is taken into service, one must use a safe routine. The following general description covers points to be observed in these situations.

1. The autoclave is filled with soft oil and agitated for at least 15 minutes to remove any deposited catalyst. The oil filling will have displaced any hydrogen that was present (Alternatively, if a normal-sized oil charge was used, the gas space may be freed from hydrogen by a nitrogen purge. As the oil is withdrawn, nitrogen is supplied to take its place. If nitrogen was not used as a purge and replacement, but hydrogen was allowed to return, the custom for some operators is, immediately after the last oil has left the vessel, to commence filling it with water to the point of overflowing, thus displacing any hydrogen present. Water is drained, being replaced by air; then go to item 2.) The oil is next withdrawn, and air is allowed to take its place.

2. Fuses must be removed from the agitator motor circuit, and a notice saying that was done must be hung on the switchgear concerned.

3. Various services must be blanked off and on no account should reliance be placed on closed valves. The pipe lines to be blanked are: hydrogen feed lines (HP or LP connections), waste gas line, vacuum line, catalyst feed line, oil feed line, and outlet.

4. The vessel is cooled, if necessary, by passing cooling water through the coils. If the coil has to be repaired/replaced, self-evidently a blank must be placed in the steam–water connection.

5. The manhole is opened and the safety valve is removed. The latter will be located very near the top of the autoclave, and the opening left assists the circulation of air. Air is next passed through the autoclave for at least an hour. The permit to work must be signed to say precautions were completed (see the "Safety and Personnel" section in this chapter).

6. After the manhole is opened, the interior of the autoclave may then be cleaned further by almost filling with hot water in which some sodium carbonate was dissolved, and then withdrawing the wash.

7. After work is completed, blanks removed, and fuses, manhole, and safety valve replaced, the autoclave is charged with oil and the gas space purged free from air with nitrogen, which is then replaced by a hydrogen purge; or, the vessel is filled completely with oil and then hydrogen is allowed to replace the oil as it is withdrawn; or, water is used instead of oil completely to fill the vessel and expel air, after which hydrogen is drawn in to replace water as it is withdrawn.

8. Permit to work is signed to return autoclave to service.

These variations were described, as the same facilities are not available in all plants.

All autoclaves and filters, like other units through which oil passes, must be effectively bonded to earth to discharge static charges; the bonding must be inspected at intervals—and particularly if the unit has sustained any movement—to see that the bonding is still effective.

Hydrogen Storage

Low-Pressure Stores

Commonly, these are of the wet-holder type fabricated from mild steel, brushed clean, and then painted inside with bituminous paint and outside with a protective paint which reflects solar heat. Large dry-holders of comparable design exist. As regarding fabric, the main risk is corrosion, signs of which are checked by regular inspection. Corrosion is likely to be reduced if the seal is filled with a 4% borax solution. Plant shutdown, such as on a weekend, allows patrol staff to log the position of holders so that allowing for temperature changes, one can discover if movement due to leaks is occurring. The safety questions in relation to LP holders which attract immediate attention are: How is control exercised on the source of hydrogen feeding them and the compressor removing hydrogen from them, in relation to whether they are nearly full or nearly empty? and What happens if the quality of the gas being supplied to them drops below the standard required? Some detail on this topic in relation to the LP storage of electrolyte hydrogen is given in the "Security" section of Chapter 4, which one should also consult. In general, the safety of an LP holder is controlled by:

1. An ultimate high-level switch to stop the incoming flow of hydrogen because the holder is almost full. This could entail some combination of ceasing hydrogen production, closing the supply pipe, and temporarily diverting surplus production to the atmosphere before it reaches the holder.

2. A high-level alarm to warn that the level is becoming high and that one must take action to prevent the rising of it further. Often worthwhile is to arrange a visible alarm as well as an audible one.

3. A low-level alarm to warn that the level is low and also to switch off the compressor withdrawing gas from the holder so that it may commence to gain. To render this alarm visible as well as audible may also be useful.

4. An ultimate low-level switch which completely cuts off all power from the compressor, since the normal low-level alarm has evidently failed to stop it.
5. A switch which is linked to some middle position of the holder and allows normal on/off control of the compressor. The more sophisticated arrangement allows the compressor to function automatically, in effect, above a low level of the LP store and below a high level of the HP store.
6. The quality of the hydrogen being supplied to the LP holder is no doubt being monitored for total hydrogen content, air, oxygen, mercury, carbon monoxide, or any other relevant impurity according to circumstances, so that in no case is a supply accepted which is beginning to approach the explosive limit. For example, the alarm condition might be set to act at 2.5% of air content or less. The automatic action attached to this alarm could then be set to ensure that at 3% of air an immediate diversion of the pipeline supply to the atmosphere via a flame arrester would occur. Do not allow air to accumulate in the system. Regarding other impurities, management should choose an alarm level which is a clear indication that an abnormal condition has been reached. One should discover from the suppliers of such monitoring equipment whether it is sensitive to temperature or moisture, and if so, to what extent. One may overcome difficulties by housing sensitive equipment in a meter house whose temperature is automatically prevented from falling below a certain level. Some hardeners receive their hydrogen by pipeline from a source miles away. In such a situation, to maintain routine telephone contact at set times is advisable (e.g., the beginning of each shift).

    Recorders are a considerable advantage in monitoring equipment, as they may easily give a very early indication that some purifying equipment in the hydrogen-supply train is now deteriorating; also, a recorder covering the operation of the low-level alarm (see 3 in the list above) can be helpful, as this is an area where accidents are more common, for example—The failure to shut down a compressor when it had exhausted its supply of hydrogen led to its drawing in air for some time and then overheating, after which the hydrogen store exploded, scattering pieces of plant well beyond the factory perimeter.

If normal oil pressure on the compressor crank shaft does not build up after a short period of running, the motor should automatically be switched off to prevent the seizure of the compressor and other troubles.

On the delivery side of the compressor, an HP switch is located in the pipe line leading to the HP store. This switch cuts out the compressor if a preset HP is reached.

Earlier emphasized was that an LP switch is always situated on the suction side of the compressor, so that if at least 0.5 in WG pressure above atmospheric pressure is not maintained, the compressor is shut down. Some LP holders include a built-in stopper which is located in the mouth of an upturned withdrawal pipe as the ultimate low level is reached.

High-Pressure Stores

Typically, HP stores may include such items as a Horton sphere of 50,000 ft^3 geometric capacity working at 10 atm to cylindrical vessels up to 6 ft in diameter and 40 ft in length working at over 30 atm. The material, design, fabrication, and final pressure testing are subject to local/national regulations following codes such as BSI, ASME, AOTC, Lloyds, etc. For cold climates, stressed-steel embrittlement needs to be considered, and in hot climates, some shade from the sun. As already stated, the delivery side of a compressor (and this covers the HP store) must be provided with an HP switch to shut down the compressor when a certain safe maximal pressure is reached. Relief valves for overpressure and drain valves to remove condensate must be provided, and only reasonably these should be at least duplicated so as to bypass malfunctioning. The relief valves must not direct gas at adjoining constructions. Both HP and LP storage units must be situated within the security zone; electric lighting in the zone may be permitted with certain precautions, such as—the lights must be sealed and be at a 3.5-m distance from the storage units with their control switches outside of the zone. These matters must always be checked against local regulations.

Hydrogen Receipt by Road/Rail

Some hardeners receive all or part of their hydrogen at 100–150 atm in steel containers mounted as a group on a vehicle; the rail vehicle may be one very large HP container. Certain basic and seemingly obvious precautions, if neglected, can cause trouble. The different HP containers on the vehicle almost certainly employ a common manifold through which to discharge, and the hydrogen supplier may also provide reducing valve(s). If the customer's HP store has been almost filled from a container and the pressure of the store is therefore virtually in equilibrium with it, this container is closed, and a second one could inadvertently be opened quickly to complete the filling. To smooth the surge in pressure quickly may be beyond the capacity of the reducing valve(s); hence, it communicates itself to the HP store relief valve(s). That valve's setting may be such that it responds, unseats, and then vents hydrogen so rapidly that it inflames. The indication is, therefore, that the stabilizing characteristics of the reducing valve(s) must be within the setting of the relief valve(s).

In a situation where the supplier has left his trailer connected to the customer's system for receiving hydrogen, and has, of course, purged his section of the delivery line, the master-connecting valve will be left shut. The customer is now expected to take hydrogen as required; then important is that the pressures in the different containers (if more than one) are equalized gradually before opening the master-supply valve. One should make an arrangement whereby the hydrogen vehicle cannot be driven away while still linked to the receiving station; for a road vehicle, this could be done by inhibiting the motor starter while discharge connections are still engaged.

General Precautions Covering Static Charges and Electrical Equipment

Mention was already made of the need to bond to earth the items of equipment which could acquire charges of static electricity, and this has an obvious additional importance in areas where hydrogen might be present, presumably by leakage. The rapid movement of dry oil is the obvious source of static; thus, filter units and vessels being filled with oil, such as autoclaves, call for particular attention both at the time of installation and whenever maintenance and modification may have disturbed the effectiveness of the bonding.

As regarding equipment, a popular answer for a long time was to group items which could spark in an enclosed area into which hydrogen could not leak (i.e., it was slightly air-pressurized, or the equipment itself was pressured). Now several countries—Germany, the United States, and the United Kingdom in particular—have developed many items of electrical equipment which are incapable of transmitting ignition to an ambient explosive atmosphere. Where these standards are acceptable, to install such equipment is often very economical. The final decision must rest upon the advice of qualified electrical engineers; in the mining industry, a large fund of practical expertise exists.

Chapter 12

Quality and Control

H. B. W. Patterson

Classification of Tests

This chapter attempts to give a brief explanation of the significance and, as appropriate, the limitations of analytical tests commonly encountered in fats and oils hydrogenation. For the methodology of the tests, analytical textbooks (*Analysis of Oils and Fats,* 1986; *Analysis of Oilseeds, Fats and Fatty Foods,* 1991; Boekenoogen, 1968; Christie, 1973; Cocks & van Rede, 1966) and detailed official handbooks (*DGF Standard Methods of Analysis*; *Methods of Analysis of Oils and Fats,* 1958; *Official and Tentative Methods of the American Oil Chemists' Society*) must be consulted. Tests are grouped together on the basis of their relating particularly to one or moreof the basic features of fats and oils, such as melting behavior, amount, and type of unsaturation. Some tests are obsolescent, as they were overtaken by instrumental techniques which save labor, time, or relate more directly to the characteristic being measured. Others are empirical but were useful and convenient for a long time; these, no doubt, will disappear only slowly from common use. Several reviews which consider briefly the impact of a number of new techniques on the analysis of fats and oils have appeared as symposia lectures, etcetera (AOCS Short Course, 1966: (i) Dutton, (ii) Erickson, (iii) Lantondress, (iv) Sherwin, (v) Sullivan, (vi) Wiedermann; Patterson, 1989; Sleeter, 1983, 1985; Taylor, 1973); one should read these and similar reviews, with their references, for greater detail and examples of application in research, quality, and process control.

Saponifiable Matter

Free Fatty Acid (%)

This is the percentage by weight of uncombined fatty acid in the material as it is. If a considerable amount of nonfatty matter is in this material (e.g., water), a useful and common practice is to measure the total fatty matter (TFM) present and then express the FFA as a percentage of the TFM—making it clear this was done.

Since the FFA is measured by titration against a standard caustic-soda solution, the conversion of the titration to a weight of fatty acid depends on the molecular weight of the fatty acid concerned. Most often oleic acid (mol wt 282) is taken as representing the average molecular weight of the acids being estimated. If the actual molecular weight is less than or greater than 282, the true result ought to be less or greater in proportion. Thus, if fatty acids other than oleic acid are considered to be more representative, the molecular weights could be:

- For FFA% expressed as lauric acid, mol wt = 200
- For FFA% expressed as palmitic acid, mol wt = 256
- For FFA% expressed as erucic acid, mol wt = 338

(Being all monocarboxylic acids, the molecular weight is the equivalent weight.) When calculating the efficiency of refining an oil, this consideration is important [AOCS Short Course, 1966: (ii) Erickson, (iv) Sherwin, (v) Sullivan].

Acid Value (AV)

Acid value (AV) is the milligrams of potassium hydroxide required to neutralize the FFA in one gram of fat. It is independent of molecular weight and, coincidentally, is almost twice the FFA% when the latter is expressed on an oleic-acid basis.

Saponification Value (SV)

Saponification value (SV) is the milligrams of potassium hydroxide required to saponify completely 1 gram of fat. This covers neutral fat and FFA present, and obviously relates to the molecular weights of the fatty acids involved. Where these molecular weights are low, the one gram will require more potassium hydroxide (coconut oil SV = 250), and where they are high, it will require less (rapeseed oil SV = 175). [Note: The weight of fat saponified by a 1-gram equivalent of potassium hydroxide (56.108 g.) is the saponification equivalent of the fat.]

Ester Value (EV)

Ester value (EV) represents combined (with glycerol) fatty acids in a fat, and therefore, when expressed in milligrams of potassium hydroxide, it is the difference between SV and AV.

Unsaponifiable Matter (%)

As far as natural and uncontaminated fats and oils are concerned, this is simply the unsaponifiable matter—such as hydrocarbons, sterols, etcetera—which is present and is expressed as a weight percentage. Typical ranges are quoted for the various fats and oils, and may show substantial variation according to maturity and species; amounts are usually less than 1.5%. In reporting on the quality of a sample, however, the unsaponifiable (%) will cover other unsaponifiable organic matter which may have reached the triglyceride as a contaminant, such as mineral oil. If present, the latter needs further assessment (Cocks & van Rede, 1966, pp. 78, 123; Mehlenbacher, 1960, pp. 157–162).

Unsaturation

Iodine Value (IV)

Iodine value (IV) is the percentage of its own weight—with which a fat will combine—of halogen calculated as iodine, the conditions of the test being indicated. The most popular procedure is known as the Wijs method, but all methods suppose that a normal carbon-carbon double bond takes up two atoms of iodine. Apart from the number of double bonds present in the molecule, the IV is also influenced by the molecular weight, thus,

- IV oleic acid = (2 × 127/282) × 100 = 90.1
- IV glyceryl = [6 × 127/(41 + 3 × 281)] × 100 = 86.2 triolein similarly
- IV linoleic acid =181, trilinolein = 173.2
- IV linolenic acid = 273.5, trilinolenin = 261.6.

Where several fatty acids are present, the IV alone cannot give the detailed picture as to how much high IV—and therefore potentially reactive-unstable oil—is present in the mixture [AOCS Short Course, 1966: (ii) Erickson]; therefore, additional information relative to this question remains to be gained by the several techniques available. According to the "Hydrogen Requirements" section in Chapter 5, the following is noted:

- 1000 kg of oil dropping 1 unit of IV requires 0.8835 m^3 of hydrogen (0°C, 760 mm of Hg)
- 1000 kg of oil dropping 1 unit of IV requires 0.9319 m^3 of hydrogen (15°C, 760 mm of Hg)
- 2240 lb of oil dropping 1 unit of IV requires 31.69 ft^3 of hydrogen (15°C, 760 mm of Hg).

Accelerated versions of the IV test are convenient for end-point checks on the hardening process (Cocks & van Rede, 1966, pp. 109–113).

Polybromides

Triunsaturated fatty-acid groups readily form hexabromides insoluble in diethyl ether at 0°C, and more unsaturated groups add correspondingly more bromine. If the precipitate which formed at 0°C disappears again when warmed to c. 37°C, probably, only tetrabromides are present (*Methods of Analysis of Oils and Fats,* 1958, p. 86).

Trans *Index*

The "Fatty-Acid Chain Length and Unsaturation" section in Chapter 1 explains how double bonds exist in two geometrical forms, the *cis* and the *trans*. The *cis* form greatly predominates in natural fats, and because it distorts the fatty-acid chain more than the *trans* form, this emphasizes soft or liquid characteristics. The "Isomerization" section in Chapter 1 explains how, in hydrogenation, *cis* isomers often change to *trans* isomers; how, in effect, they migrate along the chain; and how, in their new position, they may also change to the *trans* form. Since the *trans* isomers have the higher melting point, this kind of change is especially interesting to the hardener. The "Low-Temperature Hydrogenation" and the "Iso- or *trans*-Suppressive Hydrogenation" sections in Chapter 2 describe the means of discouraging *trans* formation, and the "Iso- or *trans*-Promoting Hydrogenation "section in Chapter 2 describes the ways of encouraging it.

One of the most convenient ways of checking how much *trans* formation has taken place at any point during the hydrogenation of an oil is to make use of

TABLE 12.1
The Melting Points of Four Isomers of Oleic Acid Illustrate the Importance of *trans* Formation

16°C		42°C
Oleic	*Geometric*	Elaidic
Octadeca-*cis* 9-enoic	*isomers*	Octadeca-*trans* 9-enoic
Positional isomers		*Positional* isomers
30°C		51.9°C
Petroselenic	*Geometric*	Petroselaidic
Octadeca-*cis* 6-enoic	*isomers*	Octadeca-*trans* 6-enoic

infrared spectrophotometry. Double bonds in unsaturated fats are not usually part of a conjugated system (see the "Fatty-Acid Chain Length and Unsaturation" section in Chapter 1), but are more isolated. Isolated *trans* double bonds show a characteristic outstanding absorption in the infrared at 10.36 µm (Cocks & van Rede, 1966; Taylor, 1971). Since complete glycerides and FFAs would also exhibit some degree of absorption in this region, the methyl esters of the fatty acids from the sample of hardened oils are prepared and their absorbance compared with that shown by methyl elaidate (methyl *trans* 9 octadecenoate) of at least 99% purity, which is taken as a standard. The sample under testing ought to have less than 5% of conjugated double-bond unsaturation in it. Further, some di- and triunsaturated fatty acids may be present in various arrangements, all or partly *trans*; these will show separate peaks or shoulders on a peak at recognizable positions close to 10.36 µm (Taylor, 1971), and can be taken into separate account if they also are relevant to the object of the investigation. When it comes to relating the *trans* index (percentage of *trans* isomers compared to methyl elaidate) to texture, remember that the position in the chain of the *trans* double bond also has a marked effect on how the melting point, and hence texture, is affected. Therefore, the significance of the *trans* index must not be exaggerated as though it were the only criterion. For fats which were hydrogenated according to the "Low-Temperature Hydrogenation" and the "Iso- or *trans*-Suppressive Hydrogenation" sections in Chapter 2, a final *trans* index of less than 25% is indicative of considerable success. Comments on an in-line control application are given by Frankel (1981) and Metcalfe (1979).

Alkali Isomerization

Normal *cis, cis,* 9,12 linoleic acid and *cis, cis, cis,* 9,12,15 linolenic acid (where the double bonds are in the skipped or 1:4 situation) may be isomerized to a conjugated arrangement—e.g., 9,11 and 10,12,14 unsaturation (among others)—by warming in a potassium hydroxide–glycol solution. Conjugated dienes have a characteristic absorption in the ultraviolet at 232 nm, and conjugated trienes at 268 nm. The change makes possible the formation of an estimate of the amount of linoleic and linolenic acids originally present (Cocks & van Rede, 1966, pp. 150–156; *Official and Tentative Methods of the American Oil Chemists' Society*). When fats possessing a reactive methylene group are oxidized, the hydroperoxide formed entails a double-bond movement into a conjugated arrangement, and if the hydroperoxide breaks down to a hydroxide from which the elements of water are next removed, the conjugated system is then extended. Truly, a conjugated triene system of sorts exists when a carbonyl group is located next to conjugated diene (i.e., -CH=CH-CH=CH-CH=O). This "pseudo triene" interferes with the measurement at 268 nm. Evidently, therefore, ultraviolet spectrophotometry is not only useful in identifying unsaturated structures, but also it is able to give very useful information on how far an oil was damaged. If other information is available, possibly an assessment of some mishandling of the oil or of attempts to disguise the deterioration can occur. Some standards of quality based on E. 1.1.232 and E. 1.1.268 may be empirical, but are useful, and they were used in the legislation of some countries. So far as hydrogenation is concerned, alkali isomerization is mainly used to investigate structure and, hence, reaction mechanism.

Refractive Index (n_D^t) (Berl, 1960)

The refractive index, n, is a measure of the relative velocities of light in air, v—in some cases, vacuum—and the medium under testing, v^1, such as oil or fatty acid. When a beam of light strikes the surface of oil, the angle it makes with a line perpendicular to the surface is the angle of incidence, i; the beam is bent or refracted at the surface so that the angle it now makes with the perpendicular in the oil is smaller and is the angle of refraction, r (Eqs. 12.1,2).

$$v \sin r = v^1 \sin i \qquad \text{[Eq. 12.1]}$$

$$n = v/v^1 = \sin i/\sin r \qquad \text{[Eq. 12.2]}$$

For different i, the ratio sin i/sin r remains the same. It is influenced, however, by wavelength, temperature, density, and constitution. The difference in n for monochromatic light of sodium D line (5890 Å) and mercury G line (4358 Å) or any other two wavelengths is known as the dispersion. For oil tests, the sodium D line is used. In the literature on oils and fats, considerable variation occurs in the temperature at which the refractive index is reported between 20 and 65°C. In this book, 60°C was used, and where necessary, readings at other temperatures converted to it. British Standard 684:1958 (*Methods of Analysis of Oils and Fats,* 1958) quotes a variation of 0.00035 units in refractive index for oils per degree centigrade, a rise in temperature causing a drop in refractive index and vice versa (Cocks & van Rede, 1966, p. 99). This is the conversion used in this book to suit the convenience of the reader. The AOCS (*Official and Tentative Methods of the American Oil Chemists' Society*) (*Handbook of Soy Oil Processing and Utilization,* 1980, p. 36) quotes a factor of 0.000385. When dealing with hardened oils, obviously necessary is to employ a fairly high temperature, since melting points up to c. 52°C are commonly encountered (Cocks & van Rede, 1966). Above this melting point, the end of hydrogenation frequently has to be checked to confirm that it is below a certain maximal IV, and when this is achieved, melting points in the range of 58–68°C follow as a matter of course. In reading the refractometer, for practical purposes all that can be sought is at best a personal error of ±0.0001, and in work practice, ±0.0002 can easily be met. This means the personal error is about equivalent to 0.3°C; hence, sensibly one controls the temperature of the instrument to ±0.1°C as a means of restricting personal variation. Such thermostatic control is provided by instrument makers by the circulation of warm water, but to shield the refractometer from drafts in the control room where it is situated is still important. Not only should samples be filtered clear, they must also be dry, and the oil introduced between the prisms of the refractometer must be allowed to reach the temperature of the instrument, which takes about two minutes, and then a constant reading is obtainable. All of these are simple precautions; most unfortunately, their neglect may lead to a distrust on the part of some operators of the accuracy of the method. To overcome this, a reading should be taken by three or four persons, one of whom is a supervisor, and noted privately. This is repeated on three or four different samples. All results are then compared. An easily agreed figure should then emerge; possibly, one person in the group will read consistently high or low, but confidence in the method is achieved. As explained earlier (see the "Process Control" section in Chapter 8), now available refractometers exist which measure to the fifth decimal place (c. 0.1 IV) and display the digital result. This eliminates observer error. The above advice remains valid, however, for all those operating the older manual system.

Continuously-recording refractometers are now on the market. With continual frequent use, the prism surfaces may become coarsened and will have to be repolished by the maker. Many organic liquids show a variation of about 0.0004 units per degree centigrade, like oils, but water shows a corresponding variation of 0.0001 and carbon disulfide 0.0008 units (Berl, 1960). Various progressive changes in refractive index are noted for different changes in constitution, such as the prolongation of the fatty-acid chain length, and these were noted in detail (Swern, 1964, p. 128), but the one of immediate interest to the hardener is the steady fall in refractive index as the IV falls during hydrogenation. Although various equations were proposed to relate refractive index of lipids to their constitution, this may remain for an individual hardener to decide that in the context of his operation a fall of a set amount in refractive index best represents obtaining a hardened oil of a certain IV; as a first approximation, it may be taken that a fall of 10 IV is indicated by a fall in n of 0.00116. Of course, a fallacy is to suppose that different fatty materials of the same IV will have closely similar refractive indices. This is an appropriate point at which to say that the conjugation of existing double bonds causes an immediate increase in the refractive index, and that if polymers then form, the upward movement in refractive index is consolidated as far as they are concerned. To defeat such a tendency, the procedure explained in the "Cyclization and Polymerization" section of Chapter 2 is employed as may prove necessary.

Melting

Natural fats are not single compounds, and therefore different possibilities exist for packing the molecules together in crystals. Even the same triglyceride may exist in different crystal modifications. Hence, the melting behavior of a sample depends on the temperature at which it was held previous to the test (i.e., which crystal form was adopted). Therefore, one can easily understand why, prior to a melting-point test, a sample is often melted and then solidified according to a set procedure which may take minutes or hours. Of course, sometimes the case is that a melting-point report is required on a fat just as it stands. Further, the temperature of transition from one form to another and the rate at which this takes place vary from fat to fat, so preconditioning procedures are not necessarily the same. A more complete picture of a fat's softening and melting behavior is given by a series of measurements [solid-fat content (SFC)] over a temperature range. Explanatory notes are given below concerning different melting tests which have gained popularity in different parts of the world or different parts of the fats industry.

Slip Melting Point

About 1 cm of melted fat is drawn into a capillary tube open at both ends, and cooled for a set time to solidify the fat. The tube (or duplicates) is then immersed to a certain depth in water, well below the anticipated melting point, and the temperature of the water is raised at a steady rate with a gentle agitation of the bath. When the fat has softened or melted sufficiently for it to be forced by hydrostatic pressure to slide up the tube, this is noted as the slip melting point, sometimes referred to as the "rise point" (Cocks & van Rede, 1966, p. 88). Also, "slip point" in American practice may refer to the temperature at which solid fat, held in a little cup submerged in a brine bath which is heated steadily, melts and suddenly rises to the surface (*Official and Tentative Methods of the American Oil Chemists' Society*).

Incipient Fusion

Incipient fusion is the point during melting when the fat has softened so far as to form an obvious meniscus at its surface in the tube.

Complete Fusion, Complete Melting Point, Clear Point, and FAC Melting Point

When the fat has melted sufficiently to be free from obvious unmelted particles, these terms—complete fusion, complete melting point, clear point, and FAC melting point—are sometimes applied in American practice, and the capillary may be sealed at its lower end [AOCS Short Course, 1966 (ii) Erickson and (iii) Lantondress]; Boekenoogen, 1964, Vol. 1; *Official and Tentative Methods of the American Oil Chemists' Society*; Take care to avoid confusing the abbreviation for complete melting point (CMP) with °C MP—this can be expensive.

Wiley Melting Point

A tube of c. a 3.5-cm internal diameter is half-filled with boiled water and then cautiously topped-off with boiled alcohol so that the latter remains above the water with which it forms an interface. A cold disc of fat is then added to the tube so that it floats at the interface, after which the temperature is steadily raised. When the disc softens enough to become a spherical globule, this is the Wiley melting point.

Wiley melting points may be 1.5°C lower than the clear point or FAC melting point mentioned in the "Complete Fusion, Complete Melting Point, Clear Point, and FAC Melting Point" section above [AOCS Short Course, 1966, (ii) Erickson and (iii) Lantondress; Boekenoogen, 1964; *Official and Tentative Methods of the American Oil Chemists' Society*].

Cloud Point

The cloud point is the temperature at which a melted fat first becomes clouded when gradually cooled (Boekenoogen, 1964; *Methods of Analysis of Oils and Fats,* 1958, Sec. 1.5; *Official and Tentative Methods of the American Oil Chemists' Society*).

Congeal Point

A melted sample of fat in a beaker is stirred gently with a titer thermometer while being cooled until it becomes cloudy. It then remains undisturbed in air at 68°F (20°C) while the latent heat of crystallization elevates the temperature to a peak and then declines. The maximal temperature attained is the congeal point [AOCS Short Course, 1966, (ii) Erickson and (iii) Lantondress; *Official and Tentative Methods of the American Oil Chemists' Society*]. This is commonly used in American practice along with IV for the control of hardening.

Titer

Titer is very similar to the congeal point, but applies to fatty acids rather than fats (AOCS Short Course, 1966, (ii) Erickson; Cocks & van Rede, 1966; *Methods of Analysis of Oils and Fats,* 1958, Sec. 1.6; *Official and Tentative Methods of the American Oil Chemists' Society*).

Drop Point, Flow Point, and Pour Point

Drop point, flow point, and pour point are, in general for each, the lowest temperature at which the effect described is obtained within the conditions of the test. For drop point and flow point, see Boekenoogen, 1964; *Methods of Analysis of Oils and Fats,* 1958, Sec. 1.4; and *Official and Tentative Methods of the American Oil Chemists' Society*; for pour point, see *Official and Tentative Methods of the American Oil Chemists'* Society.

Dilatations and Solid-Fat Index

Over a substantial temperature range, fats are mixtures of solid crystals and liquid, the solid being the more dense.

By following the expansion–contraction of the fat with a change in temperature when it is entirely liquid, one may predict what volume a given amount of fat would occupy if it remained liquid at cooler temperatures. In fact, at cooler temperatures, some of the fat changes to solid which occupies less volume; thus, the contraction is enhanced. The more solid that forms, the greater the difference between the actual volume and what it would have been had the fat remained liquid. If this difference is expressed in mm^3 for 25 grams of fat at any particular temperature, this gives an empirical indication of the proportion of solid present and is the dilatation value for that temperature. A series of such readings gives a picture of the melting behavior of the fat (*Methods of Analysis of Oils and Fats,* 1958, Sec. 1.12). Obviously, when completely liquid, the dilatation is zero; when completely solid, the indicated value is nearing 2500. As a matter of convenience, customarily one works with about 5 grams of fat and multiplies up to a 25-gram basis. The solid-fat index (SFI) is a closely similar test, but numerically related to 1 gram of fat. This was popular in the United States and other countries (*Official and Tentative Methods of the American Oil Chemists' Society*). Both dilatation (D) and SFI were mostly replaced by percentage of SFC (SFC%), which is an absolute rather than an empirical type of measurement, and now determined by pulse nuclear magnetic resonance (pNMR). Conversion from dilatation to SFC% is done by the use of tables (van den Enden et al., 1978, 1982, 1986). Descriptions of dilatation and SFI are retained here for the information of those who may never have used them. Further comment on the empirical nature of these tests is given in the "Dilatations" section of the Glossary and in the literature (Boekenoogen, 1968, p. 119; Cocks & van Rede, 1966, p. 91).

Solid Fat Content (SFC) as Determined by Nuclear Magnetic Resonance (NMR)

When the protons in a fat molecule are in a magnetic field and a suitable radio-frequency field is applied, the protons held in a solid react differently from the less constrained protons of the liquid, and emit a different signal. If the signal when the fat is entirely liquid was measured, the difference between this and that from the solid–liquid mixture enables the percentage of solids to be estimated. This is the so-called "wide-line" or "broad-line" NMR technique (Cocks & van Rede, 1966). A later development employs the application of the radio-frequency field as a series of pulses; in this case, the overall response of the protons in the solid, being more rapid, is easily distinguished. The pNMR has gained ground as an analytical technique, as it is appreciably less cumbersome (van Putte et al., 1975). In the early days of NMR use, c. 1966 (AOCS Short Course, 1966: (ii) Erickson and (vi) Wiedermann; Van Den Enden et al., 1978; van Putte et al., 1975), the accuracy of the conversion of D or SFI values to SFC% and vice versa was acceptable for the 25–35°C temperature range, but less acceptable below 25°C. This meant NMR soon came to be preferred for higher (c. 60%) solid contents. So-called "direct" and "indirect" methods of wide-line NMR were capable of good agreement (Van Den Enden et al., 1978; Waddington), but the extra work in the direct method (Mills & van de Voort, 1981) gave no corresponding advantage. The pNMR is the one now in general use. When used in combination with other techniques, including radio carbon, NMR can be used to determine molecular structure, such as the distribution of fatty acids within a triglyceride (Metcalfe, 1979). Analytical literature gives more detailed comment on the workings of NMR (Haighton et al., 1971, 1972; Mills & van de Voort, 1981; Van Den Enden et al., 1978, 1986; van Putte et al., 1975; Waddington; Walker & Bosin, 1971).

Differential Thermal Analysis (DTA)

In forming crystals, triglycerides may take up several packing arrangements. The least stable, probably the one obtained on rapid cooling and itself the lowest melting, is the alpha (α) form. More stable is the beta prime (β′) form, and some nonsymmetrical triglycerides adopt this: for simple saturated triglycerides, the most stable is the β form. Fats tend toward the β′ form in the natural state. How many more forms exist and how transitory is their stability are still matters of conjecture. The change from one form to another is accompanied by the release or taking up of heat. To detect the temperatures at which these changes occur, a sample of melted fat under testing is poured into a small cylindrical cell and rapidly chilled so that the α form is produced. Into a similar cell, a reference material—which is known not to release or absorb heat by structural change over the temperature range in question—is packed. Both cells are placed in close-fitting cavities in a copper block, a thermometer is inserted in a third cavity, and thermocouples are placed in the sample cell and the reference cell; the whole is cooled some 20°C below what is anticipated to be the lowest transition point. The block is then programmed to be heated gradually. When the sample under testing undergoes a change in form which liberates or absorbs heat, this is recorded as a difference from the temperature of the reference material. DTA identifies the temperature at which changes occur, but other means such as infrared and X-ray analysis must be employed to identify the respective forms (Directive 76/621/EEC, 1976, Vol. 1—see also Anon., 1978; Taylor, 1971). Walker and Bosin (1971) compared differential scanning calorimetry (DSC) (Bentz & Breidenbach, 1969) with SFI and NMR in detail, for which these and related original papers (Bentz & Breidenbach, 1969; Mills & van de Voort, 1981; Van Den Enden et al., 1978; van Putte et al., 1975; Waddington; Walker & Bosin, 1971) must be consulted.

Oxidation and Stability

Before considering various tests—oxidation and stability—in this field, some brief explanation is desirable of their relevance to various needs. To forecast how well a fat will retain its flavor or to assess how severely it was mishandled may be advantageous. As far as flavor and odor are concerned to remember that the chemical assessment of the degree of oxidation is only a part of the question is essential; the equally important considerations remain of what is oxidizing and what type of oxidative reaction is proceeding (see the "Linseed Oil" section in Chapter 8). Beyond this, again, is the consideration that as far as sensory perception is concerned, the concentration of the flavor or odor may markedly alter its appreciation by the observer, even to the extent of identifying it as two different materials. Finally, what to some is very desirable is to others most disagreeable. Like the estimations of texture, many oxidative tests are empirical and were invented as relatively simple ways of forecasting future behavior with an acceptable degree of reliability. Oxidation also affects color by removing it or fixing it at an undesirable level; it may be promoted to achieve texture and stability, as in the case of drying oils for varnishes. Oxidation is promoted by light, warmth, moisture, prooxidant trace metals, and some microorganisms.

Mechanism

Oxygen may add to a reactive methylene group (-CH_2-) in a fatty-acid chain to give, in the first place, a hydroperoxide:

```
–CH –
 |
O – OH
```

This may lose oxygen giving a hydroxyl compound:

```
–CH –
 |
 OH
```

which in turn with a neighboring –CH_2- loses the elements of water to give another double bond in the chain, =CH-. The oxygen may add directly to a double bond giving a peroxide:

```
–CH – CH –
  |    |
  O————O
```

or epoxides may form:

```
–CH – CH –
     \ /
      O
```

and eventually, at the rupture of the chain, -CHO and =CO aldehydes and ketones, which are the most potent flavors and odors. In the end come the carboxyl groups -COOH, generally described as rancidity, but whose threshold value (minimal concentration of possible organoleptic detection) is far above the off-flavor type of aldehydes just mentioned.

Significance of Oxidation Tests

Hydroperoxides may have no flavor, and aldehydes a strong one. Nevertheless, a point useful to know is how far the oxidation of a particular parcel has progressed in comparison with other shipments of the same oil on which a well-documented record was kept. This is using empirical knowledge to the greatest advantage. Some initial forms of oxidation, such as peroxides and conjugated dienes, may diminish not because the oil has improved, but because the deterioration has progressed to aldehydes, fatty acids, or conjugated trienes. A pragmatic approach can yield highly-useful results. If, for example, a simple oxidative test check is maintained for a period of two or three weeks at three or four points between the crude hardened oil leaving the filtered oil tank and the deodorized oil 48 hours after deodorization, what may appear is that parcels which showed a PV above a certain level shortly after hydrogenation (e.g., PV 7 after 48-hour storage as filtered crude hardened oil) are frequently, if not invariably, the ones which exhibit poor flavor stability within a few days of posttreatment. This knowledge helps management to decide where and how far to provide protection from oxidation during the posttreatment sequence, and even to estimate how far the complete exclusion of air is likely to be cost-effective. In view of this complexity, to understand the proliferation of empirical tests is easy [AOCS Short Course, 1966: (iv) Sherwin].

Specifications of Present Oxidative State and Resistance to Further Oxidation

The specification of the processed oil may state that it shall not already be oxidized beyond a certain level [PV and AnV (anisidine value) tests], and that it shall have a certain minimal resistance to further oxidation (Swift, Quality, E232/E268 tests). Side-by-side with these tests must go the realization that a minute concentration of off-flavor precursor can cause an easily detected bad flavor following a very modest degree of further oxidation. If absent, the possibility that considerable oxidation may occur before the evil flavor due to rancidity is recognizable, since if due to fatty acid alone, this may reach 0.3% w/w before becoming objectionable. Deodorization reduces FFA% to one-tenth of this level initially.

Peroxide Value (PV)

Peroxide value (PV) states the milliequivalents of peroxide oxygen combined in a kilogram of oil and able, under testing, to liberate iodine from potassium iodide; the iodine is next estimated using a standard sodium–thiosulfate solution (*Methods of Analysis of Oils and Fats,* 1958; *Official and Tentative Methods of the American Oil Chemists' Society*). Some tests, such as the Lea value, refer to the millimols of peroxide oxygen present, and the result will therefore be numerically one-half that expressed as milliequivalent of O_2. The size of the result is an indication of how well-represented in the present oil the first stage of oxidation still is. Oxidation may not have gone very much further; in any case, the amount of off flavor able to develop depends on the fatty-acid groups in which the peroxide is present [AOCS Short Course, 1966: (iv) Sherwin].

Anisidine Value (AnV)

The *p*-anisidine, $CH_3O\text{-}C_6H_4\text{-}NH_2$-, combines with aldehydes to form substances (Schiff's bases) which possess a strong absorbance in a 350-nm band, but the peak absorbance varies somewhat from one aldehyde to another. As the oxidative deterioration of different oils gives rise to differing groups of aldehydes, the absorbances are only comparable within each type of oil, and are empirical in the sense that the numerical value has to be related to flavor stability in the light of previous experience with that kind of oil. The reagent is 1 mL of 0.25% of *p*-anisidine in glacial acetic acid, and this is made up to 100 mL with isooctane, 1 gram of fat being dissolved in the mixture and allowed to react for 10 minutes at room temperature before the absorbance is read in a 1-cm^3 cell. The AnV, by convention, is 100 times the observed absorbance at 350 nm (Holm & Ekbom-Olsson, 1972). This is an empirical measure of secondary oxidation. From this, an expression was derived known as the "oxidation value" (OV). Since a peroxide group has twice the oxygen of an aldehyde group, the convention states:

$$OV = 2\ PV + AnV \qquad [Eq.\ 12.3]$$

This OV is used as an indication of the extent to which an oil may, in effect, have suffered oxidative damage, but it does not profess to measure chemically all manner of oxidative deterioration. This being the case, unfortunately the OV has come to be described by some as the "totox value." Read this section in conjunction with the "Alkali Isomerization" section in this chapter (Evans et al., 1973; Going, 1968; Jackson, 1981).

Swift Test; Swift Life; Active Oxygen Method

This is probably the most widely popular method of estimating oxidative stability (Cocks & van Rede, 1966; *Methods of Analysis of Oils and Fats,* 1958, sec. 2.25; *Official and Tentative Methods of the American Oil Chemists' Society*); it also provides a means of discovering the extent to which an antioxidant added to a fat can exert its influence (Stuckey & Osborne, 1965). A stream of dry air is sparged steadily (usually 10 liters/hour) through each of several oil samples maintained at 98 ± 0.5°C, and the ascent of PV is measured with the passage of time. Many fats and oils show little increase at first, and then a comparatively rapid ascent. The time taken to reach a certain PV is the Swift Life; alternatively, the requirement may be put in the form that in a certain number of hours, a particular PV shall not be exceeded. The complete record of the ascent is the most informative [AOCS Short Course, 1966: (iv) Sherwin].

An automated version of the Swift Test was devised by Pardun and Kroll (1972).

Rancimat Test

The Rancimat Test, which may well replace the Swift Test as a method of estimating relative oxidative stability, involves passing pure dry air through a measured quantity of hot oil (a temperature of 120°C is commonly used) and then through a measured volume of deionized water into which is inserted a platinum electrode. A sharp change in the conductance of the water indicates that the oil is becoming oxidized, and the time taken to reach this "induction point" is noted and can be compared against either a standard oil or a required specification.

Quality Test

The Quality Test is run along the same lines as the Swift Test, except that the temperature is 130°C and the time for hardened oils is fixed at 4 hours (Cocks & van Rede, 1966). If a PV of 20 is not then exceeded, the oil is accepted as adequately stable. The actual PV obtained is sometimes referred to as the "Quality Test."

Oven or Schaal Test

In the Oven or Schaal Test, samples of fat are maintained at 63°C in air for days until a persistent rancid odor is noted when lifting the cover from the beaker in which the sample is held [AOCS Short Course, 1966: (iv) Sherwin].

Sylvester Test; Barcroft-Warburg Test; Eckey Method

A set volume of fat (e.g., 300 mL) is maintained at 100°C in a closed, round, one-liter flask connected to a pressure recorder and vibrated at 200 oscillations/minute for 24 hours. The progressive drop in pressure indicates the absorption of oxygen from the air in the flask, and hence, the rate of deterioration of the fat. A standard may be set whereby no more than a certain pressure drop is acceptable after so many hours [AOCS Short Course, 1966: (iv) Sherwin; *Official and Tentative Methods of the American Oil Chemists' Society*]. Eckey devised a simpler manometric version of this test (Hunter, 1951).

2-Thiobarbituric Acid (TBA) Test

This reagent reacts with fat-oxidation products and, supposedly, especially with aldehydes to give a reddish-orange color which may be employed as a qualitative test for oxidation, or quantitatively, via spectrophotometry (AOCS Short Course, 1966: (iv) Sherwin; Tarladgis et al., 1964).

Spectrometric Test

The Spectrometric Test refers to the visible spectrum. The light of a given wavelength in passing through some thickness of a particular medium is partly attenuated or absorbed. If the intensity of the beam before entering (I_0) is compared with the intensity after leaving (I) by a photoelectric cell, the relationship $\log_{10} I_0/I$ is known as the extinction, being written simply as E. Within reasonable limits, the absorbance of a fat dissolved in a solvent varies with the molecular concentration of the fat, the thickness of the light path through the test cell, and (slightly) with the solvent used, with the understanding that no chemical reaction occurs between the fat and the solvent. The molar extinction is:

$$(\varepsilon) = \text{E/cell thickness (cm)} \times \text{molar concentration} \qquad \text{[Eq. 12.4]}$$

Since the precise molecular weight is often not known, the convention is adopted of using a 1% w/w concentration in a cell of a 1-cm thickness. In this case, the values now obtained are known as the absorptivity, and this is written as E1% 1 cm nm for the given wavelength (in nm), or briefly as E1% 1 cm, taking the wavelength for granted and spoken of as the "E value." As light beams of changing wavelength are transmitted through the cell, the attenuation or absorptivity varies, and a peak value may be found which is characteristic of the type of fat being examined (e.g., for palm oil at 458 nm). If the fat were oxidized, the absorptivity at this wavelength is likely to change noticeably, and the peak value may also drift to another wavelength (e.g., for palm oil, from 458 nm to 450 nm). What readily will be appreciated is that this test is a very empirical procedure, and its usefulness lies in having a collection of E values relating to the same type of fat with corresponding data as to performance.

Smoke, Flash, and Fire Points

The smoke point is the temperature under the conditions of the test at which the fat or oil sample commences to emit a discernible steady stream of smoke; the flash point is the temperature at which the brief flash effect is first obtainable during heating; and the fire point is the temperature at which the sample can first be induced to burn for a minimum of 5 seconds when the flame specified in the test procedure is applied [AOCS Short Course, 1966: (ii) Erickson; *Methods of Analysis of Oils and Fats,* 1958, Sec. 1.8; *Official and Tentative Methods of the American Oil Chemists' Society*; Swern, 1979, p. 451]. As the proportion of FFA in the sample increases, the above temperatures decrease markedly; traces of flammable solvents naturally affect the results (Swern, 1979).

Miscellaneous Tests

Besides the mention of some older tests, this section includes short notes on some of the analytical techniques that have greatly accelerated and improved the investigation of the structures of fats and oils among a host of quite different materials since the 1950s; and this, in its turn, has made possible the more detailed tracing of the course of the hydrogenation reaction (AOCS Short Course, 1966: (i) Dutton; Coenen, 1978; Dutton, 1972; Gunstone, 1976).

Hydroxy Fatty Acids

Castor oil provides the best-known example of a natural oil which contains a hydroxy fatty acid in its composition—in this case, usually 84–94% of ricinoleic (12 hydroxy-*cis* 9-octadecenoic) acid. A hydroxyl value is the number of milligrams of a potassium-hydroxide equivalent to the acetic acid needed to acetylate 1 gram of oil. This is a little different from an acetyl value, which is the number of milligrams of a potassium-hydroxide equivalent to the acetic acid liberated when 1 gram of acetylated oil is saponified. In the case of castor oil of good quality, the hydroxyl value is c. 160; the specification of fully-hardened castor oil may call for a minimal hydroxyl value of 150.

Reichert-Meissl, Polenske, and Kirschner Values

Respectively, these are the milliliters of 0.1 N of a sodium–hydroxide equivalent to:

- Steam-volatile water-soluble fatty acids (C6 and shorter);
- Steam-volatile water-insoluble fatty acids (mainly C8 and C10);
- Steam-volatile water-soluble fatty acids whose silver salts are soluble in water (mainly C4) obtained from 5 grams of fat. This well-known trio of classic analyses was overtaken by physical techniques. Their usefulness is indicated (Cocks & van Rede, 1966, p. 362) by Table 12.2.

Fuel OD Contamination

Hydrogenation and marine oils were associated since the very earliest days of the oil-hardening industry. The hazard of contamination of triglyceride oil by mineral oil in ships' tanks was recognized for a long time. As little as 0.01% of heavy fuel oil is reputed to be the minimal amount of contamination needed to lower the category of good whale oil by darkening its color, and this was considered to carry through to the crude hardened oils. Although the general situation has greatly altered, experience gained in assessing this class of contamination is still of interest. In ultraviolet radiation, heavy fuel oil fluoresces blue–purple and triglyceride green–yellow to light blue. If the unsaponifiable matter is prepared (Cocks & van Rede, 1966, p. 78) and acetylated, fuel oil remains as dark-brown streaks of liquid which, when taken up in a little acetic anhydride, show again the typical blue ultraviolet fluorescence. Although this procedure may be made semi-quantitative, thin-layer chromatography has now been brought into service (Cocks & van Rede, 1966, p. 78) and is generally more satisfactory.

Chromatography

Since Martin and James first developed the chromatographic separation of fatty-acid material in 1952, the technique has grown immensely. The reason it is described here in the miscellaneous section is that its application is now so widespread that it would be misleading to place it under other headings, such as Unsaturation, Stability, etcetera, along with many of the older classic analytical techniques, some of which it is supplanting in one or more of its various forms. Important variations exist within these separate procedures, and no doubt several more will be added

TABLE 12.2
Typical Test Values

	Reichert-Meissl	Polenske	Kirschner
Butter fat	20–30	2.3–3.3	19–27
Coconut oil	8	15–20	1.9
Palm kernel oil	5	10–12	1.0
Other (nonlauric) oils	0.5	0.5	0.1–0.2

within a few years. The notes give no more than an appreciation of the basic features of each technique.

Gas Chromatography (GC) or Gas–Liquid Chromatography (GLC)

Gas chromatography (GC) or gas–liquid chromatography (GLC) has identified the fatty acids by chain length and degree of unsaturation; it was further developed to achieve the separation of positional and geometric isomers. To have an adequate set of standards against which to make comparisons of investigational test runs is important. When a volatile material (solute) is carried across the surface of a fixed solvent (stationary phase) by an inert gas (mobile phase or carrier), its rate of progress is governed by its solubility: that is, the progress is inversely proportional to the partition coefficient between the fixed solvent and gas. Many solvents were especially useful for particular tasks [polyethylene glycol adipate (PEGA) and 1,4-butane diol succinate (BDS) are but two examples]; the solvent is held as a thin-layer spread on the surface of an inert finely divided solid, often a siliceous support meeting set standards of purity and physical texture. The prepared solid is packed into a narrow glass tube several meters long which can be maintained at a constant temperature, which itself can be altered automatically according to a pre-set program if desired This is the GLC column. The heated carrier gas flows along the tube; then, at the point of entry, a very small sample of solute is injected onto the packing. The solute spreads down the column at a rate inversely proportional to its partition coefficient. If more than one substance is present in the solute, they are likely to have different partition coefficients, and hence, move at different rates. They separate and emerge in succession at different intervals to be detected by sensitive means, such as their combustion in a flame, and recorded in amplitude and duration so that, when corrected, the area under the peak indicates the proportion of that particular component in the solute if the sum of all peak areas is made equal to 100%. The record is known as a chromatogram.

In the case of glycerides or fatty acids, particularly convenient is to prepare the methyl esters of the fatty acids, which, being volatile and not unduly reactive, are ideal subjects for this kind of separation. Not only does separation occur according to chain length, but also according to unsaturation; so the sequence of separation for one chain length is saturates, monoenes, dienes, trienes, etcetera. A standard chromatogram showing the times of emergence of the different members of a homologous series—such as the methyl esters of lauric, myristic, palmitic, and stearic acids—is prepared under the same test conditions. From this, the retention times can easily be read and their $\log_{10}$ values plotted against their respective chain lengths or so-called carbon numbers. This plot or line then relates to a succession of saturated fatty acids. A succession of monounsaturated fatty acids would give a similar line, but at higher retention times, di- and triunsaturated fatty acids have corresponding lines at still higher values, granted that the column is packed with appropriate polar solvent on the support. When a mixture of methyl esters of unknown fatty acids is passed through the column and its chromatogram prepared, then one can possibly discover which of its peaks correspond with those of the standard regarding chain length and the degree of unsaturation.

A popular practice has been to take the retention time of a well-known and well-suited material, such as methyl myristate or even stearic acid itself, as unity, and to express the retention times of other comparable materials on this basis, always maintaining the same working conditions of solvent, gas flow, temperature, and, if possible, even the same apparatus, so as not to introduce extraneous variables into so sensitive a situation. A set of precise times for a complete range of various fatty materials may then be obtained, and hence, if an unknown within this range is discovered to have a corresponding retention time with one of these, it is identified with it. For example, methyl esters of C6,9,12 linolenic acids and its isomer C9,12,15 have relative retention times of 1.56 and 1.74 on a basis of stearic acid being 1.0. As is apparent from what already was said in the "*Trans* Index" section in this chapter, changes in constitution regarding number, position, and the configuration of double bonds alter physical behavior so that differing retention times are to be expected. An increase in chain length and an increase in saturation increase retention times; overlapping between greater chain length and lesser chain length of greater unsaturation is therefore foreseeable. Being aware of the hazard, caution is used and any interpretations of such a chromatogram are confirmed by other techniques, some of which fortunately have come forward during the same period. To assist in the process of ultimate precise identification, newly investigated components may be allotted an "equivalent carbon number" such as 16.20, 17.80, etcetera, meaning that retention times are those of a methyl ester of a saturated unbranched fatty acid of that hypothetical number of carbon atoms obtained in the same working conditions. That the numbers are fractional merely demonstrates their empirical nature until further precision is obtained. If highly polar-liquid stationary phases, such as cyano propyl silicone (Silar C5, C10, 9CP, etc.) are used, packed on chromosorb, *cis* and *trans* isomers of fatty acids and amines can be separated (Metcalfe, 1979). Apparently, up to now, we concerned ourselves with fatty acids in the triglyceride via their methyl esters, and therefore, the distribution of the fatty acids over various groupings of triglycerides and the arrangement of possibly different fatty acids within a group (mixed triglyceride) have still to be determined. Nonetheless, the knowledge of the constitution, and hence, the understanding of a reaction like hydrogenation, is substantially advanced.

One means of detection used with GLC is burning the effluent in a very small flame, thus increasing ionization in the flame, and therefore the potential on an anode so situated as to respond. The effluent need not be burned; it can be collected as different fractions. These, in turn, can be subjected to other analyses. This is the technique of preparative GC or GLC (Cocks & van Rede, 1966; Dutton, 1972; Frankel, 1981; Metcalfe, 1979; Perkins et al., 1977; Taylor, 1971).

Liquid Chromatography (LC) or Liquid–Liquid Chromatography or Partition Chromatography

Here the bonded stationary phase is a liquid, such as octadecylsilane, and the mobile phase, such as acetonitrile, is also liquid. Now triglyceride components of triglyceride mixtures can be separated according to polarity, and hence, carbon number and unsaturation. Because it is faster and simpler, liquid chromatography (LC) has gained in popularity on GC (otherwise known as GLC) (Frankel, 1981).

Liquid–Solid or Adsorption Chromatography
In this, the oldest chromatographic separation, much improvement was made by a tighter specification of the solid stationary phase on which the solute is adsorbed from the liquid mobile phase. Thus, various tocopherols were separated directly from their parent oil on a Porasil column with 1–5% of isopropyl alcohol in hexane as a solvent, and a spectrophotometer operating at 295 nm as a detector. Again, this technique can be many times quicker than GC or classical methods (Frankel, 1981).

Gel Permeation Chromatography (GPC)
Here, the column contains a gel into which large or polymerized molecules penetrate with difficulty, and therefore in the limit the largest molecules are washed through the void spaces between gel globules without being adsorbed—the so-called exclusion limit. This technique is used to assess the degree of polymerization resulting from oxidation or heating (Frankel, 1981).

Glass Capillary Chromatography
In the 1970s, coating the inside of fine glass capillaries with very polar stationary-phase cyano propyl silicone for up to 50 m became practical. Positional, geometric, and branched-chain isomers are separable by this means (Perkins et al., 1977). Short capillary columns have achieved rapid separations suitable for quality control (D'Alonzo et al., 1981).

High-Performance Liquid Chromatography
Improvements in columns, derivatives of compounds to be separated, and means of detection have extended the usefulness of LC (Gunstone, 1976; Metcalfe, 1979; Perkins et al., 1977). One example quoted (Jungawala et al., 1976) is the separation within 15 minutes of lecithin from sphingomyelin on a 10-µm particulate silica gel using a mixed solvent of acetonitrile–methanol–water (65:21:14).

Thin-Layer Chromatography (TLC)
Popular because sensitive, simple, and cheap, this form of chromatography basically consists of a glass plate coated with a thin absorbent layer of silica gel or similar material, near one end of which are spotted some micrograms of sample. This end of the plate is then stood in a little solvent, which at once makes its way up the plate by capillary action. Components in the sample spot travel with the solvent, but at different rates according to their characteristics (solubility, polarity), so that when the solvent reaches the top of the plate, they have attained different levels. The plate is then removed from the solvent and "developed" by charring or treatment with a reagent to make the different spots stand out. As with other forms of chromatography, the results need to be interpreted by comparison with a standard (Cocks & van Rede, 1966; Metcalfe, 1979).

Mass Spectrometry (MS)

The basis of mass spectrometry (MS) was developed about the time of World War I, and was improved greatly during World War II. Molecules, when bombarded with electrons, have one or even two of their electrons stripped from them, leaving behind the main bulk with a positive charge. The molecule is thus ionized, and if these ions are accelerated in an electric field, they will be deflected upon entering a magnetic field, so as to take up a circular path. As far as an ion is concerned, the character (radius) of the path depends on the ratio of the ion's mass to its charge, and as far as the fields are concerned, the deciding factors become the magnetic field strength and the electrical potential. Since the ion's charge and the field strengths are known, the mass number of the ion can be found. Where molecules of an element are concerned, this leads to an accurate knowledge of the element's mass number and its isotopes. Of more immediate interest to the organic chemist is the fact that the ionized molecules usually fragment into neutral and positively charged pieces, which then follow paths which together give a pattern typical of the parent molecule. Thus, the molecular weight and the molecular makeup may be deduced sometimes so completely that no other kind of analysis may be needed to decide on the structural formula. Important is that the breakup of the molecule arises from electronic bombardment and not thermal disruption. Techniques for ensuring this with less volatile substances have also advanced (Cocks & van Rede, 1966; Metcalfe, 1979; Taylor, 1971).

Gas Chromatography–Mass Spectrometry (GC–MS)

Mass spectrometry (MS) is most easily interpreted when applied to single substances, some of which may have molecular weights well over 2000. If components of a mixture arising from gas chromatography (GC) are fed individually into a mass spectrograph, the data then rapidly acquired can help decide the structure of the components and take us a long way to deciding on the structure of the original substance which provided the mixture for the gas chromatogram. Starting from the original substance, such as a triglyceride, the preparation of the methyl esters of the fatty acids could be accomplished in very few hours; the gas chromatograph, in minutes; the mass spectrograph of a peak, in a second; and then the final data fed to a computer for assessment, and if possible comparison, with a data bank for identification by comparison. Older methods of analysis would require several man-years (Cocks & van Rede, 1966; Metcalfe, 1979; Taylor, 1971).

Laser Pyrolysis

Laser radiation is used to induce the pyrolysis of organic molecules so that large, nonvolatile molecules can be fragmented in a controlled manner and the fragments next identified by GC and MS. The low molecular-weight fragments give information concerning functional groups, and the larger fragments information on the main carbon skeleton (Perkins et al., 1977).

Atomic Absorption (AA) Spectrophotometry

The estimation of trace metals in fats and oils is of general interest, and for fat hardeners, the estimation of traces of catalyst which may have escaped post-treatment to remove the catalyst is especially important. If a solution of the oil in question is very finely dispersed and introduced into a flame, the solvent and oil are combusted and the trace metal vaporized so that the characteristic absorption lines of the metals concerned are then measured by a spectrophotometer. In the latest apparatus, efficiency and, hence, sensitivity are much improved by injecting the sample into the hollow interior length of a carbon rod kept under observation by the spectrophotometer. When this tiny carbon furnace is progressively heated, first the solvent evaporates, next the oil is destroyed, and then, for an interval of 3–4 seconds, a much better concentration of virtually stationary atomized metal vapor lies in the path of light shining through the rod. The limit of detection for nickel by flame atomic absorption (AA) is quoted as 0.008 ppm, but for flameless AA, 0.015 ppb (Frankel, 1981).

Color Values

Numerous methods were in use for many years, all based on the comparison by the human observer with some standard range. Attempts at numerical conversion from one to another are notoriously misleading and will not be made here. Table 12.3 shows the results on four different neutralized and bleached (nb) oils as determined on three old, popular, and yet different scales.

Numerous attempts were made to eliminate personal error in judging color by using spectrophotometric methods, which yet gave results which would not clash unduly with the traditional results of human observation. Progress is being made. Photometric measurements of absorbance [see the "2-Thiobarbituric Acid (TBA) Test" section in this chapter] at four different wavelengths across the visible spectrum, each then being multiplied by a separate factor, are made and summed. The results are in harmony on most, if not all, occasions with the findings of the Lovibond red scale (*Bailey's Industrial Oil and Fat Products,* 1982). Other systems are making their bid for popular acceptance (Frankel, 1981), so that, in the end, likely an impersonal assessment will become more general. In the Automatic Lovibond Tintometer AF960, three beams of colored light are first passed through the oil and then observed by a photoelectric cell. One of the beams acts as a reference to

TABLE 12.3
Typical Color Readings for Four Oils

nb Oil	Lovibond Y	(51/4") R	Iodine Scale	Bichromate Scale
Palm oil	30	14	47	20
Hardened fish oil	25	2.5	5.5	4.5
Sunflower oil	10	1.0	2.5	2.0
Coconut oil	10	1.0	2.4	1.8

compensate for changes in light intensity, presence of dirt, etcetera, while the other two beams evaluate yellow and red. Levels of carotene and chlorophylls can also be assessed in this instrument. Rossell (*Analysis of Oils and Fats,* 1986) comments that if the reference beam over- or undercompensates for blue or green, this could affect the accuracy of yellow and red readings, although this feature is likely to be significant in the case of some crude oils rather than in the generalities of processed oils. The Color in Oils Committee of the AOCS (Berner, 1991) found that the AF 960 tended to give rather higher readings than the manual tintometer, and that this should be investigated and corrected before AF 960 was granted full approval in AOCS methodology. Color evaluations for the full range of fats and oils are also discussed by Patterson (1993).

Glossary of Hydrogenation and Related Technical Terms

A glossary of technical terms related to hydrogenation is intended to provide immediate access to key information for the reader of this and other books on the subject. In some instances, the opportunity is taken to expand on the treatment given in earlier chapters; an undue interruption of the main text for many readers already familiar with these terms is also thus avoided. Conversely, reference is made to previous chapters as appropriate, especially when these give literature references leading to greater detail.

Activity

Catalyst

The tests in general commercial use are quite empirical. They may refer to the ability of a set weight of nickel as a catalyst dose to achieve a set drop in iodine value (IV) on a triglyceride oil not worse than a set quality standard (defined or assumed) in a set time at set temperature and pressure, with hydrogen not less than a minimal standard of purity dispersed at a set flow rate in standard test equipment. Advisably, carry out an activity test on a standard catalyst as the first test of each day's work in a laboratory section devoted to catalyst testing to ensure continuity of standard from one day to the next. Sesame oil, being notably free from catalyst poisons, was and still is a popular medium for activity assessment. A set dose was used to hydrogenate the oil (usually at 180°C and atmospheric pressure) for 30 minutes. In these conditions, a satisfactory standard was represented by a drop in the refractive index of 100 units (0.0100), known among users as the "sesame drop" or "sesame activity." As catalysts became more active, the dose used as the basis of comparison was lowered (e.g., 0.07% of nickel/oil), but the drop remains in the 80–120-unit range. The next obvious step is to base the comparison not on an oil but against the performance of another catalyst (Allen, 1978) which can be supplied in the reduced state (but protected probably in hydrogenated vegetable oil), or as a siliceous nickel base ready to be reduced in the laboratory at whatever interval is prudent to check on the day-to-day standard currently in use. Such a standard serves for years. Over a limited range in a refractive- index drop of perhaps 60–80 units, a direct linear relationship exists between the amount of nickel used and the drop obtained in the standard conditions. Hence, within these limits, the amount of nickel under testing can be converted by simple proportion to its equivalent weight of standard. To compare activities, one must compare weights of nickel, which will give the same hardening result. This could entail a second test if the drops from the first two differ widely.

Activity of Oil, Hardening Capacity, or Quality of Oil

The test procedure remains as above, but the oil changes and the catalyst dose remains the same. Apart from checking a new shipment of oil, this procedure is very useful for evaluating whether a more rigorous pretreatment is worthwhile in terms of quicker hardening, lower solid-fat index (SFI) or solid-fat content (SFC), better product color, reduced catalyst requirement, etcetera.

Activity of Hydrogen, Gas Quality

If catalyst and oil (as well as other conditions) used in the tests are made the same, evaluating the performance of a new supply of hydrogen becomes quite simple. Generally speaking, this would be made available as samples of gas compressed in small cylinders for laboratory test purposes.

Activity Apparatus Standardization

As already indicated, each day's program of catalyst testing should be preceded by a test with a catalyst, oil, and hydrogen of known quality. When no interference by other factors occurred with the test, the routine standard drop in 30 minutes should not vary by more than 0.0003 refractive-index units.

Activity at Low Temperature

Some catalysts do not have their performance diminished by operating at 100–120°C as much as others; therefore, when a program of such work exists, one is interested in discovering if catalysts recommended for it give the required results, and to what extent they may be reused. In this case, the test procedure is repeated, but at the appropriate temperature. Probably, apart from taking the drop in refractive index into account, some other feature, such as SFI or SFC% at 20°C, needs to be checked to ensure it is not too high.

Purchase

Because of the highly empirical nature of catalyst testing, one must evaluate the performance of a catalyst in tests which correspond reasonably to the tasks which will encounter in the production scale. No reason can explain why a user should not establish his own standard and empirical tests, taking into account what is explained here, in the "Catalyst" and in the (end of) the "Estimation of Selectivity" sections in this chapter and Chapter 1, respectively, and in Chapter 7 under "*Raney Nickel and Other Nickel Catalysts*" and "*Copper Catalysts*," and "*Selectively*" sections. Catalyst suppliers are likely to be cooperative in this regard.

Chromatography

Basically, chromatography secures the separation of components in a mixture when they partition themselves differently between two media, one of which is caused to flow over the other held stationary. Variants of the procedure can involve gas, liquid, and solid media. In its modern application, gas–liquid chromatography achieved a historic success in the separation of fatty acids, leading to a highly successful technique for their identification. This has continued to be refined so that structural, positional, and geometric isomers are being separated and identified with increasing success. This, in its turn, makes possible the tracing of the course of hydrogenation of a polyunsaturated fatty acid group with five and six double bonds to an extent never before possible, even if not all of the variations are clearly quantifiable.

Dilatations

In the past, dilatation values were useful, although empirical. They are now replaced by the absolute value of SFC% as determined by pulsed nuclear magnetic resonance (pNMR). As explained in the Preface, when SFC% values were available, they were used in the text. When the conversion of dilatation values to SFC% was necessary, use was made of the tables published for this purpose (van den Enden et al., 1978, 1982, 1986). To allow an explanation of the dilatation system to remain in this edition seemed useful for the information of readers who may never have had experience with it and are interested in its limitations. Most oils and fats expand upon heating and contract upon cooling. Exceptions exist, like fully hardened sunflower oil, but not many. An oil cooled below its melting point is a supercooled liquid, and this allows us to produce a complete range of volumes which a set weight of oil would occupy were it to remain liquid. In practice, if the oil is not supercooled, its various component triglycerides separate as crystalline solids as the temperature falls until the oil has completely solidified. "Solidified" in the popular sense means it has ceased to flow, yet this merely means the solid portion dominates behavior but not composition; much liquid may remain, until, at a far lower temperature, all crystallizes. Solid being more dense than liquid oil, the volume at any point is less than it would have been had the oil remained liquid; the more solid, the greater the difference in volume. By one convention, this difference in volume is expressed in cubic millimeters and is related to 25 grams of fat, thus giving an index or arbitrary measure of the solid present at some particular temperature, say, 20°C. This is called the dilatation value and is written as D20. A similar procedure relates the volume difference to 1 gram of fat and is called the SFC for some temperature—say, 21.1°C—and is written as SFI 21.1. More important is the fact that both methods are empirical; solids also contract and expand with a change in temperature, and they may also change their crystalline form in passing from one polymorph to another, both of which have characteristic densities. Therefore, to adhere to whatever temperature routine is laid down for the test to render results comparable with one another is very necessary. The numerical values of D or SFI do not give with complete accuracy the percentage by weight of solids present in the fat. Nevertheless, by accepting the empirical nature of the test, one can put it to good practical use without having to make corrections of various kinds. A further point is that some dilatation ranges may be expressed at intervals of 5°C and others at 10°C intervals. Both are valid procedures, but as one represents a rather different tempering regime from the other, the numerical results at corresponding temperatures are likely to be a little different, although acceptable within their context. A fat usually flows or melts when the dilatation value is ca 100 or the SFI is c. 4. One is mistaken to suppose that SFI or dilatation values are exactly numerically additive. When a formulation covers a number of components which need to vary in their proportion while still giving the same texture for the lowest current cost, one must make a framework of experimentally determined results within which to operate. This enables the effective contribution of the different components to be assessed. Also, hydrogenation is now such a flexible technique that a domestic margarine or shortening texture can very often be produced, either by the partial hydrogenation of one oil alone or a blend of two components, themselves derived from that oil.

Not only can one calculate what volume a weight of oil would occupy if cooled to well below its melting point, but one may also calculate what volume the solid fat would occupy if it remained solid at above its melting point. The difference between these two results for a given weight of fat at any particular temperature is the so-called melting dilatation. The percentage of solid phase present for that temperature is represented by the proportion of the actual dilatation to the melting dilatation (Boekenoogen, 1964; Stansby, 1967).

Fatty Acids

Nomenclature

A hydrocarbon chain terminating in a carboxyl group (-COOH) is classified as a fatty acid [commencing with butyric acid $CH_3\ (CH_2)_2COOH$], since these, in combination with glycerol, make up the natural fats or triglycerides.

Most fatty acids consist of a straight chain, but minor variants were identified containing a short branch, hydroxyl, keto, or terminal ring group. In the systematic name, the total number of carbon atoms in the straight chain is shown by a Greek numerical prefix:

- octanoic $CH_3\ (CH_2)_6COOH$
- decanoic $CH_3\ (CH_2)_8COOH$
- octadecanoic $CH_3\ (CH_2)_{16}COOH$
- eicosanoic $CH_3\ (CH_2)_{18}COOH$
- docosanoic $CH_3\ (CH_2)_{20}COOH$

Chains of an even number of carbon atoms greatly predominate in the natural fatty acids, and the position of a carbon atom is most often shown by numbering from the carboxyl as 1. Sometimes a useful purpose is served in numbering from the opposite end, the terminal CH_3—then being 1. This is done when discussing the families of biologically important fatty acids (some being known as essential fatty acids), how they are built up in organisms, and how they may continue to be elaborated biochemically. Obviously, the context should make clear if any system other than the usual numbering from the carboxyl is being used.

Saturation and Unsaturation

Fully saturated fatty acids contain the suffix "anoic" in the systematic name, such as "pentanoic" derived from "pentane."

Unsaturated fatty acids contain the suffix "enoic" to indicate the presence of one or more double bonds in the straight chain; how many and their positional number in the chain often are indicated at the same time if those are known: hence, "diethenoid," "triethenoid," etcetera:

$CH_3\ (CH_2)_4CH{=}CH{-}CH_2{-}CH{=}CH{-}(CH_2)_7{-}COOH$ 9, 12 octadecadienoic (linoleic acid).

Where two (and often more) double bonds are present, the fatty acid is "polyunsaturated."

Some fatty acids containing triple bonds occur in nature; the suffix then changes to "ynoic," indicating an "ethynoic" or "acetylenic" structure for example:

$CH_3{-}(CH_2)_{10}{-}C^{\circ}C{-}(CH_2)_4{-}COOH$ 6 octadecynoic (tariric acid).

Where double and triple bonds occur in the same chain, the nomenclature is dealt with as in this example:

$CH_2{=}CH{-}(CH_2)_4{-}C^{\circ}C{-}C^{\circ}C{-}(CH_2)_7{-}COOH$ octadec 17-en-9, 11-diynoic (isanic acid).

A much abbreviated form of nomenclature for polyunsaturated fatty acids (PUFAs) has become popular in discussing structure and reaction mechanisms during hydrogenation, biosynthesis, etcetera. Thus, the number immediately following C indicates the length of chain; the next number separated by a colon the number of unsaturated links in the chain; and then, in brackets, the position of the carbon atom associated with the link, and a letter to show whether ethylenic (ethenoid, olefinic, or double bond) e, acetylenic (triple bond) a, *cis* double bond c, or *trans* double bond t, for example:

$CH_3{-}(CH_2)_4CH{=}CH{-}CH_2{-}CH{=}CH{-}(CH_2)_7{-}COOH$ C18:2(9*c* 12*c*) (linoleic acid).

Generally expected in discussing natural fatty acids is that double bonds are separated or interrupted by single reactive methylene—CH_2—groups as is clear in this example of linoleic acid. In some texts, the Greek letter delta (Δ) is used to indicate a double bond at carbon atom numbers which follow.

PUFAs appear to arise in "families" characterized by the distance in number of carbon atoms, of the first double bond from the terminal methyl group, which is then called the "end carbon chain" and indicated by the Greek prefix omega (ω). Thus, C18:3 ω3 signifies 18 carbon atoms and three double bonds, the first being located three atoms from the terminal number 18 carbon atom in the CH_3—group, i.e., on carbon number 15:

$$CH_3CH_2\text{-}CH{=}CH\text{-}CH_2\text{-}CH{=}CH\text{-}CH_2\text{-}CH{=}CH\text{-}{<}CH_2)7\text{-}COOH\ 9,\ 12,\ 15$$

octadecatrienoic (linolenic acid).

If ω3 indicates the linolenic family, ω6 indicates the linoleic family, and ω9, the oleic family.

When indicating the position of any double bond, not necessarily the first, or indeed, any feature relating to a particular carbon atom, the number of that carbon atom counting from the terminal CH_3—as 1, may be indicated as n-3, n-8, etcetera. Thus, erucic acid, which has a *cis* double bond at number 13 carbon atom (δ13, i.e., counting from carboxyl) in a 22-carbon atom chain (C22), can also be described as an ω9 or n-9 fatty acid. This brings us to the important group known as essential fatty acids.

Essential Fatty Acids

Where fat rich in saturates provides above about 30% of the total calories in food intake, the cholesterol content of the human bloodstream shows a substantial increase. High-cholesterol levels are associated with heart disease. By increasing the proportion of PUFAs in the diet, the blood-cholesterol level is lowered. Linoleic acid is the most plentiful polyunsaturate group in oils and is readily available. A minimal ratio of PUFAs to saturated fatty acids (SFAs) of 1.2:1 for dietary fat products was recommended to help offset the SFA in the diet as a whole, and some health margarines aim at 1.5:1 or more. Now one can understand why hardeners are concerned to leave as much linoleate untouched while more unsaturated groups are being reduced to improve oxidative stability. This is not all. In 1930, Burr and Burr (Taylor, 1968) demonstrated that a deficiency in particular unsaturated fatty-acid groups in the diet led to a variety of disorders, particularly in the very young. These groups were named "essential fatty acids" (EFAs) (Taylor, 1968). While the human metabolism is capable of converting some EFAs to others, it is not capable of producing them from non-EFA material. To be more specific, linoleic acid can be converted to arachidonic acid, and thence, to prostaglandins, the latter being very much involved with the workings of nerve fibers, blood pressure, male fertility, uterine contraction, etcetera. The body cannot make its own linoleic acid. By all estimates, the amount of EFA needed to overcome deficiency is only about one-quarter of the amount of PUFA needed to counteract excess SFA. EFAs are all *cis*-type unsaturated fatty acids with the double bonds separated by active methylene groups. As mentioned above, on many occasions, what is useful in the particular case of the EFAs is to describe their structure by numbering the carbon atoms from the terminal CH_3-group as 1 rather than from the carboxyl end, which is the usual system. Resemblances are then made evident (see Table G.1).

Noticeably, all important EFAs have an ECC (equivalent carbon chain)of 6. Other PUFAs, including those of some marine oils, show a little EFA potency. Evidently, hydrogenation could diminish EFA activity not merely by removing

TABLE G.1
Essential Fatty Acids

Chain Length	Double Bonds (*cis*) at n-	ECC (ω)	Fatty Acid	
C18	6,9	6	linoleic	important EFAs
C18	6,9,12	6	gamma linolenic	important EFAs
C20	6,9,12,15	6	arachidonic	important EFAs
C18	3,6,9	3	linolenic	less important EFAs
C20	6,9	6	eicosanoic (6c9c)	less important EFAs

double bonds, but by causing some to migrate or change to the *trans* form. Approximate percentages of PUFA and EFA for various vegetable oils are as follows (PUFA is listed first and EFA is in parentheses): groundnut 21 (21), cottonseed 52 (51), soybean 61 (53), sesame 43 (42), sunflower 63 (62), and safflower 77 (76).

Fatty-Acid Isomerism

Structural Isomerism

The most common case of structural isomerism is the branched chain (e.g., pentanoic acid and 4- methyl butanoic acid).

Positional Isomerism

This may reasonably be regarded as no more than a type of structural isomerism, and in the fatty-acid context, it refers especially to different possible positions for one or more double bonds. But it applies equally to different locations of a hydroxyl or other group along the chain.

Geometric Isomerism

These are the alternative shapes, *cis* and *trans*, of double bond (see the "Fatty-Acid Chain Length and Unsaturation" sections in Chapters 1 and Glossary, respectively). Alternative positions about a triple bond are not possible.

Optical Isomerism

This exists in both fatty acids and triglycerides when the minimal condition of an asymmetric carbon atom present is satisfied, but the effect is generally feeble. Ricinoleic acid (see the "Castor Oil" section in Chapter 8) is an example (Anderson et al., 1970; Hilditch & Williams, 1964).

Cyclization and Polymerization

Highly unsaturated fatty acids and their methyl, glyceryl, and pentaerythrityl esters can be caused to polymerize in a variety of ways, the promotion of which is an important study in the paints and varnishes industry (Swern, 1979). As far as fatty-acid dimers and trimers are concerned, it was already noted (see the "Fatty Acids" section in Chapter 8) that their natural viscosity requires operating temperatures up to 240°C during hydrogenation, presumably to assist the dispersion of the hydrogen. In the case of edible products, every effort is made to avoid any kind of polymerization of the triglycerides and the attendant risk of cyclization, since the control of texture is lost and the products of such reactions are biologically undesirable. The cyclization is likely to produce an aromatic ring in the fatty-acid chain (AFA). The "Cyclization and Polymerization" section in Chapter 2 describes the precautions to be adopted and the reasons for them.

Hardening

Reaction Rate

Usually, this is simply expressed as the drop in IV per minute or per hour. It is sometimes noted as the rate of hydrogen uptake, in which case, temperature and pressure correction must not be forgotten; also, the need exists to check for loss or buildup within the system of uncombined hydrogen. A continuously recording refractometer would, of course, give a good indication of the hardening rate. Brisk hydrogenation reactions, such as the hydrogenation of fish oils in conventional batch autoclaves, easily reach a drop of 2 IV/minute at the outset; then, this slows steadily so that after 1–3 hours, the drop rate is under 1 IV/minute. Some slowing down for successful end-point control by the operator is needed in any case.

Order of Reaction

Order of reaction is an indication of the number of molecular species taking part in the reaction; hence, radioactive decay is first order because one kind of molecule is reacting or breaking apart. Although hydrogen and triglycerides are involved in hydrogenation, which would seem to qualify it as a second-order reaction, hydrogen is continuously renewed so that its disappearance counts for nothing. In effect, the reaction rate then depends on the unsaturated triglycerides, and most of the time, it behaves as a first-order reaction. However, at the start, especially with a highly-unsaturated oil, so many double bonds are available that the disappearance of a few from the many also has no noticeable effect for a time, and this part of the process becomes a pseudo zero-order reaction. Note that where a molecular species must first associate with a catalyst surface which can be regarded as constantly available (although eventually poisoned), the circumstances of a first-order reaction are created (see also the "Order of Reaction" section in Chapter 1).

Single-Stage Hardening

Single-stage hardening is an expression used to indicate that the process is performed virtually under the same operating conditions throughout, particularly regarding temperature. Thus, for example, remaining true is that gassing may commence at 140°C, but the temperature is given every chance to rise to 180 ± 5°C, at which it is controlled to the end: this would be single-stage hardening. The operating temperature could, of course, be low (110°C), medium (150°C), or "normal" (180 °C). For some unsaturated oils, such as marine and linseed, "single stage" would not strictly apply, since a temporary restriction is at 150°C before allowing a further rise (see the "Cyclization and Polymerization" section in Chapter 2).

Two-Stage Hardening

Two-stage hardening is an expression used to indicate that, at some stage, operating conditions such as temperature and pressure are suddenly changed to favor a different effect in the type of hydrogenation. Strictly, the pause at 150°C in the hydrogenation of marine oils and very unsaturated vegetable oils such as linseed oil (see the "Cyclization and Polymerization" section in Chapter 2) has to be regarded as creating a two-stage hydrogenation. More drastic, however, is the decision to carry out the first part of a hydrogenation at, say, 110°C and 3 atm, then to change abruptly to encourage the temperature to rise to 180°C and to drop the pressure to 1 atm. The reasoning for this is explained in the "Two-Stage Hydrogenation" section in Chapter 2 and is quite empirical in practice.

Switch Point

Switch point is an expression used to indicate at what point (such as a refractive index or IV) during hydrogenation the conditions of temperature and pressure are no longer controlled at the same level, but every effort is to be made to reach a new level. To make the change expeditiously, steam heating may be used to supplement the heat of reaction in raising the temperature, while the hydrogen pressure may be dropped to (i) slow the hydrogenation and enjoy the assistance of the steam heating, and to (ii) favor more selective hydrogenation thereafter by favoring a lower concentration of hydrogen at the catalyst surface. The choice of switch point is a compromise and empirical. Degrees of freedom before the switch point entail keeping down the proportion of *trans* isomers by employing low temperature and normal (3 atm) to moderately high pressure (3–5 atm) for as long as will still allow some subsequent stage the chance to exert its effect. After the switch point, the choice entails operating for sufficiently long at sufficiently high (say, 160–180°C) temperature and low pressure to ensure the selective removal of oxidation-prone triunsaturates, and meanwhile having to accept the acceleration in *trans* formation. The second stage, from the time the controls are altered to the end point, may be expected to cover a total further drop in IV of not less than 20 units, but this is entirely a matter of trial, and in any case, will be influenced also by quality and dose of catalyst.

Oils change during the hydrogenation; therefore, to change operating parameters to obtain the best advantage is permissible, even necessary. More than two stages could be advantageous; in the end, the "flight of steps" becomes a "ramp." Any instrumentation installed to secure this kind of control would have to justify its cost in quality, improvement, and time saved.

Hardening Techniques

Hardening techniques are described in Chapter 2, especially in "Ultralight, Touch, Brush, or Flash Hydrogenation," "Low-Temperature Hydrogenation," "Iso- or *trans*-Suppressive Hydrogenation," "Normal Hydrogenation," "Cyclization and Polymerization," "Two-Stage Hydrogenation," "Iso- or *trans*-Promoting Hydrogenation," and "Higher-Melting and Fully-Saturated Hardened Oils" sections, and are designed to maximize the chance to obtain the most-desired texture with acceptable stability.

Hardening Flavors

Origin

In practical effect, pure triglycerides have no flavor or odor, but when left to oxidize, rancidity develops. Precursors of off flavors, such as hydroperoxides, are also flavorless, but are recognized to be a first step on the way to the breakdown of the fatty-acid chains to release shorter-chain aldehydes and ketones. These more volatile substances, in many instances, have both odor and taste. Minor nonfatty components in a fresh, crude natural oil can confer a characteristic odor/flavor which is removed by hydrogenation. Although hydrogenation removes the original odor and flavor, it may produce *trans* isomers in the triglycerides which act as the precursors of new and very potent off flavors. Even if the resistance of the oil as a whole to oxidation may have increased notably, the oxidative breakdown of these *trans* isomers to aldehyde fragments can establish an off flavor which needs, in extreme cases, to be present at a concentration of only about 10^{-10} to be detected. This risk is present when linolenic or more unsaturated groups were hydrogenated, whereas a thousand times that amount of oxidation of oils whose only fatty acid was oleic acid would be possible without creating noticeable bad flavor. Some off flavor is considered also to arise during partial hydrogenation in the unsaponifiable matter of certain oils, such as marine, soybean, and rapeseed. All the conditions which minimize the production of *trans* isomers in the triglycerides appear to be beneficial also in avoiding or minimizing off flavors in the unsaponifiable matter, and these are itemized later.

A notorious off-flavor precursor arises during the partial hydrogenation of oils which contain an appreciable amount of linolenic acid or fatty acids of higher unsaturation, such as occur in marine oils. After oxidation, the linolenic hardening flavor (LHF), identified by Keppler et al. in 1965 (see the "Hydrogenation" section in Chapter 8) as being principally 6 *trans* nonenal, becomes evident; other aldehydes of about the same chain length were recognized about the same time as having distinct off flavors. Later, Yasuda et al. (1975), working in their case from a soybean oil which was hydrogenated to 58.5 IV (51.5% *trans* isomers) and after deodorization allowed to oxidize at 85°C for three weeks, identified 48 volatile compounds with the help of GC, IR, and MS (see the "Miscellaneous Tests" section in Chapter 8). The aldehyde 2-*trans*-6-*trans* octadienal was prominent among those contributing to this type of off flavor (Keppler's starting point was linseed oil). The type of degradation reactions proposed by Hoffmann (1962) would account for some of these compounds. Meijboom and Stroink (1972) identified 2-*trans*-4-*cis* decatrienal as being the fishy off flavor occurring in strongly autoxidized oils containing linolenic acid or any fatty acids having the ω3, ω6, ω9 grouping of unsaturation (see the "Nomenclature" and "Essential Fatty Acids" sections in this chapter). This, they further pointed out, was one of the four aldehydes to result from the decomposition of four of the more important isomeric monohydroperoxides obtained by Frankel et al. (1961) in the auto-oxidation of linolenic acid.

Avoidance of Off Flavors
When an oil is known to be a possible source of the hardening of off flavors because of the degree of unsaturation of some of its components, and at the same time one desires to hydrogenate it to gain stability without having to exceed the mid-thirties centigrade in melting point, certain conditions adopted for the hydrogenation will be helpful. These include the use of fresh or nearly fresh nickel-pure hydrogen and a low (under 120°C) or medium (150°C maximum) temperature. A well-pretreated oil which has had most of its catalyst poisons removed is a further advantage. A final brief spell of hydrogenation above 160°C (see the "Two-Stage Hydrogenation" section in Chapter 2) may, on balance, secure the removal of the remaining very unsaturated fatty acids and cause the migration of double bonds in off-flavor precursors so that they become less potent. This is a matter of trial, and may be limited by texture considerations. Evidently, whether the off flavor arises from the triglyceride or some minor component, hydrogenation beyond a certain point has so altered the precursor that the development of an off flavor in the ultimate deodorized oil is greatly diminished. Unfortunately, this degree of hardening may well make the texture so firm that its incorporation in blends is more limited, and a substantial economic advantage is lost. This effect appears around a 40°C mp. Occasionally, refiners note that after deodorization, an oil, whether hydrogenated or not, will improve in flavor after 12–24 hours by becoming more bland; then, after 48–72 hours, deterioration commences as it is stored in contact with air. This effect seems to be consistent with a limited amount of readily oxidizable off flavor and its precursor being destroyed; then, appreciably later, the oxidation of the fat as a whole makes itself apparent. This phenomenon, however, is too irregular to allow its exploitation in production.

The following precautions are generally helpful in conserving the quality of the hardened oil from the time it leaves the autoclave. The contact with air during filtration, storage, and subsequent processing steps is kept to a minimum when these procedures are not actually protected by a nitrogen blanket. The oil is likely to be handled and stored in darkness and to be dry when stored, both circumstances being favorable. Storing hardened oils at no more than 10°C above the melting point is usually possible. Oxidation rates accelerate steeply with a rise in temperature; a margin of 10°C allows the oil to be handled and remain liquid. Pro-oxidant traces of metals such as iron and copper are held in check by doses of up to 0.01% of citric acid added as a sequestering agent. The post-bleaching stage should reduce nickel to less than 0.2 ppm, and this effect is enhanced, if necessary, when a small amount (up to 0.1% maximum of citric acid/oil) is added at the bleaching stage (see the "Filtration" section in Chapter 4). Regarding the addition of antioxidants (as distinct from sequestrants), one must remember that the threshold values for the detection of the most potent off flavors are below the effective action of these agents. The addition of a pleasant flavor may be effective in masking the presence of a hardening off flavor. The simple general rule for oil handling remains—keep it dark, dry, cool, clean, and air-free.

Iodine Value (IV)

The iodine value (IV) of an oil or fatty acid is the weight of iodine with which it will combine, expressed as a percentage of its own weight. Since each carbon-carbon double bond in the molecule of oil combines with two atoms of iodine if n double bonds are present, the IV is:

$$(2n \times 127)/(\text{mol wt oil}) \times 100$$

Notably, the size of the molecular weight has an effect on the IV; hence, if the IV of oleic acid is 90.1, that of glyceryl trioleate is 86.2. Although hedged around with restrictions as to its accuracy in various circumstances, the test in its normal or rapid form remains useful and popular.

Nickel Consumption/Poisoning

Numerous ways are available in which this concept of poisoning and catalyst consumption may be expressed and, as before, the usefulness to hardeners is the first consideration. In a particular investigation, oil may be hardened in otherwise identical conditions with a few incremental doses of nickel, and the refractive-index drops obtained in a set time plotted against nickel concentration. On extrapolating back to nil effect, this is seen to be equivalent to, say, 0.015% of nickel/oil or some much greater dose if the oil is rich in catalyst poison. Then, evidently, in deciding the response to an increased nickel dose, this concentration of 0.015% or other figure has to be regarded as the base line. If the reaction to the poisoning of one catalyst is to be compared with that of another, then to produce the above type of plot for each is not difficult. Again, some catalysts promote rapid hydrogenation at the start (having evidently overcome any poison present), and then markedly diminish their performance if the hydrogenation is pursued to around 40°C mp. This seems to be the case for some wet-reduced unsupported nickel catalysts. If some particular hydrogenation task is seen as a major part of a production program and the one most likely to exhaust the catalyst being used, then devising a test which establishes the minimal catalyst dose (% nickel/oil) which will secure the required IV or refractive-index drop in the maximal tolerable time (e.g., 4 hours of gassing, at appropriate temperature and pressure) is not difficult. This type of test can also be made the means of comparing the durability of one catalyst with another or a standard. Important to recognize is that durability is not the same as activity, and certainly not a measure of selectivity.

Selectivity

Selectivity is a preference for first hydrogenating a double bond in the more unsaturated groups before commencing to hydrogenate any in the less unsaturated ones. Followed ideally, this would entail molecules with three double bonds all being reduced to two before any with two were reduced to one, and so on.

Various classes of selectivity are recognized, and were fully explained in the "Selectivity," "Estimation of Selectivity," "Operation of Selectivity," and "Combination of Factors Affecting Hydrogenation" sections in Chapter 1; brief definitions only are given here:

- Selectivity I (S_I) expresses the rate at which groups with two or more double bonds are hydrogenated to one, compared with the rate with which those with one are hydrogenated to nil.
- Selectivity II (S_{II}) expresses the rate at which groups with more than two double bonds are reduced to two, compared with the rate with which those with two are reduced.
- Selectivity triglyceride (S_T) expresses the extent to which hydrogenation is shared over all triglyceride molecules containing similar unsaturated groups, compared with the extent to which it is concentrated over a minority, thereby giving them a distinctly lesser degree of unsaturation than the remainder.
- Specific isomerism (S_i) expresses the proportion of double bonds converted to the *trans* form to those eliminated by hydrogenation.

No reason exists why Selectivity III and IV should not be defined if they served a useful purpose and could be measured. As explained in the "Selectivity" section in Chapter 1, the early use of the term "selectivity" was made more precise, and has now received a commonly accepted interpretation. As in the case of the concept of catalyst activity, the users may be justified in designing an empirical test to suit their particular purposes, provided, naturally, that they are not founded on false premises so that one result is barely comparable with another (see the "Estimation of Selectivity" section in Chapter 1).

Triglycerides

Positional Isomers

The "Triglycerides" section in Chapter 1 explains how if the number 1, 2, and 3 positions in a triglyceride are occupied by exactly the same kind of fatty acid, the result is a simple triglyceride, but if even one fatty acid differs, it is classified as mixed. The inner or 2 position is obviously differently situated from the outer 1 and 3 positions, which for a long time were considered to be completely equivalent. Now convincing evidence exists that, because of the relative arrangements in space of groups attached to these two end positions, inherent asymmetry appears in the glycerol molecule which allows a distinction between the 1 and 3 positions (Taylor, 1973). Biochemical experiments have contributed much to making this clear; thus, the enzyme diglyceride kinase will promote the phosphorylation of the 3 hydroxyl of a 1,2 diglyceride, but will not act with the 1 hydroxyl of a 2,3 diglyceride. This kind of distinction is a very fine one which, as we have seen, may be important in biological systems, but as far as is known does not play any significant role in hydrogenation. Evidence shows, on the other hand, that the end positions 1 and 3 are marginally more vulnerable to hydrogenation than the inner 2 position (see the "Hydrogenation" section of Chapter 8 and Ackman et al., 1964 (Drozdowski, 1977; Kaimal & Lakshminarayans, 1979). Apparently, the 2-carbon atom of the glyceryl chain is well-situated to become an asymmetric center if all three fatty acids attached are different—e.g., palmitic, oleic, stearic (POS)—or if two similar ones are attached nonsymmetrically—e.g., palmitic, stearic, stearic (PSS). This, in fact, is the case, and very weak optical isomerism is in many natural fats, which is stronger where a considerable difference in chain length exists between a 1- or 3-position fatty acid and the other two (Anderson et al., 1970; Hilditch & Williams, 1964).

As far as physical characteristics such as melting behavior are concerned, a very marked difference on many occasions is apparent according to how different fatty acids are arranged over the three triglyceride positions [e.g., POS (a major component of cocoa butter) and OPS (a major constituent of lard)]. This is no more than would be expected, since crystals grow according to the arrangements in which their components can be fit together. If the location of fatty-acid chains—themselves of different length and possibly bent (*cis* and *trans* double bonds) at differing locations—is changed around, this will itself create different possibilities for packing; nor is this the end of possibilities for change.

Polymorphs

The four valencies of carbon are represented as, in effect, representing themselves according to a tetrahedral pattern. When the structure of a particular triglyceride comes to be considered, all three branches can lie in one plane like the arms of a tuning fork (after which this pattern is named). However, also possible is that one chain could lie in a plane vertical to the others, and so on. The bending of the chain adds to the complications. Even within one pattern as described, the triglyceride molecules may pack at different angles of tilt within the crystal, and the layers of identically packed molecules may be somewhat differently oriented to one another. These different species or forms of the one substance are polymorphs. The least stable form—which is also the one with the lowest melting point, least density, and lowest heat of fusion—is the so-called alpha (α) form; simple (i.e., unmixed) saturated triglycerides assume as their most stable form the tuning-fork arrangement, and this, the beta (β) form, has the highest melting point, greatest density, and highest heat of fusion. Another form, the so-called beta prime (β'), is a most stable form for some (not all) mixed saturated triglycerides, and also for nonsymmetrical monounsaturated triglycerides (e.g., OSS, OPS), as well as being the form adopted by natural fats in the organism.

One or more intermediate forms were identified for various fats. The least stable (α) is most easily made by the rapid cooling of the oil to just below the melting point; if the temperature is then gradually raised, it melts, and given the chance, will recrystallize in an intermediate form, possibly a β'. Further heating, melting, and recrystallization lead in the end to the most stable form (Taylor, 1973).

Transition from one form to the next may be exothermic. For the great majority of fats, this series of changes can only be induced in one direction toward greater stability, and to go back to the α form, one must melt the fat completely and suitably recool. This ability to change in only one direction is described as monotropic; fats which change in either direction are enantiotropic. As the movement to more stable forms occurs, more orderly and closer-packed crystal lattices are found. X-ray diffraction, NMR, and infrared spectroscopy have played vital parts in identifying the structure of polymorphs (Taylor, 1971).

References

A

Achaya, K.T. Indian oilseed processing. *Chem. Ind.* **1994,** 3, 95.

Acosta, G.M.; Smith, Jr., R.L.; Arai, K. High-pressure PVT behavior of natural fats and oils: Trilaurin, triolein, and n-tridecane from 303 K to 353 K from atmospheric pressure to 150 MPa, *J. Chem. Eng. Data,* **1996**, 41, 961-969.

Adams, D.J.; Dyson, P.J.; Tavener, S.J. *Chemistry in Alternative Reaction Media.* West Sussex, England, Wiley, 2004.

Ahmed, M.M.; T.M. Priestley; J.M. Winterbottom. Palladium catalyzed hydrogenation of soybean oil. Ibid. **1979,** *56,* 571.

Aiken, G.R.; Poliakoff, M. A critical look at reactions in class I and II gas-expanded liquids using CO_2 and other gases. **2009**, *Green Chem. 11*, 1083-1100.

Albright, L.F. Commercial processes for hydrogenating fatty oils. *Chem. Eng.* **1967,** *9(Oct.),* 249.

Albright, L.F. Quantitative measure of selectivity of hydrogenation of triglycerides. *JAOCS* **1965,** *42,* 250.

Albright, L.F. Theory and chemistry for the hydrogenation of fatty oil. *Chem. Eng.* **1967,** *11(Sept.),* 200, 201.

Albright, L.F.; J. Wisniak. Selectivity and isomerization during partial hydrogenation of cottonseed oil and methyl oleate; effect of operating variables. *JAOCS* **1962,** *39,* 14.

Allen, R.R. Calculation of the linolenic selectivity ratios of hydrogenation catalysts. *JAOCS* **1967**, *44*, 466-467.

Allen, R.R. Hydrogenation. Ibid. **1981,** *58,* 166.

Allen, R.R. Hydrogenation: Principles and Catalysts, AOCS Short Course on Processing and Quality Control of Fats and Oils, East Lansing, Michigan, USA, 1966.

Allen, R.R. Ibid. Vol. 2, p. 19.

Allen, R.R. Principles and catalysts for hydrogenation of fats and oils. *JAOCS,* **1978,** *55,* 792.

Allen, R.R.; A.A. Kiess. Isomerization during hydrogenation: Linoleic acid. Ibid. **1956,** *33,* 355.

Allen, R.R.; A.A. Kiess. Isomerization during hydrogenation: Oleic acid. Ibid. **1955,** *52,* 400.

Allen, R.R.; J.E. Covey. Hydrogenation of vegetable oils with sulphur treated catalysts. *JAOCS,* **1980,** *57,* 115A, item 124.

Allen, R.R. and J. Covey. The effects of process variables on formation of *trans* unsaturation during hydrogenation. *JAOCS,* **1970**, *47*, 494-496.

An, W. The Design and Testing of a Solid Polymer Electrolyte Reactor for the Electrocatalytic Hydrogenation of Edible Oils. Ph.D. Dissertation, Tulane University, New Orleans, LA, **1998.**

An, W.; Hong, J. K.; Pintauro, P.N. Current efficiency for soybean oil hydrogenation in a solid polymer electrolyte reactor. *J. Appl. Electrochem.*, **1998b**, *28*, 947-954.

An, W.; Hong, J. K.; Pintauro, P. N.; Warner, K.; Neff, W, The electrochemical hydrogenation of edible oils in a solid polymer electrolyte reactor. II. Hydrogenation selectivity studies. *JAOCS,* **1999**, *76*, 215-222.

An, W.; Hong, J. K.; Pintauro, P. N.; Warner, K.; Neff, W. The electrochemical hydrogenation of edible oils in a solid polymer electrolyte reactor. I. Reactor design and operation. *JAOCS,* **1998a**, *75*, 917925.

Analysis of Oils and Fats; Hamilton, R.J.; Rossell, J.B., Eds.; Elsevier: London and New York, 1986.

Analysis of Oilseeds, Fats and Fatty Foods. Rossell, J.B.; Pritchard, J.L.R., Eds.; Elsevier: London and New York, 1991.

Anderson, B.A.; C.A. Sutton; M.J. Pallansch. Optical activity of butterfat and vegetable oils. Ibid. **1970,** *47,* 15.

Anderson, C.; R. Larsson. Catalytic hydrogenation of soybean oil with Rh (1) dihydride complexes as catalysts. *JAOCS,* **1981,** *58,* 54.

Andersson, K.; Hell, M.; Löwendahl, L.; Schöön, N.-H. Diffusivities of hydrogen and glyceryl trioleate in cottonseed oil. *JAOCS,* **1974**, *51*, 171-177.

Andersson, M.B.O., King, J.W., and Blomberg, L.G. Synthesis of fatty alcohol mixtures from oleochemicals in supercritical fluids. *Green Chem.* **2000**, *2*, 230-234.

Anon. (2003) Food Labeling: *Trans* fatty acids in nutrition labeling, final rule. Fed. Reg. *68* 41433-41506.

Anon. ADM chooses a trouble free process for *trans* free fats. *Oil Mill Gazetteer*, **2004** *109* 2-5.

Anon. (2008) www.ers.usda.gov/amber waves/foodlabels.htm.

Anon. (2009) Fats and Oils Production, Consumption and Stocks. www.census.gov.

Anon. (2009) www.census.gov.

Anon. The thinking chemist. *Chemistry in Britain* **1978,** *14,* 427.

AOCS Short Course, Processing and Quality Control of Fats and Oils. Michigan State University, East Lansing, Michigan, USA, 1966, (i) Dutton, H.J. *Determination of Fat Composition.* (ii) Erickson, D.R. *Finished Product Testing.* (iii) Lantondress, E.G. *Hydrogenation and Bleaching Control Procedures.* (iv) Sherwin, E.R. *Methods for Stability and Antioxidant Measurement.* (v) Sullivan, F.E. *Refining of Oils and Fats.* (vi) Wiedermann, L.H. *Margarine Oil Formulation and Control.*

Arai, Y.; Sako, T.; Takebayashi, Y. (Eds.). *Supercritical Fluids – Molecular Interactions, Physical Properties, and New Applications.* Berlin, Springer-Verlag, 2002.

Aro, A.; Jauhiainen, M.; Partanem, R.; Salminen, I.; Mutanen, M. Stearic acid, *trans* fatty acids, and dietary fat: Effects on serum and lipoprotein lipids, apolipoproteins, lipoprotein (a), and lipid transfer proteins in healthy subjects. *Am. J. Clin. Nutr.*, **1997**, *65*, 1419-26.

Ascherio A; Willett W. C. Health Effects of Trans Fatty Acids. *Am. J. Clin. Nutr.*, **1997**, 66, 1006S-1010S

Ascherio, A.; Katan, M. B.; Zock, P. L.; Stampfer, M. J.; Willett, W. C. *Trans* fatty acids and coronary heart disease. *N. Engl. J. Med.* **1999**, *340*, 1994-1998.

Asherio. A., M.B. Katan, P.L. Zock, M.C. Stampfer, W. Willet. *Trans* fatty acids and coronary heart disease. *New Eng. J. Med.*, **1999,** *340* 1994-1998.

Available from Unichema International, D4240 Emmerich, Germany. (a) Pricat product data sheets for: 9900, 9906, 9908, 9910, 9912, 9932; (b) *Selective Hydrogenation of Edible Oils, Edible Oil Hydrogenation*, Pricat 9910 and 9920, *Hydrogenation of Corn Oil*; (c) Pricat 9920, *The Versatile Catalyst for Vegetable Oil Hydrogenation. Influence of Catalyst Structure on Hardening of Edible Oils and Fatty Acids*; (d) Pricat 9912, *Nickel Catalyst for Fatty Acid Hydrogenation, Fatty Acid Hydrogenation (1990), Hydrogenation of Soapstock Fatty Acids*: (e) *Energy Recovery and Catalyst Utilization in Fatty Acid Hydrogenation*, W.L.P. de Bruijn and A.G. Hinze, reprint from *Fat Sci. Technol.* **1987,** *89,* 512; Dates of publication of above—up to 1992.

Ayodogan, S.; Koseoglu, S.; Akman, U.; Hortascu, O. Double-bond depletion of soybean oil triglycerides with $KMnO_4/H_2O$ in dense carbon dioxide, **2006**, *23*, 704-713.

B

Bailey, A.E. Additional notes on the kinetics and theory of fatty oil hydrogenations. *JAOCS,*

1949, *26,* 644.

Bailey, A.E. Theory and mechanics of the hydrogenation of edible fats. *JAOCS,* **1949**, *26*, 596-601.

Bailey, A.E.; Fisher, G.S. Modifications of vegetable oils. V. Relative reactivities toward hydrogenation of the mono- di- and triethenoid acids in certain oils. *Oil & Soap* **1946**, *23*, 14-18.

Bailey's Industrial Oil and Fat Products; 4th ed.; D. Swern, Ed.; John Wiley & Sons: New York, USA, 1982; Vol. 2, 293, 433.

Baltes J. Process for the selective hydrogenation of fats and fatty acids. US Patent 3,687,989, 1972.

Baltes, J. Selektive Härtung von Fetten und Fettsäuren mit Sulfiden der Übergangsmetalle als Katalysatoren. *Fette Seifen Anstrichm.* **1970**, *72*, 425-432.

Basu, H.N.; Chakrabaty, M.M. Ibid. **1966,** *43,* 119; Chakrabaty, M.M. et al., Ibid. **1972,** *49,* 510; Chakrabaty, M.M.; D. Bhattacharyya; A. Basu. Conjugated Hydrogenation of Fatty Oil Containing Conjugated Acid, IX ISF Congress, Rotterdam, 1968, III-1.

Beal, R.E. and Lancaster, E.B. Effect of agitation on selectivity in the hydrogenation of soybean oil. *JAOCS,* **1954**, *31,* 619-625.

Beck, F. Electrochemical and catalytic hydrogenation: common features and differences. *Int. Chem. Engr.*, **1979**, *19*, 1-11.

Beckman, E.J. Supercritical and near-critical CO_2 in green chemical synthesis and processing. *J. Supercrit. Fluids,* **2004** *28*, 121–191.

Beckmann, H.J. Hydrogenation practice. Ibid. **1983,** *60,* 282.

Beers, A.E.W. Low *trans* hydrogenation of edible oils. *Lipid Techn.* **2007**, *19*, 56-58.

Beers, A.E.W.; Mangnus, G. Hydrogenation of edible oils for reduced *trans*-fatty acid content. *inform* **2004**, *15*, 404-405.

Beers, A.E.W.; Ariaansz, R. F.; Okonek, D. V. *Trans* isomer control in hydrogenation of edible oils In *Trans Fatty Acids,* Dijkstra, A.J., Hamilton, R.J., Hamm, W. Eds.; Blackwell Publishing: Oxford, UK, 2008, 147-180.

Beers, A.M.; Berben, A.; Goen, P.; Jaqta, C.; Lazar, R.; and Mangus, G. Lowering the *trans* content of edible oils. Third Euro. Fed. Lipid Congress. Edinburgh, Scotland, 2004.

Bek-Nielson, B. *Technical and Economic Aspects of the Oil Palm Fruit Processing Industry*; United Nations Publication (UNIDO) 1D/123: New York, USA, 1974.

Bentz, A.P.; Breidenbach, B.G. Evaluation of differential scanning calorimetric method for fat solids. *JAOCS,* **1969,** *46,* 60.

Berben, P.H.; Borninkhof, F.; Recsink, B.H.; Kuijpers, E. (1994) Production of low *trans* isomer containing products by hydrogenation. Presented at American Oil Chemists'

Society Meeting, Atlanta, GA. Text available from Englehard/BASF, Beachwood, OH.

Berben, P.H.; Bloom, P.; and Sollie, J.C. Palladium and platinum catalyzed oil hydrogenation: Effects of reaction conditions on *trans* isomer and saturate formation. Technical Report, Englehard/BASF, Beachwood, OH.

Berl, W.G. (Ed.) *Physical Methods in Chemical Analysis,* 2nd ed. New York, Academic Press, 1960, 411–462.

Bern, L.; J.O. Lidefelt; N.H. Schoon. Mass transfer and scale up in fat hydrogenation. *JAOCS,* **1976,** *53,* 463.

Bernardini, E. *The New Oil and Fat Technology;* Technologie s.r.l., Rome, 1973.

Berner, D.L. Color in Oils, paper presented at the AOCS 81st Annual Meeting, Chicago, USA, AOCS Press: Champaign, Il, May 1991.

Bertucco, A.; Canu, P.; Devetta, L; Zwahlen, A.G. Catalytic hydrogenation in supercritical CO_2: kinetic measurements in a gradientless internal-recycle reactor, *Ind. Eng. Chem. Res.*, **1997**, *36*, 2626-2633.

Beyens, Y.; Dijkstra, A. J. Positional and triglyceride selectivity of hydrogenation of triglyceride oils In *Fat Science 1983, Proceedings of 16th ISF congress, Budapest,* Holló, J. Ed.; Akadémiai Kiadó, Budapest: 1983, 425-432.

Bezanehtak, K.; Combes, G.B.; Dehghani, F.; Foster, N.R.; Tomasko, D.L. Vapor-liquid equilibrium for binary systems of carbon dioxide + methanol, hydrogen + methanol, and hydrogen + carbon dioxide at high pressures. *J. Chem. Eng. Data,* **2002**, *47*, 161-168.

Bezanehtak, K.; Dehghani, F.; Foster, N.R. Vapor-liquid equilibrium for the carbon dioxide + hydrogen + methanol ternary system. *J. Chem. Eng. Data*, **2004**, *49*, 430-434.

Bhalchandra, B.; Ikushima, Y.; Shirai, M.; Arai, M. The selective formation of unsaturated alcohols by hydrogenation of α, β-unsaturated aldehydes in supercritical carbon dioxide using unpromoted Pt/Al_2O_3 catalyst. *Catl. Lett.*, **1999**, *62*, 175-177.

Biernoth, G.; Rost, H.E. *Removal of Polycyclic Aromatic Hydrocarbons from Edible Oils.* IX ISF Congress, Rotterdam, 1968, III-15.

Biernothl, G.; Rost, H.E. *Arch. Hyg. Bakt.* **1968,** *152,* 238.

Boekenoogen, H.A. (Ed.) *Analysis and Characterization of Oils, Fat, and Fat Products.* London, Wiley Interscience, Vol. 1; 1964, Vol. 11, 1968.

Boelhouwer, C.; Snelderwaard, J.; Waterman, H.I. Problems of selectivity in the hydrogenation of linoleic acid esters. *JAOCS,* **1956**, *33,* 143-146.

Braae, B. Degumming and refining practices in Europe. *JAOCS,* **1976,** *53,* 353.

Brands, D.S.Pontzen, K.; Poels, E.K.; Dimian, A.C.; Bliek. Solvent-based fatty alcohol synthesis using supercritical butane: flowsheet analysis and process design. *JAOCS,* **2002**, *79,* 85-91.

Brands, D.S; Poels, E.K; Dimian, A.C.; Bliek, A. Solvent-based fatty alcohol synthesis using supercritical butane. thermodynamic analysis. *JAOCS,* **2001**, 75-83.

Brown, C.A.; H.C.J. Brown. *Am. Chem. Soc.* **1963,** *85,* 1003

Brownlie, D. *Ind. Eng. Chem.* **1938,** *30,* 1139.

Brunner, G. *Gas Extraction.* New York, Springer, 1994.

Brunner, G.; Saure, C.; Buss, D. Phase equilibrium of hydrogen, carbon dioxide, squalene, and squalane. *J. Chem. Eng. Data,* **2009**, *54,* 1598-1609.

Bushell, W.J.; Hilditch, T.P. The course of hydrogenation in mixtures of mixed glycerides. *J. Chem. Soc.* **1937**, 1767-1774.

Butterfield, R. et al. Analog computers and kinetics of hydrogenation. *JAOCS,* **1964,** *41,* 29.

C

Carr, R.A. Degumming and refining practices in the U.S. Ibid. **1976,** *53,* 347.

Chahine, M.H.; Cousins, E.R.; Feuge, R.A. Positional isomers formed during the hydrogenation of cotton seed oil. Ibid. **1958,** *35,* 396.

Chahine, M.H.; Guice, W.A.; Feuge, R.A. Positional isomers formed during the hydrogenation of methyl linoleate under various conditions. Ibid. **1959,** *36,* 24.

Charlesworth, P.L. The recovery of hydrogen from industrial gas mixtures. *Chem. Eng.,* April **1965**: CE87. Petrocarbon Developments Ltd., Wythenshaw, UK, specializes in this field.

Charlesworth, P.L.; Schmidt, G. The production of hydrogen to satisfy small industrial demands. Ibid., October **1965**: CE259.

Chen Y.-L; Chou, T.-C. Metals and alloys bonded on solid polymer electrolyte for electrochemical reduction of pure benzaldehyde without liquid supporting electrolyte. *J. Electroanal. Chem.*, 1993, 360, 247-259.

Chen, A.H.; McIntyire, D.D.; Allen, R.R. Modeling of reaction rate constants and selectivities in soybean oil hydrogenation. *JAOCS,* **1981,** *58,* 816.

Chen, Y.L.; Chou, T.C. Electrochemical reduction of benzaldehyde using Ag/Nafion as an electrode. *Ind. Eng. Chem. Res.*, **1994**. *33,* 676-80.

Chiba, T., Okimoto, M.; Nagai, H.; Takata, Y. Electrocatalytic reduction using raney nickel. *Bull. Chem. Soc. Jpn.*, **1983**, *56*, 719-23.

Chouchi, D.; Gourgouillon, D.; Courel, M; Vital, J.; Nunes da Ponte, M. The influence of phase behavior on recations at supercritical conditions: The hydrogenation of pinene. *Ind. Eng. Chem. Res.*, **2001**, *40*, 2551-2554.

Christianson, D.D.; Friedrich, J.P. Production of food- grade corn germ product by supercritical fluid extraction, U.S. Patent #4,495, 207,1985.

Christie, W.W. *Lipid Analysis*; Pergamon Press: Oxford, 1973.

Churchley, A.R. Sources of safety information. *Chem. Ind.* **1977,** *2* July, 524.

Cipollina, A.; Anselmo, R.; Scialdone, O.; Filardo, G; Galia, A. Experimental P–T–ρ measurements of supercritical mixtures of carbon dioxide, carbon monoxide, and hydrogen and semiquantitative estimation of their solvent power using the solubility parameter concept. *J. Chem. Eng. Data*, **2007**, 52, 2291–2297.

Ciriminna, R.; Carraro, M.L.; Campestrini, S.; Pagliaro, M. Heterogeneous catalysis for fine chemicals in dense phase carbon dioxide. *Adv. Synth. Catal.*, **2008**, *350*, 221-226.

Clegg, A.J. Composition and related nutritional and organoleptic aspects of palm oil. *JAOCS,* **1973,** *50,* 321.

Cocks, L.V.; van Rede, C. *Laboratory Handbook for Oil and Fats Analysts.* Academic Press, New York, 1966.

Coconut Processing Technology (7 parts); Asian and Pacific Coconut Community: Jakarta, Indonesia, UNIDO/IOD, 377, September 1980.

Codex Alimentarius. CODEX STAN. 30-1981, Suppl. 1, 1983.

Coenen, J. W. E. Catalytic hydrogenation of fatty oil. *Ind. Eng. Chem. Fundam.*, **1986**, *25*, 43-52.

Coenen, J.W.E. *Actes. Congr. Int. Catalyse. 2(e)* **1960,** Paris 1, 2705.X16.

Coenen, J.W.E. Catalytic processes in the field of industrial fatty acid products, Part I. *F.S.A.* **1975,** *77,* 341.

Coenen, J.W.E. Hydrogenation of edible oils. *JAOCS,* **1976**, *53*, 382-389.

Coenen, J.W.E. *Margarine Today.* International Federation of Margarine Associations, E.J. Brill, Leiden, 1970.

Coenen, J.W.E. *The Mechanism of Heterogeneous Catalysts.* J.H. de Boer, Ed.; Elsevier, Amsterdam, 1960.

Coenen, J.W.E. The rate of change in the perspective of time. Sixth Hilditch Memorial Lecture. *Chem. Ind.* **1978,** *18(Sept.),* 709-722.

Coenen, J.W.E.; Boerma, H. Absorption der Reaktionspartner am Katalysator bei der Fetthydrierung. *Fette Seifen Anstrichm.* **1968**, *70*, 8-14.

Coenen, J.W.E.; Wieske, Th.; Cross, R.S.; Rinke, H. Occurrence, detection and prevention of cyclization during hydrogenation of fatty acids. *JAOCS,* **1967,** *44,* 344.

Cowan, J.C.; Evans, C.D.; Moser, H.A.; List, G.R.; Koritala, S.; Moulton, K.J.; Dutton, H.D. Flavor evaluation of copper hydrogenated soybean oils. *JAOCS,* **1970,** *47,* 470.

Critchfield, C.E.M. Safety in solvent extraction plants—Europe. Ibid. **1976,** *53,* 295.

Cude, I.L. Dispersion of gases vented to atmosphere from relief valves. *Chem. Eng.,* **1974.**

D

D'Alonzo, R.P.; Kozarek, W.J.; Wharton, H.W.. Analysis of processed soy oil by gas chromatography. Ibid. **1981,** *58,* 215.

Dafler, J.R. Nickel crystallite size and net activity of hydrogenation catalysts. *JAOCS,* **1977,** *54,* 17.

Daum, J.P. Determining the costs of hydrogen. *inform,* **1993,** *4,* 1394.

de Jonge, A.; Coenen, J.W.E.; Okkerse. C. Selective hydrogenation of soybean oil: XI. Trialkyl silane-activated copper catalysts. *Nature,* **1965,** *206,* 573.

de la Fuente, J.C; Fornari, T; Brignole, E.; Bottini, S. Phase equilibria in mixtures of triglycerides with low-molecular weight alkanes. *Fluid Phase Equilibria,* **1997**, 128, 15, 221-227.

De Simone, J. and Tumas, W. (Eds.). *Green Chemistry Using Liquid and Supercritical Carbon Dioxide.* Oxford University Press, New York, 2003.

de Vries, B. *JAOCS,* **1963,** *40,* 184.

de Vries, B.; G. Jurriens. *F.S.A.* **1963,** *65,* 725.

den Boer, F.C.; Wosten, W.J.. *The Mechanism of Isomerization and Hydrogenation Reactions of Octadecenoates on Nickel Catalysts*; IX ISF Congress, Rotterdam, 1968, III-4.

Devetta,L.; Canu, P.; Bertucco, A.; Steiner, K. Modeling of a trickle-bed reactor for a catalytic hydrogenation in supercritical CO_2. *Chem. Eng. Sci.*, **1997**, *5,* 4163-4169.

DGF Standard Methods of Analysis; Deutsche Gesellschaft fur Fettwissenschaft e.V., Munster Westfalen: West Germany.

Dijkstra, A.J. Hydrogenation revisited. *inform* **1997,** *8,* 1150-1158.

Dijkstra, A.J. Hydrogenation In *The Lipid Handbook,* 3rd ed. Gunstone, F.D., Harwood, J.L., Dijkstra, A.J. Eds. Taylor & Francis Group, LLC, Boca Raton, FL, j2007, 266-287.

Dijkstra, A.J. On the mechanism of the copper-catalysed hydrogenation; a reinterpretation of published data. *Eur J Lipid Sci Technol.* **2002**, *104*, 29-35.

Dijkstra, A.J. Revisiting the formation of *trans* isomers during the partial hydrogenation of triglyceride oils. *Eur J Lipid Sci Technol.* **2006**, *108*, 249-264.

Dijkstra, A.J. Selectivities in partial hydrogenation. *JAOCS,* **2010**, *87*, 115-117.

Drawing No. 8-228. Chemetron Process Equipment: Louisville, Kentucky, USA.

Drozdowski, B. Effect of the unsaturated acyl position in triglycerides on the hydrogenation rate. *JAOCS,* **1977,** *54,* 600.

Drozdowski, B.; I. Gonaj-Moszora. Reactivation of partially poisoned nickel catalyst by bleaching earth. Ibid. **1980,** *57,* 171.

Drozdowski, B.; M. Zajac. Effect of concentration of some nickel catalyst poisons in oils on the course of hydrogenation. Ibid. **1977**, *54,* 595.

Dunford, N.T., King, J.W. and List, G.R. Supercritical fluid extraction in food engineering. In: C.Tzai and G. Liadakis (Eds.). *Extraction Optimization in Food Engineering.* New York, Marcel Dekker, 2003, 57-93.

Dutton, H.J. Hydrogenation of fats. In Holman, R., Ed. *Progress in the Chemistry of Fats and Other Lipids.* Pergamon Press, Oxford, 1968.

Dutton, H.J. Some new approaches in lipid research. Third Hilditch Memorial Lecture. *Chem. Ind.* **1972,** 2 September, 665.

Duveen, R.F.; Leuteritz, G. Der BUSS-Schleifenreaktor in der Öl- und Fetthärtungsindustrie. *Fette Seifen Anstrichm.* **1982**, *84*, 511-515.

E

Eckel, R.H., S. Borra, A. Lichtenstein and S. Piazza. Understanding the complexity of *trans* fatty acid reduction in the American diet. *Circulation,* **2007**, *115,* 2231-2246.

Eckert, C. A.; Hsieh, C. K.; McCabe, J. R. Molecular thermodynamics for chemical reaction design. *AIChE J,* **1974**, *20,* 20.

Eckman, D.P.; I. Lefkowitz. Optimizing control of a chemical process. *Control Engineering,* September 1957.

El Shattory, Y.; de Man, L.; de Man, J.M. Evaluation of hydrogenation catalyst activity. *JAOCS,* **1980,** *57,* 402.

Eldib, I.A.; Albright, L.F. Operating variables in hydrogenation cottonseed oil. *Ind. Eng. Chem.* **1957,** *49,* 825.

Eller, F.; List, G.R.; Teel, J.; Steidley, K.R.; and Adlof, R.O. Preparation of spread oils meeting food and drug administration requirements for *trans* fatty acids. *J. Ag. Food Chem.,* **2005**, *53,* 5982-5983.

Ellis, S.B. *The Hydrogenation of Oils,* 2nd ed.; 1919; 22, 46, U.S. Patent 1,081,182 (1913); 1,026,156 (1912).

Erickson, D.R., and Erickson, M.D. Hydrogenation and base stock formulation procedures. In Erickson, D.R. (Ed.). *Practical Handbook of Soybean Processing and Utilization.* AOCS Press, Champaign, Il, 1995.

Evans, C.D.; List, G.R.; Moser, H.A.; Cowan, J.C. Long-term storage of soybean and cottonseed salad oils. Ibid. **1973,** *50,* 218.

Evans, C.D.; Beal, R.E.; McConnell, D.G.; Black, L.T.; Cowan, J.C. Partial hydrogenation and winterization of soybean oil. *JAOCS,* **1964,** *41,* 260.

F

Farr, W. Hydrogenation. In *Bailey's Industrial Oil and Fat Products,* 6th edition, vol. 5, 2005.

Farr, W. and List, G.R. Hydrogenation techniques. In *Nutritionally Enhanced Edible Oil and Oilseed Processing.* AOCS Press: Champaign IL, 2004.

Farr, W.E. Hydrogenation In *Proceedings of the World Conference on Oilseed Processing and Utilization,* Wilson, R.F. Ed.; AOCS Press: Champaign IL, 2001, 27-30.

Fatty Acid Technology; Lurgi AG: Frankfurt/M., 1991.

Flöter, E., VanDujin, G. *Trans*-free fats for use in foods. In *Modifying Lipids for Use in Flow Sheet SFS-164*, EMI Corp.: Des Plaines, IL, 2006.

F. Gunstone, Ed. *Modifying Lipids for Use in Foods.* Woodhead Publishing, Cambridge, UK, 2006, 173-200;429-443.

Frankel, E.N. Homogeneous hydrogenation of unsaturated fats. Applications and implications. *Rev. Fr. Corps Gras.* **1977,** *24,* 241.

Frankel, E.N.; C.D. Evans; D.G. McConnell; E. Selke; H.J. Dutton. *J. Org. Chem.* **1961,** *26,* 4663.

Frankel, E.N.; J.P. Friedrich; T.R. Bessler; W.F. Kwolek. Hydrogenation of methyl sorbate and soybean esters with polymer-bound metal catalysts. *JAOCS,* **1980,** *57,* 349.

Frankel, G. Methods of analysis of processed soy oil. *JAOCS,* **1981,** *58,* 210.

Freier, H.A., et al. Hydrogenation of sunflower oil in a batch type industrial plant with recycled oil and hydrogen to obtain vegetable fats for use in margarine manufacture. **May 1967**, *Rev. Fr. Corps Gras.*

Friedrich, J.P. C_{18}-saturated cyclic acids from linseed oil: a structural study. *JAOCS,* **1967**, *44,* 244-248.

Friedrich, J.P. Supercritical extraction of lipids from lipid-containing materials. U.S. Patent # 4,466,923 (1984).

Friedrich, J.P.; Eldridge, A.C. Production of defatted soybean products by supercritical fluid extraction. U.S. Patent #4,493,854 (1985).

Fukuzumi, K.; H. Kato. Conjugated Hydrogenation of Methyl Linoleate Isomers. IX ISF Congress, Rotterdam, 1968, III-2.

G

Gander, K.F. Margarine oils, shortenings and vanaspati. *JAOCS,* **1976,** *53,* 417.

Garcia-Verdug, E.; Liu, Z.; Ramirez, E.; Garcia-Serna, J.; Fraga-Dubreuil, J.; Hyde, J.R.; Hamley, P.A.; Poliakoff, M. In-situ generation of hydrogen for continuous hydrogenation reactions in high temperature water. *Green Chem.*, **2006**, *8*, 359-364.

Gas Volume Correction Factors Including Properties of Gases; Peabody Holmes Ltd.: Huddersfield, UK.

Gering, H. Steam jet pumps as vacuum pumps in the edible oil industry. *JAOCS,* **1980,** *57,* 115A, item 279; and from Korting Hannover AG, Hannover, West Germany.

Gerstenberg, P. Food products without *trans* fatty acids. In Dijkstra, A.J., Hamilton, R.J., Hamm, W. Eds., *Trans Fatty Acids.* Blackwell Publishing: Oxford, UK, 2008, 219-234.

Goebel, E.H. New concepts of bleaching soy oil. Ibid. **1981,** *58,* 199.

Going, L.H. Oxidative deterioration of partially processed soybean oil. Ibid. **1968,** *45,* 632.

Gray, J.I.; Russell, L.F. Hydrogenation catalysts—Their effect on selectivity. Ibid. **1979,** *56,* 34.

Grinberg, V. A.; Zhuravleva, V. N.; Vasilev, Y. B.; and Kazarinov, V. E. Electrolytic synthesis reaction with the participation of protons generated by ionization of hydrogen in a system with a solid polymer electrolyte. *Soviet Electrochemistry* (*Electrokhimiya)*, **1984**, 20, 1158-1162.

Guardo, A.; Casanovas, M.; Ramirez, E.; Recasens, F.; Magana, D.; Martinez, D.; Larrayoz, M.A. CFD modeling on external mass transfer and intra-particle diffusional effects on the supercritical hydrogenation of sunflower oil. *Chem. Eng. Sci.*, **2007,** *62*, 5054-5061.

Gunstone, F.D. Natural and unnatural unsaturated fatty acids. Fifth Hilditch Memorial Lecture. *Chem. Ind.* **1976,** *20,* March, 243.

Guo, S.; Liew, K.Y.; Li, J. Catalytic activity of ruthernium nanoparticles supported on carbon nanotubes for hydrogenation of soybean oil, *JAOCS*, **2009**, *86*, 1141-1147.

Gupta, M.K. Hydrogenation. In *Practical Guide to Vegetable Oil Processing*. AOCS Press, Champaign, IL, 2008, 145-191.

Gutsche, B.; Johannisbauer, W.; Roessler, H.; Topphoff, M. Three stage processes for the separation of supercritical or near-critical mixtures, U.S. Patent 7,084, 313, 2006.

H

Hack, K.M.; Hall, B.B. The production of high purity hydrogen from butane with special reference to materials of construction and operating problems. *Chem. Eng.* November **1965**: CE282

Haighton, A.J. Blending, chilling and tempering of margarines and shortenings. *JAOCS,* **1916,** *53,* 397.

Haighton, A.J.; von Putte, K.; Vermaas, L.F. Determination of the solid contents of fats by wide-line nuclear magnetic resonance: The signal of liquid oils. Ibid. **1972,** *49,* 153.

Haighton, A.J.; Vermass, L.F.; den Hollander, C. Determination of the solid-liquid ratio of fats by wide-line nuclear magnetic resonance. Ibid., **1971,** *48,* 7.

Handbook of Soy Oil Processing and Utilization; American Soybean Association and AOCS Press, Champaign, IL, 1980; Chapter 6, 18 ff.; Chapters 9, 17; 145; 277-296..

Härröd, M.; Möeller, P. Hydrogenation of fats and oils at supercritical conditions, In P.R. van Rohr and C. Trepp, Eds. *High Pressure Chemical Engineering.* Elsevier Science, B.V., Amsterdam, 1996, 43-48.

Härröd, M.; Möeller, P. Hydrogenation of substrate and products manufactured according to the process, WO 96/01304 (1996); U.S. patent # 5, 962, 711, 1999.

Härröd, M.; van den hark, S.; Macher, M.-B.; Möeller, P. Hydrogenation at supercritical single phase conditions, In A. Bertucco and G.Vetter, Eds., *High Pressure Process Technology.* Elsevier, Amsterdam, 2001, 496-508.

Härröd, M; Möeller, P. Partially hydrogenated fatty substances with a low content of *trans* fatty acids, U.S Patent #6,265,596 B1 (2001).

Hasman, J. *Trans* suppression in hydrogenated oils. *Inform,* **1995,** *6,* 1206-1213.

Hastert, R. C. Hydrogenation. In Hui, Y.H., Ed., *Bailey's Industrial Oil and Fat Products*, 5th edition. John Wiley and Sons, New York, 1996, 213-300.

Hastert, R. C. Hydrogenation. In O'Brien, R.; Farr, W.; Wan, P., Eds. *Introduction to Fats and Oils Technology*, 2nd edition. AOCS Press, Champaign, IL, 2000, 179-193.

Hastert, R.C. Hydrogenating fatty acids, *World Conference on Oleochemicals: Into the 21st Century*; Kuala Lumpur: Malaysia, 1990

Hastert, R.C. Hydrogenation of fatty acids. *JAOCS*, **1979,** *56,* 732A.

Hastert, R.C. Practical aspects of hydrogenation and soybean salad oil manufacture. *JAOCS,* **1981,** *58,* 169.

Haumann, B. F. Tools: Hydrogenation, interesterification. *Inform* **1994**, 68-678.

Heertje, I.; Koch, G.K.; Wosten, W.J. *Catalysis* **1974,** *32,* 337.

Heintz, K. Fatty acids in body weight and obesity. *Inform,* **2009,** *20* 459.

Heldal, J. A.; Moulton, K. J. Sr.; Frankel, E. N. Fixed-bed continuous hydrogenation of soybean oil with palladium-polymer supported catalysts. *JAOCS,* **1989**, 66, 979-982.

Higgins, N.W. Low-*trans* stereoisomer shortening systems. U.S. Patent 7,149,430, 2007.

Hilditch, T.P. Die Hydrierung der Fette In *Chemie und Technologie der Fette und Fettprodukte,* Schönfeld, H. Ed.; Verlag von Julius Springer: Wien, 1937, 98-135.

Hilditch, T.P.; Jones, E.C. *J. Chem. Soc.* **1932,** 805–820.

Hilditch, T.P.; Williams, P.N. *The Chemical Constitution of Natural Fats,* 4th ed.; John Wiley and Sons, New York, 1964, 11–62.

Hitzler, M.G.; Poliakoff, M. Continuous hydrogenation of organic compounds in supercritical fluids. *Chem. Commun.*, **1997**, 1667-1668.

Hitzler, M.G.; Smail, F.R.; Ross, S.K.; Poliakoff, M. Selective catalytic hydrogenation of organic compounds in supercritical fluids as a continuous process. *Org. Process Res. Develop.* **1998,** *2,* 137-146.

Hobbs, H.R.; Kondor, B.; Stephenson, P.; Sheldon, R.A.; Thomas, N.R; Poliakoff, M. Continuous kinetic resolution catalysed by cross-linked enzyme aggregates, 'CLEAs', in supercritical CO_2. *Green Chem.,* **2006**, *8,* 816–821.

Hoffmann, G. 1 Octen-3-ol and its relation to other oxidative cleavage products from esters of linoleic acid. *JAOCS* **1962,** *39,* 439.

Holm, U.; K. Ekbom-Olsson. *Anisidine Value and Oil Quality*; Abstracts ISF Congress: Goteborg, Sweden, 1972.

Holser, R., List, G.R, King, J.W., Holliday, R.L. and Neff, W.E. Modeling of hydrogenation kinetics from triglyceride compositional data. *J. Agric. Food Chem.*, **2002**, *50,* 7111-7113.

Horiuti, I.; Polanyi, M. Exchange reactions of hydrogen on metallic surfaces. *Trans. Faraday Soc.* **1934**, *30,* 1164.

Hughes, J.P. Hydrogenation of fatty oils. *JAOCS,* **1953**, *30,* 506-515.

Human serum by balancing saturated and unsaturated fatty acids. U.S. Patent 5,843,497.

Hunter, I.R. Oxygen absorption apparatus for measuring induction period of fats. Ibid. **1951,** *28,* 160.

Hutchenson, K. W. Organic chemical reactions and catalysis in supercritical fluid media, In Y-P Sun Ed., *Supercritical Fluid Technology in Materials Science and Engineering.* Marcel Dekker, Inc. New York, 2002, 87-188.

Hutchenson, K.W.; Scurto, A.M.; Subramaniam, B. Gas-Expanded Liquids and Near Critical Media, ACS Symposium Series #1006, American Chemical Society, Washington, DC, 2009.

Hutchings, G. Catalyst synthesis using supercritical carbon dioxide: a green route to high activity materials, *Top. Catal,* **2009**, *52*, 982-987.

Hydrogenation Plant with Post-Treatment, Technical Handout, Alfa-Laval AB: Tumba, Sweden.

I

Ikariya,T and Noyori, R. Carbon dioxide as a solvent and reactant in catalysis, In: J. M. De Simone, J. W. Tumas, Eds. *Green Chemistry Using Liquid and Supercritical Carbon Dioxide.* Oxford University Press, New York, 2003, 48-63.

Ilsemann, K.; Mukherjee, K.D. Continuous hydrogenation of fats and fatty acids at short contact times. *JAOCS,* **1978,** *55,* 892.

Inaba, M.; Ogumi, Z.; Takehara, Z. Application of the solid polymer electrolyte method to organic electrochemistry. XIV. Effects of solvents on the electroreduction of nitrobenzene on copper, platinum-nafion. *J. Electrochem. Soc.*, **1993,** 140, 19-22.

International Electrotechnical Commission. *Classification of Hazardous Areas;* Recommendation No. 79, Part 10, IEC: Geneva, 1972–1973.

J

Jackson, H.W. Techniques for flavor and odor evaluation of soy oil. *JAOCS,* **1981,** *58,* 227.

Jackson, M.A.; King, J.W. Lipase-catalyzed glycerolysis of soybean oil in supercritical carbon dioxide. *JAOCS,* **1997**, *74*, 103-106.

Jackson, M.A., List, G.R., and Palmquist, D. Low-*trans* spread and shortening oils via a catalyst switching strategy. *JAOCS,* **2008,** *85* 481-486.

Jacobsberg, B.; Jacqmain D. Palm oil quality. Appreciation and forecast of stability of crude palm oil during transport and storage. *Oleagineux* **1973,** *28,* 25.

Jessop, P.G. Asymmetric catalysis in supercritical fluids. In P. York, U.B. Kompella, B.Y.Shekunov, Eds. *Supercritical Fluid Technology for Drug Product Development.* Marcel Dekker, Inc., New York, 2004, 472-480.

Jessop, P.G.; Leitner, W., Eds. *Chemical Synthesis Using Supercritical Fluids.* Wiley-VCH, Weinheim, Germany, 1999.

Jessop, P.G.; Subramaniam, B. Gas-expanded liquids, *Chem. Rev.*, **2007**, *107*, 2666-2694.

Jessop, P.G; Ikariya, T.; Noyori, R. Homogeneous catalysis in supercritical fluids. *Chem. Rev.*, **1999**, *99*, 475–494.

Johansson, G. *Chem. Ind.*, **1975,** 1 November.

Johansson, L.E.; Lundin, S.T. Copper catalysts in the selective hydrogenation of soybean and rapeseed oils. I. The activity of copper chromite catalysts. *JAOCS,* **1979,** *56,* 974.

Jörissen, J. Ion exchange membranes as solid polymer electrolytes (SPE) in electro-organic syntheses without supporting electrolytes. *Electrochim. Acta*, **1996**, 41, 553-562.

Jungawala, F.B.; Evans, I.E.; McCluer, R.H. *Biochem. J.* **1976,** *155,* 55.

K

Kaimal, T.N.B.; G. Lakshminarayana. Effect of the position of linoleic acid in triacylglycerols on its hydrogenation rate. *JAOCS,* **1979,** *56,* 578.

Katalysatoren der Süd-Chemie AG; Technisehe Informationen, Süd-Chemie AG: 8000 München 2, Germany.

Katz, D.L.; Rzasa, M.J. *Bibliography for Physical behavior of Hydrocarbons Under Pressure and Related Phenomena.* J.W. Edwards, Inc., Ann Arbor, MI, 1946.

Kaufmann, H.P. *Analyse der Fette und Fett Producte;* Springer-Verlag: Berlin, 1958; p. 763

Kaufmann, H.P. et al. Zum nachweis von raffiniertem smalz. *F.S.A.* **1956a,** *58,* 505.

Kaufmann, H.P. et al. Der nachweis der raffination von schweineschmalz 11: Das UV-absorptionsspektrum. *F.S.A.* **1956b,** *58,* 995.

Kaufmann, H.P. Wilhelm Normann, zum Gedächtnis. *Fette u. Seifen* **1939**, *46*, 259-264.

Keppler, J.G.; Schols, J.A.; Feenstra, W.H.; Meijboom, P.W. Components of the hardening flavor present in hardened linseed oil and soybean oil. Ibid. **1965,** *42,* 246.

Keppler, J.G.; Horikx, M.M.; Meijboom, P.W.; Feenstra, W.H. Isolinoleic acids responsible for the formation of the hardening fats. *JAOCS,* **1967,** *44,* 543.

King, J.W. Analytical supercritical fluid techniques and methodology: Conceptualization and reduction to practice. *J. Assoc. Off. Anal. Chem. Int*, **1998**, *81*, 9-17.

King, J.W. Coupled processing options for agricultural materials using supercritical fluid carbon dioxide. 104-129. In A. Gopalan, C. Wai, & H.J. Jacobs (Eds.). *Supercritical Carbon Dioxide: Separations and Processes*, ACS Symposium Series #860, American Chemical Society, Washington, DC, 2003.

King, J.W. Critical fluid technology for the processing of lipid-related natural products. *Compt. rendus Chimie*, **2004**, *7*, 647-659.

King, J.W. Supercritical fluid processing of nutritionally functional lipids, 99-126, Chapter 6. In Casimir C. Akoh and O.M. Lai (Eds.). *Healthful Lipids*. AOCS Press, Champaign, IL, 2005.

King, J.W. Supercritical fluid technology for lipid extraction, fractionation and reactions. 663-687. In: T.M. Kuo and H.W. Gardner (Eds.). *Lipid Biotechnology*. Marcel Dekker, New York, 2002.

King, J.W.; List, G.R. (Eds.), *Supercritical Fluid Technology in Oil and Lipid Chemistry.* AOCS Press, Champaign, IL, 1996.

King, J.W.; Srinivas, K. Multiple unit processing using sub- and supercritical fluids. *J. Supercrit. Fluids,* **2009**, *47*, 508-610.

King, J.W.; Holliday, R.L., List, G.R. and Snyder, J.M. Hydrogenation of vegetable oils using mixtures of supercritical carbon dioxide and hydrogen. *JAOCS.,* **2001**, *78*, 107-113.

King, J.W.; Johnson, J.H.; Eller, F.J. Effect of supercritical carbon dioxide pressurized with helium on solute solubility during supercritical fluid extraction. *Anal. Chem.* **1995**, *67*, 2288-2291.

King, J.W.; Holliday, R.L.; List, G.R. , Hydrolysis of soybean oil in a subcritical water flow reactor, *Green Chem.,* **2000**, *1*, 261-264.

Kirschner, E.; E.R. Lowrey. Positional and geometrical isomerization during partial hydrogenation of trilinolein. A comparison of copper and nickel catalysts *JAOCS,* **1970,** *47,* 237-241.

Kodali, D.; G.R. List, Eds. Trans *Fat Alternatives.* AOCS Press, Champaign, IL, 2005.

Koestler, D.; A.Z. Meisel. *Physik. Chem.* **1966,** *231,* 183.

Koetsier, W.T. Hydrogenation of edible oils, technology and applications In Gunstone, F.D., Padley, F.B. Eds. *Lipid Technologies and Applications.* Marcel Dekker, Inc., New York, Basel, Hong Kong, 1997, 265-303.

Kontala, S.; Dutton H.J. Selective hydrogenation of soybean oil with sodium borohydride-reduced catalysts. *JAOCS*, **1966,** *43,* 86.

Koritala, S.; Dutton, H.J. Selective hydrogenation of soybean oil with sodium borohydride-reduced catalysts. *JAOCS*, **1966,** *43,* 86-89.

Koritala, S. Selective hydrogenation of soybean oil VII Poisons and inhibitors for copper catalysts *JAOCS.* **1975,** *52,* 240-268.

Koritala, S. Selective hydrogenation of soybean oil III. Copper exchanged molecular sieves and other supported catalysts *JAOCS,* **1968,** *45,* 197-200;

Koritala, S. Selective hydrogenation of soybean oil U.S. Patent 3,515,678 (1970).

Koritala, S. Selective hydrogenation of soybean oil: XI. Trialkyl silane-activated copper catalysts. *JAOCS.* **1981,** *58,* 701-702.

Koritala, S.; Friedrich, J.P.; Mounts, T.L. Selective hydrogenation of soybean oil. X. Ultra high pressure and low pressure, *JAOCS*, **1980**, *57*, 1-5.

Koritala, S.; Dutton, H.J. Selective hydrogenation with sodium borohydride-reduced catalysts V. **1966,** *43,* 86-89.

Koritala, S.; Dutton, H.J. Selective hydrogenation of soybean oil IV. Fatty acid isomers formed with copper catalysts *JAOCS,* **1969,** *46,* 245-248;

Koritala, S.; Scholfield, C.R. Selective hydrogenation with copper catalysts: I. Isolation and identification of isomers formed during hydrogenation of linoleate *JAOCS,* **1970,** *47,* 262-265.

Koritala, S.; Butterfield, R.O.; Dutton, H.J. Selective hydrogenation with copper catalysts: II. Kinetics *JAOCS,* **1970,** *47,* 266-268

Koritala, S. Selective hydrogenation with copper catalysts: III. Hydrogen addition and isomerization *JAOCS,* **1970,** *47,* 269-272.

Koritala, S.; Selke, E. Selective hydrogenation with copper catalysts: IV. Reaction of stearolate, oleate and conjugated esters with deuterium *JAOCS,* **1971,** *48,* 222-227

Koseoglu, S.S.; Weise, M.K. Soybean oil hydrogenation via a unique reactor. *inform,* **1991,** *2,* 631.

Krishnan; Bernam, S. Bleachability test method for palm oil. *JAOCS,* **1975,** *52,* 23.

Kuberka, K.A.; Weise, M.K.; Messina, C.A. Hydrogen supply options and issues. Ibid. **1989,** *66,* 1289.

Kusdiana D; Saka S. Kinetics of transesterification in rapeseed oil to biodiesel fuels as treated in supercritical methanol. *Fuel,* **2001**, *80,* 693.

L

Lally, K.S. A315-Axial Flow Impeller for Gas Dispersion, Technical Report, Mixing Equipment Co. Inc.: Rochester, New York, October 1987.

Langstraat, A. Characteristics and composition of vegetable oil bearing materials. *JAOCS,* **1976,** *53,* 241.

Larsson, R. Hydrogenation theory: Some aspects. Ibid. **1983,** *60,* 275.

Latondress, E. (1980) Shortenings and margarines: Basestock preparation and formulation. In *Handbook of Soybean Oil Processing and Utilization.* AOCS Press, Champaign, IL, 1980, 145-154.

Leuteritz, G. Die kontinuierliche hydrierung von olen und fetten. *F.S.A.* **1969,** *71,* 441, 445; available in English: *The Continuous Hydrogenation of Oils and Fats*; BUSS AG: Basel, Switzerland.

Leuteritz, G. Liquid phase hydrogenation plants for the chemical industry. *Chemie-Anlagen und Verfahren 3* (1971); available in English from BUSS AG: Basel, Switzerland; see also these references in this listing: Leuteritz; Urosevic, 1988; van Dierendonck and Leuteritz, 1988.

Licence, P.; Ke, J.; Sokolova, M.; Ross, S.K.; Poliakoff, M. Chemical reactions in supercritical carbon dioxide: from laboratory to commercial plant, *Green Chem.*, **2003**, *5*, 99-104.

Lightning Mixers Ltd., Technical Handout, Stockport: UK.

Linsen, B.G. Thesis Delft University (Netherlands), 1964; see also reference in this listing: Wisniak and Albright, 1961.

List, G. R.; King J. W. Hydrogenation of lipids for use in food. In: Gunstone, F. D., Ed. *Modifying Lipids for Use in Foods.* Woodhead Publishing: Cambridge, 2006, 173-200.

List, G. R.; Warner, K.; Pintauro, P.; Gil, M. Low-*trans* shortening and spread fats produced by electrochemical hydrogenation, *JAOCS*, **2007**, *84*, 497-501.

List, G.R.; Jackson, M.A. Giants of the past: Battle over hydrogenation (1903-1920). i*nform*, **June 2007,***18* (6) 403-405 and **June 2009**, *20* (6) 395-397.

List, G.R.; Jackson, M.A. The battle over hydrogenation (1903-1920) Part II, Litigation. *inform* **2009,** *20* 395-397.

List, G.R.; Kritchevsky, D.; Ratnayake, N. (Eds.) *Trans Fats in Foods*. Champaign, IL, AOCS Press, 2007.

List G.R.; Jackson, M.; Eller, F.; Adlof, R.O. Low trans spread and shortening oils via hydrogenation of soybean oil. Letters to Editor. *JAOCS* **2007**, *84* 609-612.

List, G.R.; Neff, W.E., Holliday, R.L., King, J.W., Holser, R. Hydrogenation of soybean oil triglycerides: Effect of pressure on selectivity. *JAOCS*, **2000**, *77*, 311-314.

List, G.R.; Friedrich, J.P.; King, J.W. Supercritical CO_2 extraction and processing of oilseeds. *Oil Mill Gazetteer*, **1989**, *95* (6), 2834.

Lok, C.M.; Joosten, R.A. Influence of Catalyst Structure on Hardening of Edible Oils and Fatty Acids, paper presented at the AOCS Annual Meeting, Chicago, May 1991; available also from Croda.

Loncin, M. Refining of palm oil. *JAOCS* **1975,** *52,* 141A.

Loncin, M.; Jacobsberg, B.; Everard, G. *Palm Oil, A Major Tropical Product*; Tropical Product Sales, Unilever: London, 1970.

Luddy, F.E. Physical properties of fatty acids. *JAOCS* **1979,** *56,* 759A.

M

Macher, M.-B. Supercritical Hydrogenation of Vegetable Oils, Ph.D. Thesis, Chalmer's University, Gothenburg, Sweden, 2001.

Macher, M.-B.; Hogberg, J.; Moeller, P.; Harrod, M. Partial hydrogenation of fatty acid methyl esters at supercritical conditions. *Fett/Lipid*, **1999,** *101*, 301-305.

Macher, M.-B.;Holmquist, A. Hydrogenation of palm oil in near-critical and supercritical propane, *Eur. J. Lipid Sci. Technol.*, **2001**, *103*, 81-84.

Mahony, L.H. Open Impeller Agitation System, Lightning Technical Reprint, Mixing Equipment Co. Inc.: New York, 1964; see also references in this listing: Lally, 1987; Oldshue, 1983, 1987.

Maisey, H.R. Gaseous and dust explosion venting. *Chem. Process Eng.* **1965,** October, 527.

Malaysian Palm Oil. Technical Bulletin No. 1, 1974, No. 2, 1975, Malaysian Palm Oil Producers' Association: Kuala Lumpur, Malaysia.

Margarit, De Nordiske Fabriker: Oslo, 1913; Arbeiten aus dem Pharmazeutischen Institut der Universitat Berlin, 1914.

Mayorga, M.J.; Ahrweiller, C.; Cuevas, D.; Recasens, F. Effect of reactor type on trans fatty acid and stearate formation in fat hydrogenation in SCF solvent. *J. Supercrit. Fluids*, **2009**, *48*, 21-32.

McHugh, M.; Krukonis, V. *Supercritical Fluid Extraction*. Butterworths, Boston, 1986.

Meara, M.L.; Weir, G.S.D. *Rev. Ital. Sostanze Grasse* **1976,** *53,* 178.

Mehlenbacher, V.C. *The Analysis of Fats and Oils*. The Garrard Press, Champaign, IL, 1960.

Meijboom, P.W.; Jongenotter, G.A. Flavor thresholds of unbranched isomeric non-enals. *JAOCS,* **1971,** *48,* 143.

Meijboom, P.W.; Stroink, J.B.A. 2 *trans* 4 *cis* 7 *cis*-Decatrienal the fishy off-flavor occurring in strongly autoxidized oils containing linolenic acid or ω3, 6, 9 etc., fatty acids. Ibid. **1972,** *49,* 555.

Mensink, R. P.; Katan, M. B. Effect of dietary trans fatty acids on high-density and low-density lipoprotein cholesterol levels in healthy subjects. *N. Engl. J. Med.*, **1990**, *323*, 439-445.

Meric, P.; Man, K.; Yu, K.; Tsang, S.C. Molecular guided catalytic hydrogenation by micelle-hosted Pd nanoparticle in supercritical CO_2. *Catalysis Lett.*, **2004**, *95*, 39-43.

Metcalfe, L.D. Recent advances in the analytical chemistry of fatty acids and derivatives. Ibid., **1979,** *56,* 819A.

Methods of Analysis of Oils and Fats, BS684:1958, British Standards Institution: London.

Milewska, A.; Osuna, A.A.B.; Fonesca, I.M.; Nunes da Ponte, M. Biphasic hydrogenation of α-pinene in high pressure carbon dioxide. *Green Chem.*, **2005**, *7*, 726-732.

Miller, L.L.; Christensen, L. Electrocatalytic hydrogenation of aromatic compounds. *J. Org. Chem.*, **1978**. *43*, 2059-2061.

Mills, B.L.; van de Voort, F.R. Comparison of the direct and indirect wide-line nuclear magnetic resonance methods for determining solid fat contents. *JAOCS,* **1981,** *58,* 776.

Mills, V.; Sanders, J.H.; Hawley, H.K. Partial Hydrogenation of Unsaturated Glyceride Oils. U.S. Patent 2,520,423, 1950; Continuous Hydrogenator, U.S. Patent 2,520,424, 1950.

Minet, R.G. Hydrogen production by automation. Ibid. **1979,** *56,* 185A, item 75.

Moore, C.W. *JSCI* **1919,** *38,* 320–325.

Moulton, K.J.; Moore, D.J.; Beal, R.E. Ibid. **1969,** *46,* 662.

Moulton, K.J.; Beal, R.E.; Griffin, E.L. Ibid. **1973,** *50,* 450.

List, G.R.; Evans, C.D.; Beal, R.E.; Black, L.T.; Moulton, K.J.; Cowan, J.C. Ibid. **1974,** *51,* 239.

Mounts, T. L. *Bailey's Industrial Oil and Fat Products*, 5th edition. New York, John Wiley, 1996, Vol. 4.

Mounts, T.D.; Warner, K.; List, G.R.; Wilson, R.F. (USDA–ARS). Stability of low linolenic acid soybean oils. *Inform,* **1992,** *3,* 900.

Mounts, T.L. Chemical and physical effects of processing fats and oils. Ibid. **1981,** *58,* 51A.

Mounts, T.L. Codex Committee on Fats and Oils. Ibid. **1987,** *64,* 686.

Mounts, T.L.; Dutton, H.J. Micro vapor-phase hydrogenation monitored with tandem chromatography-radioactivity. II Evaluation of catalyst selectivity for linolenate. *JAOCS* **1967**, *44*, 67-70.

Mounts, T.L.; S. Koritala; J.P. Friedrich; H.J. Dutton. Selective hydrogenation of soybean oil: IX. Effect of pressure in copper catalysts. Ibid. **1978,** *55,* 402.

Mukherjee, K.D.; I. Kiewitt; M. Kiewitt. Stationary catalysts for the continuous hydrogenation of fats. *JAOCS* **1975,** *52,* 282.

N

Newton, C.L. Hydrogen production and liquefaction. *Chem. Process Eng.* December **1967**: 51.

Nickel Catalyst KE-NF 20 and KE-FS40; Technical Handout, Süd-Chemie AG: Munich, West Germany; see also reference in this listing: *Katalysatoren der Süd-Chemie AG.*

Nishiguchi, T.; Tagawa, T.; Imai, H.; Fukuzumi, K. Transfer hydrogenation and transfer hydrogenolysis. X. Selective hydrogenation of methyl linoleate by indoline and isopropyl alcohol. *JAOCS* **1977**, *54,* 114.

NIST Reference Fluid Thermodynamic and Transport Properties Database (REFPROP): Version 8.0, National Institute of Standards and Technology, Boulder, CO, 2009.

Normann W. Process for converting unsaturated fatty acids or their glycerides into saturated compounds. British Patent 1 515, assigned to Herforder Maschinenfett- und Ölfabrik Leprince und Siveke, 1903.

O

O'Brien, R.D. (2008) Soybean oil modification. In *Soybean Chemistry, Production, Processing and Utilization.* L. Johnson; P. White; R. Galloway, Eds. AOCS Press, Champaign, IL, 409-439.

O'Brien, R.D. (2009) Hydrogenation. In *Fats and Oils: Formulating and Processing for Applications*, 3rd edition. CRC Press, New York, 110-122.

Oag, R.N.; King, P.J.; Mellor, C.J; George, M.W.; Ke, Jie; Poliakoff, M.; Popov, V.K.; Bagratasvill, V.N, Determining phase boundries and vapor/liquid critical points in supercritical fluids: a multi-technique approach, *J. Supercrit. Fluids*, **2004**, *30*, 259-272.

Official and Tentative Methods of the American Oil Chemists' Society; AOCS: Champaign, Illinois, USA.

Ogumi, Z.; Nishio, K.; Yoshizawa, S. Application of the spe method to organic electrochemistry - ii. electrochemical hydrogenation of olefinic double bonds, *Electrochim. Acta*, **1981**, *26*, 1779-1782.

Oils and Oilseeds; Tropical Products Institute: London, 1971.

Oilseed Processing. Eds. N.T. Dunford and H.B. Dunford. AOCS Press, Champaign, IL, 211-218.

Okkerse, C.; de Jonge, A.; Coenen, J.W.E.; Rozendaal, A. Selective hydrogenation of soybean oil in the presence of nickel catalysts. *JAOCS* **1967,** *44,* 152.

Okonek, D., Hasman, J.M. and Mag, T.K. (1995) Catalysts. Inform 6 1196-1204.

Okonek, D.V. Nickel-Sulphur Catalysts for Edible Oil Hydrogenation; AOCS Short Course, Hydrogenation—A Colloquium, May 1986.

Oldshue, J.Y. *Fluid Mixing Technology;* McGraw-Hill: New York, 1983.

Oldshue, J.Y. *Hydrogenation: Proceedings of AOCS Colloquium;* Champaign, Illinois, USA, 1987; Chapter 7.

Osinga, Th.J.; Balemans, A.C.F. Re-use of catalyst in fish oil hydrogenation. *Fat Sci. Technol.* **1987,** *89,* 512 (Listed in these references under Available from Unichema).

Ostwald, W. *Grundlinien der anorganischen Chemie*; Verlag von Theodor Steinkopff: Dresden and Leipzig, Germany: 1912.

Ottesen, I.; Jensen, B.H. Influence of remaining phosphatides during hydrogenation of soybean oil. *JAOCS* **1980**, item 126.

Ottesson, I. Hydrogenation Inhibiting Substances in Refined Oils, the Importance and the Influence on the Hydrogenation Catalyst. Scandinavian Fats and Oils Symposium, Denmark, 1975.

P

P & S Filtration Ltd., Technical Handout, Haslingden, UK.

Pardun, H.; E. Kroll. Bestimmung der Oxydations stabilitat von Olen und Fetten mit Hilfe einer automatischen Version des SWIFT-Testes. *F.S.A.* **1972,** *74,* 366.

Park, K.; Pintauro, P. N.; Baizer, M.M.; Nobe, K. Flow reactor studies of the paired electro-oxidation and eelectroreduction of glucose. *J. Electrochem. Soc.*, **1985**, 182, 1850-5.

Patterson, H.B.W. *Bleaching and Purifying Fats and Oils: Theory and Practice*; AOCS Press: Champaign, Il, 1993.

Patterson, H. B. W. *Hydrogenation of Fats and Oils: Theory and Practice*, AOCS Press, Champaign, IL, 1994.

Patterson, H.B.W. Bleaching practices in Europe. *JAOCS* **1976,** *53,* 339.

Patterson, H.B.W. Clarity of refined edible oils. *Filtration and Separation*, January 1979, p. 68.

Patterson, H.B.W. *Handling and Storage of Oilseeds, Oils, Fats, and Meal*; Elsevier Applied Science Publishers, London and New York, 1989; p. 130; Chapter 6.

Patterson, H.B.W. Raffination tienscher Fette. *F.S.A.* *77* **1975,** 330.

Patterson, H.B.W. *The Hydrogenation of Vegetable Oils and the Production of Vegetable Ghee;* United Nations Publication (UNIDO) 1D/124: New York, USA, 1974.

Patterson, H.B.W. The refining of animal fats. *Chem. Ind.* **1976**, *18* September 771.

Paul, R.; Buisson, P.; Joseph, N. *Compt. Rend.* **1951,** *232,* 627.

Paulose, M.M.; Mukherjee, K.D.; Richter, I. *Chem. Phys. Lipids* **1978,** *21,* 187.

Pereda, S; Bottini, S.B.; Brignole, E.A., Gas-liquid reactions under supercritical conditions-phase equilibria and thermodynamic modeling, *Fluid Phase Equilibria,* **2002**, *493*, 194-197.

Perkins, E.G.; Meaus, J.C.; Picciano, M.P. Recent advances in the instrumental analysis of lipids. *Rev. Fr. Corps Gras.* **1977,** *24,* 13.

Pillai, U.R.; Sahle-Demessie, E. Hydrogenation of 4-oxoisophorone over a Pd/Al_2O_3 under Supercritical CO_2 medium, *Ind. Eng. Chem. Res.*, **2003**, 42, 6688-6696.

Pintauro P. N.; An W. Synthesis of a Low Trans-Content Edible Oil, Non-Edible Oil, or Fatty Acid in a Solid Polymer Electrolyte Reactor, U.S. Patent No. 6,218,556, April 17, 2001

Pintauro, P. N. Electrochemical Process for the Hydrogenation of Edible and Non Edible Oils and Fatty Acids, U.S. Patent No. 5,225,581, July 6, 1993.

Pintauro, P. N.; Gil, M. P.; Warner, K.; List, G.; Neff, W. electrochemical hydrogenation of soybean oil with hydrogen gas. *Ind. & Eng. Chem. Res.*, **2005**, 44, 6188-6195.

Pintauro, P. N.; Johnson, D. K.; Park, K.; Baizer, M. M.; Nobe, K., The paired electrochemical synthesis of sorbitol and gluconic acid in undivided flow cells. *J. Appl. Electrochem.*, **1984**, 14, 209-220.

Pintauro, P.N.; Bontha J., The role of supporting electrolyte during the electrocatalytic hydrogenation of aromatic compounds. *J. Appl. Electrochem.* **1991**, 21, 799-804.

Poliakof, M; Swan, T.M; Tacke, T; Hitzler, M.G.; Ross, S.K; Wieland, S., Supercritical hydrogenation, WO/1997/038995 (1997); U.S. patent #6, 156,9333 (2000).

Popescu, O.; S. Koritala; H.J. Dutton. *JAOCS,* **1968,** *46,* 97.

Priddy, M.H. New developments in hydrogen gas generation molecular sieves. Ibid. **1971,** *48,* 46A.

Proceedings of World Conference on Oilseed Processing, vol. 1. Ed. S. Koseglu, K. Rhee and R.F. Wilson. AOCS Press, Champaign, IL. 77-91.

Puri, P. (1980) Hydrogenation of oils and fats. JAOCS 57 850a-853a.

R

Ramirez, E. Contribution to the Study of Heterogenous Catalytic Reactions in SCFs: Hydrogenation of Sunflower Oil in Pd Catalysts at Single-phase Conditions, Ph. D Thesis, Universitat Politecnica de Catalunya, Barcelona, Spain, 2005.

Ramirez, E.; Recasens, F.; Fernandez, M; Larrayoz, M.A. Sunflower oil hydrogenation on Pd/C in SC Propane in a continuous recycle reactor, *AIChE J.*, **2004,** 7, 1545-1555.

Ramirez, E.; Zgarni, S.; Larrayoz, M.A.; Recasens, F. Short compilation of published reaction rate data for catalytic hydrogenations in supercritical fluids, *Eng. Life Sci.*, **2002**, 2, 257-264.

Rao, N.N.; Lutz, S.; Wurges, K.; Minor, K. Continuous biocatalytic processes, *Org. Process Res. Dev.*, **2009**, *13*, 607–616.

Ray, J.D. Behavior of hydrogenation catalysts, I. Hydrogenation of soybean oil with palladium, *JAOCS,* **1985**, 62, 1213-1217.

Raynor, C.M.; Oakes, R.S.; Sakakura, T.; Yasuda, H. Supercritical carbon dioxide, In: *Green Reaction Media in Organic Synthesis*, Blackwell Publishing, Oxford, 2002, 125-178.

Reeitff, D.D. Liquefaction and storage of hydrogen. *Chem. Eng.* **1965,** 13 September, 191.

Reimann, P. Hydrogenation catalytique d'huiles, graisses et acides gras. *Oleagineux* **1978,** *33,* 87–90.

Removal of Metal Traces from Edible Oils; Technical Handout, Süd-Chemie AG: Munich, West Germany,.

Rice, E.E. Material of construction in the fatty acid industry. *JAOCS* **1979,** *56,* 754A.

Roonev, J.J.; Gault, F.G.; Kemball, C. *Proc. Chem. Soc.* **1960,** 407.

Rost, H.E. Influence of thermal treatments of palm oil on the content of polycyclic aromatic hydrocarbons. *Chem. Ind.* **1976,** 17 July, 612.

Ruckenstein, E; Shulgin, I, Simple equation for the solubility of a solid in a supercritical fluid cosolvent with a gas or another supercritical fluid, *Ind. Eng. Chem. Res.*, **2003**, *42,* 1106–1110.

Rutledge, D.N.; El-Khaloui, M.; Ducauze, C.J. Contribution à l'étude d'une méthode de contrôle rapide de la qualité des margarines par RMN-IBR. *Rev. Franç. Corps Gras* 1988, 35, 157-162.

Rylander, P.N. Hydrogenation of natural oils with platinum metal group catalysts. *JAOCS* **1970,** *47,* 482.

S

Sagredos, A.N.; Sinha-Roy, D.; Thomas, A. Zur bestimmung, zum vorkommen und zur zusammensetzung von polycyclischen aromatischen kohlenwasserstoffen in ölen und fetten. *Fat Sci. Technol.* **1988,** *90,* 76.

Saka, S. Recent progress in supercritical fluid science for biofuel production from woody biomass, *Forestry Studies in China*, **2006**, 8, 1008-1321.

Sandrea, R.; Nielsen, R. Dynamics of Petroleum Reservoirs Under Gas Injection, Gulf Publishing Company, Houston, TX, 1974.

Santana, A.; Fernandez, X.; Larrayoz, M.A.; Recasens, F. Vegetable fat hydrogenation in supercritical-fluid solvents: melting behavior analysis by DSC and NMR., *J. Supercrit. Fluids*, **2008**, 46, 322-328.

Santana, A.; Larrayoz, M.A.; Ramirez, E.; Nistal, J.; Recasens, F. Sunflower oil hydrogenation on Pd in supercritical solvents: kinetics and selectivities, *J. Supercrit. Fluids,* **2007**, 41, 391-403.

Sarrazin J.; Tallec, A. Use of ion exchange membranes in preparative organic electrochemistry: new processes for electrolysis, *J. Electroanal. Chem. and Interfacial Electrochem.*, **1982**, 137, 183-188.

Savage, P. E.; Gopalan, S.; Mizan, T. I.; Martino, C. J.; Brock, E. E.**, Reactions at supercritical conditions: applications and fundamentals.** *AIChE J.,* **1995,** 41, 1723-1778.

Scarpiello, D.A. Sulphur promoted nickel hydrogenation catalysts. Ibid. **1979,** *56,* 176A, item 176.

Schenk Filterbau GmbH, D7076 Waldstetten, Germany, and 414 Ridge View Court, Arnold, MD, USA.

Schilling, K. Der Reaktionsverlauf bei der Hydrierung von Triglyceriden. Simultan-hydrierung von Tri- und Monolinolenin. *Fette Seifen Anstrichm.* 1978, 80, 312-314.

Schilling, K. Über den Verlauf der Hydrierung von Triolein an Raney-Nickel. *Fette Seifen Anstrichm.* 1968, 70, 389-393.

Scholfield, C.P.; Jones, E.P.; Nowakowska, J.; Selka, E.J.; Dutton, H.J. Hydrogenation of linolenate. II. Hydrazine reduction. Ibid. **1961,** *38,* 208.

Scholfield, C.R.; Butterfield , R.O.; Duff, H.J. Calculation of catalyst selectivity in vegetable oil. *JAOCS* **1979,** *56,* 664.

Sebedio, J.L.; Langman, M.F.; Eaton, C.A.; Ackman, R.G. Alteration of long chain fatty acids of herring oil during hydrogenation on nickel catalyst. Ibid. **1981,** *58,* 41.

Seifried, B; Temnelli, F. Unique properties of carbon dioxide-expanded liquids, *INFORM*, **2010**, 21, 9-12, 59.

Seki, T.; Grunwaldt, J-D; Baiker, A. Heterogeneous catalytic hydrogenation in supercritical fluids: potential and limitations, *Ind. Eng. Chem. Res.*, **2008**, 47, 4561-4585.

Semi-continuous Hydrogenation System; Technical Handout, Sullivan Systems Inc.: Tiburon, USA.

Serfass, E.J.; Silman, H. Hydrogen production and purification by the diffusion process. *Chem. Eng.,* October **1965**: CE266.

Sleeter, R.T. Instrumental analysis for quality control and quality assurance. *Bailey's Industrial Oil and Fat Products,* 4th ed.; T.H. Applewhite, Ed.; 1985; Vol. 3, p. 182.

Sleeter, R.T. Instrumental analytical methods for edible oil processing: present and future. *JAOCS* **1983,** *62,* 343.

Snyder, J.M.; Scholfield, C.R.; Mounts, T.L. Laboratory-scale continuous hydrogenation. Ibid. **1979,** *56,* 506.

Snyder, J.M.; Dutton, H.J.; Scholfield, C.R. Laboratory-scale continuous hydrogenation. Ibid. **1978,** *55,* 383.

Society Meeting, Atlanta, GA. Text available from Englehard/BASF, Beachwood, OH.

Song, Y.; Pintauro, P.N. The electrochemical synthesis of aminonitriles. I. h-cell studies with adiponitrile and azelanitrile. *J. Appl. Electrochem.*, **1991**, 21, 21-27.

Stahl, E.; Quirin, K.W.; Gerard, E. *Dense Gases for Extraction and Refining.* Springer-Verlag, Berlin, 1987.

Stansby, M.E. *Fish Oils—Their Chemistry, Technology, Stability, Properties and Uses*; AVI Publishing Co. Inc.: Westport, 1967.

Stingley, D.V. and R.J. Wroebel. (1961) The effects of *trans* isomers on the physical properties of hydrogenated oils, *JAOCS*, **1961**, *28*, 201-205.

Stuckey, B.N.; Osborne, C.E. A review of antioxidant analysis in food products. Ibid. **1965,** *42,* 228.

Sundram K., Perlman, D. and Hayes, R.C. (1998) Increasing the HDL level and the HDL/LDL ratio in

supercritical carbon dioxide and hydrogen. *JAOCS* 78 107-113.

Swern, D. (Ed.) *Bailey's Industrial Oil and Fat Products,* 3rd ed.; John Wiley and Sons: New York, 1964; p. 882.

T

Tacke, T.; Wieland, S.; Panster, P.; Bankmann, M.; Brand, R.; Magerlein, H., Continuous hydrogenation of unsaturated fats, fatty acids, and esters by passing a mix in a supercritical solvent of starting material and hydrogen over a solid bed catalyst, WO 95/22591 (1997).

Tacke T.; Wieland, S.; Panster, P. Selective and Complete Hardening of Edible Oils and Free Fatty Acids in Supercritical Fluids, Proceedings of the 4th International Symposium on Supercritical Fluids, May 11-14, Sendai, Japan, 1997, 511-514.

Tacke, T. Wieland, S.; Panster, P. Selective and complete hydrogenation of vegetable oils and free fatty acids in supercritical fluids, In: W. DeSimone, J.M.; Tumas, Eds. *Green Chemistry Using Liquid and Supercritical Carbon Dioxide*. Oxford University Press, Oxford, 2003.

Tacke, T.; Wieland, S.; Panster, P., Hardening of fats and oils in supercritical CO_2, In *High Pressure Engineering*, P.R. von Rohr and C. Trepp (Eds.), Elsevier Science B.V., Amsterdam, 1996, 17-21.

Tacke, T.; Wieland, S.; Panster, P.; Bankmann, M.; Brand, R.; Magerlein, H. Hardening of unsaturated fats. Fatty acids or fatty acid esters, U.S. Patent # 5, 734, 070 (1998); WO 95/2259 (1995).

Tagawa, T.; Nishiguchi, T.; Fukuzumi, K. Transfer hydrogenation and transfer hydrogenolysis. XII. Selective hydrogenation of fatty acid methyl esters by various hydrogen donors. *JAOCS* **1978,** *55,* 332.

Takenouchi, S.; Kennedy, G. C. The binary system H_2O-CO_2 at high temperatures and pressures, *Am. J. Sci.*, 1964, 262, 1055-1074.

Tarladgis, B.G.; Pearson A.M.; Dugan, L.R. *J. Sci. Fd. Agric.* **1964,** *15,* 602.

Tarrago-Traini, M.T.; Phillips, K.M.; Lamar, L.E. and Holden, J.M. (2006) New and existing oils and fats with reduced *trans* fatty acid content. J. Am. Dietetic Assoc. 106 867-880.

Taylor, R.J. *Essential Fatty Acids—A Review;* Unilever Information Division: London, 1968.

Taylor, R.J. *The Chemistry of Glycerides*; Unilever Educational Booklet, Advanced Series No. 4, Unilever Information Division: London, 1973.

Taylor, R.J. *The Physics of Chemical Structure*; Unilever Educational Booklet, Advanced Series No. 2, Unilever Information Division: London, 1971.

Thomas, A.E.; Paulika, F.R. Solvent Fractionated Fats (Glidden Durkee Div., SCM International); Symposium on Recent Advances in Chemistry and Technology of Oils and Fats, Liverpool Polytechnic, 1975.

Toor H. van; Rossum G.J. van; Kruidenberg M. Low trans-fatty acid compositions; low-temperature hydrogenation, e.g. of edible oils. US Patent Application Publication 2005/0027136 A1, assigned to Cargill Incorporated, 2005.

Trabelsi, F.;Stüber, F.; Abaroudi, K.; Larrayoz, M..A.; Recasens, F.; Sueiras, J.E. Coking and ex situ catalyst reactivation using supercritical CO_2: a preliminary study, *Ind. Eng. Chem. Res.*, **2000**, *39*, 3666–3670.

Twist, D.R.; K.J. Sagar. Design variables and production costs in the large scale manufacture of hydrogen. *Chem. Eng.*, October **1965**: CE252.<Q-Vol. #? Page numbers?>

U

Ultrahydrophytosqualene: New Processes for the Generation of Squalene by Supercritical Fluid Extraction from Waste of Olive Oil Production and Hydogenation of Squalene, FAIR –CT95-1075 Report, **2003**, www.nf-2000.org/secure/fair/F348.htm.

Unilever NV. Dutch Patent Application 295,863 (1963).<Q-Patent title?>

Urosevic, D. BUSS Technology for Oils and Fats and Fatty Acid Derivatives, ISF–JOCS World Congress: Tokyo, 1988; also from BUSS AG Basel.

Ustundug, O.G.; Temelli, F. Correlating the solubility behavior of fatty acids, mono- di-, and triglycerides, and fatty acid esters in supercritical carbon dioxide, *Ind. Eng. Chem. Res.*, **2000,** 39, 4756-4766.

V

Van Den Enden, J.C.; Haighton, A.J.; van Putte, K.; Vermaas, L.F.; Waddington, D.A. A method for the determination of the solid phase content of fats using pulse nuclear magnetic resonance. *F.S.A.* **1978,** *80,* 180.

Van den Enden, J.C.; Haighton, A.J.; van Putte, K.; Vermass, L.F.; Waddington, D. A method for the determination of the solid phase content of fats using pulse magnetic resonance. *Fette Seif Ans.* **1978,** *80,* 180, also in Bulletins 19 and 20 of the International Association of Fish Meal Manufacturers: Potters Bar, EN6 3AR. England, 1986.

Van den Enden, J.C.; Rossell, J.B.; Vermaas, L.F.; Waddington, D. Determination of the solid fat content of hard confectionery butters. *JAOCS* **1982,** *59,* 433.

Van den hark, S. The Use of Supercritical Fluids to reduce the Number of Phases in Catalytic Hydrogenation: The Reaction of Fatty Acid Methyl Esters to Fatty Alcohols, Ph.D. Thesis, Chalmer's University, Gothenburg, Sweden, 2000.

Van den Hark, S.; Harrod, M. Fixed-bed hydrogenation at supercritical conditions to form fatty alcohols: the dramatic effects caused by phase transitions in the reactor, *Ind. Eng. Chem. Res.*, **2001**, 40, 5052-5057.

Van den hark, S.; Harrod, M.; Moeller, P. Hydrogenation of fatty acid methyl esters to fatty alcohols at supercritical conditions, *J. Am. Oil Chem. Soc.*, **1999**, 76, 1363-1370.

Van den Hark, S; Harrod, M. Hydrogenation of oleochemicals at supercritical single-phase conditions: influence of hydrogen and substrate concentrations on the process, *Appl. Catal. A: General*, **2001**, 210, 207-215.

Van der Planck, P.; Van Oosten, H.J.; Van Dijk, L. Non-metallic palladium-on-resin: A very active and selective catalyst for the hydrogenation of di-unsaturated fatty acid esters. *JAOCS* **1979,** 56, 45; Van der Planck, P., Van Oosten, H.J. Hydrogenation of monounsaturated and ethylene- and methylene-interrupted di-unsaturated fatty acid esters on non-metallic palladium on resin. II. Ibid., **1979,** 56, 50; van der Planck P.; Van Ooster, H.J. Mechanism of hydro-isomerization of methylene-interruped dienes on non-metallic palladium-on-resin catalysts. III. Ibid., **1979,** 56, 54.

Van der Planck, P.; Van Ooster, H.J.J. *Catalysis* **1975,** *38,* 223.

Van der Plank, P.; Van der Ent, A.; Onder de Linden, A.L. Hydrogenation of dienes with cationic rhodium complexes. Ibid., **1980,** *57,* 343.

van Dierendonck, L.L.; Leuteritz, G. Scale Up of G-L Reactions Made Simple with Loop Reactors, 6th European Conference on Mixing, Pavia, May 1988. Associazione Italiana di Ingegnena Chimica.

van Putte, K.L.; Vermaas, L.; Van Den Enden, J.; Den Hollander, C. Relations between pulsed NMR, wide-line NMR and dilatometry. *JAOCS* **1975,** *52,* 179.

van Stuyvrengerg, J.H. (Ed.) *Margarine: An Economic, Social and Scientific History, 1869–1969*; Liverpool University Press: Liverpool, 1969.

Van Dujijn, G. (2000) Technical aspects of *trans* reduction in margarines. *Oleagineux corps gras lipids* 7 95-98.

VanToor H., VanRossum, G.J. and Kruidenberg, M. (2005) Low *trans*-fatty acid compositions; low-temperature hydrogenation, e.g. of edible oils. U.S. Patent Application 0027136 A1.

Veldsink, J.W.; Bouma, M.J.; Schöön, N.-H.; Beenackers, A.A.C.M. Heterogeneous hydrogenation of vegetable oils: A literature review. *Catal. Rev. Sci. Eng.* 1997, 39, 253-318.

Vles, R.O.; Bijster, G.M.; Kleinekoort, J.S.W.; Trimmer, W.G.; Zaalberg, J. Nutritional status of low-erucic rapeseed oils. *F.S.A.* **1976,** *78,* 128.

W

Waddington, D. *Determination of the Solid Phase Content in Fats Using the BRUKER Minispec P. 20i;* Bruker-Physik AG, Germany.

Walker, R.C.; Bosin, W.A. Comparison of SFI. DSC and NMR methods for determining solid-liquid ratios in fats. *JAOCS* **1971,** *48,* 50.

Wandeler, R.; Baiker, A. Supercritical fluids – Opportunities in heterogenous catalysts. *CATTECH*, **2004**, 4, 128-143.

Wandeler, R.; Kunzle, N.; Schneider, M.S.; Mallat, T.; Baiker, A. Continuous platinum-catalyzed enantioselective hydrogenation in "supercritical" solvents. *Chem. Commun.* 673-674, 2001.

Warner, K.; Neff, W.; List, G. R.; Pintauro, P. N. Electrochemical hydrogenation of edible oils in a solid polymer electrolyte reactor. Sensory and compositional characteristics of low *trans* soybean oils. *J. Am. Oil Chem. Soc.* **2000**, 77, 1113-1117.

Waterman, H.I. *Hydrogenation of Fatty Oils*; Elsevier Publishing Corp.: Amsterdam, 1951.

Weidner, E.; Brake, C.; Richter, D.Thermo- and fluid-dynamics aspects of the Hydrogenation of triglcyerides and esters in presence of supercritical fluids. In *Supercritical Fluids as Solvents and Reaction Media*, G. Brunner (ed.), Elsevier, B.V., Amsterdam, 2004, 269-294.

Weise, M.; B. Delaney. Plants using new hydrogenation agitator design. *INFORM* **1992,** *3,* 817.

Wendt, H.H.R.H. Use of activated carbon for refining edible oils. *JAOCS* **1980,** *57,* 115A, abs. 439.

Wiebe, R; Gaddy, V.L, The solubility of carbon dioxide in water at various temperatures from 12 to 40° and at pressures to 500 atmospheres. critical phenomena, *JAOCS*, **1940**, *6,* 815–817.

Willett, W. C.; Ascherio, A. Trans fatty acids: Are the effects only marginal ? *Am. J.Public. Health*, **1994**, 84, 722-724.

Wisniak, J.; Albright, L.F. Hydrogenating cottonseed oil at relatively high pressure. *Ind. Eng. Chem.* **1961,** *53,* 375.

Wisniak, J.; Stein, S. Hydrogen solubility in jojoba oil. *JAOCS* 1974, 51, 482-485.

Wu, T.; Jiang, T.; hu, B.; He, J.; Zhou, X. Cross-linked polymer coated Pd nanocatalysts on SiO_2 support: very selective and stable catalysts for hydrogenation in supercritical CO_2., *Green Chem.*, **2009**, DOI: 10. 1039/b82298300a.

Wyatt, V.T; Haas, M.J. Production of fatty acid methyl esters via the in situ transesterification of soybean oil in carbon dioxide-expanded methanol, *JAOCS,* **2009**, 86, 1009–1016.

Y

Yasuda, K.; Peterson, R.J.; Chang, S.S. Identification of volatile flavour compounds developed during storage of a deodorized hydrogenated soybean oil. *JAOCS* **1975,** *52,* 307.

Yusem, G. J. Reactor and Process Design Analyses for the Electrocatalytic Hydrogenation of Soybean Oil on Powdered Raney Nickel, *Ph.D. dissertation*, Tulane University, New Orleans, LA, **1994**.

Yusem, G.; Pintauro P. N.; Cheng, P.-C.; and An, W, electrocatalytic hydrogenation of soybean oil in a radial flow-through raney nickel powder reactor. *J. Appl. Electrochem.*, **1996**, 26, 989-997.

Yusem, G; Pintauro, P. N. The electrocatalytic hydrogenation of soybean oil, *J. Am. Oil Chem. Soc.* **1992**, 69, 399-404.

Z

Zhang, X.; Zong, B.; Qiao, M, Reactivation of spent Pd/AC catalyst by supercritical CO_2 fluid extraction. *AIChE J.*, **2009**, 55, 2382-2388.

Zhang, Y.; Erkey, C. Preparation of supported nanoparticles using supercritical fluids: a review," *J. Supercrit. Fluids.*, **2006**, 38, 252.

Zhang, Z.; King, J.W. The effect of dissolved helium on the density and solvation power of supercritical carbon dioxide. *J. Chromatogr. Sci.*, **1997**, 35, 483-488.

Zosel, K. Process for the separation of mixtures of substances, U.S. Patent 3,969,196 (1976).

Zosel, K. Process for the simultaneous hydrogenation and deodorisation of fats and oils, U.S. Patent 3, 969, 382 (1976).

Zwahlen, A.G.; Bertucco, A. CSTR-system for kinetic investigation for hydrogenation reactions, In: *High Pressure Chemical Engineering*, P.R. van Rohr and C. Trepp (Eds.), Elsevier Science B.V., Amsterdam, 1996, 37-42.

Index

D

E

F

G

I

K

L

R

S

T

CPSIA information can be obtained at www.ICGtesting.com
Printed in the USA
LVOW091916311011

252925LV00001B/3/P

9 781893 997936